AF809513

Michael Glykas (Ed.)

Fuzzy Cognitive Maps

Studies in Fuzziness and Soft Computing, Volume 247

Editor-in-Chief

Prof. Janusz Kacprzyk
Systems Research Institute
Polish Academy of Sciences
ul. Newelska 6
01-447 Warsaw
Poland
E-mail: kacprzyk@ibspan.waw.pl

Further volumes of this series can be found on our homepage: springer.com

Vol. 230. Bhanu Prasad (Ed.)
Soft Computing Applications in Business, 2008
ISBN 978-3-540-79004-4

Vol. 231. Michal Baczynski, Balasubramaniam Jayaram
Fuzzy Implications, 2008
ISBN 978-3-540-69080-1

Vol. 232. Eduardo Massad, Neli Regina Siqueira Ortega, Laécio Carvalho de Barros, Claudio José Struchiner
Fuzzy Logic in Action: Applications in Epidemiology and Beyond, 2008
ISBN 978-3-540-69092-4

Vol. 233. Cengiz Kahraman (Ed.)
Fuzzy Engineering Economics with Applications, 2008
ISBN 978-3-540-70809-4

Vol. 234. Eyal Kolman, Michael Margaliot
Knowledge-Based Neurocomputing: A Fuzzy Logic Approach, 2009
ISBN 978-3-540-88076-9

Vol. 235. Kofi Kissi Dompere
Fuzzy Rationality, 2009
ISBN 978-3-540-88082-0

Vol. 236. Kofi Kissi Dompere
Epistemic Foundations of Fuzziness, 2009
ISBN 978-3-540-88084-4

Vol. 237. Kofi Kissi Dompere
Fuzziness and Approximate Reasoning, 2009
ISBN 978-3-540-88086-8

Vol. 238. Atanu Sengupta, Tapan Kumar Pal
Fuzzy Preference Ordering of Interval Numbers in Decision Problems, 2009
ISBN 978-3-540-89914-3

Vol. 239. Baoding Liu
Theory and Practice of Uncertain Programming, 2009
ISBN 978-3-540-89483-4

Vol. 240. Asli Celikyilmaz, I. Burhan Türksen
Modeling Uncertainty with Fuzzy Logic, 2009
ISBN 978-3-540-89923-5

Vol. 241. Jacek Kluska
Analytical Methods in Fuzzy Modeling and Control, 2009
ISBN 978-3-540-89926-6

Vol. 242. Yaochu Jin, Lipo Wang
Fuzzy Systems in Bioinformatics and Computational Biology, 2009
ISBN 978-3-540-89967-9

Vol. 243. Rudolf Seising (Ed.)
Views on Fuzzy Sets and Systems from Different Perspectives, 2009
ISBN 978-3-540-93801-9

Vol. 244. Xiaodong Liu and Witold Pedrycz
Axiomatic Fuzzy Set Theory and Its Applications, 2009
ISBN 978-3-642-00401-8

Vol. 245. Xuzhu Wang, Da Ruan, Etienne E. Kerre
Mathematics of Fuzziness – Basic Issues, 2009
ISBN 978-3-540-78310-7

Vol. 246. Piedad Brox, Iluminada Castillo, Santiago Sánchez Solano
Fuzzy Logic-Based Algorithms for Video De-Interlacing, 2010
ISBN 978-3-642-10694-1

Vol. 247. Michael Glykas (Ed.)
Fuzzy Cognitive Maps, 2010
ISBN 978-3-642-03219-6

Michael Glykas (Ed.)

Fuzzy Cognitive Maps

Advances in Theory, Methodologies, Tools and Applications

Springer

Editor

Dr. Michael Glykas
Managing Director
Aegean Technopolis
(The Technology Park of the Aegean region in Greece)
Chalkios Chios 82100
Greece

and

Assistant Professor
Department of Financial and Management Engineering
University of the Aegean
Kountouriotou Street
Chios 82100
Greece

ISBN 978-3-642-03219-6 e-ISBN 978-3-642-03220-2

DOI 10.1007/978-3-642-03220-2

Studies in Fuzziness and Soft Computing ISSN 1434-9922

Library of Congress Control Number: 2010929474

© 2010 Springer-Verlag Berlin Heidelberg

This work is subject to copyright. All rights are reserved, whether the whole or part of the material is concerned, specifically the rights of translation, reprinting, reuse of illustrations, recitation, broadcasting, reproduction on microfilm or in any other way, and storage in data banks. Duplication of this publication or parts thereof is permitted only under the provisions of the German Copyright Law of September 9, 1965, in its current version, and permission for use must always be obtained from Springer. Violations are liable to prosecution under the German Copyright Law.

The use of general descriptive names, registered names, trademarks, etc. in this publication does not imply, even in the absence of a specific statement, that such names are exempt from the relevant protective laws and regulations and therefore free for general use.

Typeset & Cover Design: Scientific Publishing Services Pvt. Ltd., Chennai, India.

Printed on acid-free paper

9 8 7 6 5 4 3 2 1

springer.com

To my wife Athina for her support all these years
and to Markos, Aggeliki and Alexandros the three pillars of my life

Foreword

This important edited volume is the first such book ever published on fuzzy cognitive maps (FCMs). Professor Michael Glykas has done an exceptional job in bringing together and editing its seventeen chapters. The volume appears nearly a quarter century after my original article "Fuzzy Cognitive Maps" appeared in the *International Journal of Man-Machine Studies* in 1986. The volume accordingly reflects many years of research effort in the development of FCM theory and applications—and portends many more decades of FCM research and applications to come.

FCMs are fuzzy feedback models of causality. They combine aspects of fuzzy logic, neural networks, semantic networks, expert systems, and nonlinear dynamical systems. That rich structure endows FCMs with their own complexity and lets them apply to a wide range of problems in engineering and in the soft and hard sciences. Their partial edge connections allow a user to directly represent causality as a matter of degree and to learn new edge strengths from training data. Their directed graph structure allows forward or what-if inferencing. FCM cycles or feedback paths allow for complex nonlinear dynamics. Control of FCM nonlinear dynamics can in many cases let the user encode and decode concept patterns as fixed-point attractors or limit cycles or perhaps as more exotic dynamical equilibria. These global equilibrium patterns are often "hidden" in the nonlinear dynamics. The user will not likely see these global patterns by simply inspecting the local causal edges or nodes of large FCMs.

This feedback structure also distinguishes FCMs from the earlier forward-only acyclic cognitive maps and from modern AI expert-system search trees. Such tree structures are not dynamical systems because they lack edge cycles or closed inference loops. Nor are trees closed under combination. Combining several trees does not produce a new tree in general because cycles or loops tend to occur as the number of combined trees increases. But combining FCMs always produces a new FCM. The combined FCM naturally averages the FCMs and their corresponding causal descriptions as well as much of their dynamics. The user can combine any number of weighted FCMs combine into a single averaged FCM by the simple artifice of adding their scaled and augmented (zero-padded) adjacency edge matrices. The strong law of large numbers then ensures that the sample average of even quantized or rounded-off FCMs will converge with probability one to the underlying but often unknown FCM that generates these matrix realizations. So FCM edge-matrix combination improves with sample size. FCM knowledge representation likewise tends to improve as the user combines more FCMs from an ever larger pool of domain experts.

These same FCM properties pose structural challenges to the user. One example is that the nonlinear nature of FCM nodes makes backward-chaining or why-based inference extremely difficult. This shortcoming cries out for at least an approximate solution. Another example is that a FCM's global feedback can easily destabilize an input pattern or make it difficult or impossible to encode a desired equilibrium pattern. A related problem is the detection of missing concept nodes. Causal learning algorithms have focused almost exclusively on adjusting the edges in rough analogy to adjusting the synapses in an artificial neural network. These algorithms do not show how the same data can predict new nodes in the FCM directed graph. These are just some of the reasons why FCMs will remain an active and fruitful area of research as well as of applications.

This volume covers a wide range of FCM applications in its seventeen chapters. Theoretical advances include new causal learning laws and the use of contraction mappings to establish the existence and uniqueness of many FCM equilibria. The domain applications show how FCMs can naturally combine factual and engineering concepts with value and even policy concepts. These applications range from software reliability and finance to computer vision and the allocation of scarce water resources among many others. The result is a landmark in FCM research and in the history of knowledge engineering.

Bart Kosko
Professor of Electrical Engineering and Law
University of Southern California

Introduction

The aim of this book is to present recent advances and state of the art in Fuzzy Cognitive Mapping. But what is it all about? A person with no previous exposure in this field would very rightly ask.

In order to explain what a fuzzy cognitive map (FCM) means we need to explain the three words it is composed of. The last two (Cognitive Maps) refer to the theory upon which fuzzy cognitive mapping stems from. These two words on their own apparently define another discipline.

The credit of the creation Cognitive Maps is quite rightly awarded to Edward Tolman back in 1948. Cognitive maps have been studied and used in various fields, such as psychology, education, archaeology, planning, geography, architecture, landscape architecture, urban planning and management. However, the theory of cognitive maps was fully developed from 1976 onwards. Thousands of articles and books have been written on this subject for the interested reader. Its main aim was and still is the representation of (causal) relationships among "concepts" also known as "factors" or "nodes".

Concepts could be assigned values. Causal relationships between two concepts could be cf three types: positive, negative or neutral. Increase in the value of a concept would yield a corresponding positive or negative increase at the concepts connected to it via relationships.

The third word Fuzzy was introduced to cognitives maps in 1986 by Bart Kosko in a famous article in which he introduced the notion of fuzziness to cognitive maps and created the theory of Fuzzy Cognitive Maps (FCMs).

More specifically Kosko introduced the notion of a "fuzzy weight" which in simple terms means that the relationship between two concepts, also called nodes, can take a value in the interval [-1,1]. Since then he is considered the "father" of cognitive mapping.

For nearly a quarter of a century extensive research in the theory of FCMs has been performed that provided major improvements and enhancements in its theoretical underpinning. New methodologies and approaches have been developed. FCMs have also been applied to many different sectors. New software tools have been developed that automate FCM creation and management.

All these theories, methodologies, applications and tools exist to date scattered in journal papers, conference and workshop proceedings. This book tries to present advances in FCM in a concrete and integrated manner.

The audience of this book is both the academic and industrial-business community that has an interest in using FCMs either as a theoretical framework or as a methodology and tool for applied research, industrial or business applications etc.

Book Chapters

The number of submitted chapters was exceptionally high showing that there is a strong interest in FCMs by contemporary researchers. Unfortunately only a limited number could be published in the book based on the decision of the review committee even though a much higher number of publishable papers were submitted.

In chapter one professor Groumpos presents the challenging problem of modeling and controlling complex systems using Fuzzy Cognitive Maps (FCMs). A special focus was devoted on the issue of modeling the supervisor of large complex systems. The method was validated in a manufacturing case study and a successful application of FCM theory was applied to a health sector business case. The contribution of this chapter to the theoretical underpinnings of FCMs is considered very significant as it presents: a mathematical description of FCM models, new construction methods and a new algorithm.

In chapter 2 Wojciech Stach, Lukasz Kurgan, and Witold Pedrycz give a thorough insight on the issue of learning FCMs. They stress the lack of automated or semi-automated methods that would replace or support designers and define learning FCMs which correspond to the construction of connection matrices based on historical data presented in the form of multivariate time series. Existing automated learning methods are based either on the Hebbian learning or they apply evolutionary algorithms. In their research they formulate the task of learning FCMs and describe the corresponding design challenges. The leading learning methods are described and analyzed both analytically and experimentally with the help of a case study. They finaly compare and contrast computational approaches with expert-based methods and outline future research directions.

In chapter 3 Elpiniki Papageorgiou presents a novel approach for the construction of augmented Fuzzy Cognitive Maps based on data mining and knowledge-extraction methods. The issue of designing decision support systems based on fuzzy cognitive maps has been explored using fuzzified decision trees and other knowledge-extraction techniques. Fuzzy theoretical techniques are used to fuzzify crisp decision trees in order to soften decision boundaries at decision nodes inherent in this type of trees. Comparisons between crisp decision trees and the fuzzified decision trees suggest that the later fuzzy tree is significantly more robust and produces a more balanced decision making. The new approach proposed in this chapter could incorporate any type of knowledge extraction algorithm. The proposed approach is implemented in a well known medical decision making problem to preview the effectiveness.

In chapter 4 Jose and Contreras present an FCM tool in which the values of concepts and relationships can change during execution. Using the tool users can design and follow the evolution of an FCM or modify previously saved FCMs. The tool allows the classical casual relationship to be defined in a variety of forms according to the problem modeled, i.e static or dynamic, fuzzy or not etc. The authors present the data structures, the interfaces, and the classes that compose the tool and their application to real life case studies.

In chapter 5 Kottas, Boutalis and Christodoulou present their research on Fuzzy Cognitive Networks (FCNs) which constitute an operational extension of FCMs. FCNs are capable of capturing steady state operational conditions of the system they describe and associate them with input values and appropriate weight sets. Acquired knowledge is stored in fuzzy rule based data bases, which can be used in determining subsequent control actions. This chapter presents basic theoretical results related to the existence and uniqueness of equilibrium points in FCN, adaptive weight estimation based on system operation data, the fuzzy rule storage mechanism and the use of the entire framework to control unknown plants. The results are validated using well known control benchmarks.

In chapter 6 Hurtado describes the use of a Fuzzy Expert System used for the modeling of operative loss exposure of the Allowances and Retirement Funds. It handles the issue of risk especially in conditions in which quantitative information is limited or in cases where information about risk factors are associated to expert's knowledge thus making the dependence on modeling in formal statistical tools very difficult. The system aims to provide a structured vision of the sources of operational risk and indicate to the managers in charge where they should concentrate their efforts in order to diminish exposure. This Fuzzy Expert System can help to complement the operative risk analysis carried out with quantitative methods like Extreme Value Theory or Montecarlo simulation.

The issue of risk is also handled by Xirogiannis, Glykas and Staikouras in chapter 7. The authors address the problem of designing an "intelligent" decision support methodology tool to act as a back end to financial planning and handling risk in profit and loss accounts (P&L). The methodology and tool proposed create a novel approach to supplementing typical financial strategy formulation projects by utilizing the fuzzy causal characteristics of FCMs to generate a hierarchical and dynamic network of interconnected P&L concepts. By using FCMs, the mechanism simulates the efficiency of complex hierarchical financial models with imprecise relationships and external stimuli while quantifying the impact of strategic changes to the overall P&L status. Generic maps that supplement the decision making process demonstrate a roadmap for integrating hierarchical FCMs into the P&L model of typical financial sector enterprises. Preliminary experiments indicate that *ex ante* reasoning of the impact of strategic changes (actual or hypothetical) to the status of financial performance can be effective and realistic, without employing detailed P&L numerical calculations.

In chapter 8 Salmeron handles the issue of Information Technology (IT) risk. He argues that as firms have spent billions of dollars in IT projects, IT risk management has by default become a critical issue. The applied efforts for bug-free IT implementation should be accompanied by mechanisms for managing implementation risks. The goal is to reduce the risk of implementation failure. Through this proposal, it is possible to observe which are the most relevant risks, and, above all, which have a greater impact on IT projects.

Chytas, Glykas and Valiris in chapter 8 also elaborate on the use of FCMs in IT. They have developed a model for software reliability prediction based on FCMs. Which helps testers to predict and manage software reliability. Despite the availability of various approaches developed in the field of software reliability, there are still issues that require further research in order to succeed in supporting the decision making process and improving software quality. FCMs capture information in the relationships between concepts, are dynamic, express hidden relationships, and are combinable and tunable. Preliminary experiments indicate that the proposed mechanism forms a sound support aid for software reliability modelling.

In chapter 10 Kottas, Karlis and Boutalis propose the use of Fuzzy Cognitive Networks (FCN) as a tool to maximize the efficiency photovoltaic (PV) power generation. They argue that as the output power of the photovoltaic modules depends on solar radiation and temperature of the solar cells it is necessary to track the maximum power point of the PV array and make the array operate near it. Maximum power operation is a challenging problem, since it requires that the system load is capable of using all power available from the PV system at all times. Fuzzy Cognitive Networks (FCN) have been proposed as an operational extension of Fuzzy Cognitive Maps (FCM), which work in continuous interaction with the system they describe and may be used to control it. In this chapter FCN is used to construct a maximum power point tracker (MPPT), which may operate in cooperation with a fuzzy MPPT controller. The proposed scheme outperforms other existing MPPT schemes of the literature giving very good maximum power operation of any PV array under different conditions such as changing insulation and temperature. Moreover it has the ability to adapt to different changes that might happen during the life cycle of the PV module, such as a destroyed cell of the PV array.

In chapter 11 Pajares, Guijarro, Herrera, Ruz and de la Cruz, have applied FCMs to computer vision tasks. Computer vision is an emerging area which is demanding solutions for solving different problems. The data to be processed are bi-dimensional (2D) images captured from the tri-dimensional (3D) scene. FCMs have been satisfactorily used in several areas of computer vision including: pattern recognition, image change detection or stereo vision matching. In their research they establish the general framework of FCM use in the context of 2D images and describe three applications in the three mentioned areas of computer vision.

In Chapter 12 Papakostas and Koulouriotis focus on the use of Fuzzy Cognitive Maps (FCMs) in classifying patterns, as an alternative to the traditional classifiers such as neural networks or even as collaborators, in achieving better classification capabilities. An FCM can simulate a typical classifier that maps a set of inputs to specific output values.

The classification capabilities of the FCM classifiers are studied in several pattern classification problems, while the ability of the FCM to store knowledge about the problem in hand is investigated in conjunction to the nodes' type of activation function and the inference law used. Appropriate experiments have been undertaken, in order to analyze the behavior of the FCM-based classifiers, in well known benchmark problems.

In chapter 13 Jose presents a Dynamical Fuzzy Cognitive Map (DFCM) in which casual relationships are based on fuzzy rules. The structure of the map changes during the phase of execution (runtime). He argues that DFCM is ideal to build supervision systems for multiagent systems (MAS), in order to study the behavior of the agents community when they fail, utilize a lot of resources, etc. DFCM is used to build a supervision system for a faults management system based on multiagent systems. Analysis of the results obtained prove that the use of these maps as supervisor of multiagent systems is very reliable.

In Chapter 14 Papageorgiou, Markinos and Gemtos investigate the use of FCMs in the yield and yield variability prediction in cotton crop. Cotton crop management is a complex process with interacting parameters like soil, crop and weather factors. FCMs were used for modeling and representing experts' knowledge. The methodology was evaluated in approximately 360 cases measured over 2001, 2003 and 2006 in a 5 ha cotton field. The results were compared with some benchmarking machine learning algorithms, which were tested for the same data set, with encouraging results. The authors concluded that the main advantage of FCM is the simple structure and the easy handling of complex data.

The use of FCMs in farming is the focus of chapter 15. Ortolani, McRoberts, Dendoncker and Rounsevell, use FCMs for the analysis of farmers behaviour and decision making. One important complicating factor in the process of policy evaluation and optimisation is that the people who generate the outcomes of policy (*i.e.* the farmers) are not the people who make the policy, and the farmers' individual and collective objectives may differ from those of the policy makers and, indeed, from each others. Under these circumstances the policy optimisation problem becomes a hierarchical (bi-level) problem and standard mathematical programming techniques are unlikely to produce optimal solutions. The authors use agent based modelling (ABM) to simulate farmers' responses to potential policy designs. This approach has the advantage of allowing for differences among famers' objectives while also retaining the hierarchical structure of the real world. However, in order to implement an ABM it is necessary to know the rules by which the virtual agents will behave in the model. FCMs provide a means to describe some of those rules in a formal and structued way, as well as providing additional useful information to the analyst. The methodology developed was tested in the context of specific case study of agri-environment measure uptake by Belgian farmers.

In chapter 16 Kafetzis, McRoberts and Mouratidou present the use of FCMs in two separate case studies concerned with water use and water use policy. One is a study concerning public participation in the Water Framework Directive (WFD) of the European Union (EU) and focuses on data collected in the Pinios river basin in Greece. The other is based on previously unpublished research by the authors on transboundary river issues in the Maritza river basin shared between Bulgaria, Greece and Turkey. Apart from the obvious similarities of location and their focus on water resources, the two studies are linked by some underlying factors. The authors present their argumentation on the choice of the authors FCMs as the best analytical method for this type of problems.

In chapter 17 Giordano and Vurro elaborate on the use of FCMs as a support mechanism for conflict analysis in drought management. Empirical investigations in scientific literature have highlighted the differences between the stakeholders' perceptions of a given drought phenomenon's severity and the results of scientific – technical evaluation. This means that there can be several perceptions of the phenomenon, and the scientific models used to assess the drought's severity do not consider these differences. Therefore, an in depth analysis of the potential conflicts and the definition of effective negotiation strategies could be really useful. The authors propose a methodology based on a FCMs to support the elicitation and the analysis of stakeholders' perceptions of drought, and the analysis of potential conflicts. The method was applied to a drought management process in the Trasimeno Lake area (Umbria Region) in order to analyze potential conflict.

Research Areas and Applications Addressed

As an amalgamation of the work presented in all chapters we can conclude that:

1. FCM research focuses on the following themes:

- **Theoretical mathematical underpinnings** (chapters 1,2,3,4,5,10).
- **Learning algorithms**: methods for analysing FCM historical data (Chapters 2, 5, 10).
- **FCM extensions**: leading to augmented fuzzy cognitive maps (chapters 3, 5, 10).
- **FCM tools**: that allow the design, simulation, and result reporting of fuzzy cognitive maps. (chapters 13, 7).
- **Expert Systems for Decision support**: methodologies and tools that allow pattern recognition, inferencing, classification and provided expert advice (chapters 3, 12).
- **Multiagent Systems**: use of FCMs in multiagent systems (chapters 13, 15)

2. FCM Applications focus on the following sectors:

- **Healthcare and Medical Decision Making** (chapters 1,3).
- **Risk management**: Operational, banking and information technology and other types of risk (chapters 6,7, 8).
- **Information technology** (chapters 8, 9).
- **Banking and Management** (chapter 7, 13).
- **Computer vision** (chapter 11).
- **Controllers and supervisory agents of complex systems** (chapters 5,10, 13).
- **Farming**: Cotton yield predictions, farmers' decision making etc. (Chapter 14, 15).
- **Water resource management** (Chapter 16).
- **Drought management** (Chapter 17).

It is certain that in the years to come we will experience many more advances in theory, methodologies and tools which will be applied, validated and tested in a big number of cases studies.

Strategic Directives and Vision

We were deeply honoured when Bart Kosko accepted our invitation to contribute the preface in this book.

We were all deeply honoured and thankful when we received his preface. In his preface Bart Kosko provided a strategic insight and a new vision to the field. If we are allowed his kind permission to rephrase his statements to a minor extent we regard this book as a milestone (Bart Kosko used the word "landmark") "in FCM research and in the history of knowledge engineering" but most importantly an inspiration "for many more decades of FCM research and applications to come".

And is exactly these two goals that all researchers in the field and practitioners that use FCMs should concentrate on from now on. In order to achieve these two goals we need to make fuzzy cognitive mapping a distinctive area of research and highlight its strengths and complementarity in comparison to other approaches like neural nets, analytical and network hierarchy process and other distinctive knowledge engineering techniques.

This book certainly helps in achieving this objective but other needed activities have also been undertaken. We have already created the "FCM Society" and we have discussed with well known journals the publications of dedicated issues. The organisation of an "FCM Society" workshop is underway. Special sessions in well known conferences will also be held.

By undertaking these activities we will put in place the necessary seed for FCM research and development to flourish in the current and next generations of researchers in the decades to come. The future, like always in life, will prove if have succeeded.

Michael Glykas

Contents

Fuzzy Cognitive Maps: Basic Theories and Their Application to Complex Systems

Peter P. Groumpos

Abstract. The challenging problem of modeling and controlling complex systems is investigated using Fuzzy Cognitive Maps (FCMs). A mathematical description of FCM models is presented, new construction methods and an algorithm are developed and extensively examined. The issue of modeling the supervisor of large complex systems is addressed and is modeled using a FCM. A manufacturing example is used to prove the usefulness of the proposed method. The problem of Decision Making process in Decision Analysis is considered and analyzed using FCM models. A successful application of FCM theory in a health problem is provided.

Keywords: Fuzzy Cognitive Maps, Modeling, Control Systems, Decision Systems.

1 Introduction

Most of today systems are characterized as complex systems with high dimension and a variety of variables and factors. It is widely recognized that conventional methods in modeling and controlling modern systems have contributed a lot in the research and on the solution of many control problems. However, their contribution to the solution of the increasingly problems associated with complex dynamical systems has proved to be limited. New methods have been proposed for complex systems that utilize existence knowledge and human experience and will have learning capabilities and advanced characteristics such as failure detection

Peter P. Groumpos
Laboratory for Automation and Robotics,
Department of Electrical and Computer Engineering,
University of Patras, 26500 Rion, Patras, Greece
Tel:.+30 61 996.449; Fax: +30 61 997309
e-mail: groumpos@ece.upatras.gr

M. Glykas (Ed.): Fuzzy Cognitive Maps, STUDFUZZ 247, pp. 1–22, 2010.
springerlink.com © Springer-Verlag Berlin Heidelberg 2010

and identification qualities. In this chapter Fuzzy Cognitive Maps (FCM) are proposed for modeling and controlling complex systems. The application of FCM may contribute to the effort for more intelligent control methods and for the development of autonomous systems. A Fuzzy Cognitive Map draws a causal picture to represent the model and the behavior of system. The concepts of an FCM interact according to imprecise rules and the operations of complex systems are simulated.

Fuzzy Cognitive Maps are symbolic representation for the description and modeling of the complex system [1]-[3],[6],[16],[17]. They consist of concepts, that illustrate different aspects in the behavior of the system and these concepts interact with each other showing the dynamics of the system. The human experience and knowledge of the operation of the system is used to develop Fuzzy Cognitive Map (FCM), as a result of the method by which it is constructed, i.e., using human experts that know the operation of system and its behavior in different circumstances. An FCM illustrates the system by a graph showing the cause and effect along concepts, and it is a simple way to describe the system's behavior in a symbolic manner, exploiting the accumulated knowledge of the complex system.

Fuzzy Cognitive Map (FCM) have been applied to a variety of scientific areas [4],[10],[12],[13],[15],[20]-[25],[30],[36]-[41]. FCMs have been used to describe and model the behavior of a system and its application in the modeling the supervisor of distributed systems. Fuzzy Cognitive Maps have been used for decision analysis and operation research. The objective of this chapter is to focus on the construction and the use of FCM in modeling complex systems. It will be shown that FCMs are useful to exploit the knowledge and experiences that human have accumulated for years on the operation of a complex system. Such methodologies are crude analogs of approaches that exist in human and animal systems and have their origins in behavioral phenomena related to these beings. So, a FCM represents knowledge in a symbolic manner and relates states, variables, events and inputs in an analogous to beings manner. This methodology can contribute to engineers' intention to construct intelligent systems, since as the more intelligent a system becomes, the more symbolic and fuzzy a representation it utilizes [7]-[12],[37],[42].

2 Basic Theories

Fuzzy Cognitive Maps (FCMs) consist of concept nodes and weighted arcs, which are graphically illustrated as a signed weighted graph with feed back. Signed weighed arcs, connecting the concept nodes, represent the causal relationship that exists among concepts. In general, concepts of a FCM, represent key-factors and characteristics of the modeled complex system and stand for: events, goals, inputs, outputs, states, variables and trends of the complex system been modeled. This graphic display shows clearly which concepts influences with other concepts and what this degree of influence is.

2.1 Fuzzy Cognitive Map Representation

Figure 1 illustrates a simple FCM consisting of five (5) concepts and nine (9) weighed arcs. Thus FCMs are directed graphs capable of modeling interrelationships or causalities existing among concepts. Concept variables and causal relations constitute the fundamental elements of an FCM. Concept variables are represented by nodes, such as C_1, C_2, C_3, C_4 and C_5 in figure 1. Causal variables always depict concept variables at the origin of arrows; effect variables, on the other hand, represent concepts-variables at the terminal points of arrows. For example, looking in figure 1, at $C_1 \rightarrow C_2$, C_1 is said to impact C_2 because C_1 is the causal variable, whereas C_2 is the effect variable. Each concept is characterized by a number A_i that represents its value and it results from the transformation of the real value of the system's variable, for which this concept stands, in the interval [0,1]. Causality between concepts allows degrees of causality and not the usual binary logic, so the weights of the interconnections can range in the interval [-1,1]. Fuzzy Cognitive Map models a system as an one-layer network where nodes can be assigned concept meanings and the interconnection weights represent causal relationships among concepts.

A Fuzzy Cognitive Map is a graph shows the degree of causal relationship - among concepts of the map knowledge expressions and the causal relationships are expressed by and fuzzy weights.

Existing knowledge on the behavior of the system is stored in the structure of nodes and interconnections of the map. Each one of the key-factors of the system. Relationships between concepts have three possible types; a) either express positive causality between two concepts ($W_{ij} > 0$) b) negative causality ($W_{ij} < 0$) and c) no relationship ($W_{ij} = 0$). The value of W_{ij} indicates how strongly concept C_i influences concept C_j. The sign of W_{ij} indicates whether the relationship between concepts C_i and C_j is direct or inverse. The direction

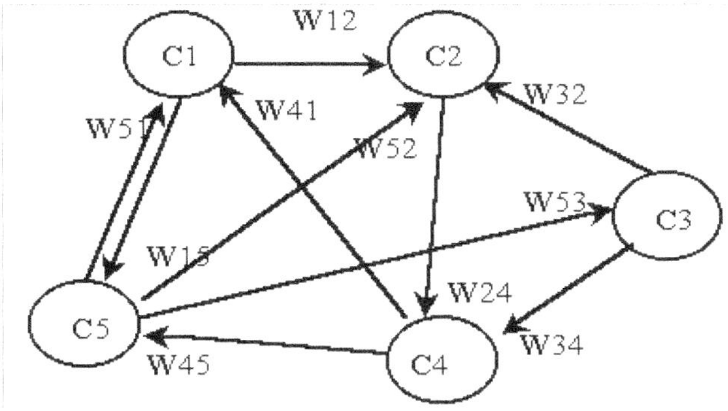

Fig. 1 A simple Fussy Cognitive Map Drawing

of causality indicates whether concept C_i causes concept C_j, or vice versa. These parameters have to be considered when a value is assigned to weight W_{ij}.

A new formulation for calculating the values of concepts at each time step, of a Fuzzy Cognitive Map, is proposed:

$$A_i^t = f(k_1 \sum_{\substack{j=1 \\ j \neq i}}^{n} A_j^{t-1} W_{ji} + k_2 A_i^{t-1}) \tag{1}$$

The k_1 expresses the influence of the interconnected concepts in the configuration of the new value of the concept A_i and k_2 represents the proportion of the contribution of the previous value of the concept in the computation of the new value and the. This new formulation assumes that a concept links to itself with a weight $W_{ii} = k_2$.

In this chapter, it is assumed that the influence of the previous value of each concept is high and it is supposed that k_1==1. This means that the previous value of each concept has a great influence in the determination of the new value. The inclusion of the previous value of each concept in the calculation rule, results in smoother variation on the values of concepts after each recalculation of their value. This will become apparent when the supervisor is modeled using a FCM of section 3. The value A_i for each concept C_i is calculated by the following rule:

$$A_i^t = f(\sum_{\substack{j=1 \\ j \neq i}}^{n} A_j^{t-1} W_{ji} + A_i^{t-1}) \tag{2}$$

Equation (1.2) is the result of equation (1) setting $k_1 = k_2 = 1$. Namely, A_i^t is the value of concept C_i at time t, A_i^{t-1} the value of concept C_i at time $t-1$, A_j^{t-1} the value of concept C_j at time $t-1$, and the weight W_{ji} of the interconnection from concept C_j to concept C_i. The function f is a threshold function and to squash the result in the interval [0,1]. Fuzzy Cognitive Maps have discrete nature, at each time step, values of all concepts are recalculated and change according to equation 2. This procedure is called a running cycle of the FCM model and it is very fundamental for the theory of FCM.

Two kinds of threshold functions are used in the Fuzzy Cognitive Map framework, the unipolar sigmoid function, where $\lambda \succ 0$ determines the steepness of the continuous function f :

$$f(x) = \frac{1}{1 + e^{-\lambda x}}$$

When nature of concepts can be negative, their values belong to the interval [-1,1], the following function is used:

$f(x) = \tanh(x)$

The simplicity of FCM model becomes apparent from its mathematical representation and operation. Suppose that an FCM is consisted by n concepts. It is mathematically represented by a 1x n state vector $\mathbf{A}$, which gathers the values of the n concepts and by an nxn weight matrix $\mathbf{W}$. Each element W_{ij} of the matrix $\mathbf{W}$ indicates the value of the weight W_{ij} between concept C_i and C_j. The diagonal of the matrix is zero since it is assumed that no concept causes itself and thus $W_{ii} = 0$.

Equation 2 can be transformed in the following equation, which describes the FCM operation with the compact mathematical model:

$$\mathbf{A}^t = f(\mathbf{A}^{t-1}\mathbf{W} + \mathbf{A}^{t-1}) \tag{3}$$

So, equation (3) computes the new state vector $\mathbf{A}^t$, which depends on the previous state of A and from the multiplication of the previous, at time $t-1$, state vector $\mathbf{A}^{t-1}$ by the weight matrix $\mathbf{W}$. The equation 3 can also be expressed as:

$$\mathbf{A}^t = f(\mathbf{A}^{t-1}\mathbf{W}^{new}) \tag{4}$$

Where, the new weight matrix $\mathbf{W}^{new}$ is the weight matrix $\mathbf{W}$ of the Fuzzy Cognitive Map with the entire diagonal elements equal to unit, which means that each concept causes itself with a weight $W_{ii} = 1$. This is a new approach, which differs from other representations of Fuzzy Cognitive Map in the literature, where it is assumed that no concept cause itself and the diagonal of the W matrix, is zero.

2.2 *Methods for Constructing Fuzzy Cognitive Maps*

The development and construction of Fuzzy Cognitive Map (FCM) have great importance for its use in the modeling of complex systems. Let us remind ourselves that FCM represent the human knowledge on the operation of the system. Experts develop FCMs, using their experience and knowledge on the complex system. Construction methodologies rely on the exploitation of experts' experience on system's model and behavior. Experts determine the number and kind of concepts that consist an FCM and the interrelationships among its concepts. Experts know the main factors that determine the behavior of the complex system, each one of these factors is represented by a concept. Experts according to their experience, determine concepts of FCM that stand for events, actions, goals. values, and trends of the complex system. Experts know which elements of the system influence other elements; for the corresponding concepts they determine the negative or positive effect of one concept to the others, with a fuzzy degree of causation. The determination of the degree of casual relationship among concepts can be improved by the application of learning rules for choosing appropriate weights for the FCM. In this way, an expert decodes his own

knowledge on the behavioral model of the system and transforms this knowledge in a weighted graph.
using

2.2.1 Assigning Numerical Weights

Knowledge on the behavior of a complex system is rather subjective and in order to construct a more accurate model of the complex system it is proposed to utilize, the experience of a group of experts. Experts are polled together and they examine the relevant factors that stand as nodes of an FCM. So, they decide the number of concepts, which consist the FCM and what characteristic of the system each concept represents. Then, the experts are individually asked to express the causal relationship among these concepts. The result of this procedure will be a collection of individual FCMs, with the same nodes but different links among concepts or/and different weights of interconnections. The individual FCMs must be combined into one collective FCM and a method to combine the individual maps. A first approach could be the summation of different weight matrixes:

$$\mathbf{W} = f\left(\sum_{1}^{N} \mathbf{W}_k\right) \tag{5}$$

Where $\mathbf{W}$ is the overall matrix, $\mathbf{W}_k$ is the individual weight matrix, which each one of the N experts has developed, and f is a threshold function, usually a type of the sigmoid function that will transform the sum of weights in the interval [-1,1].

It is accepted that experts have different experience and subjective knowledge on the system. Thus it is considered that there are experts of different credibility on the knowledge of the system, and for these experts their contributions on constructing FCMs may be multiplied by a nonnegative 'credibility' weight b_k before combining them with other expert's opinions.

$$\mathbf{W} = f\left(\sum_{1}^{N} b_k \mathbf{W}_k\right) \tag{6}$$

Where b_k is the credibility weight for the k_{th} expert, $\mathbf{W}_k$ is the weight matrix of k_{th} expert's Fuzzy Cognitive Map and N is the number of the experts. But in this case, another mechanism must be used to determine who and how credibility weights can be assigned to every expert. As an example one expert could be "penalized" with an extremely low or zero credibility weight if the expert's choice differs from other experts' average weight choice, by some predetermine rule.

A new advanced algorithm is proposed in order to assign weights for each interconnection and credibility weights for experts. Every expert constructs an individual FCM. Then for each one interconnection of the overall FCM the corresponding weight from each individual map are collected together and compared according to the following algorithm. In this algorithm, the average value of the proposed weights for each interconnection is used.

First of all, the sign of the proposed weights are examined. If the number of weights with the same sign is less than $\pi * \mathrm{N}$, this means that it is not very clear among experts the positive or negative causality between two concepts and so they are asked to reassign weights. Otherwise, the procedure continues and the proposed weights are used to determine the weight. Each expert that assigns weight for one interconnection not close enough to the average weight is penalized and the corresponding weight is partially taking into account. This mechanism is implemented using the following algorithm:

Algorithm 1

Step 1: For all the N *experts, set credibility weight* $\mathrm{b}_k = 1$

Step 2: For i,j=1 to n

Step 3: For each interconnection (C_i *to* C_j *) examine the* N *weights* W_{ij}^k *that each* k_{th} *of the* N *experts has assigned.*

Step 4: IF there are weights W_{ij}^k *with different sign and the number of weights with the same sign is less than* $\pi * \mathrm{N}$

THEN

ask experts to reassign weights for this particular interconnection and go to step3

ELSE

take into account the weights of the greater group with the same sign and consider that there are no other weights and penalize the experts who chose "wrong" sighed weight with a new credibility weight $\mathrm{b}_k = \mu_1 * \mathrm{b}_k$

Step 5: For the weights with the same sign, find their average value

$$W_{ij}^{ave} = \frac{\left(\sum_{k=1}^{N} \mathrm{b}_k W_k\right)}{N}$$

Step 6: IF $\left|W_{ij}^{ave} - W_{ij}^k\right| \geq \omega_1$ *THEN consider that there is no weight* W_{ij}^k*, penalize the* k_{th} *expert* $\mathrm{b}_k = \mu_2 * \mathrm{b}_k$ *and go to step 5*

Step 7: IF there have not examined all the $n \times n$ *interconnection go to step 2*

ELSE construct the new weight matrix $\mathbf{W}$ *which has elements the weights* W_{ij}^{ave}

END.

Example 2.1
Six experts have constructed six individual FCMs. Experts have suggested the following weights for the one interconnection from concept C_i *to concept* C_j :

$$W_{ij} = [-0.5, 0.6, 0.66, 0.7, 0.65, 0.25].$$

For this example, the requested number of weights with the same sign is $\pi = 0.8$ *and* $\omega_1 = 0.2$ *and* $\mu_1 = \mu_2 = 0.9$ According to step 3 of the algorithm, the majority of experts have assigned positive weights to that interconnection so the 1st expert is penalized with credibility weight $b_1 = \mu_1 * b_1 = 0.9b_1$ and the corresponding weight is dropping out. Then, in step 4 of the algorithm, the average weight is computing $W_{ij}^{ave} = 0.572$, and it is compared with other weights, but according to step 6 of the algorithm, the weight suggested by the 6th expert with value 0.25 is excluded from the calculation and the 6th expert is penalized. The rest of the weights are used to calculate the new average weight. In this case, the chosen weight has the value $W_{ij}^{ave} = 0.652$ for this particular weight interconnection. The same methodology is used to assign weights for all the interconnections and construct the overall Fuzzy Cognitive Map.

2.2.2 Assigning Linguistic Variables for FCM Weights

Another methodology to construct a Fuzzy Cognitive Map that is closer to fuzzy logic is proposed now. Experts are asked to describe the causality among concepts using linguistic notions. Every expert will determine the influence of one concept to the other as "negative" or "positive" and then he will describe the grade of influence with a linguistic variable such as "strong", "weak" and etc [8].

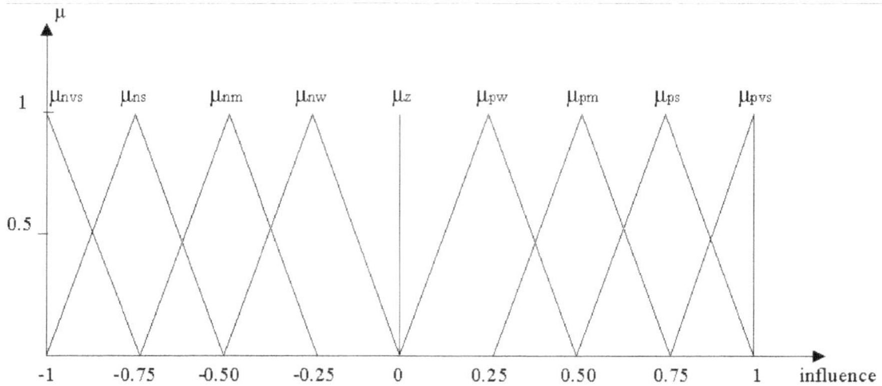

Fig. 2 Terms of the linguistic variable influence

Influence of one concept on another, is interpreted as a linguistic variable taking values in the universe *U=[-1,1]* and its term set *T(influence)* could be:

T(influence) = {negatively very strong, negatively strong, negatively medium, negatively weak, zero, positively weak, positively medium, positively strong, positively very strong}

The semantic rule *M* is defined as follows and these terms are characterized by the fuzzy sets whose membership functions are shown in Figure 2:

M(negatively very strong)= the fuzzy set for "an influence below to -75%" with membership function μ_{nvs}
M(negatively strong)= the fuzzy set for "an influence close to -75%" with membership function μ_{ns}
M(negatively medium)= the fuzzy set for "an influence close to -50%" with membership function μ_{nm}
M(negatively weak)= the fuzzy set for "an influence close to -25%" with membership function μ_{nw}
M(zero)= the fuzzy set for "an influence close to 0" with membership function μ_z
M(positively weak)= the fuzzy set for "an influence close to 25%" with membership function μ_{pw}
M(positively medium)= the fuzzy set for "an influence close to 50%" with membership function μ_{pm}
M(positively strong)= the fuzzy set for "an influence close to 75%" with membership function μ_{ps}
M(positively very strong)= the fuzzy set for "an influence above to 75%" with membership function μ_{pvs}

The linguistic variables that describe each interconnection are combined and the overall linguist variable will be transformed in the interval [-1,1]. A numerical weight for each interconnection will be the outcome of the defuzzifier, where the Center of Gravity method is used to produce this crisp weight [10]. This methodology has the advantage that experts do not have to assign numerical causality weights but to describe the degree of causality among concepts.

2.2.3 Synthesizing different Fuzzy Cognitive Maps

A distributed system is considered and for each subsystem a distinct FCM is constructed. Then all FCMs can be combined in one augmented Fuzzy Cognitive Map with a weight matrix $\mathbf{W}$ for the overal system. The unification of the distinct FCM depends on the concepts of the segmental FCM, if there are no common concepts among different maps; the combined matrix $\mathbf{W}$ is constructed according to the equation 7 (see below). In this case, there are K different FCM matrices, with weight matrices $\mathbf{W}_i$ and the dimension of matrix $\mathbf{W}$ is $n \times n$ where n equals the total number of distinct concepts in all the FCMs.

$$\mathbf{W}=\begin{bmatrix} \mathbf{W}_1 & & & \\ & \mathbf{W}_2 & \underline{0} & \\ & \underline{0} & \ddots & \\ & & & \mathbf{W}_K \end{bmatrix} \tag{7}$$

Example 2.2
It is assumed that there are two Fuzzy Cognitive Maps, F_1 with concepts C_1, C_2, C_3 and F_2 with concepts C_4, C_5, C_6. Weight matrices for F_1 and F_2 are:

$$\mathbf{W}_1=\begin{bmatrix} 0 & 0 & W_{13} \\ W_{21} & 0 & 0 \\ W_{31} & W_{32} & 0 \end{bmatrix} \text{ and } \mathbf{W}_2=\begin{bmatrix} 0 & W_{45} & W_{46} \\ W_{54} & 0 & W_{56} \\ 0 & W_{65} & 0 \end{bmatrix}$$

The augmented weight matrix will be:

$$\mathbf{W}=\begin{bmatrix} \mathbf{W}_1 & 0 \\ 0 & \mathbf{W}_2 \end{bmatrix}=\begin{bmatrix} 0 & 0 & W_{13} & 0 & 0 & 0 \\ W_{21} & 0 & 0 & 0 & 0 & 0 \\ W_{31} & W_{32} & 0 & 0 & 0 & 0 \\ 0 & 0 & 0 & 0 & W_{45} & W_{46} \\ 0 & 0 & 0 & W_{54} & 0 & W_{56} \\ 0 & 0 & 0 & 0 & W_{65} & 0 \end{bmatrix}$$

But, in most cases, the unification is used because there are common concepts among the distinct FCM and the intention is the construction of an enhanced Fuzzy Cognitive Map. In this case, there will be an overlapping between some of the diagonal elements-matrices of matrix $\mathbf{W}$ in equation 7. Overlapping represent weights of interconnections between concepts that belong to different FCMs. Then, segmental FCMs with common concepts are combined together, calculating new weights for the interconnection between common concepts. If there are more than one common concept between the Fuzzy Cognitive Maps, there will be proposed two or more weights for the same interconnection. In this case, as new weight will be the average of weights v. Then, equation 7 is implemented to construct the weight matrix of the overall Fuzzy Cognitive Map. It is consisted of n concepts that correspond to the total number of the different concepts that there are in all the segmental FCMs

Example 2.3
It is assumed that there are two Fuzzy Cognitive Maps, F_1 with concepts C_1, C_2, C_3 and F_2 with concepts C_2, C_3, C_4. C_5, Weight matrices for F_1 and F_2 are:

$$\mathbf{W}_1 = \begin{bmatrix} 0 & 0 & W_{13} \\ W_{21} & 0 & 0 \\ W_{31} & W_{32} & 0 \end{bmatrix} \text{ and } \mathbf{W}_2 = \begin{bmatrix} 0 & W_{23} & W_{24} & 0 \\ W_{32} & 0 & W_{34} & W_{35} \\ W_{42} & W_{43} & 0 & 0 \\ W_{52} & W_{53} & W_{54} & 0 \end{bmatrix}$$

The augmented Fuzzy Cognitive Map will have five concepts and its weight matrix will be:

$$\mathbf{W} == \begin{bmatrix} 0 & 0 & W_{13} & 0 & 0 \\ W_{21} & 0 & W_{23} & W_{24} & 0 \\ W_{31} & W_{32}^{ave} & 0 & W_{34} & W_{35} \\ 0 & W_{42} & W_{43} & 0 & 0 \\ 0 & W_{52} & W_{53} & W_{54} & 0 \end{bmatrix}$$

2.3 Neural Network Nature of Fuzzy Cognitive Maps

Fuzzy Cognitive Maps have been described as a hybrid methodology, because it utilizes characteristics of fuzzy logic and neural networks. The development and construction of FCMs have shown their fuzzy nature. Learning rules, used in Neural Networks theory, they are used to train the Fuzzy Cognitive Map. Parameter learning of FCM concerns the updating of connection weights among concepts.

The construction of FCM is based on experts who determine concepts and weighted interconnections among concepts. This methodology may lead to a distorted model of the system because human factor is not always reliable. In order to refine the model of the system, learning rules are used to adjust weights of FCM interconnections. The Differential Hebbian learning rule has been proposed to be used in the training of a specific type of FCMs. The Differential Hebbian learning law adjusts the weights of the interconnection between concepts t grows a positive edge between two concepts if they both increase or both decrease and it grows a negative edge if values of concepts move in opposite directions. Adjusting the idea of differential Hebbian learning rule in the framework of Fuzzy Cognitive Map, the following rule is proposed to calculate the derivative of the weight between two concepts.

$$w'_{ji} = -w_{ji} + s(A_j^{new})s(A_i^{old}) + s'(A_j^{new})s'(A_i^{old}) \tag{8}$$

Where $S(x) = \dfrac{1}{1+e^{-\lambda x}}$

Appropriate learning rules for Fuzzy Cognitive Maps need more investigation. These rules will give FCMs useful characteristics such as the ability to learn arbitrary non-linear mappings, capability to generalize to situations the adaptivity and the fault tolerance capability [4], [13], [47], [48].

3 Modeling Supervisors of Complex Systems with a FCM

Complex systems are characterized with high dimension and their dynamics are quite often unknown. A defining characteristic of complex systems in their tendency to self – organize globally as a result of many local interactions. In other words, organization occurs without any central organizing structure or entity. Therefore, conventional techniques cannot easily handle this kind of systems. The application of Fuzzy Cognitive Maps (FCM) for the modeling of the supervisor of complex systems seems to be a prospective methodology. The hierarchical structure of fig. 3 is proposed to model Large Scale complex systems. At the lower level of the structure lies the plant, which is controlled through conventional controllers. These controllers perform the usual tasks and reflect the model of the plant during normal operation conditions using conventional control techniques. The supervisor of the system is modeled as a Fuzzy Cognitive Map (FCM). The first attempt to model a supervisor of a scale complex system utilizing the concept of FCM was made early in 2000. [20]-[25]There is an amount of information that must pass from the lower level to the Supervisor-FCM. So an interface is needed, which will process, transform and communicate information from the lowerlocal controllers to the FCM on the upper level. The Fuzzy Cognitive Map will interact using equation 2; concepts of FCM will have new values that must be transmitted to the conventional controllers. So, the interface will follow the opposite direction. In this way changes on one or more concepts of the FCM could mean change in the value of one or more elements of the system.

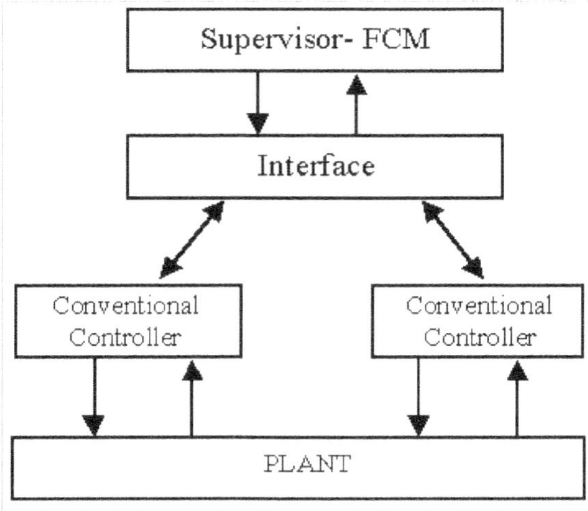

Fig. 3 An FCM – supervisor of a complex system

The model of Fuzzy Cognitive Map can be expanded to include advanced features, such as fault diagnosis, effect analysis or planning and decision making characteristics. Some of the concepts of the FCM could stand for device failure modes, their effects and causes, a subsystem's normal or irregular operation, the functionality of the system, the failures, the system mission, and the ultimate function of the system. So the FCM would represent the failure modes and their effects and the relations among them, that an expert uses to describe the functionality of the system and the failures.

A very interesting quality of FCMs is their ability in predicting and redesigning of the system. This can help the designer in evaluating what, would happen if some parameters of the system have been altered. Another useful characteristic of the FCM is its efficiency in prediction and especially to predict what would be the result of a scenario or what will be the consequences for the whole process if a state changes suddenly. This feature is especially useful for designers of systems to observe the influence of each device separately.

With Fuzzy Cognitive Maps the knowledge and human operator experience is exploited. The human coordinator of a system should know the operation of critical aspects of the whole system and uses a mental model consisted of concepts to describe it. He relates the operation of one subsystem or two different subsystems to a concept or a concept stand for a specific procedure.

FCM models the supervisor and it is consisted of concepts that may represent the irregular operation of some elements of the system, failure mode variables, failure effects variables, failure cause variables, severity of the effect or design variables. Moreover, this FCM will include concepts for determination of a specific operation of the system and it will be used for strategic planning and decision analysis. The supervisor FCM, will represent vital components of the plant and willreflect the operational state of the plant. The development of this FCM requires the integration of several experts opinions in order to construct a FCM with diagnosis and predictive capabilities. We need to point out here that conventional control methods cannot be used to model this supervisor of a complex system, a This approach is best illustrated with the following example.

Example 3.1

Four experts working on a chemical industry, which produce refreshments from water, sugar, fruit juice etc, were asked to develop a Fuzzy Cognitive Map. This FCM will be used as a supervisor of the whole plant, which will describe the operation of a process, the final product of the process and the different aspects that determine the quality of the product. Experts developed the Fuzzy Cognitive Map, which is depicted on figure 4. They decided that the most important concept is the quality of the produced product. They developed an FCM around the main concept C1, which represents the "product degradation" of the final product. Then, experts determined other concepts of the real system that influence this concept, so concept C1 depends on:

Concept C2 "the internal variation of the process",
Concept C3 "the poor quality of the input material",
Concept C4 "wear and tear machine parts",

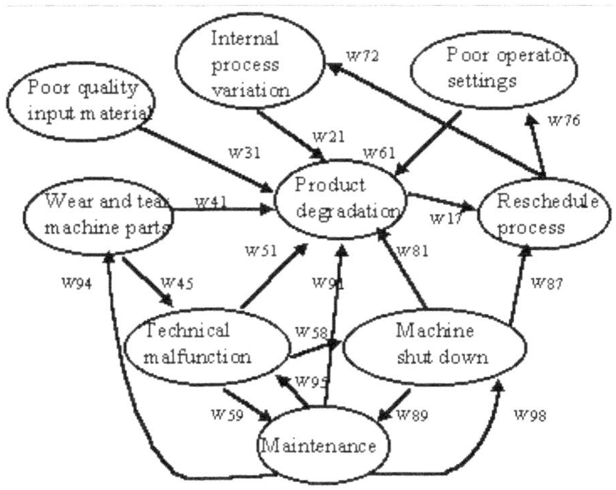

Fig. 4 Proposed Supervisor – Fuzzy Cognitive Map

Concept C5 “technical malfunction”,
Concept C6 “poor operator settings”,
Concept 7 "reschedule the process"
Concept C8 “machine shut down”.
Concept 9 "maintenance"

Then, the interrelations among concepts were determined with the following logical procedure. The value of concept 1 “degradation of product” increases the need to "reschedule the process" which is presented as concept C7.

Concept C7 decreases the value of concepts C6 “poor operator setting” and concept C2 “Internal process variation”.
Concept C4, which stand for “wear and tear machine parts”, has a positive influences on concept C5 “technical malfunction”.
Concept C5 “technical malfunction” increase the amount of concept C9 “the maintenance” and the amount of concept C8 “the machine shut down”.
Concept C9 “maintenance” decreases the amount of the following concepts: concept C5 “technical malfunction”, concept C8 “the machine shut down” and concept C4 “wear and tear machine parts”.
Concept C8 “machine shut down” increases the amount of concept C7 “reschedule process” and increases the value of concept C9 “maintenance”.

Then, experts were asked to assign values on the interconnections among concepts. Four FCMs were constructed with the same concepts, but with 4 different weights on each interconnection. Then algorithm of section 2.2.1 was implemented and an augmented Fuzzy Cognitive Map was constructed, which is depicted on figure 5 and it is used as supervisor of the plant. The experts decided on the values of weighted arcs, W_{ij}, figure 4 and is given in figure 5.

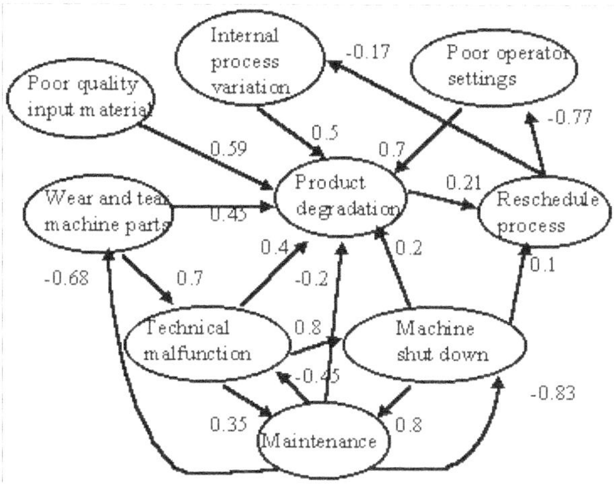

Fig. 5 Supervisor- Fuzzy Cognitive Map with weights

Table 1 The alues of concepts of the supervisor-FCM for 6 simulation steps

C1	C2	C3	C4	C5	C6	C7	C8	C9
0.24	0.48	0.20	0.10	0.15	0.40	0.00	0.07	0.61
0.65	0.50	0.30	0.40	0.45	0.50	0.51	0.40	0.53
0.74	0.48	0.30	0.41	0.51	0.50	0.54	0.48	0.62
0.73	0.48	0.20	0.40	0.51	0.40	0.55	0.47	0.64
0.72	0.48	0.20	0.39	0.50	0.40	0.55	0.47	0.64
0.72	0.48	0.20	0.39	0.50	0.40	0.55	0.47	0.64

For the constructed Fuzzy Cognitive Map, values were assigned to the concepts and the simulation of the FCM starts. Equation 2 is used to calculate the new values of concepts after each step of the FCM. Table 1 gathers the initial values of concepts and their values for six simulation cycles. FCM reaches an equilibrium point and if a new value for one or more concepts comes from the lower level then after a limited number of cycles, FCM will reach another equilibrium point.

The development of the supervisor-FCM that is dedicated to a particular plant depends on the supervisory-coordinator tasks that the user of the overall system requires. A complete Fuzzy Cognitive Map would include a decision making part and a planning part.

4 Decision Analysis and Fuzzy Cognitive Maps

Decision analysis is based on a number of quantitative methods that aid in choosing amongst alternatives. Traditional decision analysis is used to indicate decisions favouring good outcomes even though there is an uncertainly surrounding the decision itself. Furthermore, the value of each possible outcome of a decision, whether measured in costs and benefits or utility, usually varies.

Over the last years, several approaches have been investigated in the field of Decision Analysis, with the most popular one to be used that of Decision Trees (DT). Some methods combine DT with other machine learning techniques, such as Neural Networks [18] or Bayesian Networks [19]. However, very little work has been reported in combining DT with FCMs. Some research work of this combination has seen the literature the last ten years [42]-[45]. In this chapter the technique of combining a DT with a FCM model in Decision Analysis is presented.

The derived FCM model is subsequently trained using an unsupervised learning algorithm to achieve improved decision accuracy. In this chapter, the C4.5 has been chosen as a typical representative of the decision tree approach [14]. Similarly, the Nonlinear Hebbian Learning (NHL) algorithm is chosen as a representative of unsupervised FCM training.

The DT-FCM's function is briefly outlined in Figure 6. If there is a large number of input data, then the quantitative data are used to induce a Decision Tree and qualitative data (through experts' knowledge) are used to construct the FCM model. The FCM's flexibility is enriched by the fuzzification of the strict decision tests (derived fuzzy IF-THEN rules to assign weights direction and values). Finally, the derived FCM model (new weight setting and structure) is trained by the unsupervised NHL algorithm to achieve a decision.

This methodology can be used for three different circumstances, depending on the type of the initial input data: (1) when the initial data are quantitative, the DT generators are used and an inductive learning algorithm produce the fuzzy rules which then are used to update the FCM model construction; (2) when experts' knowledge is available, the FCM model is constructed and through the unsupervised NHL algorithm is trained to calculate the target output concept responsible for the decision line; and (3) when both quantitative and qualitative data are available, the initial data are divided and each data type is used to construct the DTs and the FCMs separately. Then the fuzzy rules induced from the inductive learning restructure the FCM model enhancing it. At the enhanced FCM model the training algorithm is applied to help FCM model to reach a proper decision.

The new technique has three major advantages. First, the association rules derived from the decision trees have a simple and direct interpretation and introduced in the initial FCM model to update its operation and structure. For example, a produced rule can be: If the *variable 1* (input variable) has *feature A* Then the *variable 2* (output variable) has *feature B*.

Second, the procedure that introduces the Decision Tree rules into an FCM also specifies the weight assignment through new cause-effect relationships among the FCM concepts. Third, as will be demonstrated through the experiments, this technique fares better than the best Decision Tree inductive learning technique and the FCM decision tool.

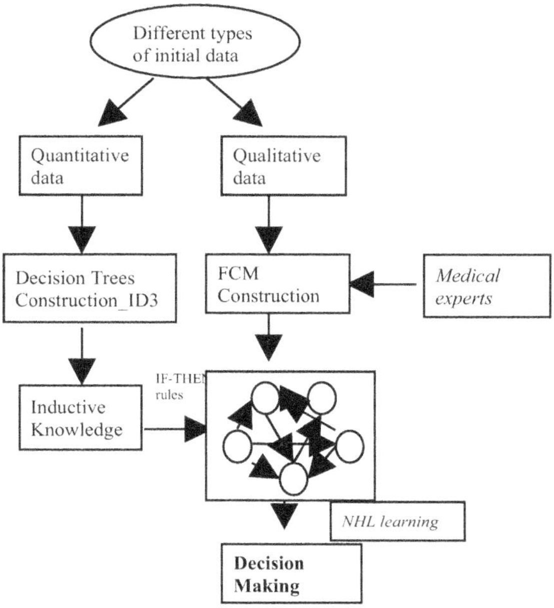

Fig. 6 The decision making system constructed by Decision Trees and Fuzzy Cognitive Maps

5 Implementation of the DT-FCM Model for Bladder Tumor Grading

The above new DT-FCM technique was used by our research team on a number of medical problems [13], [15]. Here, a representative example is presented.

Ninety-two cases of urinary bladder cancer were collected from the archives of the department of pathology of University Hospital of Patras Greece. Histopathologists-experts had diagnosed 63 cases as low-grade and 29 as high-grade using conventional WHO grading system. Following grade diagnosis, each tissue section was evaluated retrospectively, using a list containing eight well documented in the bibliography histopathological criteria essential for tumour grading. The FCM model for tumor grading had been developed and presented analytically in [13]. The FCM grading tool was able to give distinct different values for the majority of high-grade and low-grade cases using a simple Bayesian classifier for the output data. Except the experts' knowledge for determining FCM model, quantitative data for the eight main histopathological features [13,15] were also available and used for constructing DT. Then through the inductive learning procedure, a set of association rules were derived. Some of the best association rules, based on their confidence levels, are given in the Table 2. The necessary If-Then rules were induced and introduced in the FCM model enhancing its initial structure.

Table 2 Example of Association Rules derived from Decision Trees

Rules	Result/Decision Leaves
Cell-size=uniform, mitosis=absent rate	Grade Low
Cell-distribution=even, nucleoli=inconspicuous	Grade Low
Cell-distribution=clustered, cell-size=pleomorphic	Grade High
Nuclei=uniform, mitosis=absent rate	Grade Low
Cell-size=uniform, cell-number=numerous, nucleoli=inconspicuous	Grade Low

After the development of the DT-FCM model and the determination of specifications for the implementation of the NHL algorithm, the hybrid system was used to examine cases and assigned sensitivity and specificity for grading bladder tumors. The same data set that used in previously proposed FCM-TG model, were also used to evaluate the performance of the DT-FCM methodology in categorizing tumors as low grade or high grade. The results for average sensitivity and average specificity for the ninety two bladder tumour cases were 80% and 90% respectively using the DT-FCM, whereas the resulting accuracies for low grade and high grade cases were 79% and 87.5% through the FCM grading tool [15].

Our obtained results through the implementation of the proposed DT-FCM methodology are very promising and encourage us to continue our effort towards this direction.

Actually, our research group works on improving the medical diagnosis process by different means: (1) introducing a methodology based on FCMs for decision making in complex medical systems where experts' knowledge is available; (2) constructing modular FCMs for characterizing tumor grading; and (3) certain histopathological features-attributes (for example, cell distribution, nuclei, mitosis, necrosis) have been converted into discrete values, although their conceptual vagueness could be quantified by the degree of membership of a numerical value in a fuzzy set. Thus their values would be a user defined finite set of linguistic values.

A research team of the Laboratory for Automation and Robotics of the University of Patras has used extensively FCMs in modeling and analyzing medical problems and has obtained some very interesting results [13],[15],[30], [46], [47], [48].

6 Summary and Closing Remarks

In this chapter the analysis of complex systems has been investigated using the exciting and promising models of Fussy Cognitive Maps (FCM).

Fundamental mathematical theories of FCM were developed and extensively analyzed. A new algorithm was developed and used to demonstrate the usefulness of the FCM approach in modeling complex systems. Fuzzy Cognitive Map (FCM) theory, a new soft computing approach, utilizes existing experience in the operation of a complex system and combing fuzzy logic and neural networks. For such complex systems it is extremely difficult to describe the entire system with a precise mathematical model. Thus, it is more simple and useful to divide the whole plant in virtual parts and to construct an FCM for each part. The experience of different specialists who can easily judge the variables and states of a small process and then unify these to construct the final system by integrating the different Fuzzy Cognitive Maps into an augmented one have been utilized. This approach represents systems in a graphical way showing the causal relationships between states-concepts and accomplishes the unification of superposing small subsystems. FCMs offer the opportunity to produce better knowledge based on systems applications, addressing the need to handle uncertainties and inaccuracies associated with real world problems.

The issue of modeling the supervisor of a complex system was addressed and analyzed. Then, using the theory of FCM, it was modeled in a hierarchical structure where the plant was controlled using conventional controllers. A simple example from manufacturing field was given demonstrating clearly the usefulness if the proposed approach. The supervisor was modeled with nine (9) concepts and eighteen (18) weighed interconnecting arcs. The concepts of the supervisor – FCM were very interesting features of a manufacturing plant such as: machine shut down, poor operator settings, poor quality input materials, technical malfunction, maintenance and others. The inclusion of all these features as concepts in a supervisor – FCM and ability to run extensive simulation with real data can prove very useful to plant management. Another very interesting feature of this proposed approach is the ability to use several expert opinions, giving the opportunity to predict the degradation of a product. The simulation studies show that after only a few recursive steps the FCM achieves a diagnosis for the desired product. Methods for developing supervisors for manufacturing plants using the theory of FCM is needed and could be the research of future direction in this exciting field.

The basic theories of FCM were then used to address and analyze the difficult problem of Decision Trees (DT) from Medical Decision Making Models. A new integrated system has been developed to assist medical decisions. The new framework has proposed the combination of FCM and Decision Trees in a new integrated DT-FCM system. The performance of the new structure can deal with different kind of input data eliminating numerical errors. A simple medical problem, in which the case of urinary bladder cancer was considered and studied using the proposed DT-FCM structure. The simulation results were very encouraging and had given better results than existing techniques. Future work here should be directed to use extensively the proposed DT- FCM structure to further study similar health problems and compare the results with other today's techniques. Another promising research direction is to further investigate learning (both supervised and unsupervised) techniques in combination with the proposed DT-FCM structure and to be used on a number of challenging and difficult medical problems.

Closing this chapter, it is of interest to raise a basic fundamental question: what is the best way to approach the difficult issue of modeling and controlling complex systems. An issue that has been the subject of extensive investigations the last 40-50 years, especially after the Second World War.

By trying to answer this basic but generic question, a good number of future research directions can be evolved or generated. The Fuzzy Cognitive Map theories, lately have demonstrated, that can provide realistic and useful tools for addressing many problems that our society is confronted with.

References

[1] Craiger, J.P., Goodman, D.F., Weiss, R.J., Butler, A.: Modeling Organizational Behavior with Fuzzy Cognitive Maps. Intern. Journal of Computational Intelligence and Organisations 1, 120–123 (1996)

[2] Dickerson, J.A., Kosko, B.: Virtual Worlds as Fuzzy Cognitive Maps. Presence 3, 173–189 (1994)

[3] Hagiwara, M.: Extended Fuzzy Cognitive Maps. In: Proceedings of IEEE Int. Conference on Fuzzy Systems, pp. 795–801 (1992)

[4] Jang, J.S., Sun, C.T., Mizutani, E.: Neuro-Fuzzy and Soft Computing. Prentice Hall, Upper Saddle River (1997)

[5] Kim, H.S., Lee, K.C.: Fuzzy implications of fuzzy cognitive map with emphasis on fuzzy causal relationship and fuzzy partially causal relationship. Fuzzy Sets and Systems 97, 303–313 (1998)

[6] Kosko, B.: Fuzzy Cognitive Maps. Intern. Journal of Man-Machine Studies 24, 65–75 (1986)

[7] Kosko, B.: Neural Networks and Fuzzy Systems. Prentice-Hall, Englewood Cliffs (1992)

[8] Lin, C.T., Lee, C.S.G.: Neural Fuzzy Systems: A Neuro-Fuzzy Synergism to Intelligent Systems. Prentice Hall, Upper Saddle River (1996)

[9] Medsker, L.R.: Hybrid Intelligent Systems. Kluwer Academic Publishers, Norwell (1995)

[10] Nie, J., Linkens, D.: Fuzzy-Neural Control: principles, algorithms and applications. Prentice Hall Europe, Hertfordshire (1995)

[11] Schneider, M., Shnaider, E., Kandel, A., Chew, G.: Automatic construction of FCMs. Fuzzy Sets and Systems 93, 161–172 (1998)

[12] Stylios, C.D., Georgopoulos, V.C., Groumpos, P.P.: Introducing the Theory of Fuzzy Cognitive Maps in Distributed Systems. In: Proceedings of 12th IEEE Intern. Symposium on Intelligent Control, Istanbul, Turkey, pp. 55–60 (1997)

[13] Papageorgiou, E.I., Spyridonos, P., Ravazoula, P., Stylios, C.D., Groumpos, P.P., Nikiforidis, G.: Advanced Soft Computing Diagnosis Method for Tumor Grading. Artif. Intell. Med. 36, 59–70 (2006)

[14] Quinlan, J.R.: C4.5: Programs for machine learning. Morgan Kaufmann, San Mateo (1993)

[15] Papageorgiou, E.I., Spyridonos, P., Ravazoula, P., Stylios, C.D., Groumpos, P.P., Nikiforidis, G.: The Challenge of Using Soft Computing Techniques for Tumor Characterization. In: Rutkowski, L., Siekmann, J.H., Tadeusiewicz, R., Zadeh, L.A. (eds.) ICAISC 2004. LNCS (LNAI), vol. 3070, pp. 1031–1036. Springer, Heidelberg (2004)

[16] Kosko, B.: Verac, " Fuzzy cognitive maps" (1986)
[17] Dickerson, J.A., Kosko, B.: Virtual Worlds as Fuzzy Cognitive Maps. Presence 3, 173–189 (1994)
[18] D'alche-Buc, F., Zwierski, D., Nadal, J.: Trio learning: a new strategy for building hybrid neural trees. Neural Syst. 5(4), 255–274 (1994)
[19] Janssens, D., Wets, G., Brijs, T., Vanhoof, K., Arentze, T., Timmermans, H.: Integrating Bayesian networks and decision trees in a sequential rule-based transportation model. Europ. J. Operat. Research (2005)
[20] Stylios, C.D., Groumpos, P.P.: The Challenge of modeling Supervisory Systems using Fuzy Cognitive Maps. Journal of Intelligent Manufacturing 9(4), 339–345 (1998)
[21] Stylios, C.D., Groumpos, P.P.: Fuzzy Cognitive Maps: A model for Intelligent Supervisory Control Systems. Computers in Industry 39(3), 229–238 (1999)
[22] Stylios, C.D., Groumpos, P.P.: A soft computing approach for modeling the supervisor of manufacturing systems. Journal of Intelligent and Robotics Systems 26(3-4), 389–403 (1999)
[23] Stylios, C.D., Groumpos, P.P., Georgopoulos, V.C.: Fuzzy Cognitive Map Approach to Process Control Systems. J. Advanced Computational Intelligence 3(5), 409–417 (1999)
[24] Stylios, C.D., Groumpos, P.P.: Fuzzy Cognitive Maps in Modeling Supervisory Control Systems. Journal of Intelligent & Fuzzy Systems 8(2), 83–98 (2000)
[25] Groumpos, P.P., Stylios, C.D.: Modeling Supervisory Control Systems using Fuzzy Cognitive Maps. Chaos, Solitons and Fractals 11(1-3), 329–336 (2000)
[26] Stylios, C.D., Groumpos, P.P.: Modeling Complex Systems Using Fuzzy Cognitive Maps. IEEE Transactions on Systems, Man and Cybernetics: Part A Systems and Humans (IF:0,555) 34(1), 155–162 (2004)
[27] Stylios, C.D., Groumpos, P.P.: Fuzzy Cognitive Maps: A soft Computing Technique for Intelligent Control. In: Proceeding 2000 IEEE International Symposium on Intelligent Control, Patras, Greece, July 17-19, pp. 97–102 (2000)
[28] Stylios, C.D., Georgoulas, G., Groumpos, P.P.: The Challenge of Using Soft Computing for Decision Support during Labour. In: Proc. of 23rd Annual International Conference of the IEEE Engineering in Medicine & Biology Society, Istanbul, Turkey, October 25-28 (2001) (CD-ROM)
[29] Stylios, C.D., Christova, N., Groumpos, P.P.: A Hierarchical Modeling Technique of Industrial Plants Using Multimodel Approach. In: Proceeding of 10th IEEE Mediterranean conference on Control and Automation, Lisbon, Portugal, July 9-12 (2002) (CD-ROM)
[30] Christova, N., Stylios, C., Groumpos, P.P.: Production Planning For Complex Plants using Fuzzy Cognitive Maps. In: Proceeding of 7th IFAC Workshop on Intelligent Manufacturing Systems, Budapest, Hungary, April 6-8, pp. 81–86 (2003)
[31] Martchenko, A.S., Ermolov, I.L., Groumpos, P.P., Poduraev, J.V., Stylios, C.D.: Investigating Stability Analysis issues for Fuzzy Cognitive Maps. In: Proc. of 11th IEEE Mediterranean conference on Control and Automation, Rodos, Greece, June 18-20 (2003)
[32] Stylios, C.D., Groumpos, P.P.: Using Fuzzy Cognitive Maps to Achieve Intelligence in Manufacturing Systems. In: Proc. 1st International Workshop on Intelligent Manufacturing Systems, Lausanne, Switzerland, April 15-17, pp. 85–95 (1998)

[33] Stylios, C.D., Georgopoulos, V.C., Groumpos, P.P.: Decision Support System for radiotherapy based on Fuzzy Cognitive Maps. In: Int. Conference in Fuzzy logic and Technology, De Montfort University, Leicester, England, September 5-7, pp. 431–434 (2001)
[34] Glykas, M.: Workflow and Process Management in Printing and Publishing Firms. International Journal of Information Management 24(6), 523–538 (2004)
[35] Xirogiannis, G., Glykas, M.: Intelligent Modeling of e-Business Maturity. Expert Systems with Applications 32/2, 687–702 (2007)
[36] Xirogiannis, G., Stefanou, J., Glykas, M.: A Fuzzy Cognitive Map Approach to Support Urban Design. Journal of Expert Systems with Applications 26(2) (2004)
[37] Xirogiannis, G., Glykas, M.: Fuzzy Cognitive Maps in Business Analysis and Performance Driven Change. Journal of IEEE Transactions in Engineering Management 13(17) (2004)
[38] Xirogiannis, G., Chytas, P., Glykas, M., Valiris, G.: Intelligent impact assessment of HRM to the shareholder value. Expert Systems with Applications 35(4), 2017–2031 (2008)
[39] Sox, J.H.C., Blatt, M.A., Higgins, M.C., Marton, K.I.: Medical Decision Making. Butterworths, Boston (1988)
[40] Krishnan, R., Sivakumar, G., Bhattacharya, P.: Extracting decision trees from trained neural networks. Pattern Recognition 32(12), 1999–2009 (1999)
[41] Heckerman, D., Geiger, D., Chickering, D.M.: Learning Bayesian networks: the combination of knowledge and statistical data. Machine Learning 20, 197–243 (1995)
[42] Podgorelec, V., Kokol, P., Tiglic, S.B., Rozman, I.: Decision Trees: An Overview and Their Use in Medicine. Journal of Medical Systems 26(5) (October 2002)
[43] Papageorgiou, E., Stylios, C., Groumpos, P.: An Integrated Two-Level Hierarchical Decision Making System based on Fuzzy Cognitive Maps (FCMs). IEEE Trans. Biomed. Engin. 50(12), 1326–1339 (2003)
[44] Papageorgiou, E.I., Groumpos, P.P.: A weight adaption method for fine- tuning Fuzzy Cognitive Map causal links. Soft Computing Journal 9, 846–857 (2005) doi:10.10007
[45] Papageorgiou, E.I., Stylios, C.D., Groumpos, P.P.: Unsupervised learning techniques for fine-tuning Fuzzy Cognitive Map causal links. Intern. Journal of Human-Computer Studies 64, 727–743 (2006)

Expert-Based and Computational Methods for Developing Fuzzy Cognitive Maps

Wojciech Stach, Lukasz Kurgan, and Witold Pedrycz

Abstract. Development of Fuzzy Cognitive Maps (FCMs) that accurately describe a given dynamic system is a challenging task which in many cases cannot be fully completed based solely on human expertise. Some of the reasons behind this limitation include potential bias of the human experts and excessive size of the problem itself. However, due to the lack of automated or semi-automated methods that would replace or support designers, most of existing FCMs were developed using expert-based approaches. Interestingly, in the recent years we have witnessed the development of algorithms that support learning of FCMs from data. The learning corresponds to the construction of connection matrices based on historical data presented in the form of multivariate time series. Since the FCM may include feedback loops and they incorporate nontrivial transformation functions, forming these models from data is a complex task that requires searching through a large solution space. The existing automated learning methods are based either on the Hebbian learning or they apply evolutionary algorithms. This chapter formulates the task of learning FCMs and describes the corresponding design challenges. We present a comprehensive survey of the current expert-based and semi-automated/automated methods for learning FCMs. The leading learning methods are described and analyzed both analytically and experimentally with the help of a case study. We also contrast computational approaches versus expert-based methods and outline future research directions.

1 Introduction

Fuzzy Cognitive Maps (FCMs), which were introduced by Kosko (Kosko 1986) as an extension to Cognitive Maps (Axelrod 1976), are a powerful machinery for modeling of dynamic systems. They define a system as a collection of interconnected concepts where connections reflect cause-effect relationships between the

Wojciech Stach · Lukasz Kurgan · Witold Pedrycz
Department of Electrical and Computer Engineering,
University of Alberta, Edmonton, Canada
e-mail: {wstach, lkurgan, pedrycz}@ece.ualberta.ca

M. Glykas (Ed.): Fuzzy Cognitive Maps, STUDFUZZ 247, pp. 23–41, 2010.
springerlink.com © Springer-Verlag Berlin Heidelberg 2010

concepts. The relationships are represented as directed edges of the graph of the FCM and quantify the strength of causal effects between the concepts. FCMs have been recognized as a useful and flexible technique in problem solving where many decision variables are causally interrelated. In addition, they have a convenient graph representation, which consists of nodes (concepts) and weighted casual edges (relationships) representing knowledge that is easy to visualize and manipulate (Aguilar 2005). These advantages motivated application of FCMs to diverse domains including engineering (Stach et al. 2004b), medicine (Innocent and John 2004), economics (Xirogiannis and Glykas 2004), e-business (Xirogiannis and Glykas 2007), financial organizations (Glykas and Xirogiannis 2005), human management (Xirogiannis et al 2008), environmental sciences (Giordano et al. 2005), politics (Andreou et al. 2005), to name just a few. In parallel to the widespread applications, the last decade observed significant research efforts into building methodologies and tools for the development, aggregation, simulation, and analysis of the FCMs (Aguilar 2005). This chapter is entirely devoted to methods related to the development of Fuzzy Cognitive Maps.

Figure 1 shows an example process control problem which was discussed by Papageorgiou et al. (Papageorgiou et al. 2003). Two valves, *valve 1* (*V1*) and *valve 2* (*V2*), supply two different liquids into the tank. The liquids are mixed and a chemical reaction takes place. The control objective is to maintain the desired level of liquid and its specific gravity. *Valve 3* (*V3*) is used to drain liquid from the tank.

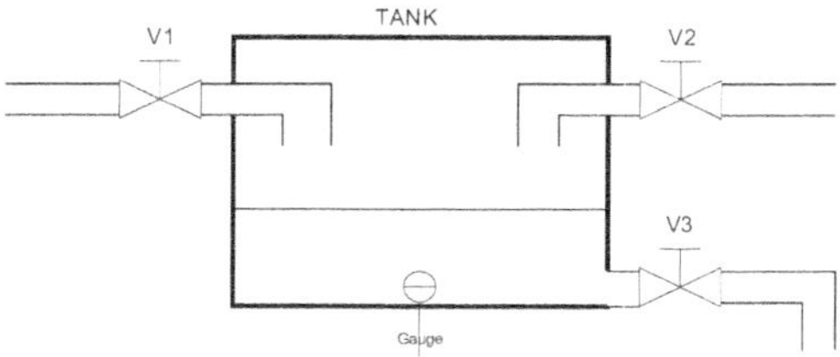

Fig. 1 Process control problem

The corresponding Fuzzy Cognitive Map that describes this system and allows for its simulation can be developed using a wide range of methods described in Sections 2 and 3. At this point, we introduce the background of FCMs based on the analysis of the FCM model that was developed for this control system by Papageorgiou et al. We also present a brief description of how an FCM model can be used to perform various tasks of analysis and simulations in order to obtain useful knowledge about the system being modeled.

The FCM model for the system presented in Figure 1 involves the following five concepts:

C1 – the amount of the liquid in the tank
C2 – the state of valve 1
C3 – the state of valve 2

C4 – the state of valve 3
C5 – the specific gravity of the liquid in the tank

These concepts are connected as illustrated in the form of a graph shown in Figure 2. The figure also shows a connection matrix which stores weights associated with directed connections between all pairs of the concepts; the matrix is equivalent to the corresponding FCM graph.

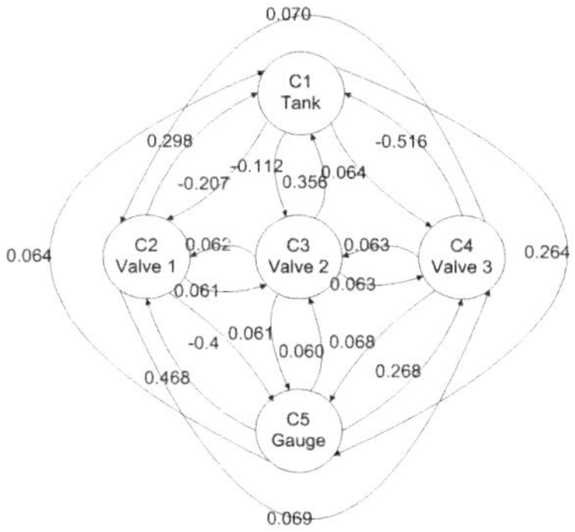

	C1	C2	C3	C4	C5
C1	0	-0.207	-0.112	0.064	0.264
C2	0.298	0	0.061	0.069	0.067
C3	0.356	0.062	0	0.063	0.061
C4	-0.516	0.070	0.063	0	0.068
C5	0.064	0.468	0.060	0.268	0

Fig. 2 FCM graph along with its connection matrix for the process control problem

Using either the graph or the connection matrix, one can perform *static analysis* of the model using techniques of graph theory, as adopted by Tsadiras et al. (Tsadiras et al. 2001). This analysis includes identification of cycles to uncover nontrivial relationships between concepts, calculation of the model density to obtain an indication of its complexity, and an analysis of importance of individual nodes. For instance, the importance can be quantified by adding up absolute values of all weights for connections both entering and leaving a given concept (node of the graph). Consequently, the corresponding values for the control process model from Figure 2 are C1=1.88, C2=1.30, C3=0.84, C4=1.18, and C5=1.32. The two concepts with the highest values are C1 and C5, which suggests they play the most important role in this system. This conclusion is consistent with the explanation of the system provided in the paper by the experts, i.e., these concepts define the control target task, which is to maintain desired liquid gravity (C5) by keeping the liquid level (C1) within a defined range. The reader may refer to the original paper or other relevant resources that describe methods for the static analysis (see e.g., Gross and Yellen 1998) for further details.

The usage of different FCM development methods may result in different maps. Static comparison of such maps, which relies solely on the values of the connection matrices, is based on the following sum of the average absolute differences reported between the corresponding weight values (Stach et al. 2004a).

$$Matrix - error = \frac{1}{N(N-1)} \sum_{i=1}^{N} \sum_{\substack{j=1 \\ j \neq i}}^{N} \left| e_{ij}^{1} - e_{ij}^{2} \right| \tag{1}$$

where e_{ij}^{1}, e_{ij}^{2} are the weights for relation from concept C_i to C_j in the FCM models 1 and 2, respectively.

The *dynamic analysis* of the FCM model allows the user to draw additional observations and conclusions concerning the underlying system, which are not available through the static analysis. The dynamic analysis is concerned with the simulation of the FCM system as a whole and the simulation of its constituent components. This provides insights into existence, interactions and dependencies between the concepts in successive iterations of the simulation. This type of analysis allows exploring "what-if" scenarios by performing simulations when imposing different initial conditions on concepts. It offers description of dynamical behavior of the underlying system, which can be used to support decision making (Stylios et al. 2008) and/or predictions about its future states (Stach et al. 2008b).

Dynamic analysis of FCMs is based on an execution model which calculates concepts *activation levels* in successive iterations. Activation levels determine degrees of presence of a given concept in the system and are represented by floating-point numbers between 0 (inactive) and 1 (active). For the control system example, the activation level of each valve determines degree to which it is open. The value of 0 means that a given valve is closed, value of 1 means that it is fully opened, and the remaining values represent the valve being partially opened. The simulation also requires defining initial values of all concepts (also called *initial condition* or initial *state vector*). To calculate successive values of all concepts (Kosko 1986), called *system state* (state vector) we use the following expression:

$$\forall j \in \{1,\ldots,N\}, C_j(t+1) = f\left(\sum_{i=1}^{N} e_{ij} C_i(t)\right) \tag{2}$$

where $C_j(t)$ is the activation level of concept j^{th} at iteration t, e_{ij} is the weight for relationship from concept C_i to C_j, and f is the transformation (transfer) function.

The transformation function is used to normalize concepts' values to the range [0,1], which allows for comparison of activation levels between different concepts. The most popular functions are continuous, although in some research work binary functions were used. The binary functions limit the dynamic analysis of concepts just to two values that correspond to two linguistic terms, inactive and active. Comparison of different transfer functions for FCMs was recently carried out by Tsadiras (Tsadiras 2008).

Figure 3 shows a result of a simulation of the control model completed when using continuous transfer function and the initial state vector suggested in the original paper, i.e., **C**[0] = [0.4, 0.708, 0.612, 0.717, 0.3].

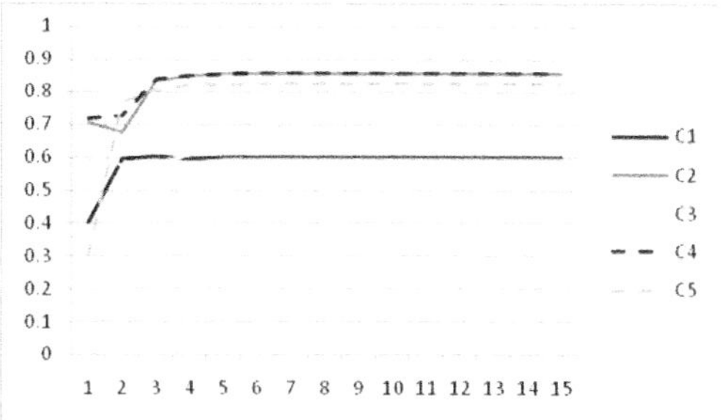

Fig. 3 Sample FCM simulation result of the control system model from Fig 2

Typically, the concept values in an FCM simulation either converge to particular state (referred to as the *fixed-point-attractor*) or they keep cycling between a fixed set of states (referred to as the *limit-cycle*). The dynamic analysis usually investigates several aspects, such as activation levels of concepts at the final state (if there is any) or cycles (intervals, concepts activation levels within the cycle), changes/trends in the activation levels throughout the simulation for either all concepts or a subset of concepts that is of interest to the user. The simulation from Figure 3 follows the fixed-point-attractor. In order to achieve the desired final values of concepts C1 and C5, when compared to their initial state, the following actions on valves are taken: the opening of the valves 1 and 3 is increased by 21% and 19%, respectively, whereas the opening of valve 2 is reduced by 1%. This leads the system to the stable state in as little as four iterations after some adjustments are made within the first three iterations. Detailed dynamic analysis of this particular model is presented in (Papageorgiou et al. 2003). A more sophisticated FCM dynamic analysis can be performed using time-series analysis (e.g. Hamilton 1994).

Similar to the *matrix-error* measure that was introduced for the static analysis, the most commonly used formula to compare two simulation results is based on the sum of the absolute average differences between the corresponding concept values normalized per concept per iteration (Stach et al. 2005b) as described below

$$Simulation - error = \frac{1}{(K-1)\cdot N}\sum_{t=1}^{K-1}\sum_{i=1}^{N}\left|C_i^1(t) - C_i^2(t)\right| \tag{3}$$

where $C_i^1(t), C_i^2(t)$ are values of ith node at t^{th} iteration obtained from simulation of the FCM models 1 and 2, respectively, K is the number of available iterations to compare (we ignore the initial state vector since it is always the same for both models), and N is the number of concepts.

Following that, we will focus on the central theme of this chapter which is the design of FCMs models, including deciding on the concepts relevant to a given system and defining relationships (weights) between the selected concepts. Generally speaking, Fuzzy Cognitive Maps models can be developed *by experts* and/or *computationally* (in either *automated or semi-automated* fashion) (Stach et al. 2005a). The expert-based approach can be classified as *deductive modeling* and involves application of human expertise in a given domain. In contrary, the computational methods can be classified as *inductive modeling* and they use available data and a learning algorithm to develop or to support development of an FCM model for a given system. In spite of their shortcomings, FCM models were developed almost exclusively using expert-based methods. The main reason for that is that the computational methods were introduced relatively recently (Stach et al. 2005a).

In this chapter, we present a comprehensive review of different approaches to develop FCMs. Next section is devoted to the expert-based methods and includes explanation of the steps that are performed by an expert to create an FCM model. We also describe how to combine multiple maps that are created by different experts for the same underlying system. Section 3 provides a systematic survey of computational methods and includes both semi-automated (which require some involvement of a human expert) and automated methods. We also demonstrate working of the considered FCM development methods using the example control system model and we provide a side-by-side comparison of these methods. Finally, conclusions and future research directions are outlined in Section 4.

2 Expert-Based Methods

2.1 Overview

Expert-based development of Fuzzy Cognitive Maps relies entirely on human expertise and domain knowledge. The relatively simple model representation makes it possible to simply manually draw the graph that corresponds to an FCM using only a pencil and a sheet of paper. The experts are also required to have a rudimentary knowledge of the FCM theory to understand the meaning of the weights and the direction of the causal effects. In order to increase credibility of the model, a group of experts instead of a single person may be involved in the development process. Experts can work together or design individual maps that represent their own understanding of a given system. In the latter case the individual maps can be combined into a single model.

2.2 Development of FCMs by a Single Expert

The expert-based development of FCMs usually consists of the following three steps (Kosko 1986, Khan and Quaddus 2004)

1. Identification of important concepts.
2. Identification of causal relationships among these concepts.
3. Estimation of the strength of the causal relationships.

In the first step, the decision which from among all available concepts should be included in the model has to be made. The most intuitive strategy is to create a list of all relevant concepts and remove the insignificant ones. In the second step, all cause-effect direct relationships between the remaining concepts have to be identified, including their directions. Usually this is accomplished by focusing on one pair of concepts at a time, since then the expert is relieved of the task of coming up with hidden or indirect cause–effect relationships. These relationships become apparent later through analyses carried out using the completed FCM. These first two steps result in a structural design which consists of a graph with nodes and directed edges.

The main challenge in expert-based development of FCMs is to accurately estimate the strength of the relationships. We note that the number of weights exhibits quadratic growth with the number of concepts, which may lead to difficulties in developing maps with several dozens of concepts. Following the original paper (Kosko 1986), each relationship strength value (weight) is expressed by a real number from the [-1,1] interval. The value of 0 denotes no relationship and is implicitly assigned at the end of the second step. Higher absolute values represent stronger relationships, whereas the sign defines the type: *promoting* (positive numbers) or *inhibiting* (negative numbers). Theoretically, each weight can take on an infinite numbers of values. Consequently, this step is potentially susceptible to subjective judgment of a given expert. A common practice to facilitate the estimation of the weight values is to first describe each relationship by a linguistic term and next to transform these terms into numerical values. The corresponding work can be divided into the following three steps (Kosko 1986, Khan and Quaddus 2004)

1. Determining the sign of each relationship.
2. Describing each relationship by means of linguistic terms, e.g. *weak, medium, strong* and *very strong*.
3. Mapping the linguistic terms to numerical values, e.g. *weak* to 0.25, *medium* to 0.5, *strong* to 0.75, and *very strong* to 1.0.

The use of the linguistic expressions to describe the degrees of causality in relationships allows the experts to avoid the difficult task of specifying the precise numerical values before a draft model is established. Additionally, analytical procedures, such as *Analytical Hierarchy Process* (Saaty 1980), may be helpful to find the numerical values used in the last step of the weight estimation procedure.

2.3 FCM Development by a Group of Experts

Fuzzy Cognitive Maps allow for a relatively simple aggregation of knowledge obtained from several experts. The aggregation should improving reliability of the final model which is less susceptible to potentially erroneous beliefs of a single expert. There are a couple of procedures for combining multiple FCMs into a single, final model. They involve simple matrix operations, such as summations and multiplications by a number (Kosko 1988), which are computed using the connection matrices developed by individual experts.

It is not uncommon that experts decide on different number of concepts. Consequently, the sizes of corresponding matrices may not be the same and/or the corresponding rows/columns may concern different concepts. In such a case, the first step towards combining the maps is to equalize their sizes. The connection matrices are augmented by including any missing concept(s), when considering all concepts in all input maps, through addition of extra rows and columns of all zeros. In other words, the omitted concepts are added "superficially" by assigning them with no incoming and outgoing relationships with other concepts. If the total number of distinct concepts over all input FCMs equals N, then each connection matrix is augmented to the matrix of $N \times N$ size (Khan and Quaddus 2004).

Assuming no additional information on the credibility of individual experts or assuming that all experts are equally credible, the simplest method for combining the maps is to calculate average of each relationship weight across all experts. Therefore, for k experts, the connection matrix of the final FCM is established by the following expression (Kosko 1988):

$$E = \frac{1}{k}\left(E_1 + E_2 + \ldots + E_k\right) \tag{4}$$

In this approach, each expert contributes equally to the final model. This basic formula can be easily modified to accommodate credibility of different experts by assigned a weight w_i that quantifies credibility of the ith expert. These weights take value from [0,1] range and their sum is usually normalized, that is . The final, combined model is calculated using the following weighted average:

$$E = \frac{1}{\sum_{i=1}^{k} w_i}\left(w_1 E_1 + w_2 E_2 + \ldots + w_k E_k\right) \tag{5}$$

As a result, the experts with higher credibility have stronger influence on the structure of the final model than those with lower credibility. More detailed discussion on the assignment of the credibility weights can be found in the literature (e.g. Taber and Siegel 1987).

2.4 Example

Figure 4 shows the FCM model proposed by experts for the process control problem defined in Figure 1 (Papageorgiou et al. 2003). The model was developed by three experts who had good understanding of the modeled system. Firstly, they jointly agreed on the set of concepts used in the model. Secondly, each expert drew the relationships between the concepts and assigned weights for each relationship. Finally, the models were merged using the technique described in Section 3; the original paper does not mention whether credibility weights were used. This model was later updated to the final map shown in Figure 2.

Static comparison between the expert-derived map and the final map from Figure 2 shows that these maps are similar. The experts' map is sparser, i.e., some of the weights that have small magnitude in the final model are rounded to zero. The weights that have larger magnitude have the same sign between the two maps. We also observe that the weight developed by experts have lower precision. The *matrix-error* between the two models is small and equals 0.09 (with the standard deviation of 0.07). Therefore, the static analysis of these two models gives very similar results. For instance, the most important concept in both models

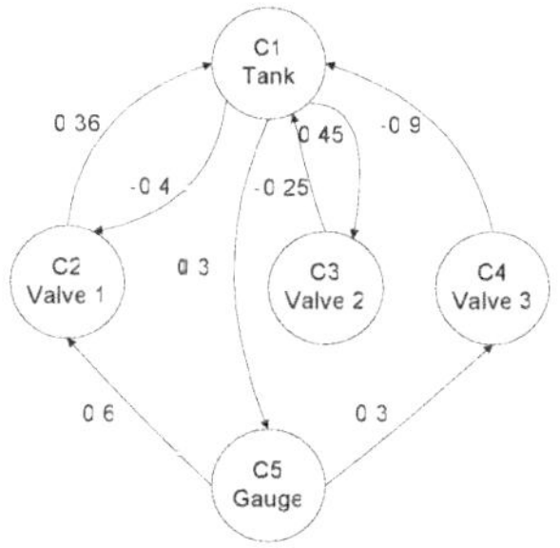

	C1	C2	C3	C4	C5
C1	0	-0.4	-0.25	0	0.3
C2	0.36	0	0	0	0
C3	0.45	0	0	0	0
C4	-0.9	0	0	0	0
C5	0	0.6	0	0.3	0

Fig. 4 FCM model of the process control problem that was developed by three experts

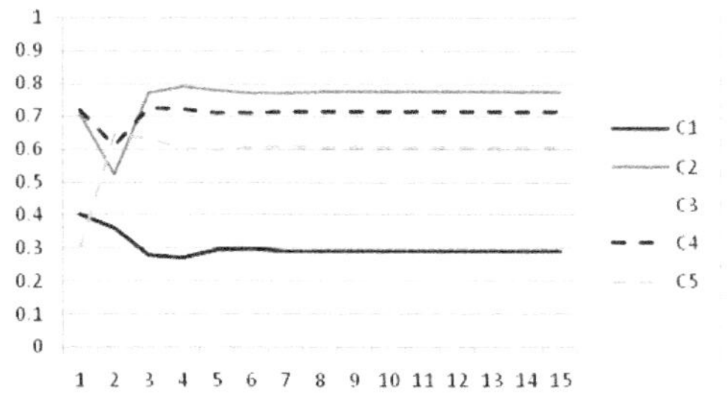

Fig. 5 Simulation result for the experts' map of the process control system from Fig. 4

is C1 with the corresponding importance measure equal 2.66 and 1.88 in the expert-based and final models, respectively. At the same time, the second and third important concept in the expert model are C2 and C5 with the values of 1.36 and 1.20, respectively, while in the final model they are C5 and C2 with the values of 1.32 and 1.30, respectively. The model proposed by the experts is structurally less complex since it has fewer non-zero connections between nodes. We observe that the expert-based models generated are usually sparsely connected (Stach et al. 2005b).

Although the expert and the final map are similar from the static analysis point of view, their simulations for the same initial conditions, which are shown in Figures 5 and 3, respectively, differ quite significantly. The *simulation-error* equals 0.19 (with the standard deviation of 0.08), which is relatively large considering that the error values range in-between 0 and 1. Therefore, the conclusions resulting from the dynamical analysis of the two models will be likely different. For instance, the amount of liquid in the tank (C1) after completing the simulation is approximately two times lower than for the final model, while the gravity of the liquid (C2) is only slightly lower. This demonstrates that models that are similar based on the static analysis may exhibit different dynamic behavior, while the experts usually do not consider the dynamic behavior in their modeling.

2.5 Summary

Expert-based methods for the development of Fuzzy Cognitive Maps are well established and have been extensively applied to real life modeling tasks in diverse domains (Aguilar 2005, Stach et al. 2005a). The popularity of the expert-based methods stems from at least two reasons. Firstly, these methods are relatively simple and straightforward. All the steps in the development process are clearly defined and described in the literature, the development does not requires sophisticated knowledge of the underlying modeling technique, and the expert knowledge is represented by a simple to comprehend graph. The second reason is the lack, until recently, of alternative methods for FCM development. These methods, which are described in the next Section, provide support to or replace experts from the model development task.

We also observe that expert-based methods aim at developing structure of the model that corresponds to expert(s) understanding of a given system. The experts virtually never simulate the model to verify whether its dynamic behavior is correct. Therefore, models created by expert(s) usually provide good static description of the system, and are better suited for the static analysis. On the other hand, the dynamic analysis (simulations) of the experts-derived maps may lead to inaccuracies when compared with the actual system.

3 Computational Methods

3.1 Overview

Computational methods utilize historical data available for a given system to establish FCM model. Semi-automated methods require a relatively limited human intervention, whereas fully automated approaches are able to compute the FCM model solely based on the historical data, i.e. without any human input. A number of algorithms for learning FCM structure from data have been recently proposed. They can be categorized into two groups based on the learning paradigm used, i.e., Hebbian-based learners and methods based on evolutionary algorithms. The following two subsections describe chronologically algorithms from the two groups.

3.2 Hebbian-Based Methods

In one of the first attempts, Dickerson and Kosko proposed a simple *Differential Hebbian Learning (DHL)* (Dickerson and Kosko 1993, Dickerson and Kosko 1994) method, which is based on Hebbian theory (Hebb 1949). During DHL learning the values of weights are iteratively updated until the desired structure is found. In general, the weights of outgoing edges for each concept in the connection matrix are modified only when the corresponding concept value changes,

$$e_{ij}(t+1)=\begin{cases} e_{ij}(t)+c_t\left[\Delta C_i \Delta C_i - e_{ij}(t)\right] & if\ \Delta C_i \neq 0 \\ e_{ij}(t) & if\ \Delta C_i = 0 \end{cases} \quad (6)$$

where e_{ij} denotes the weight for relation from concept C_i to C_j, ΔC_i represents the change in the C_i concept's activation value, t is the iteration number, and c_t is a learning coefficient. The learning coefficient is a small constant which values usually decrease as the learning progresses. The main drawback of this learning method is that the formula updates weights between each pair of concepts taking into account only these two concepts and ignoring the influence from other concepts.

An improved version of DHL learning was introduced by Huerga (Huerga 2002). The new algorithm, called *Balanced Differential Algorithm (BDA)* eliminates one of the limitations of DHL method by taking into account all the concept values that change at the same time when updating the weights. More specifically, the modified formula for $e_{ij}(t+1)$ takes into consideration changes in all concepts if they occur at the same iteration and have the same direction. Empirical comparison between DHL and BDA demonstrates that the latter method improves quality of the learned maps (Huerga 2002). On the other hand, the BDA algorithm was applied only to binary FCMs, i.e., maps with binary transfer functions, which limits its application areas.

One year later, Papageorgiou and colleagues introduced *Nonlinear Hebbian Learning (NHL)* algorithm (Papageorgiou et al. 2003). While this algorithm originates from the same learning principles, it uses a nonlinear extension to the basic Hebbian rule (Oja et al. 1991) by introducing modified weight update formula. The NHL learning method has been designed as a semi-automated approach that requires initial human intervention. Experts are required to suggest nodes that are directly connected and only these edges are updated during learning. In addition, the experts have to indicate sign of each edge according to its physical interpretation. The algorithm updates the corresponding weights while preserving their initial signs. In a nutshell, the NHL algorithm allows obtaining model that retains initial graph structure imposed by the expert(s), and therefore requires human intervention before the learning process starts. Also, the experts have to define output concepts and specify range of values that these concepts can take. The latter is used after every update of the learned model's weights to validate the model. The validation is based on checking whether the model state satisfies these constrains.

The same research group proposed *Active Hebbian Algorithm (AHL)* in 2004 (Papageorgiou et al. 2004). This approach introduces and exploits the task of determination of the sequence of activation concepts. Expert(s) determines the desired set of concepts, initial structure and the interconnections of the FCM structure as well as the sequence of activation concepts. A seven-step AHL procedure, which is based on Hebbian learning, is iteratively used to adjust the weights to satisfy defined predefined stopping criteria.

In a recent work, Stach and coworkers proposed an improved version of the NHL method (Stach et al. 2008a). The algorithm, called *Data-Driven Nonlinear Hebbian Learning (DD-NHL),* is based on the same learning principle as NHL, but it takes advantage of historical data (a simulation of the actual system) and uses output concepts to improve the learning quality. An empirical comparative study have shown that if historical data are available, then the DD-NHL method produces better FCM models when compared with those developed using the generic NHL method (Stach et al. 2008a).

3.3 Evolutionary Algorithms-Based Methods

In 2001, Koulouriotis and colleagues applied the Genetic Strategy (GS) to learn FCM's model structure, i.e., weights of relationships, from data (Koulouriotis et al. 2001). In their method, the learning process is based on a collection of input/output pairs, which are referred to as examples. The learning requires historical data consisting of multiple sequences of state vectors (multiple simulations of the system). The algorithm computes the structure of an FCM that is able to generate state vector sequences that transform the input vectors into the output vectors. The main drawback of this approach is that it requires multiple state vector sequences, which might be difficult to obtain in some of the application domains.

Particle Swarm Optimization (PSO) method, proposed by Parsopoulos and coworkers, belongs to the class of Swarm Intelligence algorithms (Parsopoulos et al. 2003). This method aims at learning FCM structure based on historical data that converge to a desired final state. PSO is a population based algorithm, which performs a search for the solution by maintaining and transforming a population of individuals. The learning requires human knowledge that is used to specify adequate constraints, which would guarantee that the relationships within the FCM model retain the physical meaning defined by the expert(s).

The next algorithm proposed by Khan and Chong aims to accomplish a different learning objective (Khan and Chong 2003). Instead of learning the structure of the FCM model, their goal was to find an initial state vector (initial condition) that leads a given model to the specified end state. Their method employed genetic algorithms to find the initial state.

A fully automated method for learning FCMs, which is based on real-coded genetic algorithms (RCGA), was introduced by Stach and colleagues in 2005 (Stach et al. 2005b). The RCGA is a floating-point extension (Herrera et al. 1998) to genetic algorithms (Goldberg 1989). This extension was used to allow finding floating point weights instead of weights that take on a limited set of values. The core of this approach is a learning module which exploits RCGA to find FCM structure that is capable of mimicking a given input historical data. This approach is flexible in terms of the input data as it can use either one or multiple sets of concepts values over successive iterations. A follow-up of this work includes analysis of the quality of the RCGA-based learning depending on the amount of the available historical data (Stach et al. 2004a). It demonstrates that the RCGA-based method can generate FCM models that are identical to models proposed by domain expert given the input data of sufficient size, and that increasing the amount of the input data improves accuracy of the learning.

Recently, the same research group introduced a parallel RCGA-based method that targets learning of large maps that consist of dozens of concepts (Stach et al. 2007). The method was reported to be up to four times faster than the sequential

RCGA learning when executed on eight processors. It allows learning maps that include several dozens of concepts in a few hours.

3.4 Example

We use the process control system from Figure 1 to investigate the quality of models learned using state-of-the-art learning methods. The data from Figure 3 were used to learn the FCM model and we tested three learning algorithms that include NHL and DD-NHL, two most recently proposed Hebbian learning based methods, and RCGA, which is the most recent genetic-based method.

Since the RCGA method is initialized with a 100 randomly generated maps, whereas the two other methods use just a single map, the experiments for both Hebbian-based methods were repeated 100 times using the 100 initial maps generated for the RCGA method. The final output was selected as the map that provides simulations with the lowest value of the *simulation-error*.

Table 1 presents a summary of the results. We computed both *matrix-error* and *simulation-error* to quantify the quality with respect to both the static and the dynamic analysis, respectively. In addition, the last column gives the learning time for each method in seconds. We report the average values together with the corresponding standard deviations (shown in brackets).

Table 1 Experimental results concerning comparison of the quality of FCM models for the process control system developed by the experts and learned using computational methods including NHL, DD-NHL and RCGA

Learning method	*Matrix-error*	*Simulation-error*	Time [s]
NHL	0.236 (0.162)	0.064 (0.053)	8
DD-NHL	0.245 (0.145)	0.056 (0.063)	9
RCGA	0.225 (0.154)	0.003 (0.004)	68
Expert-based	0.092 (0.078)	0.187 (0.077)	N/A

We show detailed results, i.e., the resulting map and its simulation, only for the RCGA method since it outperformed the two Hebbian-based methods on both quality criteria. Figures 6 and 7 present the FCM model and its simulation result for the same initial condition as in Figure 3, respectively. We note that the *simulation-error* values obtained with maps learned using the NHL and DD-NHL methods are substantially lower than the error of the expert-based method.

In spite of the relatively different connection matrix generated by RCGA (comparison with the final map, i.e., comparison of models from Figures 2 and 7, reveals that the *matrix-error* of RCGA equals 0.255 and that 25% of the weights have opposite signs), the *simulation-error* is very small (equals 0.003 and it reduces to 0.001 if we consider only the final state). This observation can be generalized, based on experiments reported in the literature (e.g. Stach et al. 2004a), to a statement that the structurally different maps can generate very similar simulations. The chances that the RCGA method will find suboptimal solutions, i.e., solutions with low *simulation-error* and relatively high *matrix*-error, decrease

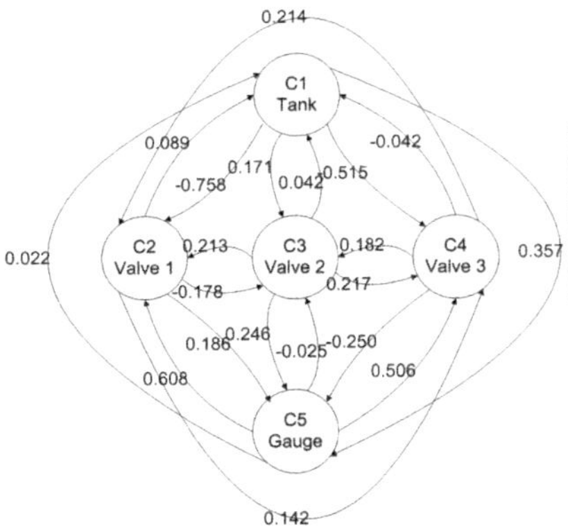

	C1	C2	C3	C4	C5
C1	0	-0.758	0.171	-0.515	0.357
C2	0.089	0	-0.178	0.142	0.186
C3	0.042	0.213	0	0.217	0.246
C4	-0.042	0.214	0.182	0	-0.250
C5	0.022	0.608	-0.025	0.506	0

Fig. 6 FCM model of the process control system learned using RCGA method

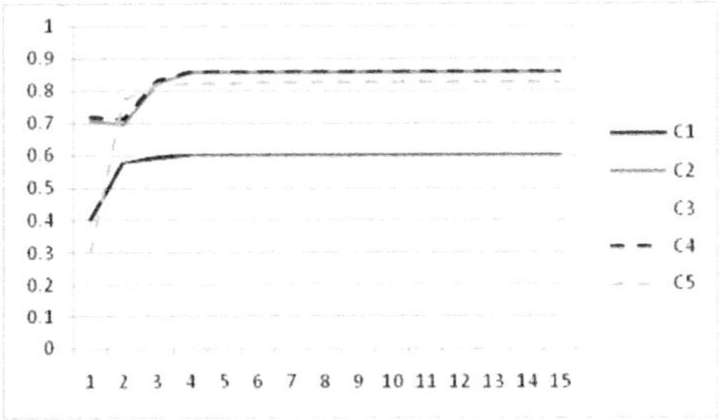

Fig. 7 Simulation result of the FCM for the process control system learned using the RCGA method

as the amount of the available input data increases (see Stach et al. 2004a for details). One has to be aware that suboptimal solutions can be generated by the fully automated learning approaches and in these cases results of the static analysis could be inaccurate.

In our example of dynamic analysis we compare the final states for the three simulations from Figures 3 (the actual system), 5 (expert-derived model), and 7 (best-performing automated computational model). The comparison of the difference between the final values of the two most important concepts, C1 and C5 (see Section 1), shows that

- For C1, the expert-derived model underestimates the final state value by 51% while the model developed with RCGA makes only 0.1% error
- For C5, the error equals 26% and 0.2% for the expert-based and RGCA-based learning, respectively.

While the RCGA-based method provides satisfactory performance, the expert-based solution may lead to problems. The incorrect liquid gravity (C5) would likely produce wrong setups of the valves. Additionally, substantial underestimation of the liquid level (C1) would inevitably lead to exceeding its maximum level by making incorrect control decision regarding the opening of the valves.

On the other hand, example of static analysis shows that the expert-derived map correctly identifies C1 as the most important concept with C2 and C5 as the second and third most important concepts when compared to the correct order of C1, C5 and C2. Similar analysis applied to the map generated by the RCGA method shows that the top three important concepts are C2, C5 and C4 the corresponding importance measures equal 2.39, 2.20 and 2.07, respectively.

3.5 Summary

Computational methods, when compared to the expert-based methods, are a relatively new branch of research devoted to the Fuzzy Cognitive Maps. They emerged to eliminate the drawbacks and limitations of the expert-based methods. Several years of the research resulted in computational methods that are summarized and compared in Table 2. The table performs a side-by-side comparison that includes several aspects, such as learning goal, involvement of a domain expert, input data type, and learning strategy used.

Table 2 Comparative analysis of computational methods used for learning FCMs

Method	Reference	Learning Goal	Expert input	Data used[1)]	Transformation function	# concepts	Learning Algorithm
DHL	(Dickerson and Kosko 1994)	Connection matrix	No	Single	N/A	N/A	Hebbian
BDA	(Huerga 2002)	Connection matrix	No	Single	Binary	5,7,9	Modified Hebbian
NHL	(Papageorgiou et al. 2003)	Connection matrix	Yes and No[2)]	Single	Continuous	5	Modified Hebbian
AHL	(Papageorgiou et al. 2004)	Connection matrix	Yes and No[2)]	Single	Continuous	8	Modified Hebbian
DD-NHL	(Stach et al. 2008a)	Connection matrix	Yes and No[2)]	Single	Continuous	5	Modified Hebbian
GS	(Koulouriotis et al. 2001)	Initial vector	No	Multiple	Continuous	7	Genetic
PSO	(Parsopoulos et al. 2003)	Connection matrix	No	Multiple	Continuous	5	Swarm
Genetic	(Khan and Chong 2003)	Initial vector	N/A	N/A	Continuous	11	Genetic
RCGA	(Stach et al. 2005b)	Connection matrix	No	Single	Continuous	4,6,8,10	Genetic
Parallel RCGA	(Stach et al. 2007)	Connection matrix	No	Single	Continuous	5,10,20, 40,80	Parallel Genetic

1) Single – historical data consisting of one sequence of state vectors, Multiple – historical data consisting of several sequences of state vectors for different initial conditions

2) Initial human intervention is necessary but later when applying the algorithm the human input is not needed

The choice of a particular computational learning method is affected by several factors. One needs to consider the type of available data since some methods require multiple simulations (state vectors). The Hebbian-based methods are faster since evolutionary optimization requires complex and time-consuming calculations. On the other hand, methods from the latter group provide better quality of the learned models in the context of the similarity of their dynamic behavior defined as the *simulation-error*. Semi-automated methods are preferred if some structural constraints can be imposed on the map by the expert. Otherwise, if the only criterion is the quality of model's dynamic behavior, then the fully automated genetic optimization seems to be the best option.

4 Conclusions and Future Directions

4.1 Conclusions

Fuzzy Cognitive Maps have gained a well-deserved attention in the recent years. Numerous successful applications in various research and industrial domains clearly imply the effectiveness of this modeling technique. The unquestionable advantages of FCMs, such as simplicity and adaptability to a given application area, encourage researchers and practitioners to apply this method. However, it seems that further development of FCMs is somewhat constrained by deficiencies that are present in their underlying theoretical framework. One of the issues that have been recently investigated is to provide a systematic approach to efficient design of FCMs.

The two categories of approaches to develop FCMs include the *expert-based* and the *computational* methods. The advantages, disadvantages and additional characteristics of these two types of methods are summarized in Table 3.

Table 3 Comparison of expert-based and computational methods for the development of FCMs

	Expert-based	Computational
Type of modeling	Deductive	Inductive
Main objective	To create a model that is structurally understandable	To create a model that provides accurate simulations
Main application	Static analysis	Dynamic analysis
Main shortcoming	Dynamic analysis could be inaccurate	Static analysis could be inaccurate and more difficult

The expert-based methods which are fairly well established have been used for a relatively long time. Their main advantage is the easiness of the representation that is used by the expert(s) to develop the maps. Nevertheless, models developed by experts are vulnerable to subjectivity of expert(s) beliefs and could be difficult to develop for large problems that involve dozens of concepts. Moreover, although maps developed by experts provide accurate static analysis of the FCM model, they may lead to inaccurate dynamic analysis. These limitations motivated researchers towards alternative learning strategies that would provide models that accurately represent the dynamics of the modeled system.

The computational methods aim at learning FCMs from data and therefore at providing models suitable to perform accurate dynamic analysis. Two main methodologies used for computational learning of FCM include approaches based on Hebbian learning rule and methods that exploit genetic algorithms. Unfortunately, fully automated computational methods may fail to provide models that allow for accurate static analysis. A partial solution to this problem is provided by the semi-automated methods, in which the experts supervise the learning. However, the existing semi-automated learning methods often provide models that are not as good for the static analysis as the models obtained from the expert-based methods, and worse in the context of the dynamic analysis when compared with the fully automated methods.

4.2 Future Directions

The progress in research towards finding an efficient approach to develop Fuzzy Cognitive Maps that has been observed within last few years provides a solid foundation for future investigations. Recent interest in computational methods suggests that this will be the main direction of future research. Even though the first step towards automation of FCM development from data was done, there are still problems that need to be overcome.

One of the main challenges is to provide solution to the main drawback of automated methods, which generate solutions that are hard or impossible to interpret and which may lead to incorrect static analysis. The ultimate solution should be fully automated as only then FCMs could be applied to model large problems such as those encountered in systems biology. At the same time, the intermediate steps will likely involve developing efficient and accurate semi-automated algorithms. Another pressing issue is that the currently most accurate automated methods which are based on genetic optimization cannot scale to work on problems exceeding several dozens of concepts. Although some work was done towards improving scalability of these methods, further research in this direction is required. The final challenge facing the researchers working on the computational methods is to popularize their work. This could be accomplished by freely providing implementations of the developed methods to the relevant research and development communities and by creation of links between the developers of the methods and the practitioners who build the FCM models.

References

Aguilar, J.: A survey about fuzzy cognitive maps papers. Int. J. Comp. Cogn. 3(2), 27–33 (2005)

Andreou, A.S., Mateou, N.H., Zombanakis, G.A.: Soft computing for crisis management and political decision making: The use of genetically evolved fuzzy cognitive maps. Soft. Comput. 9(3), 194–210 (2005)

Axelrod, R.: Structure of decision: the cognitive maps of political elites, Princeton, NJ (1976)
Dickerson, J.A., Kosko, B.: Virtual worlds as fuzzy cognitive maps. In: VRAIS 1993, pp. 471–477 (1993)
Dickerson, J.A., Kosko, B.: Virtual worlds as fuzzy cognitive maps. Presence 3(2), 173–189 (1994)
Giordano, R., Passarella, G., Uricchio, V.F., Vurro, M.: Fuzzy cognitive maps for issue identification in a water resources conflict resolution system. Phys. Chem. Earth 30(6-7), 463–469 (2005)
Glykas, M., Xirogiannis, G.: A soft knowledge modeling approach for geographically dispersed financial organizations. Soft. Comput. 9(8), 579–593 (2005)
Goldberg, D.E.: Genetic algorithms in search, optimization, and machine learning. Addison-Wesley, Reading (1989)
Gross, J.L., Yellen, J.: Graph theory and its applications. CRC Press, Boca Raton (1998)
Hamilton, J.D.: Time series analysis. Princeton University Press, Princeton (1994)
Hebb, D.O.: The organization of behavior. Wiley, NY (1949)
Herrera, F., Lozano, M., Verdegay, J.L.: Tackling real-coded genetic algorithms: operators and tools for behavioural analysis. Artif. Intell. Rev. 12(4), 265–319 (1998)
Huerga, A.V.: A balanced differential learning algorithm in fuzzy cognitive maps. In: QR 2002, poster (2002)
Innocent, P.R., John, R.I.: Computer aided fuzzy medical diagnosis. Inf. Sci. 162(2), 81–104 (2004)
Khan, M., Quaddus, M.: Group decision support using fuzzy cognitive maps for causal reasoning. Group Decis. Negotiation J. 13(5), 463–480 (2004)
Khan, M.S., Chong, A.: Fuzzy cognitive map analysis with genetic algorithm. In: IICAI 2003 (2003)
Kosko, B.: Fuzzy cognitive maps. Int. J. Man Mach. Stud. 24, 65–75 (1986)
Kosko, B.: Hidden patterns in combined and adaptive knowledge networks. Int. J. Approximate Reasoning 2, 377–393 (1988)
Koulouriotis, D.E., Diakoulakis, I.E., Emiris, D.M.: Learning fuzzy cognitive maps using evolution strategies: a novel schema for modeling and simulating high-level behaviour. In: CEC 2001, pp. 364–371 (2001)
Oja, E., Ogawa, H., Wangviwattam, J.: Learning in nonlinear constrained Hebbian networks. In: Kohonen, T., Makisara, K., Simula, O., Kangas, J. (eds.) Artificial Neural Networks. Elsevier, Amsterdam (1991)
Papageorgiou, E., Stylios, C., Groumpos, P.: Fuzzy cognitive map learning based on nonlinear Hebbian rule. In: Gedeon, T.(T.) D., Fung, L.C.C. (eds.) AI 2003. LNCS (LNAI), vol. 2903, pp. 256–268. Springer, Heidelberg (2003)
Papageorgiou, E., Stylios, C.D., Groumpos, P.P.: Active Hebbian learning algorithm to train fuzzy cognitive maps. Int. J. Approximate Reasoning 37(3), 219–249 (2004)
Parsopoulos, K.E., Papageorgiou, E.I., Groumpos, P.P., Vrahatis, M.N.: A first study of fuzzy cognitive maps learning using particle swarm optimization. In: CEC 2003, pp. 1440–1447 (2003)
Saaty, T.L.: The analytic hierarchy process: planning, priority setting, resource allocation. McGraw-Hill, New York (1980)
Stach, W., Kurgan, L., Pedrycz, W., Reformat, M.: Learning fuzzy cognitive maps with required precision using genetic algorithm approach. Electron. Lett. 40(24), 1519–1520 (2004a)

Stach, W., Kurgan, L., Pedrycz, W., Reformat, M.: Parallel fuzzy cognitive maps as a tool for modeling software development project. In: NAFIPS 2004, pp. 28–33 (2004b)
Stach, W., Kurgan, L.A., Pedrycz, W.: A survey of fuzzy cognitive map learning methods. In: Grzegorzewski, P., Krawczak, M., Zadrozny, Z. (eds.) Issues in Soft Computing: Theory and Applications. Exit (2005a)
Stach, W., Kurgan, L., Pedrycz, W., Reformat, M.: Genetic learning of fuzzy cognitive maps. Fuzzy Sets Syst. 153(3), 371–401 (2005b)
Stach, W., Kurgan, L., Pedrycz, W.: Parallel learning of large fuzzy cognitive maps. In: IJCNN 2007, pp. 1584–1589 (2007)
Stach, W., Kurgan, L.A., Pedrycz, W.: Data-driven nonlinear Hebbian learning method for fuzzy cognitive maps. In: WCCI 2008, pp. 1975–1981 (2008a)
Stach, W., Kurgan, L.A., Pedrycz, W.: Numerical and linguistic prediction of time series with the use of fuzzy cognitive maps. IEEE Trans. Fuzzy Syst. 16(1), 61–72 (2008b)
Stylios, C.D., Georgopoulos, V.C., Malandrakic, G.A., Chouliara, S.: Fuzzy cognitive map architectures for medical decision support systems. Appl. Soft Comput. 8(3), 1243–1251 (2008)
Taber, W.R., Siegel, M.A.: Estimation on expert weights using fuzzy cognitive maps. In: ICNN 1986, vol. 2, pp. 319–325 (1987)
Tsadiras, A.K.: Comparing the inference capabilities of binary, trivalent and sigmoid fuzzy cognitive maps. Inf. Sci. 178(20), 3880–3894 (2008)
Tsadiras, A.K., Kouskouvelis, I., Margaritis, K.G.: Making political decisions using fuzzy cognitive maps: The FYROM crisis. In: PCI 2001, vol. 2, pp. 501–510 (2001)
Xirogiannis, G., Glykas, M.: Fuzzy cognitive maps in business analysis and performance driven change. IEEE Trans. Eng. Manage 51(3), 334–351 (2004)
Xirogiannis, G., Glykas, M.: Intelligent modeling of e-business maturity. Expert Syst. Appl. 32(2), 687–702 (2007)
Xirogiannis, G., Chytas, P., Glykas, M., Valiris, G.: Intelligent impact assessment of HRM to the shareholder value. Expert Syst. Appl. 35(4), 2017–2031 (2008)

A Novel Approach on Constructed Dynamic Fuzzy Cognitive Maps Using Fuzzified Decision Trees and Knowledge-Extraction Techniques

Elpiniki I. Papageorgiou

Abstract. A novel approach for the construction of augmented Fuzzy Cognitive Maps based on data mining and knowledge-extraction methods has been investigated for decision making and classification tasks. Specifically, through this work, the issue of designing decision support systems based on fuzzy cognitive maps has been explored using fuzzified decision trees and other knowledge-extraction techniques. Fuzzy cognitive map is a knowledge-based technique that works as an artificial cognitive network inheriting the main aspects of cognitive maps and artificial neural networks. Decision trees, in the other hand, are well known intelligent techniques that extract rules from both symbolic and numeric data. Fuzzy theoretical techniques are used to fuzzify crisp decision trees in order to soften decision boundaries at decision nodes inherent in this type of trees. Comparisons between crisp decision trees and the fuzzified decision trees suggest that the later fuzzy tree is significantly more robust and produces a more balanced decision making. The new approach proposed in this paper could incorporate any type of knowledge extraction algorithm. Furthermore, through the knowledge extraction methods the useful knowledge from data can be extracted in the form of fuzzy rules and inserted those into the FCM, contributing to the development of a dynamic approach for decision support. The proposed approach is implemented in a well known medical decision making problem to preview the effectiveness.

Keywords: fuzzy cognitive maps, decision trees, fuzzy, neuro-fuzzy, data mining, causal paths, decision making.

1 Introduction

This chapter presents a soft computing procedure to handle different data types for decision support tasks in diverse scientific areas. The proposed methodology

Elpiniki I. Papageorgiou
Dept of Informatics and Computer Technology, Technological Educational Institute of Lamia, TEI Lamias, 3rd Km Old National Road Lamia-Athens, 35100 Lamia, Greece
e-mail: epapageorgiou@teilam.gr

M. Glykas (Ed.): Fuzzy Cognitive Maps, STUDFUZZ 247, pp. 43–70, 2010.
springerlink.com © Springer-Verlag Berlin Heidelberg 2010

establishes an advanced framework by implementing different knowledge extraction methods for Fuzzy Cognitive Mapping (FCM) decision support.

Nowadays, the knowledge acquisition and representation constitutes a major knowledge engineering bottleneck. A large number of techniques in the field of artificial intelligence used to represent knowledge: production rules, decision trees, rule-based architectures semantic nets, frameworks, fuzzy logic, causal cognitive maps, among others. The decision trees gained popularity because of their conceptual transparency. The well-developed design methodology comes with efficient design techniques supporting their construction, i.e. (Quinlan, 1986; Sestino, and Dillon, 1991; Sison, and Chong, 1994). The decision trees generated by these methods were found useful in building knowledge-based expert systems. Due to the character of continuous attributes as well as various facets of uncertainty one has to take into consideration, there has been a visible trend to cope with the factor of fuzziness when carrying out learning from examples in the case of tree induction. In a nutshell, this trend gave rise to the name of fuzzy decision trees and has resulted in a series of development alternatives; i.e. (Mitra et al., 2002; Umano et al., 1994; Olaru, 2003). The incorporation of fuzzy sets (Pedrycz, and Sosnowski, 2000; Crockett et al., 2005; Ishibuchi et al., 1995; Janikow, 1998) into decision trees enables us to combine the uncertainty handling and approximate reasoning capabilities of the former with the comprehensibility and ease of application of the latter. Fuzzy decision trees (Janikow, 1998; 1999) assume that all domain attributes or linguistic variables have pre-defined fuzzy terms for each fuzzy attribute. Those could be determined in a data driven manner. The information gain measure, used for splitting a node, is modified for fuzzy representation and a pattern can have nonzero degree of matching to one or more leaves (Yuan, and Shaw, 1995; Weber, 1992).

One of the great challenges for computational intelligence is the knowledge extraction from data, combining it with available symbolic knowledge, and refining the resulting knowledge-based expert systems. The reasoning with symbolic-logical rules is more acceptable to human users than the recommendations given by black box systems (Zurada et al., 2004), because such reasoning is comprehensible, provides explanations, and may be validated by human inspection.

Fuzzy logic and causal cognitive maps, in the other hand, are some of the main topics of artificial intelligence on representation of knowledge and approximation of reasoning with uncertainty (Kosko, 1986). The choice of a particular technique is based on two main factors: the nature of the application and the user's skills. The fuzzy logic theory, based on representation of knowledge and approximation of reasoning with uncertainty, is very close to the expert's reasoning, and it is well known as artificial intelligence-based method, especially in the field of medical decision making. An outcome of this theory is fuzzy cognitive maps (Kosko, 1992).

Fuzzy cognitive maps (FCMs) are diagrams used as causal representations between knowledge/data to represent events relations. They are modeling methods based on knowledge and experience for describing particular domains using concepts (variables, states, inputs, outputs) and the relationships between them. They can be obtained by asking human experts to define the variables of the system and to identify fuzzy causal relationships among the variables using 'if–then' rules and thus producing fuzzy weights (Stylios and Groumpos 2004; Papageorgiou and Groumpos, 2005b). Human experts who supervise a system and know its behavior under different circumstances develop a FCM model in such a way that their accumulated experience and knowledge are integrated in the causal relationships between factors/characteristics (Groumpos and Stylios, 2000). It is a very convenient, simple, and powerful tool, which is used in numerous areas of application (Aguilar, 2005, Xirogiannis and Glykas 2004, Xirogiannis et al. 2007; Xirogiannis et al. 2008; Froelich et al. 2009; Sordo et al. 2008; Wei et al., 2009).

The performance of FCMs is known to be sensitive to the initial weight setting and architecture. This shortcoming can be alleviated and the FCM model can be augmented if a fuzzy rule base as well as new fuzzy relationships is available. A number of knowledge extraction techniques (i.e. machine learning, fuzzy decision trees, association rules, Bayesian networks, neural networks, pattern recognition techniques, hybrid computational intelligent algorithms) (Nauck et al., 1997; Au and Farm, 1999; Mitra and Hayashi, 2000; Zurada and Lozowski, 1996; Chen and Wei, 2002; Wells and Niederer, 1998) could be used for the generation of a fuzzy rule base (X. Liu et al., 1997; Mitra and Hayashi, 2000). These methods can extract the available knowledge from data in the form of fuzzy rules and thus insert them into the FCM system.

In the case of medical decision systems based on fuzzy cognitive maps only a few studies have been undertaken (Papageorgiou et al, 2003; Georgopoulos and Stylios, 2005; Papageorgiou et al. 2006a; Papageorgiou et al. 2006b; Papageorgiou et al., 2007; Papageorgiou et al., 2008; Georgopoulos and Stylios, 2008; Stylios et al., 2008). Few frameworks have been proposed such as the integrated two level structure for making decisions in external beam radiotherapy (Papageorgiou et al. 2003), a learning approach and an FCM based grading tool, namely FCM-GT for characterizing tumours' malignancy (urinary bladder and brain tumors), (Papageorgiou et al. 2006a; Papageorgiou et al. 2008), an hybrid approach using complementary case based reasoning and competitive FCMs for the differential diagnosis problem from the speech pathology area (Georgopoulos and Stylios, 2008). Some appropriate FCM architectures were presented recently and applied to examples from two medical disciplines, i.e. speech and language pathology and obstetrics (Stylios et al., 2008). FCMs have been also used for pattern recognition and classification approaches (Papakostas et al., 2008, Froelich et al., 2009, Boutalis et al., 2009).

Most decision tree induction methods used for extracting knowledge in classification problems do not deal with cognitive uncertainties such as vagueness and ambiguity associated with human thinking and perception. Fuzzy decision trees represent classification knowledge more naturally to the way of human thinking and are more robust in tolerating imprecise, conflict, and missing information.

In this work, a new approach on constructing fuzzy cognitive maps combining knowledge from experts and from data using rule extraction methods is proposed. The explored knowledge extraction methods generate meaningful fuzzy linguistic weights incorporated to restructure and augment FCMs for decision support tasks.

The methodology is partly data driven and knowledge driven so some expert knowledge of the domain is required. The whole approach is applied to an FCM constructed to handle the complex problem of making decision on radiation therapy treatment.

The following sections are organized as: second section gives the necessary background information about fuzzy cognitive maps theory and development, as well as the knowledge extraction method of fuzzy decision trees for generating fuzzy rules. The third section is reserved for explanation of the developed fuzzy cognitive map based decision support system in medical informatics. Then the fourth section presents the application results of the proposed methodology for treatment planning selection of prostate cancer. Finally the last section gives the discussion for this proposed framework and outlines the conclusions.

2 Background of Proposed Approach

2.1 Main Aspects of Fuzzy Cognitive Maps

Fuzzy cognitive map is a causal knowledge-driven methodology for modeling complex decision systems, originated from the combination of fuzzy logic and neural networks (Kosko, 1986). An FCM describes the behavior of a knowledge-based system in terms of concepts; each concept represents an entity, a state, a variable, or a characteristic of the system (Kosko, 1992). FCM nodes are named by such concepts forming the set of concepts *C = {C1,C2, . . . ,Cn}*. Arcs *(Cj,Ci)* are oriented and represent causal links between concepts; that is how concept *Cj* causes concept *Ci*. Weights of arcs are associated with a weight value matrix *Wn·n*, where each element of the matrix *wji* taking values in [-1, . . .,1] . Kosko has developed a fuzzy causal algebra that describes the causal propagation and combination of concepts in an FCM. The algebra depends only on the partial ordering P, the range set of the fuzzy causal edge function *e*, and on general fuzzy-graph properties (e.g., path connectivity). Kosko notes that this algebra can be used on any digraph knowledge representation scheme.

A causal path from some concept node *Ci* to concept node *Cj*, say *Ci--~Ck1, Ckl--~... Ckn, Ckn --~Cj*, can be indicated by the sequence (i, *k*, ... *kn,j)*. Then the indirect effect of *Ci* on *Cj* is the causality *C~I* imparts to *Cj* via the path *(i, kl . . . . . kn,j)*. The total effect of *Ci* on *Cj* is the composite of all the indirect-effect causalities C~ imparts to *Cj*. If there is only one causal path from *Ci* to *Cj*, the total effect C~ imparts to *Cj* reduces to the indirect effect.

The indeterminacy can be removed with a numeric weighting scheme. A fuzzy causal algebra, and hence FCMs, bypasses the knowledge acquisition processing tradeoff.

A simple fuzzy causal algebra is created by interpreting the indirect effect operator *I* as the minimum operator (min) and the total effect operator *T* as the maximum operator (max) on a partially ordered set P of causal values (Peláez and Bowles, 1996). Formally, let ~ be a causal concept space, and let e: ~ × ~ P be a fuzzy causal edge function, and assume that there are m-many causal paths from

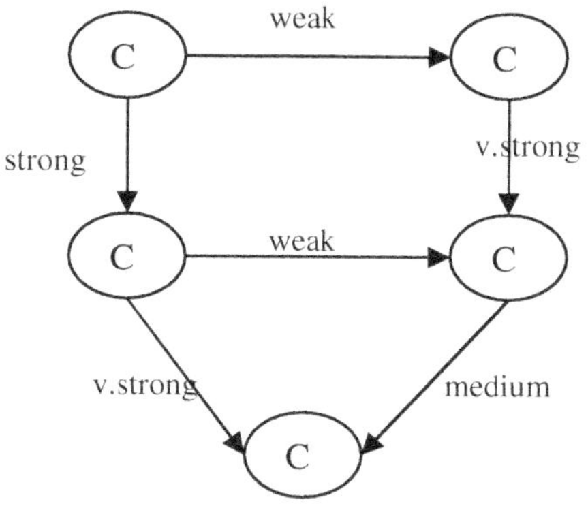

Fig. 1 A cognitive map with fuzzy labels at the edges

Ci to *Cj*: (i, k~ ..... k~, j) for 1 ~< r ~< m. Then let *Ir(Ci, Cj)* denote the indirect effect of concept *Ci* on concept *Cj* via the rth causal path, and let *T(i, Cj)* denote the total effect of *Ci* on *Cj* over all m causal paths. Then

I~(Ci,Cj)=min{e(Cp,Cp+,):(p,p+ 1) ~ (i,k~ . . . . . k,~,j) }

T(Ci,Cj)= max(Ir(Ci,Cj)) , where l <~r<~m

where p and p + 1 are contiguous left-to right path indices.

The indirect effect operation amounts to specifying the weakest causal link in a path and the total effect operation amounts to specifying the strongest of the weakest links (Kosko, 1992; Peláez and Bowles, 1996). For example, suppose the causal values are given by P = {none, weak, medium, strong, very strong} and the FCM is defined as in Figure 1. There are three causal paths from C1 to C5: path (C1, C3, C5), path (C1, C3, C4, C5) and path (C1, C2, C4, C5).

The three indirect effects of C1 on C5 are:

I1(C1,C5) = min {el3 , e35) = min {strong, v.strong} = strong
12 (C1 ,C5) = min{e13,e34,e45}= weak,
I3(C1,C5) = min{e12,e24,e45}= weak.
Hence, the total effect of C1 on C5 is:
T(C1,C5) = max {I1,(C1,C5), I2(C1,C5), I3(C1,C5) }
= max {strong, weak, weak} = strong.

– In words, C1 can be said to impart strong causality to C5.

FCMs display two distinct characteristics: firstly Nodes or concepts are also considered to be fuzzy, ie, each node is a fuzzy set, and can have an activity level to some degree from 0% to 100% and secondly, the causal relationships between nodes are fuzzified, a number is assigned to the causal link to express the degree of relationship between two concepts. The directed edge eij from the causal concept *Ci* to concept *Cj* measures how much *Ci* causes *Cj*. The edges *eij* take values from the fuzzy causal interval [-1, 1], *eij* = 0 indicates no causality; eij > 0 indicates a causal increase (ie *Cj* increases as *Ci* increases, and *Cj* decreases as Ci decreases); eij < 0 indicates causal decrease (ie, *Cj* decreases as *Ci* increases, and

Cj increases as *Ci* decreases). Word weights like '*little*' or '*somewhat*' can be used instead of numeric values (Groumpos and Stylios, 2000). An FCM can be described by a connection matrix and the activation levels of its nodes can be represented as a state vector, whereby simple vector-matrix operations allow extension to neural or dynamical systems techniques. FCMs can be subjected to an initial stimulus in the form of a state vector A representing the states of the system's variables.

At each simulation step, the value *Ai* of a concept *Ci* is calculated by computing the influence of other concepts *Cj*'s on the specific concept *Ci* following the calculation rule:

$$A_i^{(k+1)} = f(A_i^{(k)} + \sum_{\substack{j \neq i \\ j=1}}^{N} A_j^{(k)} \cdot e_{ji}) \tag{1}$$

where $Ai^{(k+1)}$ is the value of concept *Ci* at simulation step $k+1$, $Aj^{(k)}$ is the value of concept *Cj* at simulation step *k*, e_{ji} is the weight of the interconnection from concept *Cj* to concept *Ci* and *f* is a sigmoid threshold function that have been selected since the values *Ai* of the concepts, lie within [0,1]:

$$f = \frac{1}{1+e^{-mx}} \tag{2}$$

The parameter *m* is a real positive number and *x* is the value $Ai^{(k)}$ on the equilibrium point. In this work we use m=5, because this value showed best results in previous works (Bueno & Salmeron, 2009). A concept is turned on or activated by making its vector element 1. New state vectors showing the effect of the activated concept are computed using method of successive substitution, i.e., by iteratively multiplying the previous state vector by the relational matrix using standard matrix multiplication.

For the construction of FCMs, experts of the specific domain problem develop a mental model manually based on their knowledge in related area. At first, they identify key domain issues or concepts. Secondly, they identify the causal relationships among these concepts and thirdly, they estimate causal relationships strengths. This achieved graph (FCM) shows not only the components and their relations but also the strengths. In fuzzy diagrams, the influence of a concept on the others is considered as "negative", "positive" or "neutral". All relations are expressed in fuzzy terms, e.g. very weak, weak, medium, strong and very strong.

In a simple FCM, all fuzzy variables are mapped into interval [-1, 1]. A simple way is to map fuzzy expression to numerical value in a range of [-1, 1]. For example, positively weak is mapped to 0.25, negatively medium to -0.5, positively strong to 0.75 (Stylios and Groumpos, 2004). Then, all the suggested by experts linguistic variables, are considered and an overall linguistic weight is obtained, which transformed to a numerical weight with the defuzzification method of Centre of Gravity (COG) (Jang, 1997).

The above situation shows that in many cases, to develop a FCM manually becomes very difficult and experts' intervention could not resolve the problem. Therefore, a systematic way should be found in order to bridge this gap. For example, designing a new method using data mining and knowledge extraction approaches from data could eliminate the existing weakness.

2.2 Fuzzy Rules and Linguistic Weight Generation by Using Knowledge Extraction Methods

The task of decision making, especially in medicine, is difficult and complex due to the huge amount of medical data and the different sources of medical information. Thus data mining and knowledge processing systems are used in medicine for the tasks of diagnosis, prognosis, treatment planning and decision support (Fayyad and Uthurusamy, 1996; Fayyad et al., 1996).

From the literature, a large number of knowledge extraction approaches have been found (Pal and Mitra, 1999; Zurada et al., 2004). Frequently, machine learning systems can be used to develop the knowledge bases used by expert systems. Given a set of clinical cases that act as examples, a machine learning system can produce a systematic description of those clinical features that uniquely characterize the clinical conditions. This knowledge can be expressed in the form of simple rules, often used for decision making in medicine (Nauck and Kruse, 1999).

Some known rule generation algorithms existing from the literature are: Subset (Fu, 1993), MoFN (Towel, Shavlik, 1994), X2R (Liu and Tan, 1998), C4.5 (Quinlan, 1993), FuNN (Kasabov, 1993), Rulex (Andrew and Geva, 1997), NEFCLASS (Nauck, 1997), fuzzy logical MLP (Mitra and Pal, 1994), Rough fuzzy MLP (Pal and Mitra, 2003). All of these functions and methodologies tried to discover knowledge from historical data. This knowledge represented in the form of rules most of the time.

In the medical field, it is preferable not to use black box approaches. The user should be able to understand the modeler and to evaluate its results. Fuzzy rule based systems are especially suitable, because they consist of simple linguistically interpretable rules and do not have some of the drawbacks of symbolic or crisp rule based classifiers. Among the wide range of possible approaches, the fuzzy decision trees based rule generation computing method was selected to extract the knowledge exploring causal paths and useful fuzzy rules.

2.2.1 Extraction Method Using Fuzzy Decision Trees

Fuzzy decision trees are an extension of the classical artificial intelligence concept of decision trees. The main fundamental difference between fuzzy and crisp trees is that with fuzzy trees, gradual transitions exist between attribute values (Pedrycz and Sosnowski, 2000). The reasoning process within the tree allows all rules to be fired to some degree, with the final crisp classification being the result of combining all membership grades. Modifications to the ID3 algorithm have been made for developing such trees (Sison and Chong, 1994, Umano et al., 2003,

Olaru, 2003, Hayashi et al., 1998, Crockett et al., 2006). Sison and Chong (Sison and Chong, 1994) proposed a fuzzy version of ID3 which automatically generated a fuzzy rule base for a plant controller from a set of input–output data. Umano et al. (Umano et al., 2003) also proposed a new fuzzy ID3 algorithm. This algorithm generates an understandable fuzzy decision tree using fuzzy sets defined by the user. These fuzzy tree methodologies require the data to have been pre-fuzzified before the fuzzy decision trees are induced.

Recent work by Janikow involves the induction of fuzzy decision trees directly from data sets by the FID (Fuzzy Induction on Decision Tree) algorithm (Janikow, 1998; 1999). Janikow introduced the non fuzzy rules and the different kind of fuzzy rules (Janikow, 1998). In this point it is essential to refer that the data (real values) are partitioned into fuzzy sets by experts.

The following steps identify the proposed algorithm process based on FID algorithm:

Step 1: A fuzzy clustering algorithm is used for input domain partition. The supervised method takes into account the class labels during the clustering. Therefore the resulted partitions, the fuzzy membership functions (fuzzy sets) represent not only the distribution of data, but the distribution of the classes too.

Step 2: During a pre-pruning method the resulted partitions could analyze and combine the unduly overlapped fuzzy sets.

Step 3: The results of the pre-pruning step are input parameters (beside data) for the tree induction algorithm. The applied tree induction method is the FID algorithm by C. Z. Janikow.

Step 4: The fuzzy ID3 is used to extract rules which are then used for generating causal paths and the fuzzy rule base.

Step 5: While the FID algorithm could generate larger and complex decision tree as it is necessary, therefore a post pruning method is applied. The rule which yields the maximal fulfillment degree in the least number of cases is deleted.

This method provides causal paths and fuzzy rules for producing linguistic weights and thus building dynamic FCMs for decision support.

2.2.2 A Generic Example Representing the Proposed Approach

A generic example of the causal knowledge-driven FCM model consisting of eight concepts and eleven interconnections among concepts, with fuzzy labels at the edges of connections, is depicted in Figure 2. This FCM will be restructured using the proposed methodology and the available knowledge from fuzzified decision trees. Only for implementation reason, we consider that the fuzzy decision tree presented in Figure 3 has been produced by using the fuzzified decision tree above method on the available data set.

The produced tree has a number of three paths for C1 to C8, two paths for C2 to C8, and one path of each one of the other concepts to C8, thus defining new interconnections and/or update the initial ones of the FCM model.

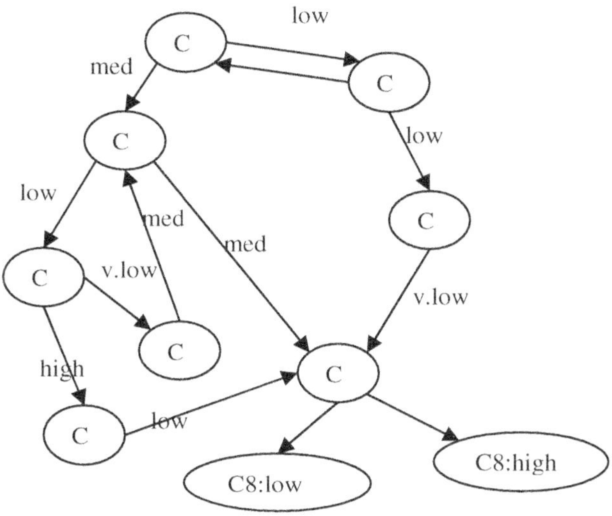

Fig. 2 Example FCM model with initial linguistic labels on interconnections (weights)

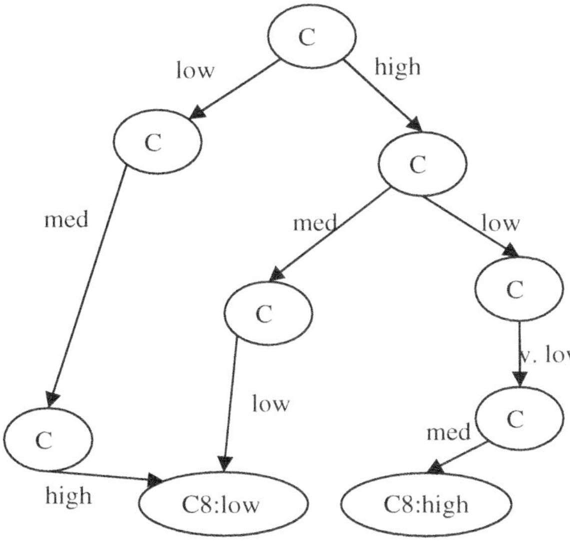

Fig. 3 Example Fuzzy decision tree induced from the data showing membership grades at each branch

Here, the causal effect of C1 to C8 is determined by taking the minimum of the attached labels of the individual paths. Let I1, I2 and I3 denote the effect of C1 to C8 through the paths 1 to 3 respectively, and e_{ij} be the label attached with edge from node ith to node jth. Then, to determine the total effect of C1 to C8, we take the maximum of paths I1 through I3 causal paths.

Path 1 from C1 to C8: c1→c3→ c6→ c8

I1(C1 to C8)=min(low, med, high)=low

Path 2 from C1 to C8:c1→c2→ c5→ c7 →c8

I2(C1 to C8)=min(high, low, v. low, med)=v. low

Path 3 from C1 to C8: c1→c2→ c4 →c8

I3(C1 to C8)=min(high, med, low)=low

Thus total effect of C1 to C8, denoted by T(C1,C8) is computed below:

T(C1,C8)=max{I1,I2,I3}=max{low, v. low, low}=*low*

– In words, C1 imparts *low* causality to C8.

To determine the total effect of C2 to C8, we take the maximum of paths I4 through I5.

Path 4 from C2 to C8: c2→ c5→ c7 →c8

I4(C2 to C8)=min(low, v.low, med)=v. low

Path 5 from C2 to C8: c2→ c4 →c8

I5(C2 to C8)=min(med, low)=low

Thus total effect of C2 to C8, denoted by T(C2,C8) is:

T(c2,c8)=max{I4,I5}=max{low, v. low}=*low*

– In words, C2 imparts *low* causality to C8.

Path 6 from C6 to C8: c6 →C8:

I6=high

To determine the total effect of C6 to C8, we take the maximum of path I6.

– In words, C6 imparts *high* causality to C8.

Path 7 from C4 to C8: c4 →c8:

I7=low

The total effect of C4 to C8 is determined by taking the maximum of path I7.

– In words, C4 imparts *low* causality to C8.

To determine the total effect of C5 to C8, we take the maximum of path I8.

Path 8 from C5 to C8: C5→ C7 →C8:

I8(C5 to C8)=min(v.low, med)=v.low

Thus total effect of C5 to C8, denoted by T(C5,C8) is computed:

T(C5,C8)=max{I8}=*v. low*

– In words, C5 imparts *v. low* causality to C8.

Summarizing, new causal paths describing the interconnections among concepts as well as some of the existed interconnections have been explored updating their initial values due to the above paths.

After the implementation of the investigating methodology, the FCM model was restructured and a new augmented FCM model was produced illustrated in Figure 4. Where each branch has fuzzy labels, fuzzy values derived from corresponding fuzzy sets as they have been initially prescribed by experts and data handle.

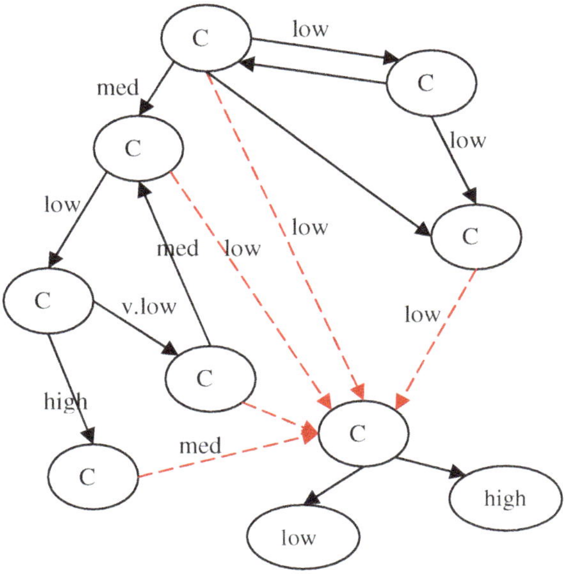

Fig. 4 The new restructured FCM model using the proposed approach

3 New Approach on Constructing Dynamic Fuzzy Cognitive Maps Using Knowledge Extraction Techniques

There is a necessity to propose a methodology and generally a framework for extracting fuzzy interconnections among attributes from available data using knowledge extraction techniques. Through the knowledge extraction methods, fuzzy linguistic interconnections could be identified to restructure the fuzzy cognitive map model thus producing a new dynamic FCM-based tool for decision support tasks. The proposed approach can incorporate any decision tree algorithm, but for the purpose of this work C4.5 has been chosen as it is a well-known and well-tested decision tree induction algorithm for classification problems (Quinlan, 2002). As it has already been stated, the central idea of the proposed method is to combine a fuzzy decision tree to extract the available knowledge of data and to generate fuzzy linguistic weights through causal paths. The resulted fuzzy relationships among leaf nodes are applied to restructure the FCM model. Among the different fuzzy inference techniques we selected for our approach the Zadeh's union and intersection parameters (see above section 2.1). The inference algorithm of FCMs remains the same and only the weight matrix multiplied with previous concept values was changed. Figure 5 illustrates the proposed process with the corresponding steps and final decision.

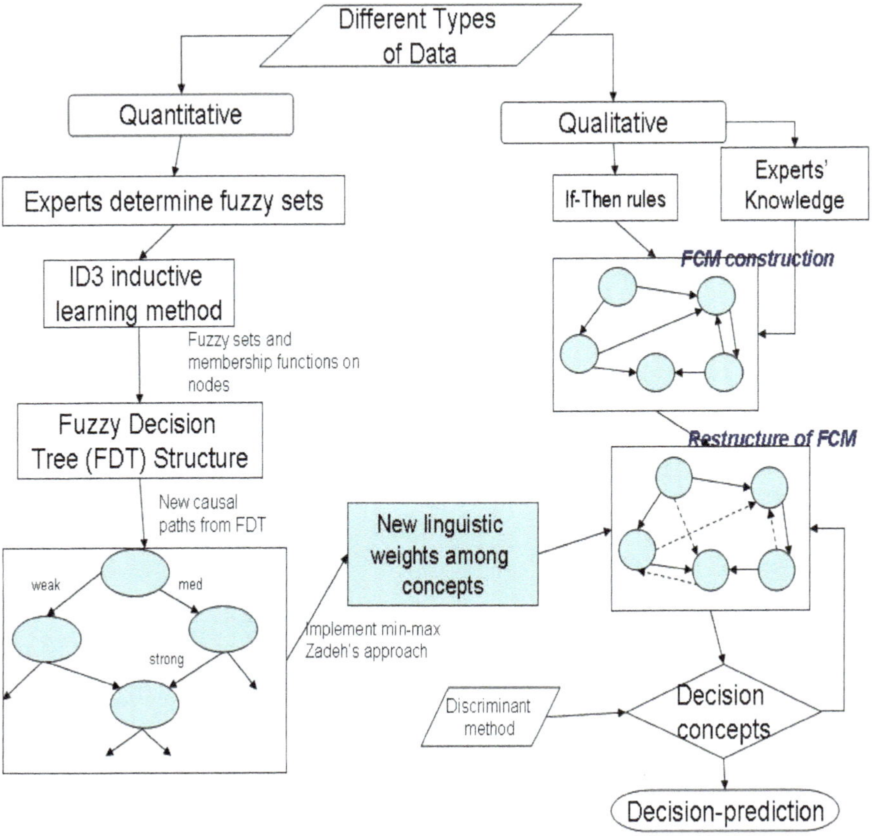

Fig. 5 New approach for constructing augmented FCMs by complementary use of fuzzy decision trees

The algorithmic approach for the restructure of FCM using fuzzy decision trees (and/or other knowledge extraction methods) is consisting on the following steps:

Step 1: For all the M experts, set credibility weight $b_k = 1$

Step 2: Each of the M experts is asked to suggest and describe each of the N concepts that comprise the FCM.

Step 3: For all the ordered pair of concepts (*Ci* and *Cj*) each kth of the M experts is asked to make the following statement (using an if-then rule):

IF the value of concept *Ci* {increases, decreases, is stable} **THEN** causes value of concept *Cj* to {increase, decrease, nothing} **THUS** the influence of concept *Ci* on concept *Cj* is *T(Ci,Cj)*

Through this step a number of linguistic weights have been assigned by experts.

Step 4: If quantitative data (numeric or symbolic) are available, the approach of using fuzzified crisp decision trees (presented in above section 2.1) is implemented into the data set to derive the available structure of fuzzy decision trees and the fuzzy labels in the branches *Di*.

Step 5: From the created fuzzy decision trees, a number of causal paths among the branches *i*, connecting leaf nodes *Di* to *Dj*, is determined. These causal paths transferred in FCM model as fuzzy rules between interconnecting concepts *Ci* to *Cj*, through a number of direct positive relationships.

Step 6: Using the fuzzy causal algebra, an indirect effect operator *I* used as the minimum operator (min) on an ordered set *P* of causal values. The simple fuzzy causal algebra is created by interpreting the indirect effect operator *I* as the minimum operator (min) on the set *P* of fuzzy values, corresponding to the respective causal paths among the FCM concepts. Then the max operator *T* is applied to the resulted effect operators *I*, and a new linguistic weight produced among *Ci* and *Cj*. The overall linguistic weight is the sum of the path products. Thus a new inferred fuzzy weight is assigned between the concepts *Ci* and *Cj*.

Step 7: Aggregate all the linguistic weights both produced by experts and knowledge extraction methods, using the SUM method where the membership function μ suggested by k^{th} expert is multiplied by the corresponding credibility weight b_k. Use the COG defuzzification method to calculate the numerical weight e_{ij} for every interconnection.

Step 8: IF there is an ordered concept pair not examined go to step 3, ELSE construct the weight matrix $\boldsymbol{E}$ whose are the defuzzified weights e_{ij}.

END.

Using the above algorithm, someone could use fuzzy decision trees (and/or other knowledge extraction techniques) to pass available knowledge into FCM reconstructed by causal paths through fuzzy rules. Experts construct fuzzy sets and fuzzy membership functions for each problem and these fuzzy sets are used into the fuzzy decision tree algorithm due to compatibility reasons. This happens in the case of FCMs to derive the respective linguistic variables and then make the necessary comparisons.

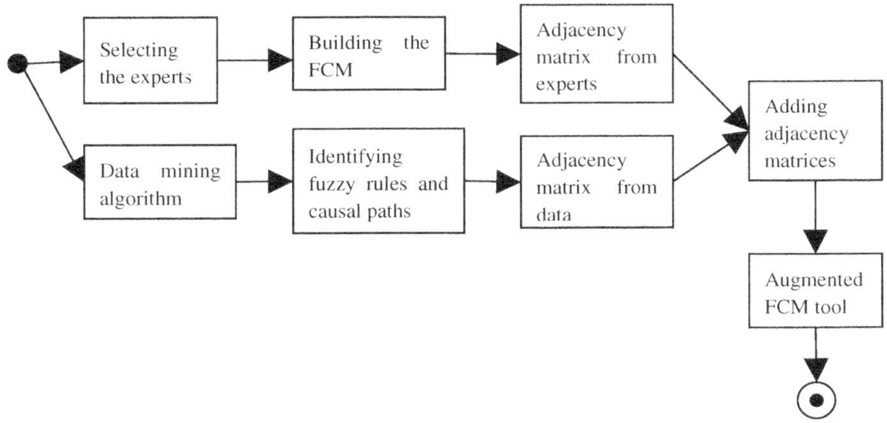

Fig. 6 Basic steps for building an augmented FCM tool

The causal paths of the leaf nodes used to determine new causal paths and weights in the FCM model. Thus the FCM model was augmented as new direct and indirect relationships among concepts determined. The basic stages for building augmented FCMs are given in Figure 6.

4 Application of the Proposed Framework into FCM-DSS in Radiotherapy

Radiotherapy is the application of ionizing radiation to cure patients suffering from cancer (and/or other diseases) and to eliminate infected cells, alone or combined with other modalities. The aim of radiation therapy is to design and perform a treatment plan on how to deliver a precisely measured dose of radiation to the defined tumor volume with as minimal damage as possible to the surrounding healthy tissue.

In a previous work, a decision making system for radiation therapy based on human knowledge and experience was developed by Papageorgiou et al., 2003. That system was consisted of a two-level hierarchical structure where an FCM in each level was created producing an advanced decision-making system. The lower-level FCM modeled the treatment planning, taking into consideration all the factors and treatment variables as well as their influences (CTST-FCM). The upper-level FCM modeled the procedure of the treatment execution and calculated the optimal final dose for radiation treatment. The upper level FCM supervised and evaluated the whole radiation therapy process. The proposed two-level integrated structure for supervising the procedure before treatment execution seems a rather realistic approach to the complex decision making process in radiation therapy.

As it has already been stated, the central idea of the proposed technique is to combine different data driven methods to extract the available knowledge from data and to generate causal paths through fuzzy rules by determining new linguistic weights. The resulted linguistic weights are applied to construct an augmented FCM-based clinical treatment simulation tool (new CTST-FCM) used for decisions in radiation treatment planning. According to the desired values of output concepts, the augmented FCM-DSS reaches a decision about the acceptance of treatment planning technique.

At this point, according to the guidelines (AAPM and ICRU protocols) and radiotherapists' opinions for the most important variables taken under consideration (in order to achieve a good distribution of the radiation on the tumor, as well as to protect the healthy tissues), as well as to minimize the complexity of the presented model, five factor concepts and eight selector-concepts were selected to determine the system performance through the calculation of output concepts. Thus, a new CTST-FCM model that represents the radiotherapy treatment planning procedure according to the test packages, guidelines and radiotherapists' opinions is designed and illustrated in Figure 7.

The number of concepts has been reduced to 16 concepts thus to avoid the complexity of the previously developed CTST-FCM model and to be more clear the proposed technique to no specialist readers. Concepts F-C1 to F-C5 are the

Factor-concepts, that represent the depth of tumor, the size of tumor, the shape of tumor, the type of the irradiation and the amount of patient thickness irradiated. Concepts S-C1 to S-C8 are the Selector-concepts, representing size of radiation field, multiple field arrangements, beam directions, dose distribution from each field, stationery vs. rotation-isocentric beam therapy, field modification, patient immobilizing and use of 2D or 3D conformal technique, respectively. The concepts OUT-C1 to OUT-C3 are the three Output-concepts. Table 1 gathers these respective concepts. The value of the OUT-C1 represents the amount of dose applied to mean Clinical Target Volume (CTV), which have to be larger than the 90% of the amount of prescribed dose to the tumor. The value of concept OUT-C2 represents the amount of the surrounding healthy tissues' volume received a dose, which have to be as less as possible, less than the 5% of volume received the prescribed dose and the value of concept OUT-C3 represents the amount of organs at risk volume received a dose, which have to be less than the 10% of volume received the prescribed dose (Khan, 1994; ICRU Report 50).

Table 1 Description and Type of new CTST-FCM concepts

Concepts	Description	Number & Type of values scaled
F-C 1	Accuracy of depth of tumor	Five fuzzy
F-C 2	Size of tumor	Seven fuzzy (very small, small, positive small, medium, negative large, large, very large)
F-C 3	Shape of tumor	Three fuzzy (small, medium, large)
F-C 4	Type of irradiated tissues-presence of inhomogeneities	Five fuzzy
F-C 5	Amount of patient thickness irradiated	Five fuzzy
S-C 1	Size of radiation field	Five fuzzy
S-C 2	Single or multiple field arrangements	Two discrete
S-C 3	Beam direction(s) (angles of beam orientation)	converted to Five fuzzy
S-C 4	Dose distribution from individual field	Two fuzzy
S-C 5	Stationery vs. rotation-isocentric beam therapy	Two discrete
S-C 6	Field modification (no field modification, blocks, wedges, filters and multileaf-collimator shaping blocks)	Five discrete
S-C 7	Patient immobilization	Three discrete
S-C 8	Use of 2D or 3D conformal technique	Two discrete
Out-C 1	Dose given to treatment volume (must be within accepted limits)	Five fuzzy
Out-C 2	Amount of irradiated volume of healthy tissues	Five fuzzy
Out-C 3	Amount of irradiated volume of sensitive organs (organs at risk)	Five fuzzy

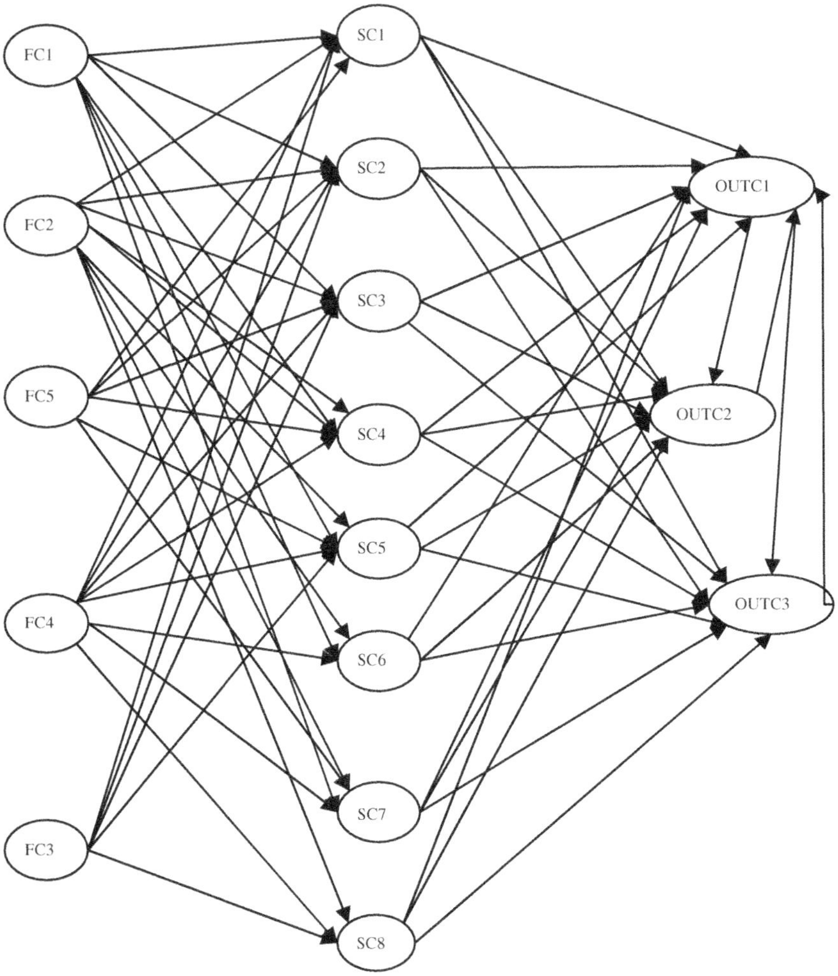

Fig. 7 The new CTST-FCM tool for decision making in radiotherapy

After the description of new CTST-FCM concepts, the design of FCM model continues with the determination of fuzzy sets for each one concept variable. The radiotherapists that work as experts specified the fuzzy membership functions for the fuzzy values of factor concepts, selector concepts and output concepts. For the factor concept F-C3 and selector concept S-C1 (size of radiation field) the experts proposed the fuzzy membership functions illustrated in Figure 8.

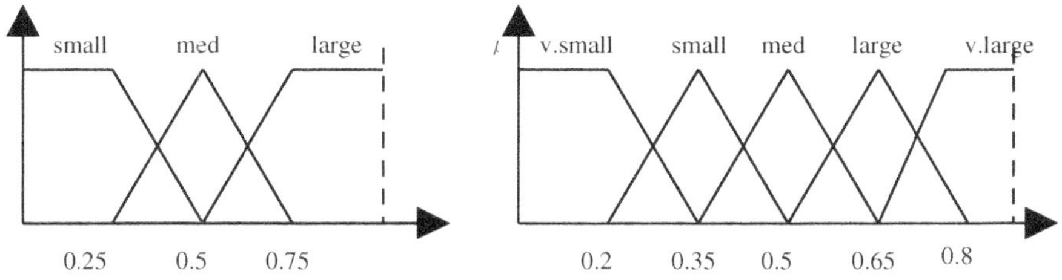

Fig. 8 Partitions (fuzzy trapezoidal membership functions) for F-C3 and S-C1 determined by a priori knowledge from radiotherapists-experts

Then, using the experimental data derived from measurements (Papageorgiou, 2000)), for the initial values of concepts and implementing the knowledge extraction method of fuzzy decision trees, a large set of fuzzy rules among the related concepts were derived. Some of the fuzzy rules that considered important to the decision making approach were selected from the fuzzy decision tree-based rule extraction technique according to the test packages and experimental data. Some example rules are presented at follows:

If F-C1 is medium Then S-C1 is high
If F-C1 is medium Then S-C2 is very high
If F-C2 is high Then S-C1 is high
If F-C2 is small and F-C3 is small Then S-C1 is very high
If S-C4 is 1 and S-C6 is medium Then F-C5 is very high
If F-C1 is small and F-C2 is small Then S-C3 is small

In this point, due to the large number of fuzzy rules produced by the fuzzy decision tree algorithm, we selected only those which differ from the initially suggested by experts and used for the reconstruction of the augmented CTST-FCM in radiation treatment planning. These rules accompanied by rules suggested by experts produce the new CTST-aFCM simulation tool for radiation therapy, which has new strengths among concepts and assigns new decisions and treatment planning suggestions.

5 Results on Implementing the Proposed Approach in Radiotherapy Process

After construction of new CTST-FCM tool for giving a decision about the acceptance or no of the radiotherapy process, a number of scenarios have been introduced and the decision making capabilities of the technique are presented by simulating these scenarios and finding the predicted outcomes according to the available data. Two scenarios for the problem of prostate cancer therapy were considered using the new CTST-FCM model, which consists of 16 concepts and 64 interconnections among concepts, in order to test the validity of the model. In each of the test scenarios we have an initial vector **A**, representing the presented events at a given time of the process, and a final vector **A_f**, representing the last state that can be arrived at.

The final vector **A_f** is the last vector produced in convergence region and the 14th, 15th and 16th value of this vector are the final values of decision concepts.

The algorithm used to obtain the final vector **A_f** is consisting on the following steps:

(1) Definition of the initial vector **A** that corresponds to the elements identified in Table 1.

(2) Multiply the initial vector **A** and the matrix **E** defined by experts by the eq. (1), as indicated by the respective Tables of each case study.

(3) The resultant vector is updating using eqs. (1)–(2).

(4) This new vector is considered as an initial vector in the next iteration.

(5) Steps 2–4 are repeated until $\mathbf{A}^{k} - \mathbf{A}^{k-1} \leq e = 0.001$.

The FCM performance is illustrated by means of simulation of the following two case scenarios in radiotherapy process.

In the first scenario, the 3-D conformal technique consisting of six-field arrangement is suggested and in the second one the conventional four-field box radiation technique. Radiotherapy physicians and medical physicists choose and specified, in our previous study, the fuzzy membership functions for the weights for each case study as well as the fuzzy rules according to their knowledge for each treatment planning procedure. The numerical values of weights between factor and selector concepts for the new CTST-aFCM are summarized after the defuzzification process and are depicted in Table 2.

Table 2 Weight values between F-Cs and S-Cs for new CTST-aFCM after defuzzification process

Factors/ Selectors	S-C1	S-C2	S-C3	S-C4	S-C5	S-C6	S-C7	S-C8
F-C1	0.6	0.62	0.4	0.4	0.6	0.6	0.2	0
F-C2	0.7	0.6	0.2	0.53	0.55	0.5	0.6	0.5
F-C3	0.6	0.63	0.45	0	0.4	0	0	0.7
F-C4	0.32	0.6	0.5	0.55	0.47	0.5	0	0.6
F-C5	0.5	0.6	0.6	0.6	0.2	0.5	0.5	0

For the first case study, the conformal radiotherapy was selected. Multiple CT-based external contours define the patient anatomy and isocentric beam therapy is used (Khan, 1994). Beam weights are different for the six fields, and blocks, wedges are used. The specific characteristics of conformal therapy determine the values of concepts and weights interconnections of new CTST-aFCM model. So, the S-C2 takes the value of six-field number; S-C1 has the value of "small-size" for radiation field that means that the influence of S-C1 and S-C2 toward OUT-Cs is great. In the same way the S-C3 and S-C4 have great influence at OUT-Cs because different beam directions and weights of radiation beams are used. The S-C5 takes the discrete value of isocentric beam therapy. Concept S-C6 takes values for the selected blocks and/or wedges, influencing the OUT-Cs. The S-C7 takes a value for accurate patient positioning and the S-C8 takes the discrete value of 3-D radiotherapy.

Considering the above and the measured experimental data, the initial values of concepts and weights of interconnections between S-Cs and OUT-Cs are suggested. Table 3 gathers the numerical weights among Factor-concepts, Selector-concepts, and Output-concepts, of new CTST-aFCM for the first case study, as they identified from combined knowledge from experts and data. The new CTST-aFCM model is presented in Figure 9, where the modified relationships are shown by red line.

Table 3 Numerical weights among F-Cs, S-Cs and OUT-Cs of new CTST-FCM for the first case, as they derived from combined knowledge from experts and data

Concepts	S-C3	S-C4	S-C5	S-C6	S-C7	S-C8	S-C9	S-C10	OUT-C1	OUT-C2	OUT-C3
F-C1	0.7	0.75	0.4	0.4	0.65	0.6	0	0	0	0	0
F-C2	0.75	0.6	0	0.6	0.55	0.5	0.6	0.5	0	0	0
F-C3	0.6	0.7	0.45	0.2	0.4	0	0	0.75	0	0	0
F-C4	0.25	0.6	0.5	0.55	0.4	0.5	0	0.4	0	0	0
F-C5	0.5	0.6	0.6	0.5	0.2	0.5	0.6	0	0	0	0
S-C1	0	0	0	0	0	0	0	0	0.4	-0.4	-0.4
S-C2	0	0	0	0.5	0	0	0	0	0.3	-0.5	-0.4
S-C3	0	0	0	0	0	0	0	0	0.4	-0.3	-0.3
S-C4	0	0	0	0	0	0	0	0	0.4	-0.4	-0.4
S-C5	0	0	0	0	0	0.7	0	0	0.3	-0.3	-0.3
S-C6	0	0	0	0	0.6	0	0	0	0.4	-0.4	-0.4
S-C7	0	0	0	0	0	0	0	0	0.5	-0.5	-0.5
S-C8	0	0	0	0	0	0	0	0	0.6	-0.5	-0.5
OUT-C1	0	0	0	0	0	0	0	0	0	-0.6	-0.5
OUT-C2	0	0	0	0	0	0	0	0	-0.7	0	0
OUT-C3	0	0	0	0	0	0	0	0	-0.6	0	0

The following initial vector is formed for this particular treatment technique:

$\mathbf{A_1}$= *[0.6 0.5 0.5 0.6 0.6 0.4 0.65 0.7 0.45 0.6 0.6 0.5 0.6 0.5 0.5 0.5].*

Through the simulation algorithm described above, the resulting CTST-FCM starts to interact and simulates the radiation procedure. New values of concepts were calculated after 8 simulation steps. Fig. 10 illustrates the values of concepts for eight simulation steps, where is concluded that after the 5th simulation step FCM reaches an equilibrium region. The following vector gives the calculated values of concepts in the equilibrium region.

A_1_f= [*0.6590 0.6590 0.6590 0.6590 0.6590 0.9420 0.9568 0.8988 0.9412 0.9515 0.9585 0.8357 0.8770 0.9813 0.0203 0.0336].*

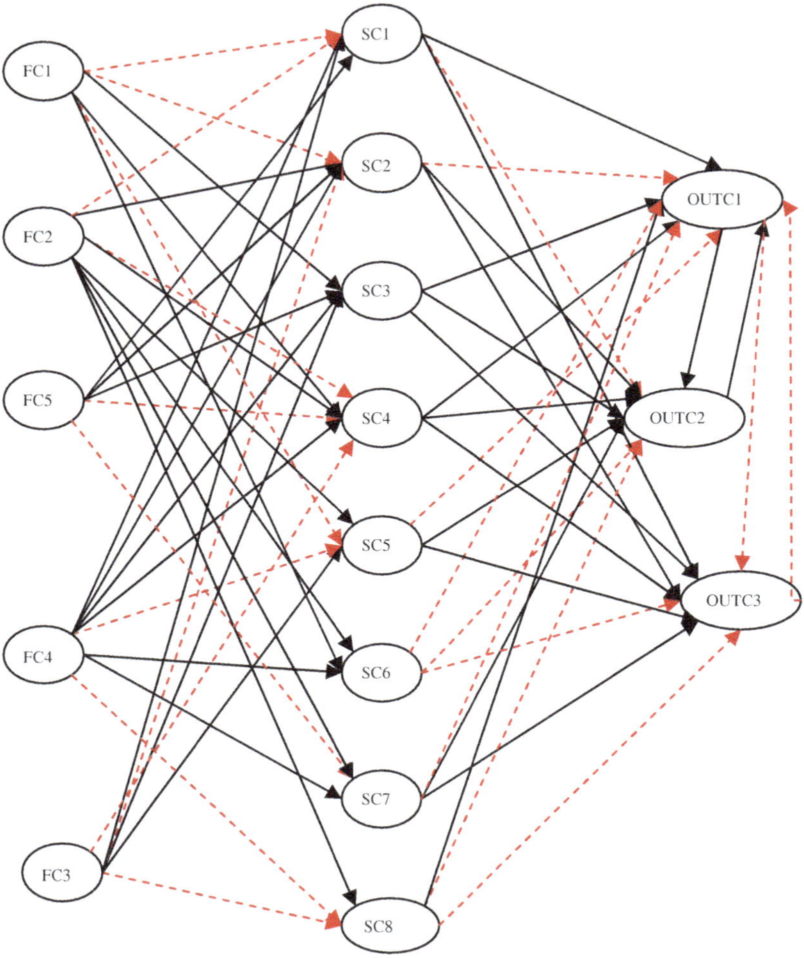

Fig. 9 The new CTST-aFCM tool for decision making in radiotherapy after combining knowledge from experts and data (the broken lines are the new weight values)

At the equilibrium point, the following values of OUT-Cs are: for OUT-C1 is 0.9813, for OUT-C2 is 0.0201 and for OUT-C3 is 0.0336. Based on the referred protocols (AAPM, 1995; ICRU Report 50), the calculated values of output concepts are accepted. The calculated value of OUT-C1 is 0.981, which means that the CTV receives the 98% of the amount of the prescribed dose, which is

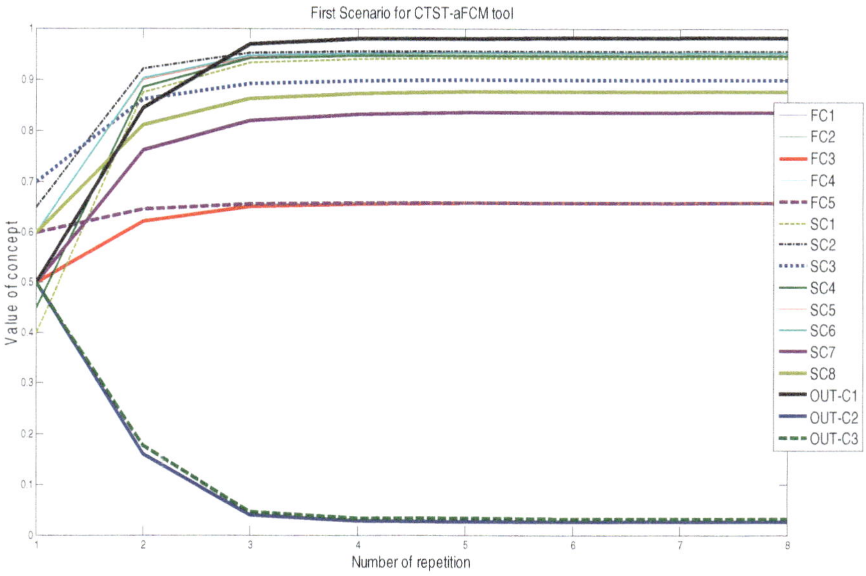

Fig. 10 Variation of values of 16 concepts for the new CTST-aFCM for the first case study for eight simulation steps

accepted. The value of OUT-C2 that represents the amount of the surrounding healthy tissues' volume received a dose was found equal to 0.0201, so the 2.01% of the volume of healthy tissues receives the prescribed dose, and the OUT-C3 was found equal to 3.36% of the dose received from organs at risk.

In the second scenario, the conventional four-field box technique is implemented for the prostate cancer treatment. This technique is consisted of a four-field box arrangement with gantry angles 0, 90, 180, and 270. For this case, the new CTST-FCM was reconstructed which means that the cause-effect relationships and weights have been reassigned not only from radiotherapists' suggestions but also from data knowledge using the proposed rule extraction technique. For this case, the Selector-concept S-C2 has the value of four-field number; S-C1 has the value of "large-size" of radiation field, which means that the influence of S-C1 and S-C2 toward OUT-Cs is very low. In the same way, the S-C3 and S-C4 have lower influence on OUT-Cs because different beam directions and weights of radiation beams are used. The S-C5 takes the discrete value of isocentric beam therapy and has the same influence on OUT-Cs as the above conformal treatment case. S-C6 has zero influence on OUT-Cs because no blocks (and/or no wedges and any filters) are selected for this treatment case. The S-C7 takes a low value for no accurate patient positioning and the S-C8 takes the discrete value of 2-D radiotherapy. The numerical weights among F-Cs, S-Cs and OUT-Cs, of new CTST-aFCM for the second case study, are given in Table 4.

Table 4 Numerical weights among F-Cs, S-Cs and OUT-Cs of new CTST-FCM for the second case

Concepts	S-C3	S-C4	S-C5	S-C6	S-C7	S-C8	S-C9	S-C10	OUT-C1	OUT-C2	OUT-C3
F-C1	0.7	0.75	0.4	0.4	0.6	0.6	0	0	0	0	0
F-C2	0.75	0.6	0	0.6	0.55	0.5	0.6	0.5	0	0	0
F-C3	0.6	0.7	0.45	0.2	0.4	0	0	0.75	0	0	0
F-C4	0.25	0.6	0.5	0.5	0.4	0.5	0	0.4	0	0	0
F-C5	0.5	0.6	0.6	0.5	0.2	0.5	0.6	0	0	0	0
S-C1	0	0	0	0	0	0	0	0	0.3	-0.4	-0.3
S-C2	0	0	0	0.5	0	0	0	0	0.25	-0.5	-0.4
S-C3	0	0	0	0	0	0	0	0	0.3	-0.3	-0.3
S-C4	0	0	0	0	0	0	0	0	0.25	-0.2	-0.2
S-C5	0	0	0	0	0	0.7	0	0	0.3	-0.3	-0.3
S-C6	0	0	0	0	0.6	0	0	0	0.2	0	0
S-C7	0	0	0	0	0	0	0	0	0.4	-0.3	-0.3
S-C8	0	0	0	0	0	0	0	0	0.4	-0.4	-0.4
OUT-C1	0	0	0	0	0	0	0	0	0	-0.4	-0.4
OUT-C2	0	0	0	0	0	0	0	0	-0.7	0	0
OUT-C3	0	0	0	0	0	0	0	0	-0.6	0	0

Using this new CTST-aFCM model, with the new modified weight matrix, the simulation of the radiotherapy procedure for this case starts with the following initial values of concepts:

$\mathbf{A_2}$ = *[0.5 0.48 0.4 0.6 0.5 0.7 0.45 0.4 0.6 0.6 0.3 0.2 0.4 0.4 0.2 0.2].*

Through the simulation algorithm described above, the resulting CTST-aFCM simulates the radiation procedure and converges to a steady point after 8 simulation steps. Fig. 11 illustrates the values of concepts for eight simulation steps, where is concluded that after the 5th simulation step FCM reaches an equilibrium region. The following vector gives the calculated values of concepts in the equilibrium region:

$\mathbf{A_2\_f}$ = *[0.6590 0.6590 0.6590 0.6590 0.6590 0.9420 0.9568 0.8988 0.9412 0.9515 0.9585 0.8357 0.8770 0.9541 0.0754 0.0910].*

The final values of OUT-Cs are as follows: for OUT-C1, 0.9541; for OUT-C2, 0.0754; and for OUT-C3, 0.0910. These values for OUT-C2 and OUT-C3 are not accepted according to related protocols and task groups (AAPM, 1995; ICRU Report 50).

The new explored CTST-aFCM model seems to be a less complex but dynamic model working efficiently with less number of concepts and weights and especially with weights identified by knowledge extracted from fuzzy decision

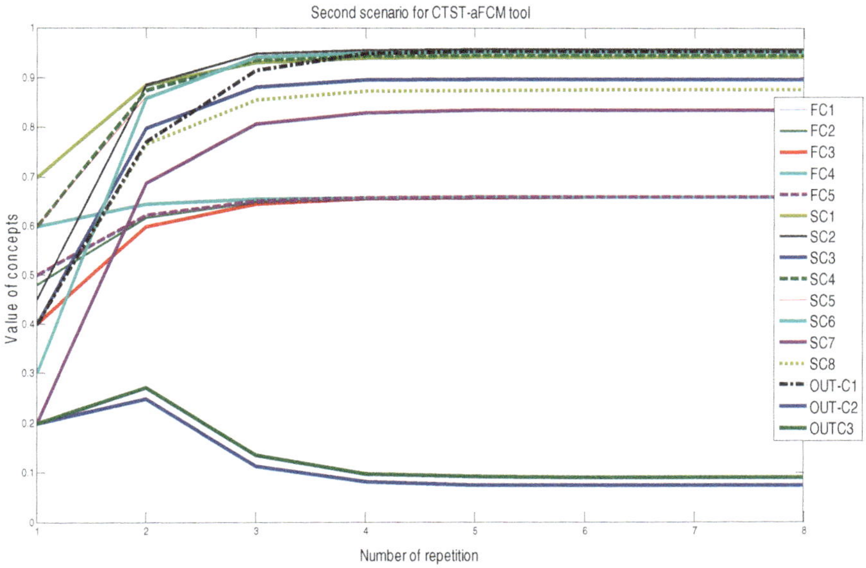

Fig. 11 Variation of values of 16 concepts of new CTST-aFCM for the second case study, with the classical treatment planning case

tree-based rule extraction technique. This tool for making decisions in radiation therapy can adapt its knowledge from available data and not only from experts' opinions. Thus, through the proposed approach, an acceptable decision is succeeded and the new CTST-aFCM tool is less time consuming and easy for use from no specialists.

6 Discussion and Conclusions

In this chapter, a novel approach on producing dynamic FCM-DSS combining knowledge extracted from the available data sets using a data mining algorithm and from guidelines has been explored and a medical application giving meaningful results has been depicted. The proposed approach was based partly on fuzzy rules derived from data using fuzzy decision tree rule extraction algorithm and on the experts' knowledge. All the available knowledge from data was used to enrich the FCM which works as a knowledge-based decision making model.

The produced CTST-aFCM was constructed combining knowledge from the available fuzzy rule base created by knowledge exploiting by experts, guidelines and data, using rule extraction methods for producing fuzzy rules. In this point, due to the large number of fuzzy rules produced by the fuzzy decision tree algorithm, we selected only those which differ from the initially suggested by experts and used for the construction of the CTST-aFCM in radiation treatment planning. Thus, some of the initial CTST-aFCM weights could be changed according to the new knowledge inserted from the fuzzy rules.

Pinpointing, the main goal of this work is to represent a different dynamic approach for construction of augmented FCM-based decision support tools rather to compare with other decision support systems. The new CTST-aFCM model with less number of concepts and weights and especially with weights not only determined by guidelines and radiotherapists-experts' suggestions but also using fuzzy rules extracted through fuzzy decision tree technique, is a dynamic and less complex model which works efficiently. Through the proposed approach was proven that the decision making tool can adapt its knowledge not only using experts' opinions and medical guidelines but also using the available data. Thus, an acceptable decision can be succeeded and the new CTST-aFCM tool is less time consuming and easy for use from no specialists.

In our opinion, revival of interest in DSS research crucially depends on new frameworks and architectures that would extend the cognitive capacities of DSS to meet the real world. We hope that our work points in this direction. Furthermore, the aim of the proposed methodology was not to achieve better accuracies or to present a better classifier, but to introduce a novel framework for FCM-DSS enhancement by fuzzy rule base constructed by efficient extraction of knowledge methods. The new decision support tool is simple, less complex, transparent and interpretable to be accepted for medical applications. The distinguishing feature of such dynamic FCM-DSS is its situations with large amount of data, not enough knowledge from experts and difficulty to handle the available knowledge from many different sources of information.

As disadvantages could be referred the following:

- If not enough information is available, the approach can not be more efficient than other decision making methods and should be complemented by other intelligent methods
- Its outcomes are dependent on the attentiveness of the analysts about the knowledge extraction methods.

Summarizing in this chapter, a novel approach for the construction of dynamic FCMs for inference and decision making is presented. The fuzzy weights and causal paths, which derived by using fuzzy decision tree induction algorithm for this data set, were incorporating to construct the new CTST-aFCM producing acceptable decisions not only based on experts' suggestions. In the future, other rule-extraction methods for decision support in medical domain will be investigated. Furthermore, the proposed methodology will be implemented in other medical problems which have been handled till today using only knowledge from data or from experts, as well as in other scientific domains with large amount of available data and information sources.

Acknowledgments

The research of E.I. Papageorgiou was supported in part by the European Commission's Seventh Framework Information Society Technologies (IST) Programme, Unit ICT for Health, project DEBUGIT (no. 217139).

References

AAPM Report No. 55, Radiation Treatment planning dosimetry verification. American Association of Physicists in Medicine. Report of Task Group 23 of the Radiation Therapy Committee. American Institution of Physics, Woodbury (1995)

Au, W.-H., Chan. K.C.C.: FARM: A data mining system for discovering fuzzy association rules. In: Proc. of the 8th IEEE International Conference on Fuzzy Systems, Seoul, Korea, August 22-25, pp. 1217–1222 (1999)

Aguilar, J.: A survey about fuzzy cognitive maps papers. International Journal of Computational Cognition 3(2), 27–33 (2005)

Alam, R., Ibbott, G.S., Pourang, R., Nath, R.: Application of AAPM Radiation Therapy Committee Task Group 23 test package for comparison of two treatment planning systems for photon external beam radiotherapy. Med. Phys. 24, 2043–2054 (1997)

Boutalis, Y., Kottas, T.L., Christodoulou, M.: Adaptive estimation of fuzzy cognitive maps with proven stability and parameter convergence. IEEE Transactions on Fuzzy Systems 17(4), 874–889 (2009)

Bueno, S., Salmeron, J.L.: Benchmarking main activation functions in fuzzy cognitive maps. Expert Systems with Applications 36(3), 5221–5229 (2009)

Chen, G., Wei, Q.: Fuzzy association rules and the extended mining algorithms. Information Sciences 147, 201–228 (2002)

Crockett, K., Bandar, Z., Mclean, D., O'Shea, J.: On constructing a fuzzy inference framework using crisp decision trees. Fuzzy Sets and Systems 157, 2809–2832 (2006)

Fayyad, U.M., Piatetsky-Shapiro, G., Smyth, P., Uthurusamy, R.: Advances in Knowledge Discovery and Data Mining. AAAI/MIT Press, Menlo Park (1996)

Fayyad, U., Uthurusamy, R.: Data mining and knowledge discovery in databases. Commun. ACM 39, 24–27 (1996)

Froelich, W., Wakulicz-Deja, A.: Predictive Capabilities of Adaptive and Evolutionary Fuzzy Cognitive Maps - A Comparative Study. In: Nguyen, N.T., Szczerbicki, E. (eds.) Intel. Sys. for Know. Management. SCI, vol. 252, pp. 153–174. Springer, Berlin (2009)

Froelich, W., Wakulicz-Deja, A.: Mining temporal medical data using adaptive fuzzy cognitive maps. In: Proceedings - 2009 2nd Conference on Human System Interactions, HSI 2009, pp. 16–23 (2009) art. no. 5090946

Fu, L.M.: Knowledge-Based Connectionism for Revising Domain Theories. IEEE Trans. on Systems, Man, and Cybernetics 23(l), 173–182 (1993)

Gath, G., Geva, A.B.: Unsupervised optimal fuzzy clustering. IEEE Transactions on Pattern Analysis and Machine Intelligence 7, 773–781 (1989)

Georgopoulos, V.C., Stylios, C.D.: Complementary case-based reasoning and competitive fuzzy cognitive maps for advanced medical decisions. Soft Computing 12, 191–199 (2008)

Georgopoulos, V.C., Stylios, C.D.: Augmented Fuzzy Cognitive Maps Supplemented with Case Base Reasoning for Advanced Medical Decision Support. In: Nikravesh, M., Zadeh, L.A., Kacprzyk, J. (eds.) Soft Computing for Information Processing and Analysis Enhancing the Power of the Information Technology. Studies in Fuzziness and Soft Computing. pp. 391–405. Springer, Heidelberg (2005) ISBN: 3-540-22930-2

Hayashi, Y., Maeda, T., Bastian, A., Jain, L.C.: Generation of fuzzy decision trees by fuzzy ID3 with adjusting mechanism of and/or operators. In: Proc. of Int. Conf. Fuzzy Syst., pp. 681–685 (1998)

ICRU Report 50, Prescribing, recording and reporting photon beam therapy. International Commission on Radiation Units and Measurements, Washington (1993)

Ishibuchi, H., Nozaki, K., Yamamoto, N., Tanaka, N.: Selecting fuzzy if–then rules for classification problems using genetic algorithms. IEEE Trans. Fuzzy Systems 3(3), 260–270 (1995)

Janikow, C.Z.: Fuzzy decision trees: issues and methods. IEEE Trans. Systems Man and Cybernetics 28(1), 1–14 (1998)

Janikow, C.Z.: Fuzzy partitioning with FID3.1. In: Proceedings of the 18th International Conference of the North American Fuzzy Information Society, pp. 467–471 (1999)

Janikow, C.Z.: Fuzzy Decision Trees Manual, free version for Fuzzy Decision Trees (1998) `http://www.cs.umsl.edu/Faculty/janikow/janikow.html`.

Jang, J.S.R., Sun, C.T., Mizutani, E.: Neuro-Fuzzy & Soft Computing. Prentice-Hall, Upper Saddle River (1997)

Jang, L.: Soft Computing Techniques in Knowledge-Based Intelligent Engineering Systems: Approaches and Applications. Studies in Fuzziness and Soft Computing, vol. 10. Springer, Heidelberg (1997)

Khan, F.: The Physics of Radiation Therapy, 2nd edn. Williams & Wilkins, Baltimore (1994)

Kosko, B.: Fuzzy Cognitive Maps. Int. J. Man-Machine Studies 24, 65–75 (1986)

Kosko, B.: Neural Networks and Fuzzy Systems. Prentice-Hall, New Jersey (1992)

Kurgan, L.A., Musilek, P.: A Survey on Knowledge Discovery and Data mining processes. The Knowledge Engineering Review 21(1), 1–24 (2006)

Lee, K.C., Kim, H.S.: A Causal Knowledge-Driven Inference Engine for Expert System. In: Proc. of the 31st Hawaii International Conference on System Science, January 6-9, vol. 1(1), pp. 284–293 (1998)

Liu, H., Tan, S.T.: X2R: A Fast Rule Generator. In: Proc of IEEE Inter. Conf. on Systems, Man & Cybernetics, Vancouver, Canada (October 1995)

Liu, X., Cohen, P., Berthold, M.R.: IDA 1997. LNCS, vol. 1280. Springer, Heidelberg (1997)

Lozowski, A., Zurada, J.M.: Extraction of linguistic rules from data via neural networks and fuzzy approximation. In: Cloete, J., Zurada, J.M. (eds.) Knowledge-Based Neurocomputing. The MIT Press, Cambridge (2000)

Miao, Y., Liu, Z.Q.: On causal inference in fuzzy cognitive maps. IEEE Transactions on Fuzzy Systems 8, 107–119 (2000)

Mitra, S., Konwar, K.M., Sankar, K.P.: Fuzzy decision tree, linguistic rules and fuzzy knowledge-based network: generation and evaluation. IEEE Trans. Syst. Man Cybern. Part C: Appl. Rev. 32(4), 328–339 (2002)

Mitra, S., Hayashi, Y.: Neuro-Fuzzy rule generation: Survey in soft computing. IEEE Trans Neural Networks 11(3), 748–760 (2000)

Nauck, D., Klawonn, F., Kruse, R.: Foundations of neuro-fuzzy systems. Wiley, Chichester (1997)

Nauck, D., Kruse, R.: Obtaining interpretable fuzzy classification rules from medical data. Artificial Intelligence in Medicin 16(2), 149–169 (1999)

Nauck, D.: NEFCLASS toolbox (1997), `http://fuzzy.cs.uni-magdeburg.de/nefclass/`

Olaru, C.W.: A complete fuzzy decision tree technique. Fuzzy Sets and Systems 138, 221–254 (2003)

Pach, F.P., Abonyi, J.: Association Rule and Decision Tree based Methods for Fuzzy Rule Base Generation. Transactions on Engineering, Computing and Technology 13 (2006) ISSN 1305-5313

Pal, S.K., Mitra, S.: Neuro-Fuzzy Pattern Recognition: Methods in Soft Computing. Wiley, New York (1999)

Papageorgiou, E., Stylios, C., Groumpos, P.: An Integrated Two-Level Hierarchical Decision Making System based on Fuzzy Cognitive Maps (FCMs). IEEE Trans. Biomed. Engin. 50(12), 1326–1339 (2003)

Papageorgiou, E.I.: A model for dose calculation in treatment planning using pencil beam kernels. MSc. Thesis, Medical University Hospital of Patras, Greece (June 2000)

Papageorgiou, E.I., Groumpos, P.P.: A weight adaptation method for fine-tuning Fuzzy Cognitive Map causal links. Soft Computing 9, 846–857 (2005a)

Papageorgiou, E.I., Groumpos, P.P.: A new hybrid learning algorithm for Fuzzy Cognitive Maps learning. Applied Soft Computing 5, 409–431 (2005b)

Papageorgiou, E.I., Spyridonos, P., Ravazoula, P., Stylios, C.D., Groumpos, P.P., Nikiforidis, G.: Advanced Soft Computing Diagnosis Method for Tumor Grading. Artificial Intelligence in Medicine 36(1), 59–70 (2006a)

Papageorgiou, E.I., Stylios, C.D., Groumpos, P.P.: A Combined Fuzzy Cognitive Map and Decision Trees Model for Medical Decision Making. In: Proceedings of the 28th IEEE EMBS Annual Intern. Conference in Medicine and Biology Society, EMBS 2006, New York, USA, 30 August-3 September, pp. 6117–6120 (2006b)

Papageorgiou, E.I., Groumpos, P.P.: Neuro-fuzzy, fuzzy decision tree and association rule based methods for fuzzy cognitive map grading process. In: Proceedings of International Conference on Computational Intelligence in MEDicine, CIMED 2007, Plymouth, UK, July 25-27 (2007) (CD-ROM)

Papageorgiou, E.I., Spyridonos, P., Glotsos, D., Stylios, C.D., Ravazoula, P., Nikiforidis, G., Groumpos, P.P.: Brain tumour characterization using the soft computing technique of fuzzy cognitive maps. Applied Soft Computing 8, 820–828 (2008)

Papageorgiou, E.I., Papandrianos, N., Apostolopoulos, D., Vassilakos, P.: Fuzzy Cognitive Map based Decision Support System for thyroid diagnosis management. In: Zurada, J.M., Yen, G.G., Wang, J. (eds.) Computational Intelligence: Research Frontiers. LNCS, vol. 5050, pp. 1204–1211. Springer, Heidelberg (2008)

Papakostas, G.A., Boutalis, Y.S., Koulouriotis, D.E., Mertzios, B.G.: Fuzzy cognitive maps for pattern recognition applications. International Journal of Pattern Recognition and Artificial Intelligence 22(8), 1461–1486 (2008)

Pedrycz, W., Sosnowski, A.: Designing decision trees with the use of fuzzy granulation. IEEE Trans. Syst. Man Cybern. A 30, 151–159 (2000)

Peláez, C.E., Bowles, J.B.: Using fuzzy cognitive maps as a system model for failure modes and effects analysis. Information Sciences 88, 177–199 (1996)

Quinlan, J.R.: Decision trees and decision making. IEEE Trans System, Man and Cybernetics 20(2), 339–346 (1990)

Quinlan, J.R.: C4.5: Programs for machine learning. Morgan Kaufmann, San Mateo (1993)

Quinlan, J.R.: Is C5.0 better than C4.5 (2002), `http://www.rulequest.com/see5-comparison.html`

Sestino, S., Dillon, T.: Using single-layered neural networks for the extraction of conjunctive rules and hierarchical classifications. J. Appl. Intell. 1, 157–173 (1991)

Sison, L., Chong, E.: Fuzzy modeling by induction and pruning of decision trees. In: IEEE Symposium on Intelligent Control, U.S.A., pp. 166–171 (1994)

Sordo, M., Vaidya, S., Jain, L.C.: An introduction to computational intelligence in healthcare: New directions. Studies in Computational Intelligence 107, 1–26 (2008)

Stach, W., Kurgan, L., Petrycz, W.: A Framework for a novel scalable FCM learning method. In: Proceedings of the 2007 Symposium on Human-Centric Computing and Data Processing (HCDP 2007), Canada, February 21 - 23, pp. 13–14 (2007)

Stylios, C.D., Georgopoulos, V.C., Malandraki, G.A., Chouliara, S.: Fuzzy cognitive map architectures for medical decision support systems. Appl. Soft Comput. 8(3), 1243–1251 (2008)

Stylios, C.D., Groumpos, P.P.: Modeling Fuzzy Cognitive Maps. IEEE Transactions on Systems, Man, and Cybernetics, Part A 34, 155–162 (2004)

Taber, R., Yager, R., Helgason, C.M.: Quantization Effects on the Equilibrium Behavior of Combined Fuzzy Cognitive Maps. International Journal of Intelligent Systems 22, 181–202 (2007)

Towell, G., Shavlik, J.: Extracting Refined Rules from Knowledge-Based Neural Networks. Machine Learning 131, 71–101 (1993)

Umano, M., Okamoto, H., Hatono, I., Tamura, H.: Generation of fuzzy decision trees by fuzzy ID3 algorithm and its application to diagnosis by gas in oil. In: Japan–U.S.A. Symposium, pp. 1445–1450 (1994)

Weber, R.: Fuzzy ID3: a class of methods for automatic knowledge acquisition. In: 2nd International Conference on Fuzzy Logic and Neural Networks, Iizuka, Japan, pp. 265–268 (1992)

Wei, Z., Baowen, S., Yanchun, Z.: Design of inference model based on activation for fuzzy cognitive map. In: 2009 International Workshop on Intelligent Systems and Applications, ISA 2009 (2009) art. no. 5072819

Wells, D., Niederer, J.: A Medical Expert System approach using Artificial Neural Networks for standardized treatment planning. Int. J. Radiat. Oncol. Biol. Phys. 41(1), 173–182 (1998)

Xirogiannis, G., Chytas, P., Glykas, M., Valiris, G.: Intelligent impact assessment of HRM to the shareholder value. Expert Systems with Applications 35(4), 2017–2031 (2008)

Xirogiannis, G., Stefanou, J., Glykas, M.: A fuzzy cognitive map approach to support urban design. Expert Systems with Applications 26(2), 257–268 (2004)

Xirogiannis, G., Glykas, M.: Intelligent Modeling of e-Business Maturity. Expert Systems with Applications 32/2, 687–702 (2007)

Yuan, Y., Shaw, M.J.: Induction of fuzzy decision trees. Fuzzy Sets Systems 69, 125–139 (1995)

Zurada, J.M., Duch, W., Setiono, R.: Computational intelligence methods for rule-based data understanding. In: Proc. of the IEEE International Conference on Neural Networks, vol. 92(5), pp. 771–805 (2004)

Zurada, J.M., Lozowski, A.: Generating linguistic rules from data using neuro-fuzzy framework. In: Proc. 4th Intern. Conf. on Soft. Computing (IIZUKA 1996), Iizuka, Fukuoda, Japan, pp. 618–621 (1996)

The FCM Designer Tool

Aguilar Jose and José Contreras

Abstract. In this work we present a tool for the development of Fuzzy Cognitive Map (FCM). In our proposed tool, the FCM is defined by concepts and relationships that can change during the execution time. Using the tool, we can design a FCM, follow the evolution of a given FCM, change FCM defined previously, etc. Our tool allows to use the classical casual relationship defined in [2, 4], or to define the casual relationships according to the problem modeled, which can be statics or dynamics, fuzzy or not. In this work, we present the data structures, the interfaces, and the classes that compose the tool. Additionally, we give some examples of utilization of our tool in problems where we need define them specific casual relationships.

1 Introduction

Fuzzy Cognitive Maps (FCM) have been present previously in different works [1, 2, 3, 4, 5]. They are a technique to model system based on concepts that describe the main aspects of the modeled system, and causal representations between the concepts. FCM use the theory of fuzzy logic to describe their structure and to infer answers of the map from input data. These maps have been used in different domain for different proposes [6, 8, 9, 11, 12, 13, 14, 15, 16].

An interesting domain of study about FCM is in the area of methodologies and tools to develop them. This area is very important because in the next years it could expand the utilization of the FCM in different domains. The definition of methodologies to guide the development of FCM in specific domains is necessary because this is a complex task and depends on the characteristics of the problem modeled. The tools help and facilitate the design and the execution of the FCM.

Aguilar Jose
CEMISID, Departamento de Computación,
Facultad de Ingeniería, Universidad de Los Andes,
Núcleo la Hechicera, Mérida 5101-Venezuela
e-mail: aguilar@ula.ve

José Contreras
Distrito Tecnológico, Petróleos de Venezuela (PDSVA),
Hosteria El Vallecito, Mérida 5101, Venezuela

M. Glykas (Ed.): Fuzzy Cognitive Maps, STUDFUZZ 247, pp. 71–87, 2010.
springerlink.com © Springer-Verlag Berlin Heidelberg 2010

Recently, different tools have been proposed to model FCM. Some of them are: the FCModeler tool displays the known using interactive graph visualization [19]. The system also models pathway interactions and the effects of assumptions using a FCM-based modelling tool. In [20] has been proposed a tool to study the casual relationships map in multiagent systems, this tool is called Multiagent-Causal Maps. Other work has been proposed by Amit Roy [21], specifically, it developed a FCM tool based on Python. In [22] has been proposed a FCM applet to be used from programs.

In this work we present the FCM Designer Tool. This is a tool in Spanish, which allows the design of the structure of the maps (the definition of the concepts and the relationships between them). In this tool, the causal relationships between concepts can be the classical defined in [2, 4], or we can establish specific casual relationships according to the problem modeled. These casual relationships can be static or dynamic. In the case of dynamic casual relationships, they can be based on logic rules, mathematical equations, fuzzy logic, among others. Additionally, the inference patterns from a given input to a map can be followed. Furthermore, using our tool we can stop the evolution of a map during its runtime, of such form to introduce new information to the map, continue a previous execution, etc. The FCM Designer Tool is a versatile tool that allows to design, and later to execute, the FCM. This work is organized as follows. In section 2 is presented the FCM Designer Tool. Section 3 presents some examples of utilization of our tool, where we need to define specific dynamic casual relationships.

2 The FCM Designer Tool

In this section we present the data structure, the interface and the organization of the tool code (classes diagram, etc.).

2.1 The Data Structure

For the representation of the map we have defined a class called *Map*. The main attributes of this class are: a set of concepts and a set of relationships (see figures 1 and 9). We can see in figure 1 that each relationship contains a reference to its antecedent and consequent concepts, this is carried out in order to avoid the utilization of an adjacency matrix, since the execution time for the adjacency matrices have a complexity equal to O(nxn), where n is the number of concepts. In our data structure, the runtime is O(m), where m is the number of relationships; in this way, our tool has a runtime of linear order instead of quadratic order.

2.2 Tool Interface

The application is divided in 4 parts:

- A work area, where we design the Map
- A thread of execution of the Maps

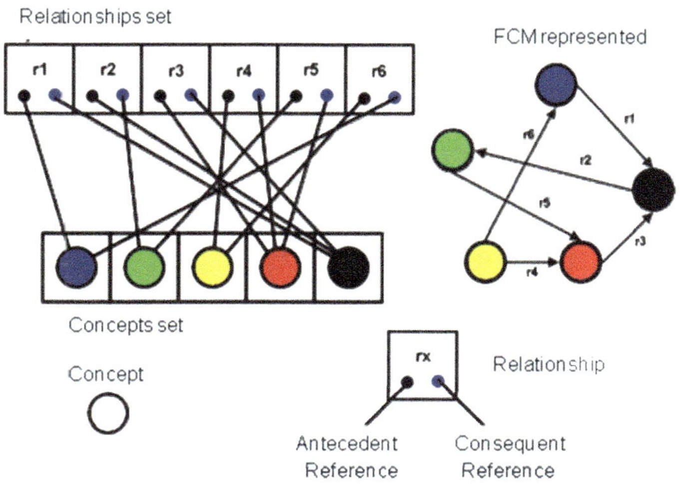

Fig. 1 Data Structure of the Map class

- A Menu Bar: to access to the general options of the tool
- A Control Panel: to access to the options of design, visualization and execution of the map.

2.2.1 The Work Area

For the visualization of the map (see figure 2), the concepts are represented by circles accompanied by their name. Each concept is defined by:

- An initial value
- A name
- A values set that represents the domain of values of the concept.
- A description of the concept

Each relationship is represented on the work area by an arrow that connects its antecedent concepts with its consequent concepts. Each relationship is defined by:

- A value or weight. When we use static casual relationships, this value does not change during the map execution. When we use dynamic casual relationships, it is the relationship weight defined by the experts. This value can be positive (the value of the Antecedent concept affects directly the Consequent concept) or negative (the value of the Antecedent concept affects inversely the Consequent concept).
- A reference to its Antecedent concept
- A reference to its Consequent concept

In order to modify the parameters of the nodes or of the relationships, we can use the mouse to generate different actions (see figure 3). These windows are

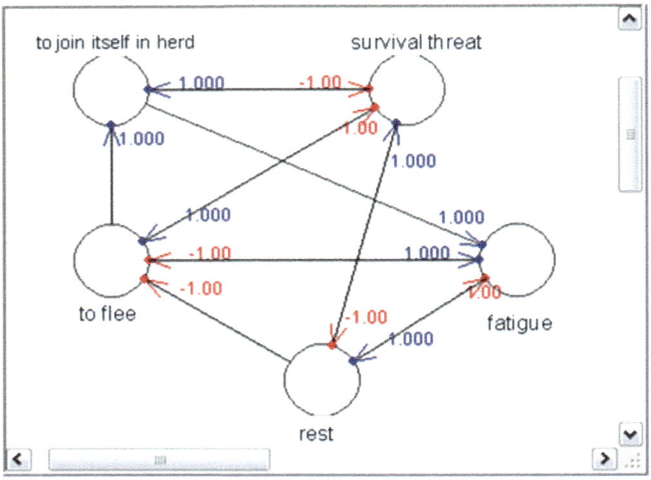

Fig. 2 Example of the work area

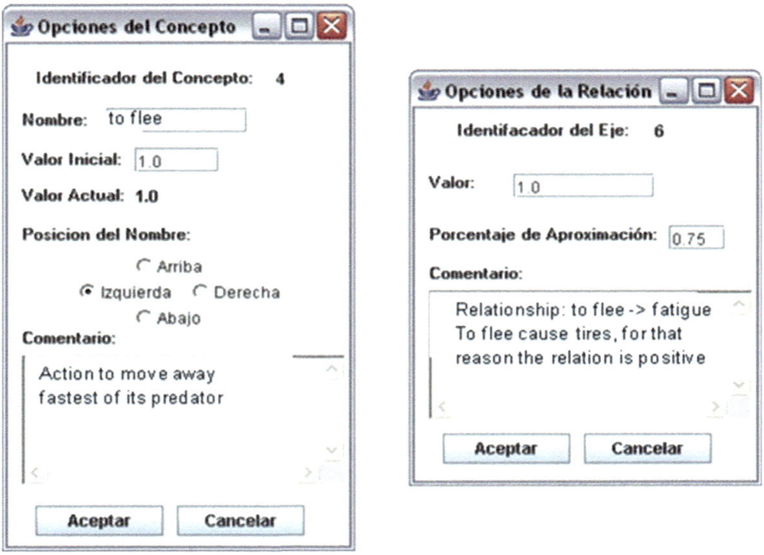

Fig. 3 Windows of the options to modify concept and relationship characteristics

created when the user pushes the right button of the mouse, when it is on a concept or on a relationship. We can see in figure 3 that for each concept is possible to modify its name, its initial value, the position of the label that contains the name, and also the description of the concept. In the case of the relationships, we can modify its value, the position of the label that identifies the relationship, and also the description of the casual relationship.

2.2.2 Thread of Execution

For the execution of the map we use a thread of execution, called RunFCM, which is in charge to carry out the different iterations of the map for a given input, until its convergence. By default, the condition to stop the execution of a map is given by a maximum number of iterations equal to 10, but the user can change this value. Also there is the option to stop the execution of the map when it is stabilized. These options are shown ahead.

Additionally, using the tool we can define specific casual relationships, according to the characteristics of the problem that we like to model. That is interesting when we like to define dynamic relationships. In the tool, they can have different forms: logic rules, mathematical equations, fuzzy rules, etc. In the tool, the type of dynamic relationship is specified in the class Rules, which allows the definition of each relationship using Java source code. The definition of the dynamic casual relationships must be codified in this class, which is compiled by the tool before the execution of the map.

2.2.3 Menu Bar

The general options of the system are: to establish like stopped condition a maximum number of iterations to be executed, to establish like stopped condition that the map stabilizes, to select the type of visualization and normalization used on the map, to save or to load a map, or its execution. These options are accessed through a menu bar (see figure 4).

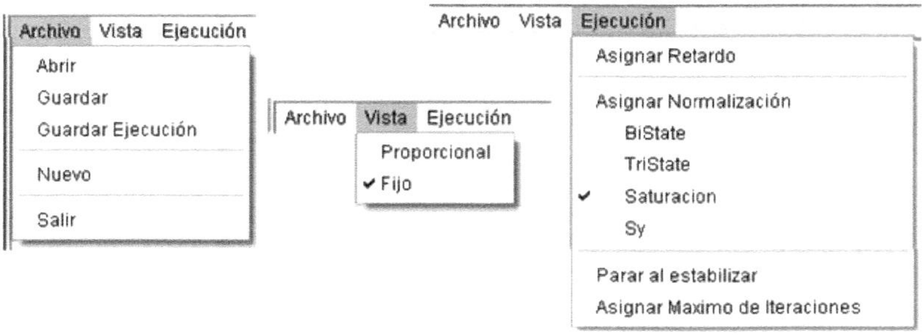

Fig. 4 General Menu of the tool

We can see in figure 4 that the "Archivo" menu offers the options: "Abrir" (this option will load a map from a specified file and will draw it in the work area); "Guardar" (this option allows saving the current map in some file (its concepts and relationships, with their characteristics)); "Guardar Ejecución" (this option allows keeping in a file the different iterations of the execution of a map, in order to save these data for their later analysis); "Nuevo" (this option cleans the work area and creates a new map to begin to work); and finally, "Salir" (this option closes the current window).

The "Vista" menu offers two options for the visualization of the concepts in the work area. The default option is "Fixed", which indicates that all the concepts were drawn in the work area with a size of 8 pixels. The option "Proporcional" would draw the nodes of proportional form to its value, with radios that go from 5 pixels for concepts of value 0 (zero), to 25 pixels for concepts of value 1 (one). The "ejecución" menu offers different types of normalization (possible output values of the concepts) during the execution of the map; they can be:

- For saturations:

$$S(x)=\begin{cases}0 & ,x<0\\ x & ,0\le x\le 1\\ 1 & ,x>1\end{cases}$$

- In the case nonlinear:

$$S(x)=\frac{1}{1+e^{-9\cdot(x-0.5)}}$$

- For bivalent states:

$$S(x)=\begin{cases}0, x<0.5\\ 1, x\ge 0.5\end{cases}$$

- For trivalent states:

$$S(x)=\begin{cases}0 & ,x\le 1/3\\ 0.5 & ,1/3<x\le 2/3\\ 1 & ,x>2/3\end{cases}$$

Where, x is the current value of the concept.

The tool also allows adding a delay among iterations of the map, during its execution, with the option "Asignar Retardo". This delay is in milliseconds, and has a value by default of 0. This option allows visualizing the iterations of the map during its execution, in a way that the user has a time interval between iterations to analyze the map. With the option "Asignar Máximo de Iteraciones", the tool allows to modify the maximum number of iterations allowed before stopping the execution of the map. The windows to modify these options are shown in the figure 5.

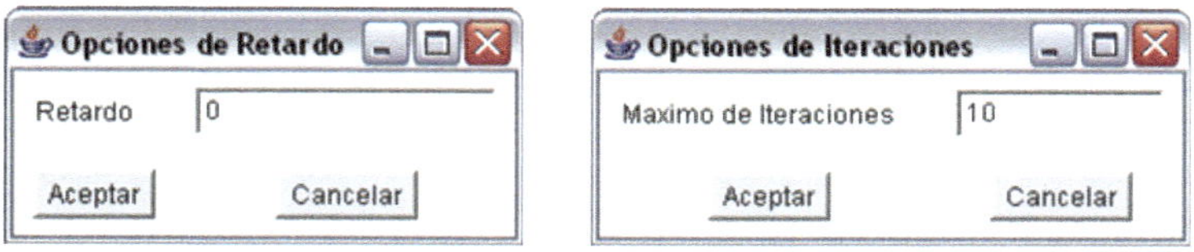

Fig. 5 Windows for the options of maximum number of iterations and delay

Another form to stop the execution of a map is using the option "Parar al Estabilizar". In this case, when the map in the current iteration has the same vector of states that in several of the more recent iterations, the tool stops the execution of the map.

2.2.4 Control Panel

The Control Panel has the different controls necessary to design, to execute and to visualize the map. Figures 6 and 7 show the components of the panel. This panel is divided in three regions, a superior region to design maps, an intermediate region of execution, and an inferior region of visualization.

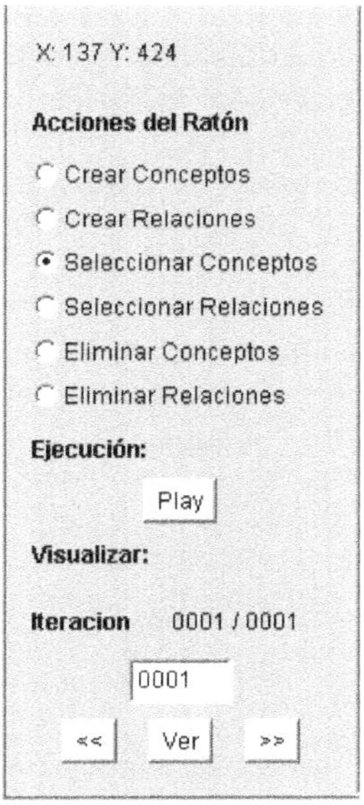

Fig. 6 Control Panel

The superior region of the panel has associate the interface for the option "CambiaAccionesDelRaton", and it is composed by 6 possibilities, which indicate the type of actions that we can assign to the mouse:

- "Crear Conceptos": it allows adding concepts to the map, pushing the left button of the mouse when we are in the work area screen. The concepts are created by default with the name "node x", where x represents the number of the concept in the map. When we push the right button on some concept, we open the window that allows modifying the different characteristics of the concept (see figure 3).
- "Crear Relaciones": it allows establishing relationships between two concepts. We need to push on the antecedent concept and later on the consequent concept, in order to establish the relationship.

- "Seleccionar Conceptos": with this option activated we can move a concept in the work area, dragging it with the mouse. Also, we can modify the values of the concept using the right button that opens the windows of the figure 3.
- "Seleccionar Relaciones": this option allows changing the value of the relationship and its position, when we push on it with the mouse. For that, it opens the window to modify the characteristics of the relationships (see figure 3).
- "Eliminar Conceptos": it allows activating this action on the mouse, so that, when it is pressed on some concept it eliminates the concept, with all the relationships associated to this concept.
- "Eliminar Relaciones": it allows deleting a relationship pushing on the label of the relationship.

The intermediate region of the panel, which is called "ejecución", has associated the option "Play". When we press this button, the tool executes the thread of execution called RunFCM (it executes the FCM that is in the work area, for a given input). This thread uses static casual relationships by default (see figure 3), but if the Rules object is not empty, it uses the casual relationships defined in the code of this object. When the map is in execution state, the button is inactive, and the label of the option is changed to "Wait", which indicate that the user cannot do nothing while the map is in this state.

The Inferior region of the Panel has associated the interface to manage the visualization of the map. In this region is the label: "Iteracion: x/y", which indicates that we are visualizing the iteration x of y possible. In order to visualize a given iteration in the map, we can use the labels: "<<" or ">>" (see figure 6), which allow to show the previous or the following iteration. Another form to visualize a given iteration of the map is to write the number of the iteration and then to press the button "ver", which will show this iteration.

2.3 Main Windows of the Tool

The main window contains all the elements of the tool. The window is divided in three parts, in the superior part is observed the menu bar, in the left part the controls, and the central part the work area. This structure is typical of many computer systems. All the elements that compose the main window belong to the MapWindows class, and can be visualized in figure 7: the menu bar, the work area, the Control Panel and the thread of execution of the map. In figure 7 we can see the complete interface of the user with the tool.

2.4 Source Code

We have placed the different classes in different packages according to the roll carried out by them (see figure 8).

We are three main packages, which are the Map package that contains all the objects that model a map, the Algorithms package that contains all the classes used for the execution of the maps, and the Interface package that contains all the classes used for the creation of the windows for the interaction with the user. The

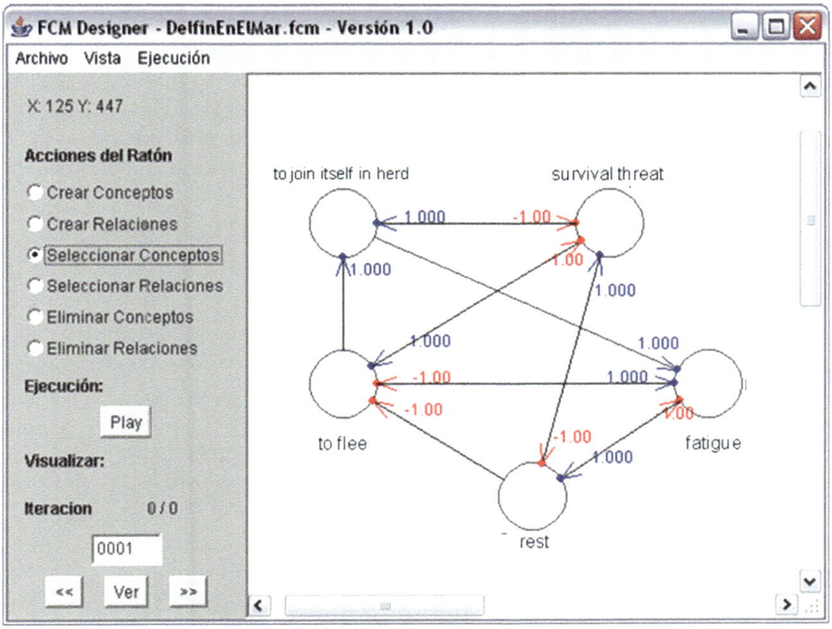

Fig. 7 Main window of the tool

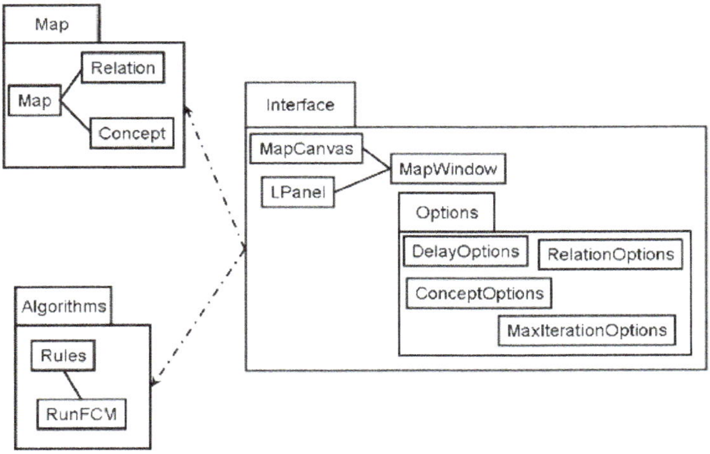

Fig. 8 Organization of the classes in packages

secondary windows used to change the different options of the tool were placed in a sub-package of the interface package, called Options.

Now, we are going to describe two classes, for the rest see [10]. They are the classes used to build the maps in the work area (see figures 2 and 9).

- MapCanvas: it defines a screen with white background, where we design the map. When this objet is instanced, it is created a ScrolledPanel with the dimensions of the work area (1024x1024 pixels). Additionally, it calls the *MovimientoRatonEnMapCanvas* and *AccionesRatonEnMapCanvas* classes, which are classes used to define the different actions of the mouse on the work area. These actions can be: to create, to modify, to eliminate or to move concepts or relationships, in the work area.
- Map: it is the class that models the map in the tool. It calls the *relation* and *concept* classes, which are the classes that describe the characteristics of the concepts and relationships that compose the map.

In a work area we can only build a map in a given moment. With these classes, our tool defines the screen of the figure 2.

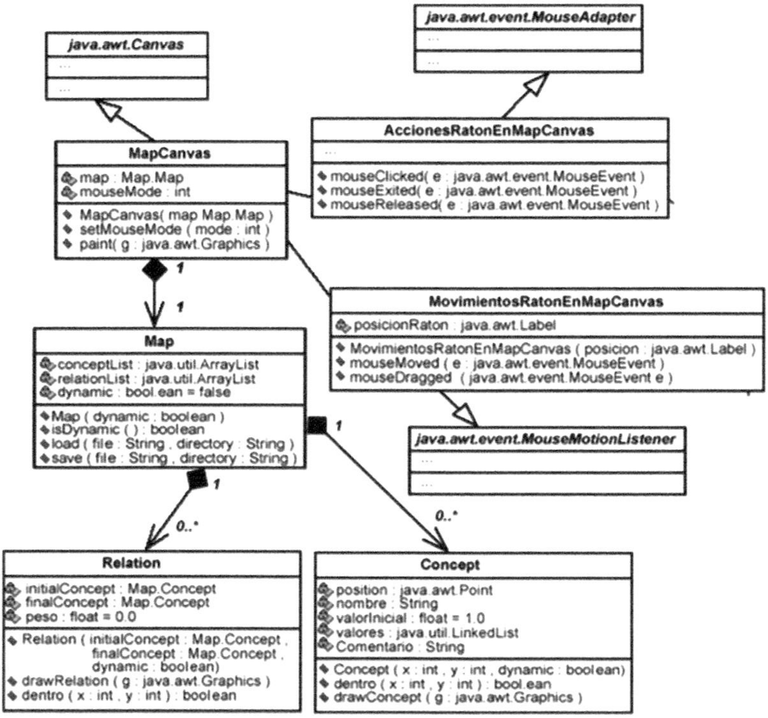

Fig. 9 UML Diagram for the MapCanvas and Map classes

Remember that we have said previously that our tool has an execution thread called RunFCM, which executes the map. It is defined in a class with the same name. It is in charge of carry out the different iterations of the map, given initial values to the concepts of the map. The source code of the thread is:

```
private void runStatic(){
 int c=1;
 do{
  java.util.Iterator <Map.Relation> itE =
          this.mapCanvas.getMap().getRelationList().iterator();
  while ( itE.hasNext() ){
   Map.Relation ref = itE.next();
   ref.getFinalConcept().valorAuxiliar +=
          ref.getInitialConcept().getLastValue()*ref.getValue();
  }
  java.util.Iterator <Map.Concept> itN =
          this.mapCanvas.getMap().getConceptList().iterator();
  while ( itN.hasNext() ){
   Map.Concept ref = itN.next();
   ref.addLastValue( this.normalization( ref.valorAuxiliar ) );
   ref.setCurrentValue( ref.getLastValue() );
   ref.valorAuxiliar=0.0f;
  }
  this.mapCanvas.update( this.mapCanvas.getGraphics() );
  try{
    this.hilo.sleep(this.retardo);
  }catch ( java.lang.InterruptedException e ){
     System.out.println ( e );
  }
  this.mapCanvas.getMap().setCurrentIteration( c++ );
  this.etiqueta.setText( this.mapCanvas.getMap().getIterationLabel() );
 }while ( this.Continue( c ) );
}
```

3 Example of Utilization

In this section we give some examples of utilization of our tool. In the first case, to build FCM where the casual relationships are mathematical equations; and in the second case, where the casual relationships are rules.

3.1 Casual Relationships Like Dynamic Equations

In the first example, we use the FCM proposed in [7] to model a system which consists of a water tank with a valve of input and a valve of output (see figure 10).

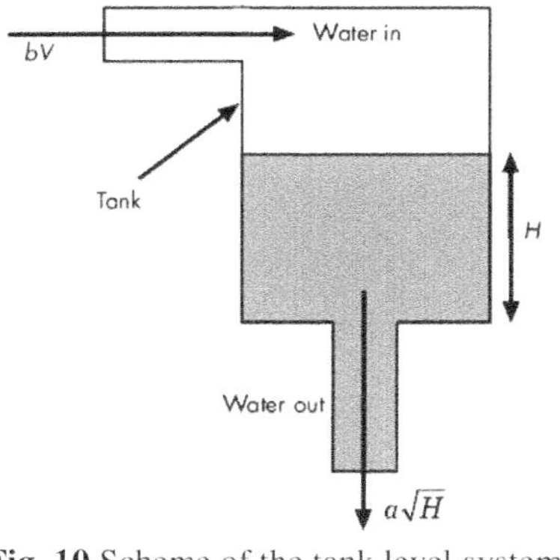

Fig. 10 Scheme of the tank level system

The water enters to the tank from above and leaves the same one by an orifice in the base. The rate of water input is proportional to voltage V applied to a pump that sends water to the tank, the rate of exit of the liquid is proportional to the root square of the level of the water in the tank. The equation for this system is given by:

$$\frac{dVol}{dt} = A\frac{dH}{dt} = b \cdot V - a \cdot \sqrt{H}$$

Where Vol is the volume of water in the tank, A is the cross-sectional area of the tank, b is the constant of the input stream, and a is the constant of the exit flow. The equation describes the height of water H, like a function of the time.

For the design of the FCM, three concepts were used in [7]: the constant of input stream, called b/A, the constant of flow of exit, called a/A, and the level of the liquid in the tank, called H. The dynamic relationships are defined of the following form [7]:

- The relationship between the constant of input stream and the level is given by the value of the applied voltage to pump V.
- The relationship between the constant of the exit flow and the level is given by the squared root of the value of the concept that represents the level.

The resulting map designed using our tool is shown in figure 11.

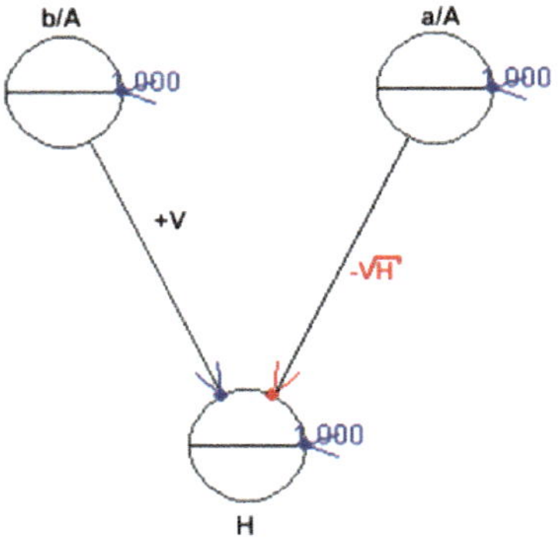

Fig. 11 FCM to model the dynamic system of the level of the liquid in a tank

We can see in figure 11 that exist relationships that feedback each concept, that is due to that these concepts have a memory property, for example, the values of the concepts b/A and a/A remain constant during all the execution. The casual relationship in this case has the following source code (this is the code of the Rule class in our tool):

```
static public float computeRelationValue ( Map.Relation  relation ){
 if ( relation.getInitialConcept().getName().equals("b/A") &&
     relation.getFinalConcept().getName().equals("H")      )
   relation.setValor ( relation.getDinamicImput() );
```

```
 else if ( relation.getInitialConcept().getName().equals("a/A") &&
        relation.getFinalConcept().getName().equals("H")        )
   relation.setValor( -1.0f * (float)Math.sqrt(
               relation.getFinalConcept().getCurrentValue() ) );
 else
   relation.setValor ( 1.0f );
 return relation.getValue();
}
```

We can see in the code that the first "If" determines if the relationship is between the concept b/A and H, the value of this relationship is given by the auxiliary variable DynamicImput, which contains the value of voltage V. The second "If" determines if the relationship is between the concepts a/A and H, if it is thus, the value of the relationship is calculated according to the squared root of the present value of concept H, the rest of the relationships have a value of 1.

3.2 Casual Relationships Like Rules

In the second example, we are going to use a FCM that have been defined in [17] to describe the emergent and self-organizing properties in a given multiagent system. In this map, a concept can have the following values: *High* (between 2/3 and 1) when it works correctly; *Medium* (between 1/3 and 2/3) when its functioning must be validated; and *Low* (between 0 and 1/3) when it does not work or work erroneously. Additionally, they have used, to describe the type of casual relationships, the set of rules proposed in [7]. Some of the rules defined in [7] are:

- If the preceding concept is **High** and the consequent one is also **High** then the relationship is **Complete**$^+$(1.0).
- If the preceding concept is **High** and the consequent one is **Low** then the relationship is **Low**$^+$ (0.25).
- If the preceding concept is **Low** and the consequent one is **Medium** then the relationship is **Medium**$^-$ (-0.5).
- If the preceding concept is **Low** and the consequent one is **Low** then the relationship is **Complete**$^-$ (-1.0).

This set of rules was codified in the Rules class in our tool. For this example, the source code of the Rules class is:

```
public class Rules {
 static private boolean isHigh ( float v ){ return ( v > 2.0f / 3.0f); }
 static private boolean isLow ( float v ){ return ( v <= 1.0f / 3.0f); }
 static private boolean isMedium ( float v ){
   return ( v > 1.0f/3.0f && v <= 2.0f/3.0f );
 }
 static public float computeRelationValue ( Map.Relation  relation ){
  float vInicial = relation.getInitialConcept().getCurrentValue();
  float vFinal = relation.getFinalConcept().getCurrentValue();

  if ( isHigh ( vInicial ) && isHigh( vFinal ) )
    relation.setValor( 1.0f );
  else if ( isHigh ( vInicial ) && isMedium ( vFinal ) )
    relation.setValor( 0.75f );
```

```
else if ( isHigh ( vInicial ) && isLow( vFinal ) )
  relation.setValor( 0.25f );
else if ( isMedium ( vInicial )&& isHigh( vFinal ) )
  relation.setValor( 0.75f );
else if ( isMedium ( vInicial ) && isMedium ( vFinal ) )
  relation.setValor( -0.5f );
else if ( isMedium ( vInicial ) && isLow( vFinal ) )
  relation.setValor( -1.0f );
else if ( isLow( vInicial ) && isHigh( vFinal ) )
  relation.setValor( -0.75f );
else if ( isLow( vInicial ) && isMedium ( vFinal ) )
  relation.setValor( -0.5f );
else if ( isLow( vInicial ) && isLow( vFinal ) )
  relation.setValor( -1.0f );
relation.setValor ( relation.getValue() * relation.getDinamicImput() );
return relation.getValue();
}}
```

In this code, the concepts are labeled according to their values, using the functions isHigh, isMedium and isLow. The value of the antecedent concept is labeled as vInicial, and the value of the consequent concept is labeled like vFinal. Once the values of the concepts are obtained, we can calculate the value of the relationship according to the rules specified previously. Once obtained the value of the relationship, it is multiplied by the DynamicImput variable (it defines the weight of the relationship established by the experts in [17]).

The specific FCM proposed in [17], designed with our tool, is:

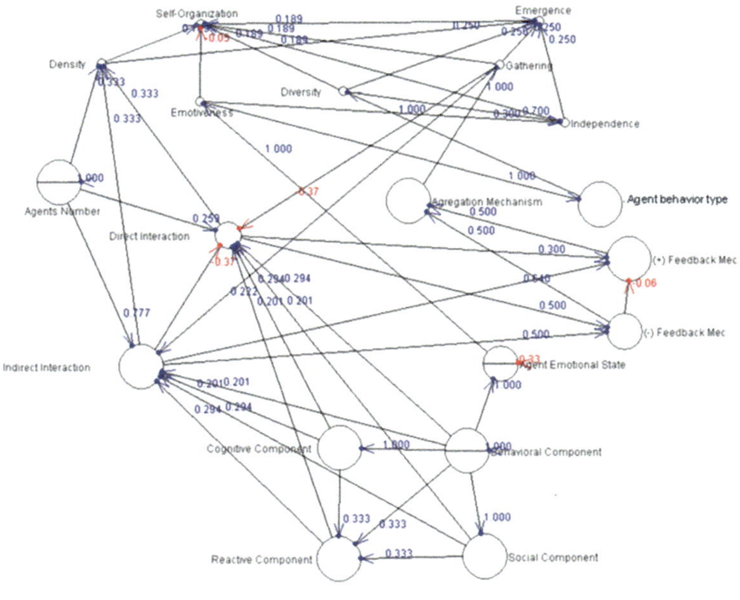

Fig. 12 The FCM proposed in [17], designed with our tool

The meanings of some of the concepts of this FCM are (for the rest, see [17]):

- Self-Organization: It measures the degree of adaptability in a multiagent system. The self-organization is an adaptive and dynamic process where systems acquire and maintain their own structure without outside control.
- Emergence: It measures the degree of system's evolution through the possibility of the appearance of emerging properties. Some of the things that could emerge are temporary and space patterns such as: new collective policies and norms, cooperation between agents, among others.
- Diversity It measures the homogeneity or heterogeneity of the society of agents. It is measured by the number of agents of different types defined within the system

And one example of the results of the evolution (iterations) of this map for one specific input set of concept values is:

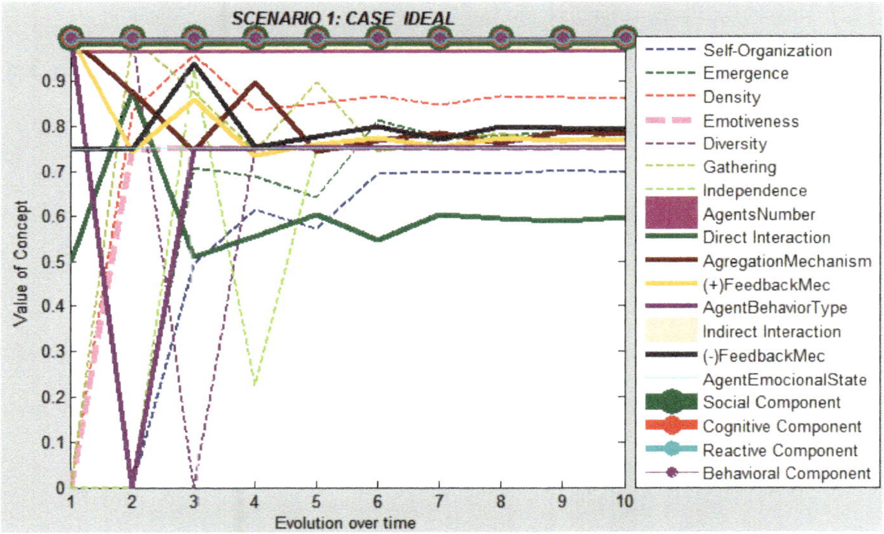

Fig. 13 Example of the execution of the FCM proposed in [17] using our Tool

In the figure 13 we can see the values evolution of the concepts during the execution of the FCM, for a given set of initial values of concepts (it is the case of an emergent and self-organized multiagent system, which is called ideal scenario). In this example, the system converges for 8 iterations.

4 Conclusions

There are several tools to built and execute FCM. Each one has a specific use (to study multiagent systems, etc.) or defines FCM with static casual relationships.

The tool proposed in this work allows the definition of different types of dynamic relationship. They are defined in a class that must be programmed by the designer on the class Rules.

The tool proposed in this work requires an expert knowledge about the problem to study in order to determine the type of dynamic relationships among the concepts, and knowledge about the tool and the language Java to program the specific relationships of the problem modeled. Future versions of the tools will include a library of dynamic casual relationships, a library of learning methods [6, 13, 14]. Also, a methodology based on this tool will be proposed for specific problems in control, economy, etc. [6, 18].

References

[1] Axelrod, R.: The Structure of Decision: Cognitive Maps of Political Elites. Princeton University Press, Princeton (1976)

[2] Kosko, B.: Fuzzy Cognitive Maps. International Journal of Man-Machine Studies 24, 65–75 (1986)

[3] Eden, C.: On the Nature of Cognitive Maps. Journal of Management Studies 29, 261–265 (1992)

[4] Kosko, B.: Fuzzy Engineering. Prentice-Hall, New Jersey (1999)

[5] Aguilar, J.: A Dynamic Fuzzy-Cognitive-Map Approach Based on Random Neural Networks. International Journal of Computational Cognition 1, 91–107 (2003)

[6] Aguilar, J.: A Survey about Fuzzy Cognitive Maps Papers. International Journal of Computational Cognition 3, 27–33 (2005) (invited paper)

[7] Aguilar, J.: Different Dynamic Causal Relationship Approaches for Fuzzy Cognitive Maps. Technical Report N 10-2009, CEMISID, Universidad de los Andes (2009) (submitted to publication)

[8] Khan, M., Quaddus, M.: Group Decision Support using Fuzzy Cognitive Maps for Causal Reasoning. Group Decision and Negotiation Journal 13, 463–480 (2004)

[9] Sharif, A., Irani, Z.: Exploring Fuzzy Cognitive Mapping for IS Evaluation. European Journal of Operational Research 173, 1175–1187 (2006)

[10] Contreras, J., Aguilar, J.: La Herramienta FCM Designer. Technical Report N 25-2008, CEMISID, Universidad de los Andes

[11] Papageorgiou, E., Stylios, C., Groumpos, P.: An integrated two-level hierarchical decision making system based on fuzzy cognitive maps. IEEE Trans. Biomed. Eng. 50, 1326–1339 (2003)

[12] Papageorgiou, E., Parsopoulos, K., Groumpos, P., Vrahatis, M.: Fuzzy Cognitive Maps Learning through Swarm Intelligence. In: Rutkowski, L., Siekmann, J.H., Tadeusiewicz, R., Zadeh, L.A. (eds.) ICAISC 2004. LNCS (LNAI), vol. 3070, pp. 344–349. Springer, Heidelberg (2004)

[13] Stach, W., Kurgan, L., Pedrycz, W.: A Survey of Fuzzy Cognitive Map Learning Methods. In: Grzegorzewski, P., Krawczak, M., Zadrozny, S. (eds.) Issues in Soft Computing: Theory and Applications, Exit, pp. 71–84 (2005)

[14] Stach, W., Kurgan, L., Pedrycz, W., Reformat, M.: Genetic Learning of Fuzzy Cognitive Maps. Fuzzy Sets and Systems 153, 371–401 (2005)

[15] Elpiniki, I., Papageorgioua, E., Stylios, C., Groumpos, P.: Unsupervised learning techniques for fine-tuning fuzzy cognitive. Int. J. Human-Computer Studies 64, 727–743 (2006)
[16] Papageorgiou, E., Spyridonos, P., Glotsos, D., Stylios, C., Ravazoula, P., Nikiforidis, G., Groumpos, P.: Brain tumor characterization using the soft computing technique of fuzzy cognitive maps. Applied Soft Computing 8, 820–828 (2008)
[17] Perozo, N., Aguilar, J., Teran, O.: A Verification Method for MA-SOES. Technical Report N 14-2009, CEMISID, Universidad de los Andes (2009) (submitted to publication)
[18] Xirogiannis, G., Glykas, M.: Fuzzy Cognitive Maps in Business Analysis and Performance Driven Change. IEEE Transactions in Engineering Management 51, 334–351 (2004)
[19] http://orion.math.iastate.edu/danwell/GET/GETFC.html
[20] http://www.damas.ift.ulaval.ca/~fabiola/recherche/
[21] http://www.artecs.net/cgi-bin/FCM.cgi
[22] http://www.ochoadeaspuru.com/fuzcogmap/software.php

Fuzzy Cognitive Networks: Adaptive Network Estimation and Control Paradigms

Thodoris L. Kottas, Yiannis S. Boutalis, and Manoli A. Christodoulou

Abstract. Fuzzy Cognitive Networks (FCN) constitutes an operational extension of Fuzzy Cognitive Maps (FCM), which assume that they always reach equilibrium points during their operation. Moreover, they are in continuous interaction with the system they describe and may be used to control it. FCN are capable of capturing steady state operational conditions of the system they describe and associate them with input values and appropriate weight sets. In the sequence they store the acquired knowledge in fuzzy rule based data bases, which can be used in determining subsequent control actions. This chapter presents basic theoretical results related to the existence and uniqueness of equilibrium points in FCN, the adaptive weight estimation based on system operation data, the fuzzy rule storage mechanism and the use of the entire framework to control unknown plants. The results are validated using well known control benchmarks.

1 Introduction

Fuzzy Cognitive Maps (FCM) have been introduced by Kosko [1] based on Axelrod's work on cognitive maps [2]. They are inference networks using cyclic directed graphs that represent the causal relationships between concepts .They use a symbolic representation for the description and modelling of the system. In order to illustrate different aspects in the behavior of the system, a fuzzy cognitive map consists of nodes where each one represents a system characteristic feature. The node interactions represent system dynamics. An FCM integrates the accumulated experience and knowledge on the system operation, as a result of the method by which it

Thodoris L. Kottas · Yiannis S. Boutalis
Democritus University of Thrace, Xanthi, V.Sofias 12, Greece
e-mail: tkottas@ee.duth.gr, ybout@ee.duth.gr

Manoli A. Christodoulou
Technical University of Crete, Chania, Crete, Greece
e-mail: manolis@ece.tuc.gr

M. Glykas (Ed.): Fuzzy Cognitive Maps, STUDFUZZ 247, pp. 89–134, 2010.
springerlink.com © Springer-Verlag Berlin Heidelberg 2010

is constructed, i.e., by using human experts who know the operation of the system and its behavior. Different methodologies to develop FCM and extract knowledge from experts have been proposed in [3] - [6].

Kosko enhanced the power of cognitive maps considering fuzzy values for their nodes and fuzzy degrees of interrelationships between nodes [1],[7]. He also proposed the differential Hebian rule [7] to estimate the FCM weights expressing the fuzzy interrelationships between nodes based on acquired data. After this pioneering work, fuzzy cognitive maps attracted the attention of scientists in many fields and have been used to model behavioral systems in many different scientific areas. Application examples can be found in political science [8], [9], in economic field [10]-[15], in representing social scientific knowledge and describing decision making methods [16] - [18]. Other applications include geographical information systems [19] - [21], cellular automata [22], pattern recognition applications [23], [24] and numerical and linguistic prediction of time series functions [25]. Fuzzy cognitive maps have also been used to model the behavior and reactions of virtual worlds [26] - [30], as a generic system for decision analysis [18], [31] and as coordinator of distributed cooperative agents.

Regarding FCM weight estimation and updating, recent publications [44] - [48] extend the initially proposed differential Hebian rule [7] to achieve better weight estimation. Another group of methods for training FCM structure involves genetic algorithms and other exhaustive search techniques [49] - [53], where the training is based on a collection of particular values of input output historical examples and on the definition of appropriate fitness function to incorporate design restrictions.

Various extensions of FCMs have been proposed in the literature [32] - [43]. Dynamic Cognitive Networks (DCN) appear in [35], the Fuzzy Causal Networks in [36] - [40], while the neutrosophic cognitive maps appear in [41], [42]. Recently Fuzzy Cognitive Networks (FCN) [43] has been proposed as a complete computational and storage framework to facilitate the use of FCM in cooperation with the physical system they describe.

Fuzzy Cognitive Networks (FCNs) and their storage mechanism assume that they reach equilibrium points, each one associated with a specific operation condition of the underlying physical system. However, the conditions under which FCMs and consequently FCNs reach an equilibrium point and whether this point is unique have not been adequately studied so far. Simple FCMs have bivalent node values and trivalent edges (weights) and are equipped with binary threshold functions or sigmoid functions with very large inclination [54]. According to Kosko [54], [55], starting from an initial state, simple FCMs follow a path, which ends in a fixed point or limit cycle, while more complex ones may end in an aperiodic or "chaotic" attractor. These fixed points and attractors could represent *meta rules* of the form "If input then attractor or fixed point". A more detailed study on the performance of FCMs has been presented recently in [56], where the inference capabilities of FCMs equipped with binary, trivalent or sigmoid functions are compared. The relation of the existence of these attractors or fixed points to the weight interconnections of the FCM and FCN has not been fully investigated. This is, however, of paramount importance if one wants to use FCNs with learning capabilities in reliable adaptive system identification and control schemes.

In this chapter, we present a study on the existence of the above fixed points of FCMs equipped with continuous differentiable sigmoid functions having contractive or at least non expansive properties. The study is based on the work presented in [57], [58] and is made by using an appropriately defined contraction mapping theorem and the non-expansive mapping theorem. It is proved that when the weight interconnections fulfill certain conditions, related to the size of the FCM and the inclination of the sigmoid functions, the concept values will converge to a unique solution regardless of their initial states, or in some cases a solution exists that may not necessarily be unique. Otherwise the existence or the uniqueness of equilibria may or may not exist, it may depend on the initial states, but it can not be assured. In case the FCM has also input nodes (that is nodes that influence but are not influenced by other nodes), the unique equilibrium does not depend solely on the weight set, as in the case of FCMs with no input nodes; it depends also on the values of the input nodes.

In the sequel, an adaptive weight estimation algorithm is proposed with guaranteed exponentially fast error convergence to zero, which uses the obtained conditions to construct appropriate weight projection rules assuring that the obtained weights do not compromise the existence of the FCM solution. In view of these results *meta rules* of the form "If weights then fixed point" are more appropriate to represent the behavior of an FCM which satisfy the above weight conditions. Fuzzy Cognitive Networks (FCN) [43], [59], [60], introduced recently as an extension of FCMs can work on the basis of such *meta rules* and provide the application framework of the obtained results. When the necessary weight conditions are fulfilled by an FCN during its updating procedure, its information storage mechanism is actually a depository of this kind of *meta rules*.

Based on this depository of information, one can devise control approaches using the FCN and following an inverse procedure to obtain the control input(s), which drives the FCN (and consequently the system it describes) in the desired condition. Two such approaches are presented in this chapter. The first approach is tested to control the well known *inverted pendulum* benchmark. It is assumed that the depository of fuzzy *meta rules* is complete and therefore the determination of the appropriate control input is based on simple inverse formulas. The second approach is applied to control a nonlinear wastewater treatment bioprocess. In this case the depository is not complete and the control inputs are determined by following an inverse procedure using a gradient technique.

The chapter is organized as follows. Section 2 describes the representation and mathematical formulation of Fuzzy Cognitive Maps. Section 3 provides the proof of the existence solution of the concept values of a Fuzzy Cognitive Map. Section 4 presents the adaptive weight estimation algorithm with proven stability and parameter convergence, while Section 5 presents the FCN framework and its operation in the two control paradigms. Finally, Section 6 concludes the work.

2 Fuzzy Cognitive Maps

A graphical representation of FCMs is depicted in Fig. 1. Each concept represents a characteristic of the system; in general it represents events, actions, goals, values and trends of the system. Each concept is characterized by a number A_i that represents its value and it results from the transformation of the real value of the systems variable, represented by this concept, either in the interval [0,1] or in the interval [-1,1]. All concept values form vector A as follows:

$$A = \left[A_1\ A_2\ \ldots.\ A_n\right]^T$$

with n being the number of the nodes (in Fig. 1 $n = 8$) . Causality between concepts allows degrees of causality and not the usual binary logic, so the weights of the interconnections can range in the interval [-1,1].

The existing knowledge on the behavior of the system is stored in the structure of nodes and interconnections of the map. The value of w_{ij} indicates how strongly concept C_i influences concept C_j. The sign of w_{ij} indicates whether the relationship between concepts C_i and C_j is direct or inverse.

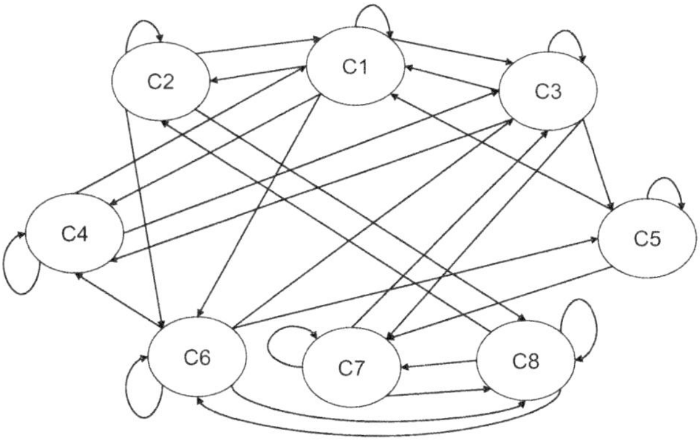

Fig. 1 An FCM with 8 nodes

For the FCM of Fig. 1 matrix W is equal to

$$W = \begin{bmatrix} d_{11} & w_{12} & w_{13} & w_{14} & 0 & w_{16} & 0 & 0 \\ w_{21} & d_{22} & 0 & 0 & 0 & w_{26} & 0 & w_{28} \\ w_{31} & 0 & d_{33} & w_{34} & w_{35} & 0 & w_{37} & 0 \\ w_{41} & 0 & w_{43} & d_{44} & 0 & 0 & 0 & 0 \\ w_{51} & 0 & 0 & 0 & d_{55} & 0 & w_{57} & 0 \\ 0 & 0 & w_{63} & w_{64} & w_{65} & d_{66} & 0 & w_{68} \\ 0 & 0 & w_{73} & 0 & 0 & 0 & d_{77} & w_{78} \\ 0 & w_{82} & 0 & 0 & 0 & w_{86} & w_{87} & d_{88} \end{bmatrix}$$

The equation that calculates the values of concepts of Fuzzy Cognitive Networks, may or may not include self-feedback. In its general form it can be written as:

$$A_i(k) = f(\sum_{\substack{j=1 \\ j \neq i}}^{n} w_{ji} A_j(k-1) + d_{ii} A_i(k-1)) \tag{1}$$

Where $A_i(k)$ is the value of concept C_i at discrete time k, $A_i(k-1)$ the value of concept C_i at discrete time $k-1$ and $A_j(k-1)$ is the value of concept C_j at discrete time $k-1$. w_{ji} is the weight of the interconnection from concept C_j to concept C_i and d_{ii} is a variable that takes on values in the interval $[0,1]$, depending upon the existence of "strong" or "weak" self-feedback to node i. Repetitive application of (1) for each node A_i will probably lead the FCN in an equilibrium point. Alternatively, it may present a limit cycle or a chaotic behavior.

Regarding the functions f used in FCMs, the following functions are usually found in the literature, allowing also different interpretation of their results:

The bivalent function [54], [56]

$$f(x) = \begin{cases} 1 & x > 0 \\ 0 & x \leq 0 \end{cases}$$

The use of this function allows the activation of each concept to either 0 or 1, leading to the development of binary FCMs, where each concept is either activated or not.

The trivalent function [54], [56]

$$f(x) = \begin{cases} 1 & x > 0 \\ 0 & x = 0 \\ -1 & x < 0 \end{cases}$$

In tis case when the concept takes the value 1, it means that this concept increases, when the activation level equals -1, it means that the concept decreases and when level equals to 0, it means that the concept is remaining stable.

The hyperbolic tangent or sigmoid function [56] with saturation level -1 and 1. In a general form it can be written as

$$f(x) = tanh(c_l x) \text{ or } f(x) = \frac{e^{2c_l x} - 1}{e^{2c_l x} + 1}.$$

where $0 < c_l \leq 1$ is a number used to adjust the inclination of the sigmoid function. This function squashes the result in the interval [-1, 1].

The sigmoid function with saturation level 0 and 1 (log-sigmoid). f is a sigmoid function commonly used in the Fuzzy Cognitive Maps, which squashes the result in the interval [0,1] and is expressed as,

$$f = \frac{1}{1+e^{-c_l x}}.$$

where $0 < c_l \leq 1$ is used to adjust its inclination.

Equation (1) can be rewritten as:

$$A(k) = f(W^T A(k-1)) \tag{2}$$

In the next Section we are deriving conditions, which determine the existence of a unique solution of (2), when continuous differentiable transfer functions f are used.

3 Existence and Uniqueness of Solutions in Fuzzy Cognitive Maps

In this Section we check the existence of solutions [57], [58] in equation (2), when a continuous and differentiable transfer function is used, such as sigmoid functions are. We know that the allowable values of the elements of FCM vectors A lie either in the closed interval $[0,1]$ or in the closed interval $[-1,1]$. This is a subset of $\Re$ and is a complete metric space with the usual l_2 metric. We will define the regions where the FCM has a unique solution, which does not depend on the initial condition since it is the unique equilibrium point.

3.1 The Contraction Mapping Principle

We now introduce the Contraction Mapping Theorem [62].

Definition 1. *Let X be a metric space, with metric d. If φ maps X into X and there is a number $0 < c < 1$ such that*

$$d(\varphi(x),\varphi(y)) \leq cd(x,y) \tag{3}$$

for all $x,y \in X$, then φ is said to be a contraction of X into X.

Theorem 1. *[62] If X is a complete metric space, and if φ is a contraction of X into X, then there exists one and only one $x \in X$ such that $\varphi(x) = x$.*

In other words, φ has a unique fixed point. The uniqueness follows from the fact that if $\varphi(x) = x$ and $\varphi(y) = y$, then (3) gives $d(x,y) \leq cd(x,y)$, which can only happen when $d(x,y) = 0$ (See [62]).

Equation (2) can be written as:

$$A(k) = G(A(k-1)) \tag{4}$$

where $G(A(k-1))$ is equal to $f(WA(k-1))$.

In FCM's $A \in [0,1]^n$ or $A \in [-1,1]^n$ and it is also clear according to (2) that $G(A(k-1)) \in [0,1]^n$ or $G(A(k-1)) \in [-1,1]^n$ depending on which squashing sigmoid function is used. If the following inequality is true:

$$d(G(A),G(A')) \leq cd(A,A')$$

where A and A' are different vectors of concept values and G is defined in (4), then G has a unique fixed point A such that:

$$G(A) = A$$

Before presenting the main theorem we need to explore the role of f as a contraction function.

Theorem 2. *The scalar sigmoid function f, ($f = \frac{1}{1+e^{-x}}$) is a contraction of the metric space X into X, were $X = [a,b], a, b$, finite, according to Definition 1, where:*

$$d(f(x), f(y)) \leq c d(x,y) \tag{5}$$

Proof. Here f is the sigmoid function, $x,y \in X$, X is as defined above and c is a real number such that $0 < c < 1$

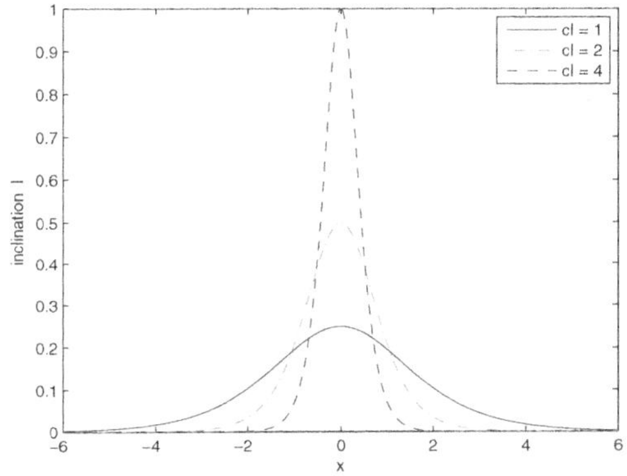

Fig. 2 Inclination of sigmoid function $f = \frac{1}{1+e^{-c_l x}}$ when $c_l = 1, 2, 4$

The inclination l of a sigmoid function f is equal to:

$$l = \frac{\partial f}{\partial x} = \frac{c_l e^{-c_l x}}{(1+e^{-c_l x})^2} = \frac{c_l}{e^{c_l x}} \left(\frac{1}{1+e^{-c_l x}} \right)^2 = \frac{c_l}{e^{c_l x}} f^2 \tag{6}$$

for $x \in X$. Equation (6) for $c_l = 1, 2, 4$ is plotted in Fig. 2. According to Equation (6) one can see that the inclination l of $f(x)$ in the bounded set X depends on c_l and x. In particular, taking into account Fig. 2 one can conclude that when $c_l < 4$ the inclination is always smaller than 1 regardless the value of x. Consequently, for the sigmoid with $c_l \geq 4$ the contraction mapping is not valid for every x. There is an interval, for which it is not valid. Figure 3 shows the inclination of the sigmoid function when $c_l = 5$. It can be seen that when $-0.1925 < x < 0.1925$ the inclination exceeds 1. In this interval, if one wants to keep the contraction property of the sigmoid used he should probably consider lowering c_l.

In deriving the results of this chapter it is assumed for simplicity that $c_l = 1$. Similar approach can be applied for other values of c_l giving analogous results. According to Fig. 2 when $c_l = 1$ it always holds that:

$$\frac{1}{4} \geq l \tag{7}$$

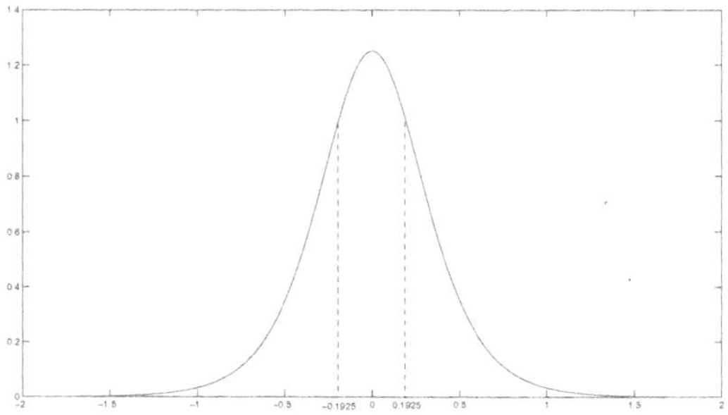

Fig. 3 Inclination of sigmoid function $f = \frac{1}{1+e^{-c_l x}}$ when $c_l = 5$

and for any x, y

$$\frac{d\left(f(x), f(y)\right)}{d\left(x, y\right)} \leq 1/4 \tag{8}$$

From (7) and (8) we get:

$$\frac{d\left(f(x), f(y)\right)}{d\left(x, y\right)} < 1 \tag{9}$$

Thus there is always a number c for which $0 \leq c < 1$, such that (9) is:

$$\frac{d\left(f(x), f(y)\right)}{d\left(x, y\right)} < c < 1 \tag{10}$$

Theorem 2 can be easily expanded for the continuous and differentiable sigmoid function $f = tanh(c_l x)$. The inclination l of $f = tanh(c_{li} x)$ is equal to:

$$l = c_l(1 - f^2)$$

and its plot for $c_l = 1$ and $c_l = 0.5$ is given in Fig. 4. According to Fig. 4 one can see that for $c_l = 1$ the inclination l of $f(x)$ in the bounded set X is always smaller than 1. Thus for the hyperbolic tangent function we get:

$$\frac{d(f(x), f(y))}{d(x, y)} \leq 1$$

and there is always a number c, such that:

$$\frac{d(f(x), f(y))}{d(x, y)} \leq c \leq 1$$

In case where $0 < c \leq 1$, the map f becomes non-expansive and the following theorem holds:

Theorem 3. *The scalar sigmoid function f, ($f = \frac{1}{1+e^{-x}}$) is a non-expansive map of the metric space X into X, were $X = [a, b], a, b$ finite, where:*

$$d(f(x), f(y)) \leq cd(x, y) \tag{11}$$

and $0 < c \leq 1$.

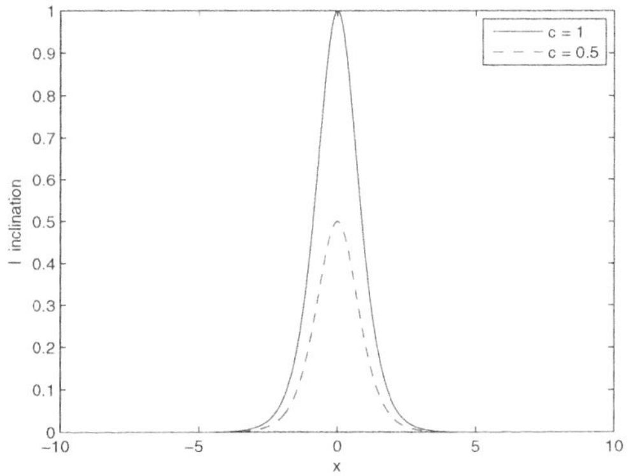

Fig. 4 Inclination of sigmoid function $f = tanh(c_l x)$ when $c_l = 1$ and $c_l = 0.5$ respectively

Proof. Use the results in the previous theorem together with the Browder-Gohde-Kirk theorem ([65], p.52) for non-expansive maps in Hilbert spaces. Here it should be noted that a solution exists but **is not unique**.

Theorem 4. *There is one and only one solution for any concept value A_i of any FCM where the sigmoid function $f = \frac{1}{1+e^{-x}}$ is used, if:*

$$\left(\sum_{i=1}^{n} \|w_i\|^2\right)^{1/2} < 4 \tag{12}$$

where w_i is the i_{th} row of matrix W^T and $\|w_i\|$ is the l_2 norm of w_i.
If

$$\left(\sum_{i=1}^{n} \|w_i\|^2\right)^{1/2} = 4 \tag{13}$$

there exists at least one solution for each concept value A_i of any FCM, which is **not necessarily unique**.

Proof. Let X be the complete metric space $[a,b]^n$ and $G : X \rightarrow X$ be a map such that:

$$d(G(A), G(A')) \leq cd(A, A') \tag{14}$$

for some $0 < c < 1$.

Vector G is equal to:

$$G = \left[f(w_1 \cdot A) \; f(w_2 \cdot A) \; \ldots \; f(w_n \cdot A) \right]^T \tag{15}$$

where n is the number of concepts of the FCM, f is the sigmoid function $f = \frac{1}{1+e^{-x}}$, w_i is the i_{th} row of matrix W^T of the FCM, where $i = 1,2,...,n$, and by $\cdot$ we denote the inner product between two equidimensional vectors which both belong in $\Re^n$.

Assume A and A' are two different concept values for the FCM. Then we want to prove the following inequality:

$$\left\|G(A) - G(A')\right\| \leq c\left\|A - A'\right\| \tag{16}$$

But $\|G(A) - G(A')\|$ according to (15) equals to:

$$\|G(A) - G(A')\| = \left(\sum_{i=1}^{n} (f(w_i \cdot A) - f(w_i \cdot A'))^2\right)^{1/2}$$

According to Theorem 2 for the scalar argument of $f(.)$, which is $w_i \cdot A$ in the bounded and closed interval $[a,b]$ with a and b being finite numbers, it is true that:

$$|f(w_i \cdot A) - f(w_i \cdot A')| \leq c'_i |(w_i \cdot A) - (w_i \cdot A')|$$

for every $i = 1,2,...,n$.
Thus

$$|f(w_i \cdot A) - f(w_i \cdot A')| \leq c' |(w_i \cdot A) - (w_i \cdot A')|$$

where $c' = max(c'_1, c'_2, ..., c'_n)$.

By using the Cauchy-Schwartz inequality we get:

$$c'|(w_i \cdot A) - (w_i \cdot A')| = c'|w_i \cdot (A - A')| \leq c'\|w_i\|\|A - A'\|$$

Subsequently, we get:

$$\|G(A) - G(A')\| = \left(\sum_{i=1}^{n} (f(w_i \cdot A) - f(w_i \cdot A'))^2\right)^{1/2}$$

$$\Rightarrow \|G(A) - G(A')\| \leq \left(\sum_{i=1}^{n} (c'\|w_i\|\|A - A'\|)^2\right)^{1/2}$$

Finally:

$$\|G(A) - G(A')\| \leq c'\|A - A'\| \left(\sum_{i=1}^{n} \|w_i\|^2\right)^{1/2}$$

A necessary condition for the above to be a contraction is:

$$c' \left(\sum_{i=1}^{n} \|w_i\|^2\right)^{1/2} < 1 \tag{17}$$

From eq. (7) we have that:

$$c' \leq 1/4$$

So that condition of eq. (17) now becomes:

$$\left(\sum_{i=1}^{n} \|w_i\|^2\right)^{1/2} < 4 \tag{18}$$

The proof of the equality according to (13) is similar, using the results of Theorem 3.

Remark 1. Using the same analysis one can prove that for the hyperbolic tangent sigmoid function $f = tanh(x)$ eq. (12) and (13) become:

$$\left(\sum_{i=1}^{n} \|w_i\|^2\right)^{1/2} < 1 \tag{19}$$

and

$$\left(\sum_{i=1}^{n} \|w_i\|^2\right)^{1/2} = 1 \tag{20}$$

respectively.

3.2 Interpreting the Results

Theorem 4 provides conditions that the weight interconnections of the FCM should fulfill in oder that it has an equilibrium condition and that condition be unique. It is clear that the above results hold only for continuous differentiable squashing functions like the log-sigmoid and the hyperbolic tangent functions. Therefore, the results are not valid for FCMs equipped with bivalent or trivalent functions; we can not conclude anything for this type of FCMs using the above analysis. Equations (12), (13) are valid for FCMs having log-sigmoid functions. Eq. (12) provides with conditions for the existence and uniqueness of an equilibrium condition, while Eq. (13) refers to condition that guarantees only the existence. Similar respective results arise from equations (19) and (20) for FCMs equipped with hyperbolic tangent functions. In case none of the conditions hold, this does not imply that an equilibrium does not exist. It may or may not exist, just there is not any guarantee for its existence.

It can be observed that the conditions provide with upper bounds for the FCM weights, which depend on the size of the FCM. Moreover, taking into account the observations made in the proof of theorem 2, it can be concluded that these upper bounds depend also on the inclination of the sigmoid used. The larger the inclination becomes the smaller these bounds are, so that in the limit no weight set can be found to guarantee the uniqueness or the existence of the equilibrium. This is however expected, because sigmoids with large inclinations tend to reach the behavior of bivalent or trivalent functions and therefore no guarantee can exist based on the above analysis. FCMs that use sigmoids with large inclinations (or in the limit bivalent or trivalent functions) tend to give mainly qualitative results, while FCMs using small inclination sigmoids give both quantitative and qualitative results (see also the relevant conclusions in [56]). This is again mathematically expected since sigmoids with large inclination have very low discriminative abilities because they produce very similar outputs for inputs that may be quite dissimilar. For FCMs with sigmoids having small inclinations it is much easier to find weight sets that fulfill the existence and uniqueness conditions.

3.3 FCM with Input Nodes

So far we have not considered the existence of input nodes. This kind of nodes is also called "*steady*" nodes in [43] in the sense that they influence but are not influenced by the other nodes of the FCM. We are showing now that the results obtained in the previous section are still valid. For the FCM of Fig. 5 C_1 is such an input (or steady) node. Its weight matrix W is equal to:

$$W = \begin{bmatrix} 0 & w_{12} & w_{13} & 0 \\ 0 & d_{22} & w_{23} & w_{24} \\ 0 & 0 & d_{33} & w_{34} \\ 0 & w_{42} & w_{43} & d_{44} \end{bmatrix}$$

while vector A containing the node values is:

$$A = \begin{bmatrix} U_1 & A_2 & A_3 & A_4 \end{bmatrix}^T$$

For the FCM of Fig. 5, matrix G in eq. (15) assumes now the following form:

$$G = \begin{bmatrix} U_1 & f(w_2 \cdot A) & f(w_3 \cdot A) & f(w_4 \cdot A) \end{bmatrix}^T$$

where U_1 is the input to FCM nodes. In a more general form matrix G can be written as:

$$G = \begin{bmatrix} U_1 & U_2 & \ldots & U_m & f(w_{m+1} \cdot A) & f(w_{m+2} \cdot A) & \ldots & f(w_{m+n} \cdot A) \end{bmatrix}^T \tag{21}$$

corresponding to vector:
$A = \begin{bmatrix} U_1 & U_2 & \cdots U_m & A_{m+1} & A_{m+2} & \cdots & A_{m+n} \end{bmatrix}^T$ where m is the number of inputs and n is the number of the other concept nodes in FCM. Under this definition eq. (4) assumes the same form:

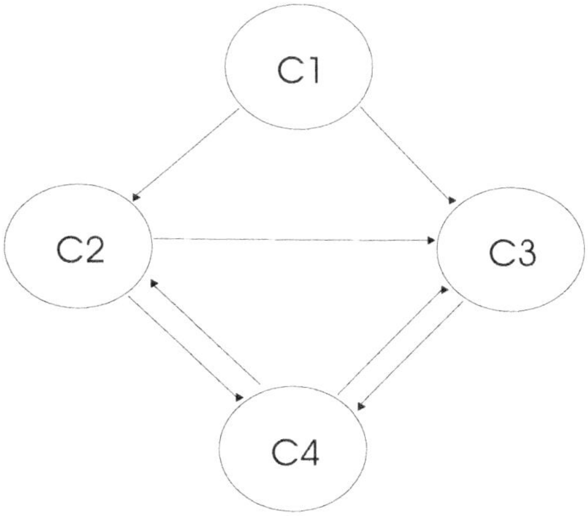

Fig. 5 FCM with one input node

$$A(k) = G(A(k-1)) \tag{22}$$

The next theorem proves that for matrix G and vector A defined above the results of theorem 4 are still valid.

Theorem 5. *For an FCM with input nodes, with its concept values driven by (22), where G is described in (21), there is one and only one solution for any concept value A_i if eq. (12) is fullfiled, that is:*

$$\left(\sum_{i=1}^{n} \|w_{m+i}\|^2\right)^{1/2} < 4$$

where w_{m+i} is the $(m+i)_{th}$ row of matrix W^T and $\|w_{m+i}\|$ is the l_2 norm of w_{m+i}. If $\left(\sum_{i=1}^{n} \|w_{m+i}\|^2\right)^{1/2} = 4$ a **not always unique** *solution exists.*

Proof. Assume A and A' are two different concept values for the FCM having one or more inputs. Then we want to prove again inequality (16), that is:

$$\|G(A) - G(A')\| \leq c\|A - A'\|$$

But since input node values are not influenced by the other nodes of the FCM $\|G(A) - G(A')\|$ according to (21) is equal to:

$$\|G(A) - G(A')\| = \left(\sum_{i=1}^{m} (U_i - U_i)^2 + \sum_{i=1}^{n} (f(w_{m+i} \cdot A) - f(w_{m+i} \cdot A'))^2\right)^{1/2}$$

where m is the number of inputs and n is the number of the other nodes in the FCM. The above equation is equivalent to the following:

$$\|G(A) - G(A')\| = \left(0 + \sum_{i=1}^{n} (f(w_{m+i} \cdot A) - f(w_{m+i} \cdot A'))^2\right)^{1/2}.$$

Therefore, $\|G(A) - G(A')\|$ assumes quite the same form with that appearing in theorem 4 leading to the same condition, that is:

$$\left(\sum_{i=1}^{n} \|w_{m+i}\|^2\right)^{1/2} < 4$$

Similarly, for the $=$ case the respective results of Theorem 4 still hold and the results obtained for the hyperbolic tangent function are still valid.

The results of the above theorem should not be misinterpreted. Similar to theorem 4, theorem 5 states that the weights have to comply with the bounds implied by its conditions in order to assure the existence or the uniqueness of the FCM equilibrium. However, when these conditions are fulfilled the unique equilibrium does not depend solely on the weight set, as in the case of FCMs with no input nodes. It depends also on the values of the input nodes. This is a quite reasonable implication because the input node values are steady and unlike the other node values they are not changing during the repetitive application of eq. (1) until the FCM reaches its equilibrium. Therefore, they act as an external excitation. Different values of the excitation will drive the FCM in different equilibria. The results obtained so far are quite significant. They are valid for a class of sigmoid functions, which may present contractive or non expansive properties and are frequently used in FCMs. Many physical systems can reach an equilibrium point after initial perturbations. If FCMs are used for the representation of such systems, it is important to know, which FCM structure may guarantee the existence and probably the uniqueness of the equilibrium. This gives rise also to the development of estimation algorithms, which can learn the structure (weights) based on real system's data. Such an algorithm will be given in the next section. Moreover, in FCMs with input nodes this unique equilibrium may be different depending on the value of the exciting input, a behavior quite similar with the behavior of stable nonlinear physical systems. Therefore FCMs could be used not only for modeling but also for the control of unknown nonlinear systems.

4 On Line Parameter Estimation of Fuzzy Cognitive Maps

Based on the results and observations of Section 3 we are now presenting a method of finding appropriate weight sets related to a desired equilibrium point of the FCM [58]. Choosing a desired state A^{des} for the FCM this is equivalent to solving the equation

$$A^{des} = f(W^T A^{des}) \tag{23}$$

in respect to W^T, where $A^{des} = \left[A_1^{des}, A_2^{des}, \cdots A_n^{des}\right]^T$ and f is a vector valued function $f : \Re^n \to \Re$, defined as follows: $f(x) = [f_1(x_1), f_2(x_2), \cdots f_n(x_n)]^T$, where $x \in \Re^n$ and $f_i(x_i) = \frac{1}{1+e^{-x_i}}$, for $i = 1, 2, \cdots, n$. Then,

$$f^{-1}(A^{des}) = W^{*T} A^{des} \tag{24}$$

where $f^{-1}(A^{des}) = \left[f_1^{-1}(A_1^{des}), f_2^{-1}(A_2^{des}), \cdots f_n^{-1}(A_n^{des})\right]^T$ and $f_i^{-1}(A_i^{des}) = W_i^T \cdot A^{des}$, with W_i being the ith column of W^*.

In order to solve the above equation we use an adaptive estimation algorithm [63] including the parameter constraints derived in Section 3. The updating scheme is illustrated in Fig. 6, demonstrating that the weight learning is performed adaptively based on the desired equilibrium point of the FCM. Taking into account that $f_i(x_i)$ is the sigmoid function weight updating laws are given as follows:

The error $\varepsilon_i(k)$ of the parametric discrete-time adaptive law is of the form:

$$\varepsilon_i(k) = \frac{f_i^{-1}(A_i^{des}) - W_i^T(k-1)A^{des}}{c + (A^{des})^T A^{des}} \tag{25}$$

while the updating algorithm is given by:

$$W_i(k) = W_i(k-1) + \alpha\varepsilon_i(k)A^{des} \tag{26}$$

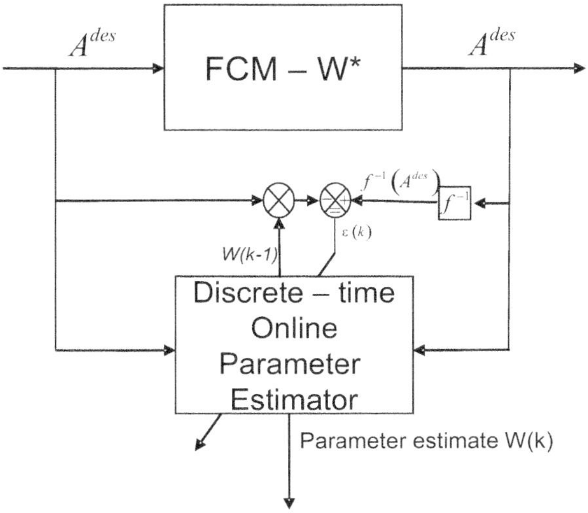

Fig. 6 Online parameter estimator designed for FCMs

where $W_i(k)$ is the ith column of $W(k)$, which is the estimator of $W^*(k)$. A^{des} is constant vector and $f_i^{-1}(A_i^{des})$ is also constant and scalar. $\alpha > 0$ and $c > 0$ are design parameters.

By using the above updating algorithm we are confident that the estimator converges to W^*. In Section 3 we proved that if inequality (12) is true then W^* corresponding to the designed FCM provide a unique solution and satisfies (23).

Proof. From (25), (26) and $\tilde{W}_i(k) = W_i(k) - W_i^*$ we obtain the error equation

$$\tilde{W}_i(k) = \left[I - \frac{aA^{des}(A^{des})^T}{c + (A^{des})^T A^{des}} \right] \tilde{W}(k-1) \tag{27}$$

$$\varepsilon_i(k) = -\frac{\tilde{W}_i^T(k-1)A^{des}}{c + (A^{des})^T A^{des}} \tag{28}$$

Consider the function

$$V(k) = \frac{1}{2a}\tilde{W}^T(k)\tilde{W}(k) \tag{29}$$

Then

$$V(k) - V(k-1) = \frac{1}{2a}\tilde{W}_i^T(k)\tilde{W}_i(k) - \frac{1}{2a}\tilde{W}_i^T(k-1)\tilde{W}_i(k-1)$$

$$= -\frac{\tilde{W}_i^T(k-1)A^{des}(A^{des})^T\tilde{W}_i(k-1)}{c+(A^{des})^T A^{des}} + \frac{\tilde{W}_i^T(k-1)A^{des}(A^{des})^T aA^{des}(A^{des})^T\tilde{W}_i(k-1)}{2(c+(A^{des})^T A^{des})^2}$$

Using $\varepsilon_i(k)(c + (A^{des})^T A^{des}) = -\tilde{W}_i^T(k-1)A^{des}$, we obtain

$$V(k) - V(k-1) = -\varepsilon_i^2(k)(c + (A^{des})^T A^{des}) \left[1 - \frac{a(A^{des})^T A^{des}}{2(c+(A^{des})^T A^{des})} \right]$$

Since $a, c > 0$ we can always choose $0 < a < 2$ such that $\frac{a(A^{des})^T A^{des}}{2(c+(A^{des})^T A^{des})} < 2$. It then follows that

$$V(k) - V(k-1) \leq -c_0 \varepsilon_i^2(k)(c + (A^{des})^T A^{des}) \leq 0 \tag{30}$$

for some constant $c_0 > 0$. From (29), (30) we have that $V(k)$ and therefore $W_i(k) \in \ell_\infty$ and $V(k)$ has a limit, i.e., $\lim_{k \to \infty} V(k) = V_\infty$. Consequently, using (30) we obtain

$$c_0 \sum_{k=1}^{\infty} (\varepsilon_i^2(k)(c + (A^{des})^T A^{des})) \leq V(0) - V_\infty < \infty$$

which implies $\varepsilon_i(k)\sqrt{(c + (A^{des})^T A^{des})} \in \ell_2$ and $\varepsilon_i(k)\sqrt{(c + (A^{des})^T A^{des})} \to 0$ as $k \to \infty$. Since $\sqrt{(c + (A^{des})^T A^{des})} \geq c > 0$, we also have that $\varepsilon_i(k) \in \ell_2$ and $\varepsilon_i(k) \to 0$ as $k \to \infty$. We have $\varepsilon_i(k)A^{des} = \varepsilon_i(k)\sqrt{(c + (A^{des})^T A^{des})}\frac{A^{des}}{\sqrt{(c+(A^{des})^T A^{des})}}$. Since $\frac{A^{des}}{\sqrt{(c+(A^{des})^T A^{des})}}$ is bounded and $\varepsilon_i(k)\sqrt{(c + (A^{des})^T A^{des})} \in \ell_2$, we have that $\varepsilon_i(k)A^{des} \in \ell_2$ and $\left\|\varepsilon_i(k)A^{des}\right\| \to 0$ as $k \to \infty$. This implies (using (26)) that $\|W_i(k) - W_i(k-1)\| \in \ell_2$ and $\|W_i(k) - W_i(k-1)\| \to 0$ as $k \to \infty$. Now

$$W_i(k) - W_i(k-N) =$$
$$W_i(k) - W_i(k-1) + W_i(k-1) - W_i(k-2) + \ldots + W_i(k-N+1) - W_i(k-N)$$

for any finite N. Using the Schwartz inequality, we have

$$\|W_i(k) - W_i(k-N)\|^2 \leq \|W_i(k) - W_i(k-1)\|^2 + \|W_i(k-1) - W_i(k-2)\|^2 + \ldots +$$
$$\|W_i(k-N+1) - W_i(k-N)\|^2$$

Since each term on the right-hand side of the inequality is in ℓ_2 and goes to zero with $k \to \infty$, it follows that $\|W_i(k) - W_i(k-N)\| \in \ell_2$ and $\|W_i(k) - W_i(k-N)\| \to 0$ as $k \to \infty$.

Remark 2. The n weight updating laws given by (26) for each column W_i, $i = 1, \ldots, n$ of W can be written in a compact form as follows

$$\underline{W}(k) = \underline{W}(k-1) + a\underline{\varepsilon}A^{des} \tag{31}$$

where $\underline{W} = \left[W_1^T, W_2^T, \ldots W_n^T\right]^T$ is a $n^2 \times 1$ column vector containing all the elements of matrix W arranged in a column, row after row. Also, $\underline{\varepsilon}$ is a $n^2 \times n$ matrix having the form $\underline{\varepsilon} = [\varepsilon_1 I_n, \varepsilon_2 I_n, \ldots \varepsilon_n I_n]^T$, where I_n is a $n \times n$ unit matrix and ε_i is given by (25).

4.1 Gradient Projection Method

In parameter identification problems, we have some a priori knowledge as to where the unknown parameter W^* is located in $\Re^n$. This knowledge could be the upper and the lower bounds for each element of W^*. In our parameter identification problem inequalities $-1 \leq w_{ij} \leq 1$ and (12) must be true. A convex projection method [63] is next illustrated to modify the adaptive laws of the previous section. A schematic representation of the orthogonal projection method is illustrated in fig. 7.

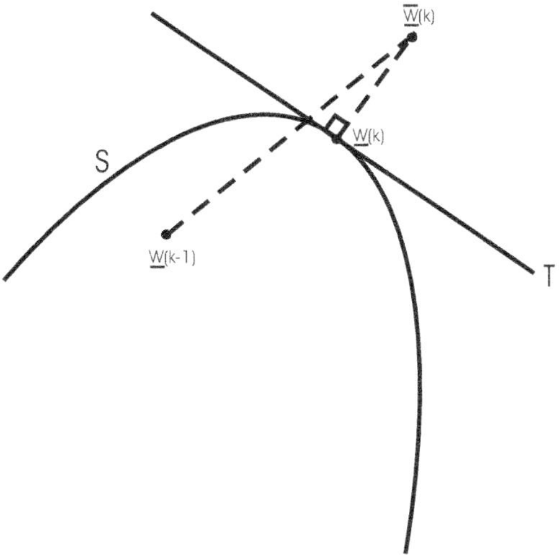

Fig. 7 Discrete Time Parameter Projection

4.1.1 Projection Method 1

The S_1 set for projection that satisfies $-1 \leq w_{ij} \leq 1$ is the convex set:

$$S_1 = \{W \in \Re^n | g(w_{ij}) \leq 0,\ g(w_{ij}) = |w_{ij}| - 1, \forall i, j \in \aleph\}$$

The updating formula of parameters W of FCM is now given by:

$$\begin{aligned} \overline{\underline{W}}(k) &= \underline{W}(k-1) + a\underline{\varepsilon}A^{des} \\ \underline{W}(k) &= \begin{cases} \overline{\underline{W}}(k), & if\, \overline{\underline{W}}(k) \in S_1 \\ \Pr(\overline{\underline{W}}(k)), & if\, \overline{\underline{W}}(k) \notin S_1 \end{cases} \end{aligned} \tag{32}$$

where
$\Pr(\overline{\underline{W}}(k)) = \perp proj$ of $\overline{\underline{W}}(k)$ on S_1.

This formula can also be element-wise written as:

$$\begin{aligned} \overline{w}_{ij}(k) &= w_{ij}(k-1) + \alpha\varepsilon_i(k)A_i^{des} \\ w_{ij}(k) &= \begin{cases} \overline{w}_{ij}(k), & if\ |\overline{w}_{ij}(k)| \leq 1 \\ -1 & if\, \overline{w}_{ij}(k) < -1 \\ 1 & if\, \overline{w}_{ij}(k) > 1 \end{cases} \end{aligned} \tag{33}$$

4.1.2 Projection Method 2

The S_2 set for projection that satisfies (12) is the convex set:

$$S_2 = \{\underline{W} \in \Re^{n^2} | g(\underline{W}) \leq 0,\ g(\underline{W}) = \|\underline{W}\| - 4\}$$

The updating equation of parameters W of FCM is now given by:

$$\overline{\underline{W}}(k) = \underline{W}(k-1) + a\varepsilon \underline{A}^{des}$$
$$\underline{W}(k) = \begin{cases} \overline{\underline{W}}(k), & if\, \overline{\underline{W}}(k) \in S_2 \\ \Pr(\overline{\underline{W}}(k), & if\, \overline{\underline{W}}(k) \notin S_2 \end{cases} \tag{34}$$

where the orthogonal projection $\Pr(\overline{\underline{W}}(k)) = \perp proj$ of $\overline{\underline{W}}(k)$ on S_2 is given by $\frac{4}{\|\overline{\underline{W}}(k)\|}\overline{\underline{W}}(k)$. Therefore, equation (34) can also be written in the following compact form:

$$\overline{\underline{W}}(k) = \underline{W}(k-1) + a\varepsilon \underline{A}^{des}$$
$$\underline{W}(k) = \overline{\underline{W}}(k) \min\left(1, \frac{4}{\|\overline{\underline{W}}(k)\|}\right) \tag{35}$$

4.1.3 Concurrent Projection Method

In the previous subsections two projection methods were presented. Our objective is to combine both projection methods. The set S which satisfies both inequalities is the intersection of sets S_1 and S_2.

$$S = S_1 \bigcap S_2$$

We replace the weight updating equation:

$$\underline{W}(k) = \underline{W}(k-1) + a\varepsilon \underline{A}^{des}$$

with

$$\overline{\underline{W}}(k) = \underline{W}(k-1) + a\varepsilon \underline{A}^{des}$$
$$\underline{W}(k) = \begin{cases} \overline{\underline{W}}(k), & if\, \overline{\underline{W}}(k) \in S \\ \Pr(\overline{\underline{W}}(k), & if\, \overline{\underline{W}}(k) \notin S \end{cases} \tag{36}$$

where $\Pr(\overline{\underline{W}}(k)) = \perp proj$ of $\overline{\underline{W}}(k)$ on S.

4.2 Bilinear Adaptive Estimation

In the analysis made so far the c_l factor of the sigmoid functions is considered constant and specifically $c_l = 1$. If c_l is allowed to take also other values then the conditions of existence and uniqueness of equilibrium points are modified. Also the weight updating laws are modified and the problem of adaptive weight estimation is not linear anymore but it becomes a bilinear adaptive estimation problem, where both the FCN weights and the c_l factor of each sigmoid are modified, using again appropriate projection methods. These results have been very recently presented in [66], they are not however given here because they go beyond the scope of this chapter.

5 Control Paradigms Using the FCN Framework

The adaptive weight estimation algorithm presented in the previous sections uses the obtained conditions to construct appropriate weight projection rules assuring that the obtained weights do not compromise the existence of the FCM solution. In view of these results *meta rules* of the form "If weights and inputs then fixed point" are more appropriate to represent the behavior of an FCM which satisfy the above weight conditions. Fuzzy Cognitive Networks (FCN) [43], [59], [60], [58] introduced recently as an extension of FCMs can work on the basis of such *meta rules* and provide the application framework of the obtained results. During its updating procedure, The FCN framework , stores these *meta rules* in a fuzzy rule database.

Using the stored information, one can devise control approaches using the FCN and following an inverse procedure to obtain the control input(s), which drives the FCN (and consequently the system it describes) in the desired condition. Two such approaches are presented in the following. The first approach is tested to control the well known *inverted pendulum* benchmark. It is assumed that the depository of fuzzy *meta rules* is complete and therefore the determination of the appropriate control input is based on simple inverse formulas. The second approach is applied to control a nonlinear wastewater treatment bioprocess. In this case the depository is not full and the control inputs are determined by following an inverse procedure using a gradient technique. In the paradigms that follow we present the FCN operation, its storage mechanism and the control of the physical system.

5.1 Control of an Inverted Pendulum

The control of an inverted pendulum system is a standard benchmark problem in the area of nonlinear control systems. In this system, an inverted pendulum is attached to a cart equipped with a motor that drives it along a horizontal track. A simple model of the inverted pendulum in Fig.8 is given by:

$$\begin{aligned} \dot{x}_1 &= x_2 \\ \dot{x}_2 &= \frac{mlx_2{}^2 sin(x_1) - g(M+m)x_1}{mlcos(x_1) - l(M+m)} + \frac{1}{mlcos(x_1) - l(M+m)} T \end{aligned} \tag{37}$$

where $x_1 = \vartheta$ is the angle of the pole, $x_2 = \dot{\vartheta}$ is the angular velocity of the pole, $g = 9.81$, $l = 1.0$ is the half length of the pole, $m = 1.0$ is the mass of the pole, $M = 2.5$ is the mass of the cart and T is the control input.

5.1.1 Prerequisites Regarding Stability Analysis

In this subsection we provide preliminaries regarding the Lyapunov's stability analysis of nonlinear systems, which will be used for the stability of the inverted pendulum. Suppose that a dynamic system is represented with

$$\dot{x}(t) = f(x(t)) \tag{38}$$

where $x \in \Re^n$ is a n vector and $f : D \to \Re^n$ with $D = \Re^n$ or $D = B(h)$ for some $h > 0$ where

$$B(h) = \{x \in \Re^n : |x| < h\}$$

is a ball centered at the origin with a radius of h and $|\cdot|$ is a norm on $\Re^n$.Assume that for enery x_0 the initial value problem

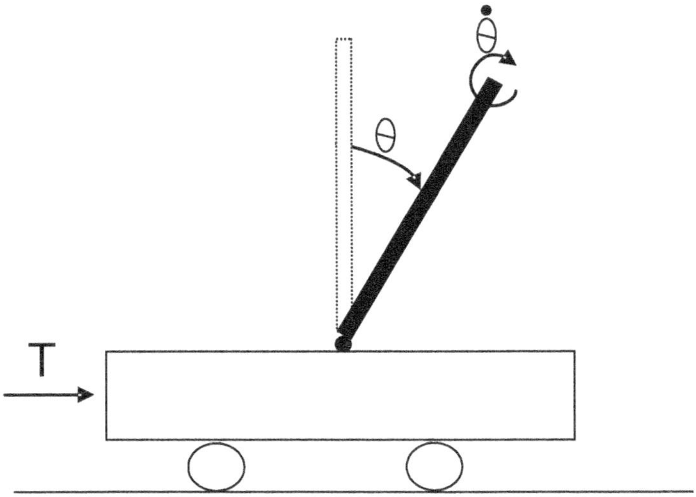

Fig. 8 Inverted Pendulum

$$\dot{x}(t) = f(x(0)), \quad x(0) = x_0 \tag{39}$$

posseses a unique solution $\varphi(t, x_0)$ that depends continuously on x_0. A point $x_e \in \Re^n$ is called an "equilibrium point" of eq. (38) if $f(x_e) = 0$ for all $t \geq 0$. An equilibrium point x_e is an "isolated eqilibrium point" if there is an $h' > 0$ such that the hyper sphere around x_e,

$$B(x_e, h') = \{x \in \Re^n : |x - x_e| < h'\}$$

The equilibrium x_e of eq. (38) is "stable" (in the sence of Lyapunov) if for every $\varepsilon > 0$ there exists a $\delta(\varepsilon) > 0$ such that $|\varphi(t, x_0)| < \varepsilon$ for all $t \geq 0$ whenever $|x_0| < \delta(\varepsilon)$.

The equilibrium x_e of eq. (38) is "asymptotically stable" if it is stable and there exists $\eta > 0$ such that $\lim_{t \to \infty} \phi(t, x_0) = 0$ whenever $|x_0| < h$.

The set $X_d \subset \Re^n$ of all $x_0 \in \Re^n$ such that $\phi(t, x_0) \to 0$ as $t \to \infty$ is called the "domain of attraction"of the equilibrium $x_e = 0$ of eq. (38). The equilibrium $x_e = 0$ is said to be "globally asymptotically stable" if $X_d = \Re^n$.

5.1.2 Lyapunov's Direct Method

The stability results for an equilibrium $x_e = 0$ of eq. (38) depend on the existence of an appropiate "Lyapunov function" $V : D \to \Re$ where $D = \Re^n$ for global results and $D = B(h)$ for some $h > 0$, for local results. If V is continuosly differentiable with respect to its arguments then the derivative of V with respect to t along the solutions of eq. (38) is

$$\dot{V}(x(t)) = \nabla V(x(t))^T f(x(t))$$

where

$$\nabla V(x(t)) = \left[\frac{\partial V}{\partial x_1}, \frac{\partial V}{\partial x_1}, \ldots, \frac{\partial V}{\partial x_n} \right]^T$$

Lyapunov's direct method is given by the following:

Let $x_e = 0$ be an equilibrium for eq. (38). Let $V : B(h) = \Re$ be a continuosly differentiable function on $B(h)$ such that $V(0) = 0$ and $V(x) > 0$ in $B(h) - 0$, and $\dot{V}(x) \leq 0$ in $B(h)$, then $x_e = 0$ is stable. If $\dot{V}(x) < 0$ in $B(h) - 0$, then $x_e = 0$ is asymptotically stable.

5.1.3 Inverted Pendulum Stability Analysis

In this subsection we will illustrate the use of Lyapunov's direct method for the stability analysis of the inverted pendulum [67]. Eq. (37) can be rewritten as:

$$\begin{aligned} \dot{x}_1 &= x_2 = f_1(x) \\ \dot{x}_2 &= \frac{mlx_2{}^2 sin(x_1) - g(M+m)x_1}{mlcos(x_1) - l(M+m)} + \frac{1}{mlcos(x_1) - l(M+m)} T = f_2(x) \end{aligned}$$

such that it is in the form of eq. (38). T is the force applied in the cart in order to control the pole. Assume that $T_{0,0} = 0$ so that the equilibrium is preserved.

We choose

$$V(x) = \tfrac{1}{2}x_1^2 + \tfrac{1}{2}x_2^2$$

so that

$$\nabla V(x) = \left[x_1, x_2, \right]^T$$

and

$$\dot{V} = \left[x_1, x_2 \right] \begin{bmatrix} x_2 \\ \frac{mlx_2{}^2 sin(x_1) - g(M+m)x_1}{mlcos(x_1) - l(M+m)} + \frac{1}{mlcos(x_1) - l(M+m)} T \end{bmatrix}$$

and we would like $\dot{V} < 0$ to prove assymptotic stability.

We have

$$x_2 \left(x_1 + \frac{mlx_2{}^2 sin(x_1) - g(M+m)x_1}{mlcos(x_1) - l(M+m)} + \frac{1}{mlcos(x_1) - l(M+m)} T \right) < -\beta$$

If for some fixed $\beta > 0$ (note that $x_2 \neq 0$) then:

$$x_1 + \frac{mlx_2{}^2 sin(x_1) - g(M+m)x_1}{mlcos(x_1) - l(M+m)} + \frac{1}{mlcos(x_1) - l(M+m)} T < \frac{-\beta}{x_2}$$

Regarding this equation we have that:

$$T < (mlcos(x_1) - l(M+m)) \left(-\frac{\beta}{x_2} - x_1 \right) - \left(mlx_2{}^2 sin(x_1) - g(M+m)x_1 \right) \quad (40)$$

on $x \in B(h)$ for some $h > 0$ and $\beta > 0$. When eq. (40) holds then this control law T ensures that asymptotic stability for the solution x_e holds.

5.1.4 An FCN Designed for the Inverted Pendulum

As shown in Section 4 the concepts values of the FCM with a specified matrix W have a unique solution as far as (12) is fulfilled. The perspective of transforming FCMs into a modeling and control alternative requires, first to update its weight matrix W so that the FCM can capture different mappings of the real system and second to store these different kind of mappings. The Fuzzy Cognitive Network (FCN) [43] has been proposed as an operational extension framework of FCM, which updates its weights and reaches new equilibrium points based on the continuous interaction with the system it describes or in the initial knowledge that some experts can give. Moreover, for each equilibrium point a fuzzy rule based storage mechanism of the form "If weights then fixed point" is provided, which facilitates and speeds-up its operation.

The FCN designed for the Inverted Pendulum is shown in Fig. 9

Node $C(T)$ represents the force applied in the cart and is in the interval:

$[0 \rightarrow -30 Newton, 1 \rightarrow 30 Newton]$

Node $C(x_1(t))$ represents the angle of the pole at time t and is in the interval:

$[0 \rightarrow -pi/2(rad), 1 \rightarrow pi/2(rad)]$

Node $C(x_2(t))$ represents the speed of the pole at time t and is in the interval:

$[0 \rightarrow -pi/2(rad/sec), 1 \rightarrow pi/2(rad/sec)]$

Node $C(\dot{x}_1(t+\Delta T))$ represents the angle of the pole at time $t+\Delta T$ and is in the interval: $[0 \rightarrow -pi/2(rad), 1 \rightarrow pi/2(rad)]$

Node $C(x_2(t+\Delta T))$ represents the speed of the pole at time $t+\Delta T$ and is in the interval: $[0 \rightarrow -pi/2(rad/sec), 1 \rightarrow pi/2(rad/sec)]$

The weight matrix of FCN is depicted below:

$$W = \begin{bmatrix} 0\,0\,0 & w_{C(T)C(x_1(t+\Delta T))} & w_{C(T)C(x_2(t+\Delta T))} \\ 0\,0\,0 & w_{C(x_1(t))C(x_1(t+\Delta T))} & w_{C(x_1(t))C(x_2(t+\Delta T))} \\ 0\,0\,0 & w_{C(x_2(t))C(x_1(t+\Delta T))} & w_{C(x_2(t))C(x_2(t+\Delta T))} \\ 0\,0\,0 & 1 & 0 \\ 0\,0\,0 & 0 & 1 \end{bmatrix}$$

5.1.5 Off Line Training of the FCN

Collecting Training Data for the FCN

From the dynamic equations described in 37 of the inverted Pendulum we collect the training data using the following procedure:

for $x_1(t)$ = -pi/2:pi/8:pi/2
  for $x_2(t)$ = -pi/2:pi/8:pi/2
    for T = -30:7.5:30

calculate $x_1(t+\Delta T)$ and $x_2(t+\Delta T)$ using eq. (37) with $\Delta T = 0.001$.

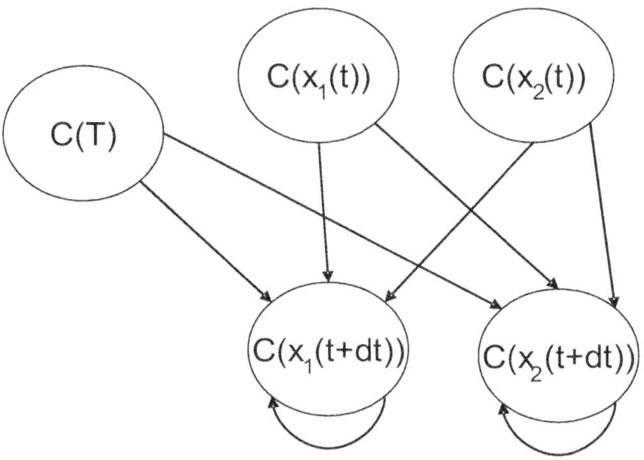

Fig. 9 An FCN designed to control an Inverted Pendulum

Also, for its sample $x_1(t)$, $x_2(t)$ we select T such that eq. (40) is fullfield. In eq. (40) we select parameter $\beta = 0.1$.

As an example, if $x_1 = -pi/4 = -0.7854$, $x_2 = pi/8 = 0.3927$ and $T = 0$ then $x_1(t+\Delta T) = -pi/4.0174 = -0.7820$ and $x_2(t+\Delta T) = -pi/10.5920 = -0.2966$. Using this procedure we collect 729 training data samples.

Translate Training Data into FCN Vectors

For node $C(T)$ the conversion coefficient is: $C(T) = \frac{T*0.5+15}{30}$, where T is the value from the fuzzy controller which is the force that is applied to the real system and $C(T)$ is the value of node $C(T)$ in the interval [01] of FCN.

For node $C(x_1)$ the conversion coefficient is: $C(x_1(t)) = \frac{x_1(t)*0.5+pi/4}{pi/2}$, where $x_1(t)$ is the value from the real system which is the angle of the pole at time t and $C(x_1(t))$ is the value of node $C(x_1(t))$ in the interval [01] of FCN.

For node $C(x_2)$ the conversion coefficient is: $C(x_2(t)) = \frac{x_2(t)*0.5+pi/4}{pi/2}$, where x_2 is the value from the real system which is the angular speed of the pole at time t and $C(x_2(t))$ is the value of node $C(x_2(t))$ in the interval [01] of FCN.

For node $C(x_1(t+\Delta T))$ the conversion coefficient is: $C(x_1(t+\Delta T)) = \frac{x_1(t+\Delta T)*0.5+pi/4}{pi/2}$, where $x_1(t+\Delta T)$ is the value from the real system which is the angle of the pole at time $(t+\Delta T)$ and $C(x_1(t+\Delta T))$ is the value of node $C(x_1(t+\Delta T))$ in the interval [01] of FCN.

For node $C(x_2(t+\Delta T))$ the conversion coefficient is: $C(x_2(t+\Delta T)) = \frac{x_2(t+\Delta T)*0.5+pi/4}{pi/2}$, where $x_2(t+\Delta T)$ is the value from the real system which is the angular speed of the pole at time $(t+\Delta T)$ and $C(x_2(t+\Delta T))$ is the value of node $C(x_2(t+\Delta T))$ in the interval [01] of FCN.

As a result, using the above conversion coefficients a training sample vector containing the values:

$x_1 = -pi/4, x_2 = pi/8$ and $T = 0$ then $x_1(t+\Delta T) = -pi/4.0174$ and $x_2(t+\Delta T) = -pi/10.5920$,

is translated to the following FCN nodes vector:

$$A^{des} = \begin{bmatrix} 0.5 \; 0.25 \; 0.6250 \; 0.2511 \; 0.4056 \end{bmatrix}^T$$

Another training sample vector containing the values:

$x_1 = pi/8, x_2 = -pi/4$ and $T = 15$ then $x_1(t+\Delta T) = pi/8.0163$ and $x_2(t+\Delta T) = -pi/4.1013$,

is translated to the following FCN nodes vector:

$$A^{des} = \begin{bmatrix} 0.7500 \; 0.6250 \; 0.2500 \; 0.6247 \; 0.2561 \end{bmatrix}^T$$

In a more general form node vector for the FCN of the Inverted Pendulum can be written as:

$$A^{des} = \begin{bmatrix} C(T) \; C(x_1(t)) \; C(x_2(t)) \; C(x_1(t+\Delta T))) \; C(x_2(t+\Delta T))) \end{bmatrix}^T$$

This training data constitutes the input to the FCN, where according to fig. 10 the parameters w of the FCN are updated according to eqs. 26, 33, 35 and stored as fuzzy knowledge into a fuzzy rule database. For these parameters w eq. (23) is also true so that one can ensure that the FCN converges to a unique equilibrium point.

Storing Knowledge from Previous Operating Conditions

This procedure modifies FCN's knowledge about the system by continuously modifying the weight interconnections and consequently the node values. During the repetitive updating operation the procedure uses input from the systems training data, producing a new weight matrix for each new equilibrium state. It is desirable to device a storage mechanism for keeping these weights for probable future use to control the Pendulum. It would be also preferable this storage to allow weight retrieval even in situations, where the equilibrium conditions were not been exactly met during the training phase. To this end we present the storing mechanism of the previous acquired operational situations in a fuzzy if-then rule database, which associates in a fuzzy manner the various weights with the corresponding equilibrium node values. The procedure is explained as follows. Suppose for example that the FCN of Fig. 9 has a unique equilibrium point:

$$A = \begin{bmatrix} 0.500 \; 0.2500 \; 0.6250 \; 0.2511 \; 0.4056 \end{bmatrix}^T$$

which is connected with the weight matrix W:

$$W = \begin{bmatrix} 0 \; 0 \; 0 & -1.0000 & -0.4849 \\ 0 \; 0 \; 0 & -0.8754 & -0.0424 \\ 0 \; 0 \; 0 & -1.0000 & -0.8561 \\ 0 \; 0 \; 0 & 1.0000 & 0 \\ 0 \; 0 \; 0 & 0 & 1.0000 \end{bmatrix}$$

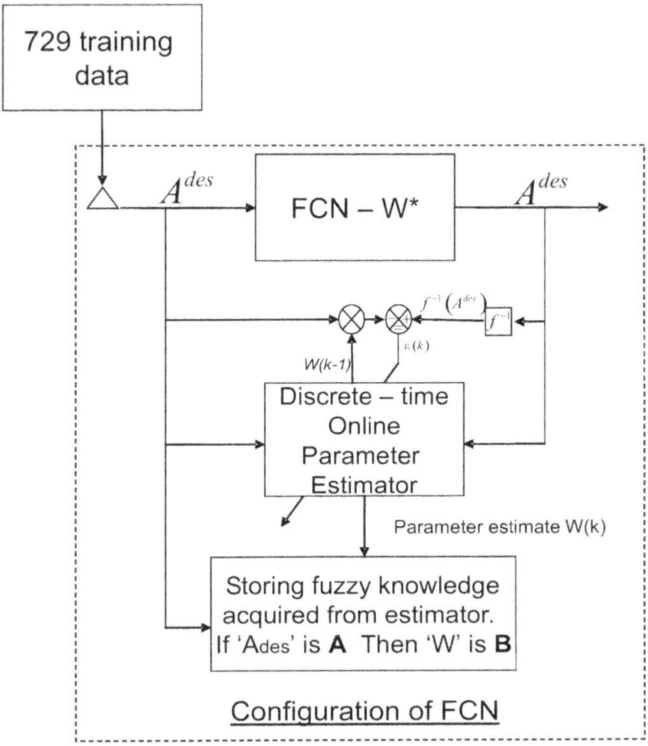

Fig. 10 Off line training of the FCN with data from the physical system

in order that A is a unique solution of (23) weight matrix W has to be such that inequality (12) is fulfilled. Indeed, for weight matrix W inequality (12) takes the form:

$$\left(\sum_{i=1}^{n} \|w_i\|^2\right)^{1/2} = 2.3950 < 4$$

where $n = 5$ is the number of concepts of the FCN.

Suppose also that the FCM in another operation point A' is related to weight matrix W' both given bellow.

$$A' = \left[0.7500\ 0.6250\ 0.2500\ 0.6247\ 0.2561\right]^T$$

$$W' = \begin{bmatrix} 0\ 0\ 0 & -0.1850 & -1.0000 \\ 0\ 0\ 0 & -0.0709 & -0.8930 \\ 0\ 0\ 0 & -0.2717 & -0.0572 \\ 0\ 0\ 0 & 1.0000 & 0 \\ 0\ 0\ 0 & 0 & 1.0000 \end{bmatrix}$$

matrix W' also fulfills (12):

$$\left(\sum_{i=1}^{5} \|w_i'\|^2\right)^{1/2} = 1.9783 < 4$$

The fuzzy rule database, which is obtained using the information from the two previous equilibrium points, is resolved as follows:

There are two rules for the description of the above two different equilibrium situations:

Rule 1

if node $C(x_1)$ is mf1 *and* node $C(x_2)$ is mf1 *and* node $C(x_1(t+\Delta T))$ is mf1 *and* node $C(x_2(t+\Delta T))$ is mf1

then $w_{C(T)C(x_1(t+\Delta T))}$ is mf1 *and* $w_{C(x_1)C(x_1(t+\Delta T))}$ is mf1 *and* $w_{C(x_2)C(x_1(t+\Delta T))}$ is mf1 *and* $w_{C(T)C(x_2(t+\Delta T))}$ is mf1 *and* $w_{C(x_1)C(x_2(t+\Delta T))}$ is mf1 *and* $w_{C(x_2)C(x_2(t+\Delta T))}$ is mf1

Rule 2

if node $C(x_1)$ is mf2 *and* node $C(x_2)$ is mf2 *and* node $C(x_1(t+\Delta T))$ is mf2 *and* node $C(x_2(t+\Delta T))$ is mf2

then $w_{C(T)C(x_1(t+\Delta T))}$ is mf2 *and* $w_{C(x_1)C(x_1(t+\Delta T))}$ is mf2 *and* $w_{C(x_2)C(x_1(t+\Delta T))}$ is mf2 *and* $w_{C(T)C(x_2(t+\Delta T))}$ is mf2 *and* $w_{C(x_1)C(x_2(t+\Delta T))}$ is mf2 *and* $w_{C(x_2)C(x_2(t+\Delta T))}$ is mf2

where each memberhip function is deriving from the two equilibrium points aquired from matrices A, W, A' and W' and presented bellow:

$$A = \left[0.5\ 0.25_{mf1}\ 0.6250_{mf1}\ 0.2511_{mf1}\ 0.4056_{mf1}\right]^T$$

$$W = \begin{bmatrix} 0\ 0\ 0 & -1.0000_{mf1} & -0.4849_{mf1} \\ 0\ 0\ 0 & -0.8754_{mf1} & -0.0424_{mf1} \\ 0\ 0\ 0 & -1.0000_{mf1} & -0.8561_{mf1} \\ 0\ 0\ 0 & 1.0000 & 0 \\ 0\ 0\ 0 & 0 & 1.0000 \end{bmatrix}$$

$$A' = \left[0.7500\ 0.6250_{mf2}\ 0.2500_{mf2}\ 0.6247_{mf2}\ 0.2561_{mf2}\right]^T$$

$$W' = \begin{bmatrix} 0\ 0\ 0 & -0.1850_{mf2} & -1.0000_{mf2} \\ 0\ 0\ 0 & -0.0709_{mf2} & -0.8930_{mf2} \\ 0\ 0\ 0 & -0.2717_{mf2} & -0.0572_{mf2} \\ 0\ 0\ 0 & 1.0000 & 0 \\ 0\ 0\ 0 & 0 & 1.0000 \end{bmatrix}$$

The number and shape of the fuzzy membership functions of the variables of both sides of the rules are gradually modified as new desired equilibrium points appear to the system during its training. To add a new triangular membership function in the fuzzy description of a variable, the new value of the variable must differ from one already encountered value more than a specified threshold. Fig. 11 depicts this procedure. The initial value, a, of the variable determines the pick of the triangular

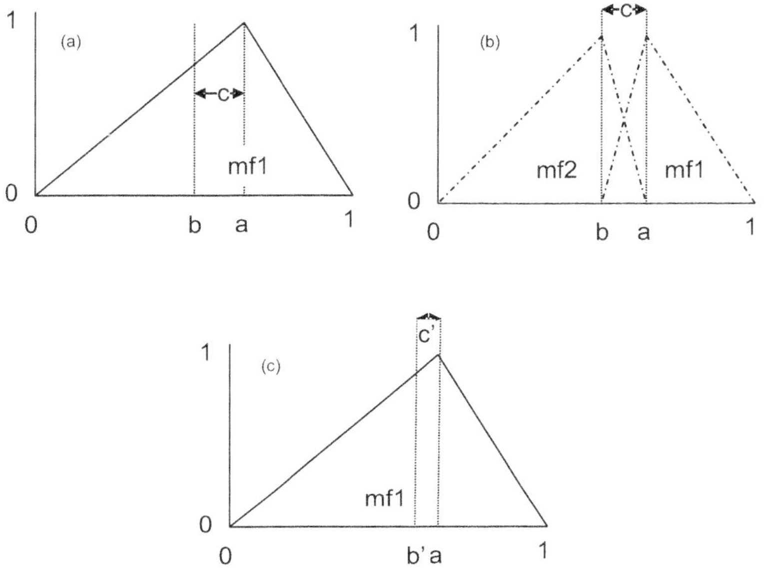

Fig. 11 Conditions for creating new triangular membership function. (a) Present triangular with pick corresponding to the value a of the variable, (b) $c > specified \quad threshold$ (creation of new triangular) and (c) $c < specified \quad threshold$ (non-creation of new triangular)

membership function ($mf1$) for the fuzzy description of the variable. When a new value, b, of the variable appears in a new equilibrium condition, it is compared with the previously encountered value a. If $|a-b|$ exceeds a specified threshold, c, then a new triangular membership function ($mf2$) is created and the initial triangular function ($mf1$) is modified as shown in fig. 11b. If $|a-b|$ does not exceed the threshold the initial fuzzy partition of the variable remains unchanged (fig. 11c). The threshold comes usually as a compromise between the maximum number of allowable rules and the detail in fuzzy representation of each variable. Instead of triangular membership functions one can use gaussian or trapezoidal memberhip functions to create the fuzzy rule database. The coefficient c in creating new triangular membership functions for nodes $C(x_1(t+\Delta T)))$, $C(x_2(t+\Delta T)))$ and for the six (6) weights in Inverted Pendulum example is set to $c = 0.01$. The maximum number of triangular membership functions that can appear during the off-line training are 100 for each node and 200 for each weight interconnection.

The created fuzzy memerbship functions of the fuzzy rule databases *Rule1* and *Rule2* are depicted in fig. 12 and 13.

5.1.6 Control of the Inverted Pendulum

After the FCN has been trained and the knowledge acquired has been stored in the Fuzzy Rule Database, it can be used to control the behaviour of the Inverted

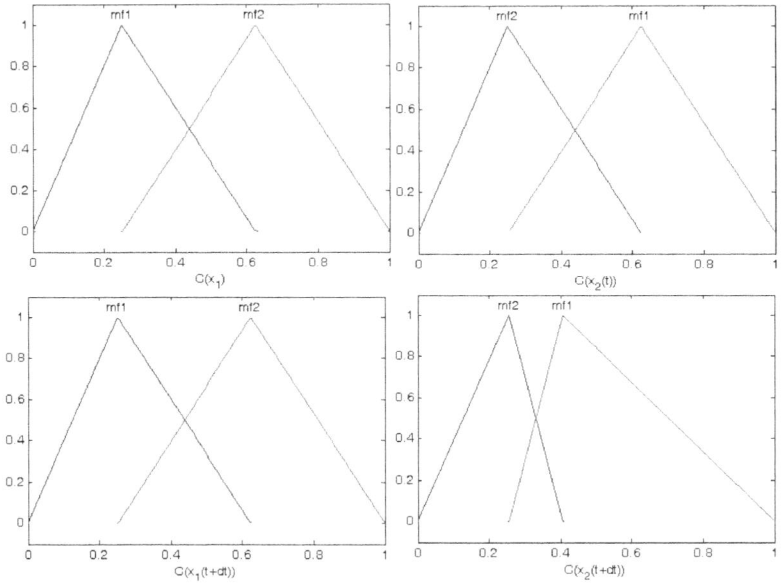

Fig. 12 Left hand side (if-part)

Pendulum. The FCN can be used either to drive the pole of the pendulum to track a desired state or simply to balance the pole around the vertical position. The control of the Pendulum is based only in the initial off-line training described above. Suppose that the desired state of the FCN is:

$$A = \left[\, \tilde{T}\ A(x_1)\ A(x_2)\ A^{des}(x_1(t+\Delta T))\ A^{des}(x_2(t+\Delta T)) \,\right]^T$$

From the Fuzzy Rule database which is of the form:

if node $C(x_1)$ is $A(x_1)$ *and* node $C(x_2)$ is $A(x_2)$ *and* node $C(x_1(t+\Delta T))$ is $A^{des}(x_1(t+\Delta T))$ *and* node $C(x_2(t+\Delta T))$ is $A^{des}(x_2(t+\Delta T))$

then $w_{C(T)C(x_1(t+\Delta T))}$ is $w1$ *and* $w_{C(x_1)C(x_1(t+\Delta T))}$ is $w2$ *and* $w_{C(x_2)C(x_1(t+\Delta T))}$ is $w3$ *and* $w_{C(T)C(x_2(t+\Delta T))}$ is $w4$ *and* $w_{C(x_1)C(x_2(t+\Delta T))}$ is $w5$ *and* $w_{C(x_2)C(x_2(t+\Delta T))}$ is $w6$

one obtains the values of the weights using convetional fuzzy inference and defuzzification procedure.

Since $C(T)$, $C(x_1)$ and $C(x_2)$ are steady nodes, then for the FCN nodes we have that:

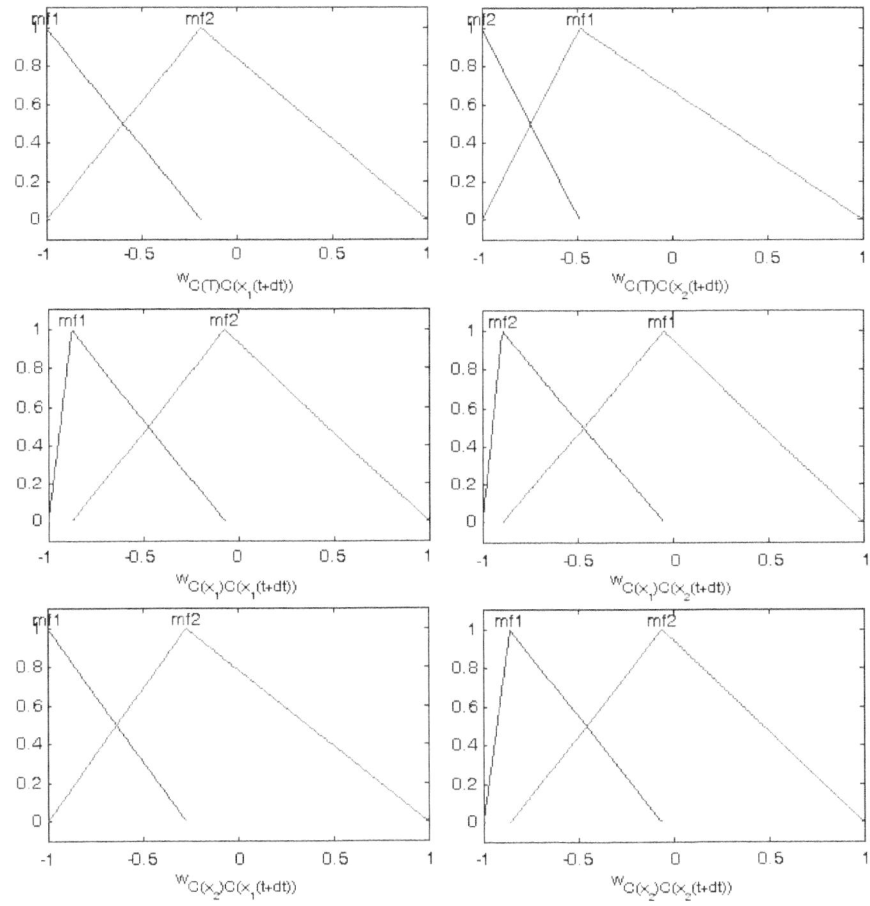

Fig. 13 Right hand side (then-part)

$$
\begin{aligned}
&\tilde{T} = \tilde{T} \\
&A(x_1) = A(x_1) \\
&A(x_2) = A(x_2) \\
&A^{des}(x_1(t+\Delta T)) = f\left(w1 * \tilde{T} + w2 * A(x_1) + w3 * A(x_2) + A^{des}(x_1(t+\Delta T))\right) \\
&A^{des}(x_2(t+\Delta T)) = f\left(w4 * \tilde{T} + w5 * A(x_1) + w6 * A(x_2) + A^{des}(x_2(t+\Delta T))\right)
\end{aligned}
$$

which can be rewritten:

$$
\begin{aligned}
f^{-1}\left(A^{des}(x_1(t+\Delta T))\right) &= w1 * \tilde{T} + w2 * A(x_1) + w3 * A(x_2) + A^{des}(x_1(t+\Delta T)) \\
f^{-1}\left(A^{des}(x_2(t+\Delta T))\right) &= w4 * \tilde{T} + w5 * A(x_1) + w6 * A(x_2) + A^{des}(x_2(t+\Delta T))
\end{aligned}
$$

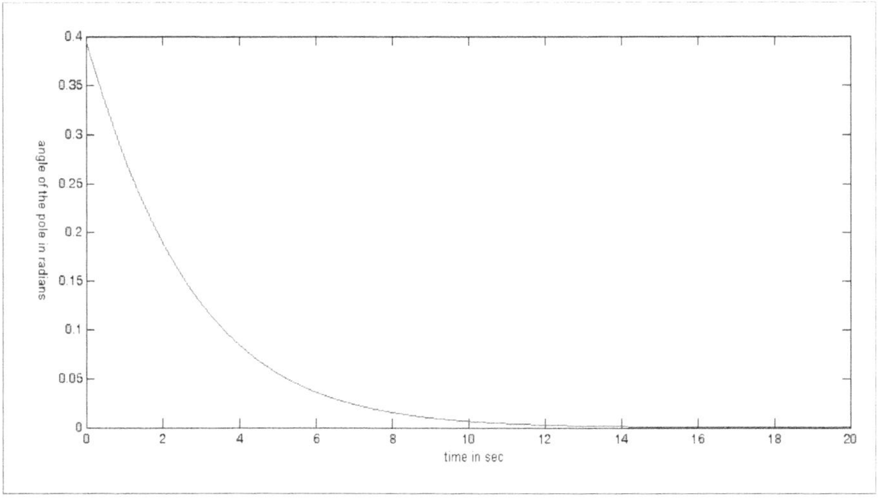

Fig. 14 State $x_1(t)$ for initial condition $x_1(0) = pi/8, x_2(0) = pi/8$

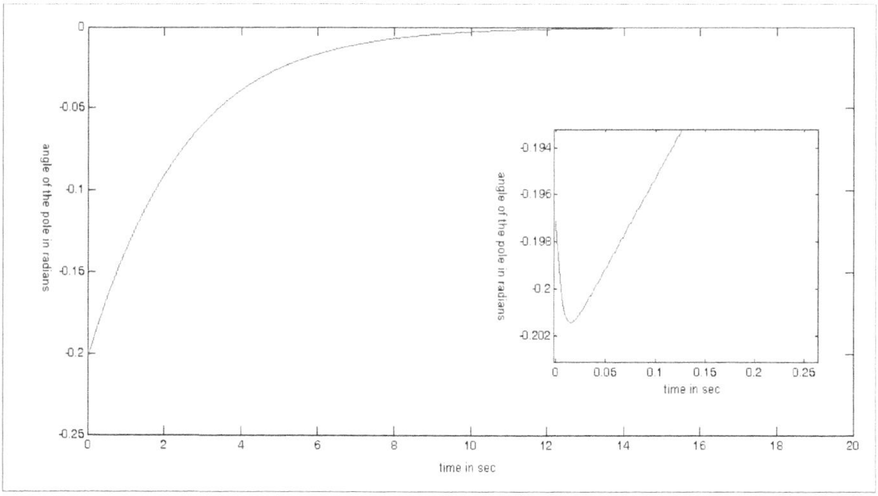

Fig. 15 State $x_1(t)$ for initial condition $x_1(0) = -pi/16, x_2(0) = -pi/4$

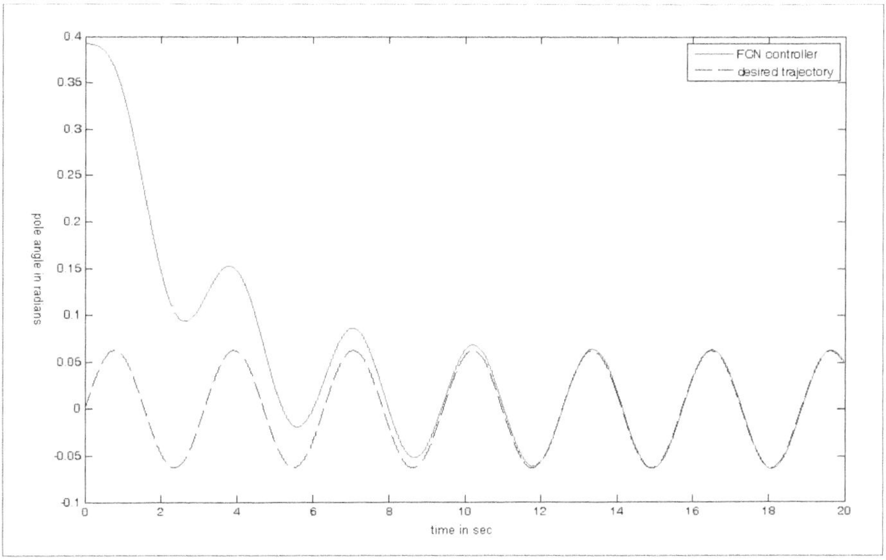

Fig. 16 State $x_1(t)$ and its desired value $\frac{sin(2*t)}{16}$ for initial condition $x_1(0) = pi/8, x_2(0) = 0$

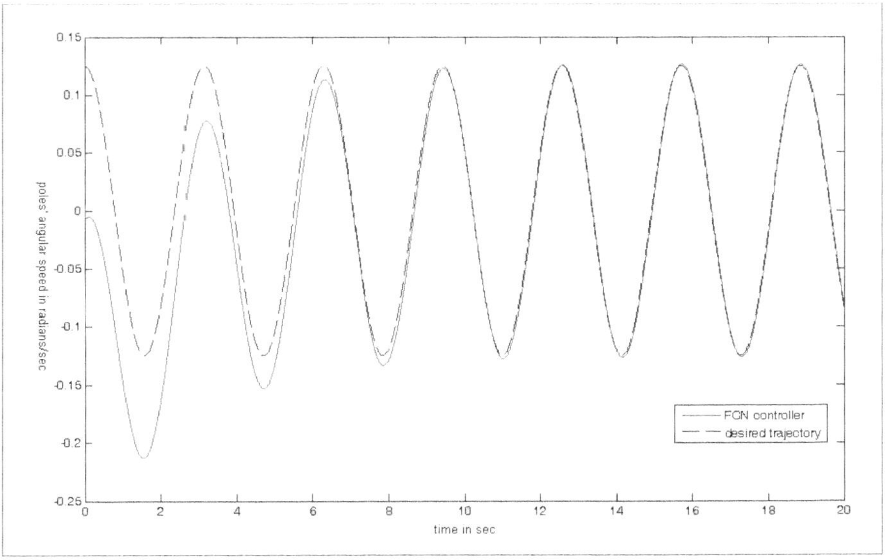

Fig. 17 State $x_2(t)$ and its desired value $\frac{cos(2*t)}{8}$ for initial condition $x_1(0) = pi/8, x_2(0) = 0$

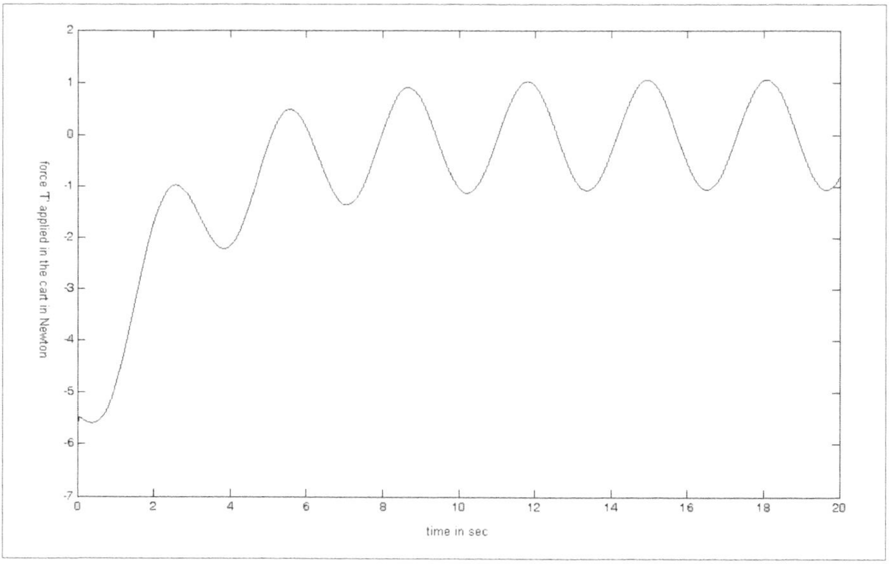

Fig. 18 Force T applied in the cart for tracking state $x_1(t)$ in the desired value $\frac{sin(2*t)}{16}$ for initial condition $x_1(0) = pi/8, x_2(0) = 0$

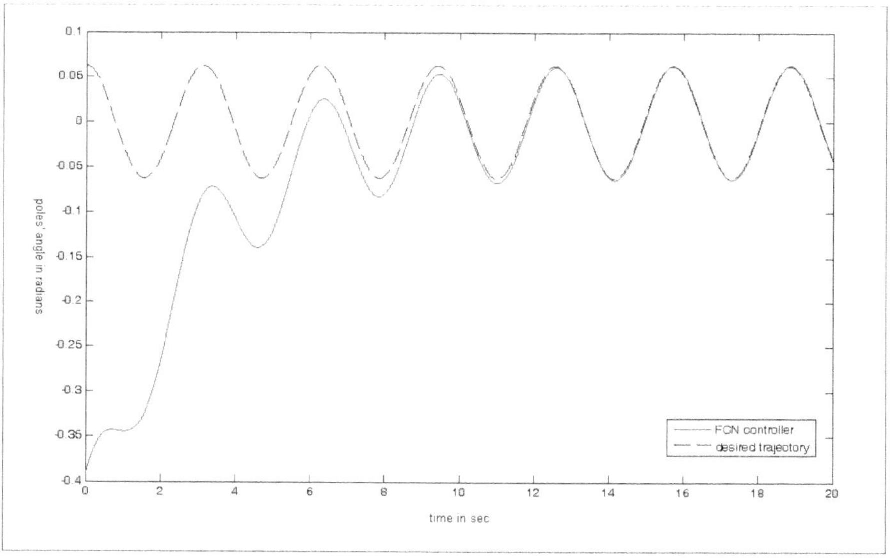

Fig. 19 State $x_1(t)$ and its desired value $\frac{cos(2*t)}{16}$ for initial condition $x_1(0) = -pi/8, x_2(0) = pi/16$

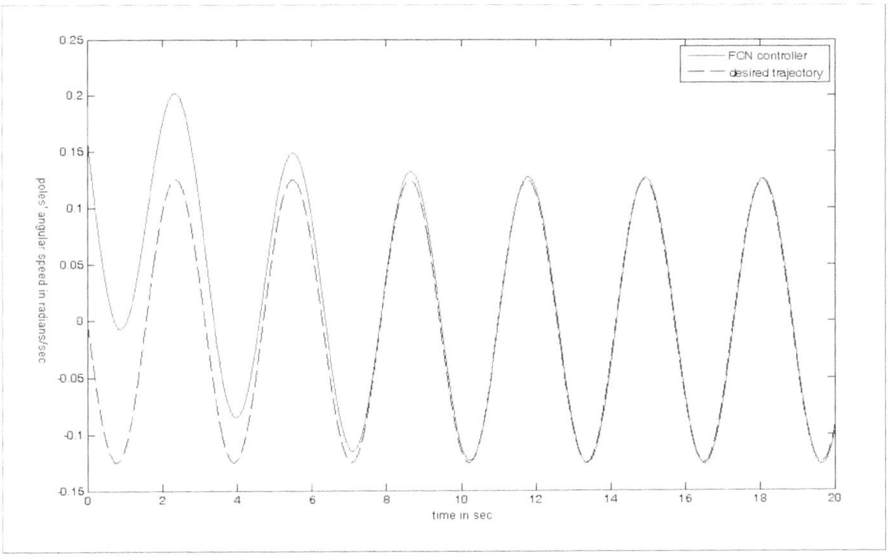

Fig. 20 State $x_2(t)$ and its desired value $-\frac{sin(2*t)}{8}$ for initial condition $x_1(0) = -pi/8, x_2(0) = pi/16$

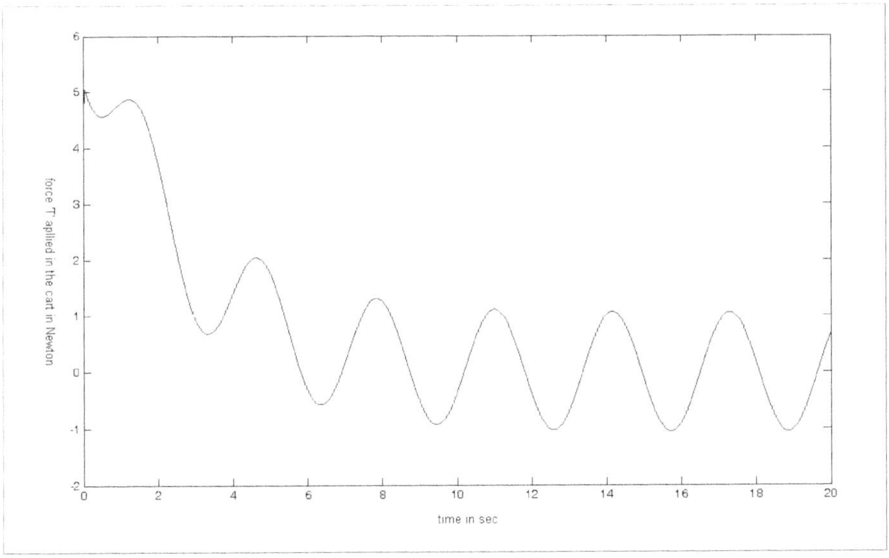

Fig. 21 Force T applied in the cart for tracking state $x_1(t)$ in the desired value $\frac{cos(2*t)}{16}$ for initial condition $x_1(0) = -pi/8, x_2(0) = pi/16$

where f is the sigmoid funtion $f(x) = \frac{1}{1+\exp^{-x}}$, $A(x_1)$ is the state of $x_1(t)$ at time t, $A(x_2)$ is the state of $x_2(t)$ at time t, $w1$, $w2$, $w3$, $w4$, $w5$, $w6$ are the estimated FCN weight values, $A^{des}(x_1(t+\Delta T))$, $A^{des}(x_2(t+\Delta T))$ are the desired state of the pole of the Inverted Pendulum at time $(t+\Delta T)$ and $\tilde{T}$ is the unknown force applied to the cart.

We can now estimate the force $\tilde{T}$ using the following formula:

$$\begin{aligned}\tilde{T} = (f^{-1}\left(A^{des}(x_1(t+\Delta T))\right) + f^{-1}\left(A^{des}(x_2(t+\Delta T))\right)\\ -w2 * A(x_1) - w3 * A(x_2) - A^{des}(x_1(t+\Delta T))\\ -w5 * A(x_1) - w6 * A(x_2) - A^{des}(x_2(t+\Delta T)))/(w1+w4)\end{aligned} \tag{41}$$

from eq. 41 all parameters are known, so the force $\tilde{T}$ is estimated in the interval $[01]$ and after it is converted in the interval $[-30 \quad 30]$ it is applied as input to the cart. It should be mentioned that eq. (41) does not guarantee that the estimated $\tilde{T}$ always fullfils (40). However, during the simulations caried out there was not any violation of (40). In case this happens one could possibly take some corrective action and modify $\tilde{T}$ accordingly.

Figures 14 and 15 show the balancing of the pole around the vertical position when the initial state values are $x_1(0) = pi/8$, $x_2(0) = pi/8$ and $x_1(0) = -pi/16$, $x_2(0) = -pi/4$ respectively. The ideal smooth convergence(without oscilations) of the state $x_1(t)$ to zero is due to the fact that in our control paradigm no restrictions are inposed regarding the position of the cart. Moreover due to simulation ideal force values can be applied at each sampling time, something which is not realistic in a real life experiment.

Figs. 16 and 17 show the tracking paradigm, where the pole starting from $x_1(0) = pi/8$, $x_2(0) = 0$ is controlled to follow state trajectories produced by $x_1^{des}(t) = (1/16)sin(2*t)$ and $x_2^{des}(t) = (1/8)cos(2*t)$, while fig. 18 impresses the force T apllied in the cart.

Similarly, fig. 19 and 20 show the tracking paradigm, where the pole starting from $x_1(0) = -pi/8$, $x_2(0) = pi/16$ is controlled to follow state trajectories produced by $x_1^{des}(t) = (1/16)cos(2*t)$ and $x_2^{des}(t) = -(1/8)sin(2*t)$, while fig. 21 impresses the force T apllied in the cart.

It can be observed that the proposed technique performs very well.

5.2 *Using FCN to Control an Anaerobic Digestion Process*

5.2.1 Description of the Plant and the FCN Graph

This chapter is based on a real experimental data. The pilot plant is a UASB reactor. The system was equipped with pH and temperature control. The reactor was used for methanization of pre-acidified food industry wastewater and operated continuously at increasing volumetric organic loading rate by increasing wastewater flowrate. The experimental unit is presented in fig. 22. The original wastewater (diluted peach pulp) was acidified using a Continuous Stirred Tank Reactor (CSTR) at Hydraulic Residence Time (HRT) equal to 6-8h. The pilot plant was monitored daily

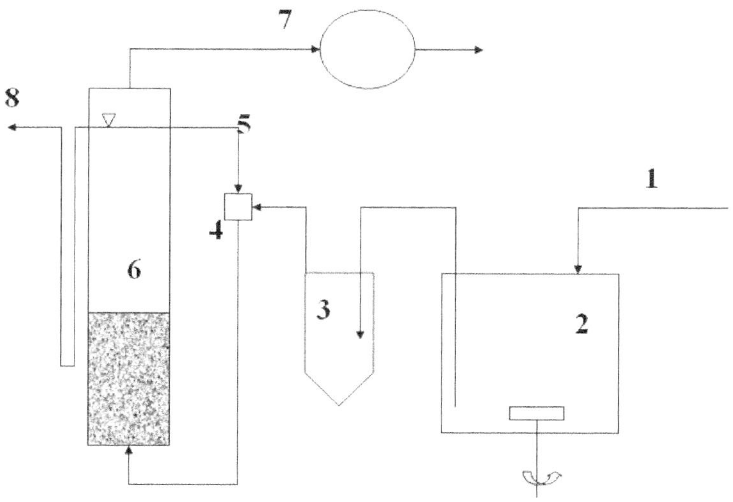

Fig. 22 Schematic representation of the pilot plant used for anaerobic wastewater treatment: 1) Raw wastewater, 2) Acidification tank, 3) Sedimentation tank, 4) pH conditioning tank, 5) Recycle stream, 6) UASB reactor, 7) Biogas measurement and analysis, 8) Treated effluent

for biogas and methane production, pH and temperature. Additionally, at steady state conditions samples were obtained from the influent and effluent of the reactor and analyzed for total and soluble COD, ethanol, acetic, propionic and butyric acid.

The graph shown in fig. 23 represents a Fuzzy Cognitive Network for the anaerobic digestion process [61]. This graph was produced based on the experience of the authors gained from the operation of the experimental unit. The graph has 11 nodes, where nodes C1, C2, C3 and C11 stand for wastewater flowrate, reactor pH, reactor temperature, and reactor Volatile Suspended Solids (VSS/SB) concentration, respectively. These nodes are steady value and at the same time control nodes, since a change of their values affects the values of the output nodes. Nodes C4, C5, C6, C7, C8, C9 and C10 are output nodes representing HRT, non-soluble COD, soluble COD, Ethanol, butyric acid, acetic acid and methane, respectively.

The FCN is trained using experimental data from the pilot plant. Details regarding the data gathering procedure can be also found in [61]. Its operating condition is accosiated throught the weight adaptation mechanism with a weight matrix. This association is stored in the fuzzy rule database using the mechanism described in the previous subsections. The fuzzy rules used are of the form:

if node 1 is mf1 and node 2 is mf1 and node 3 is mf1 and node 11 is mf1
then $W_{1,5}$ is mf1 and $W_{1,6}$ is mf1 and $W_{1,7}$ is mf1 and $W_{1,8}$ is mf1 and $W_{1,9}$ is mf1

and $W_{1,10}$ is mf1 and $W_{2,5}$ is mf1 and $W_{2,6}$ is mf1 and $W_{2,7}$ is mf1 and $W_{2,8}$ is mf1 and $W_{2,9}$ is mf1 and $W_{2,10}$ is mf1 and $W_{3,5}$ is mf1 and $W_{3,6}$ is mf1 and $W_{3,7}$ is mf1 and $W_{3,8}$ is mf1 and $W_{3,9}$ is mf1 and $W_{3,10}$ is mf1 and $W_{4,5}$ is mf1 and $W_{4,6}$ is mf1 and $W_{4,7}$ is mf1 and $W_{4,8}$ is mf1 and $W_{4,9}$ is mf1 and $W_{4,10}$ is mf1 and $W_{5,4}$ is mf1 and $W_{5,6}$ is mf1 and $W_{6,7}$ is mf1 and $W_{6,8}$ is mf1 and $W_{6,9}$ is mf1 and $W_{7,9}$ is mf1 and $W_{8,9}$ is mf1 and $W_{9,10}$ is mf1and $W_{11,5}$ is mf1 and $W_{11,6}$ is mf1 and $W_{11,7}$ is mf1 and $W_{11,8}$ is mf1 and $W_{11,9}4$ is mf1 and $W_{11,10}$ is mf1

while matrix W of the FCN of fig. 23 is:

$$W = \begin{bmatrix} 0\,0\,0 & 0 & W_{1,5} & W_{1,6} & W_{1,7} & W_{1,8} & W_{1,9} & W_{1,10} & 0 \\ 0\,0\,0 & 0 & W_{2,5} & W_{2,6} & W_{2,7} & W_{2,8} & W_{2,9} & W_{2,10} & 0 \\ 0\,0\,0 & 0 & W_{3,5} & W_{3,6} & W_{3,7} & W_{3,8} & W_{3,9} & W_{3,10} & 0 \\ 0\,0\,0 & 0 & W_{4,5} & W_{4,6} & W_{4,7} & W_{4,8} & W_{4,9} & W_{4,10} & 0 \\ 0\,0\,0 & W_{5,4} & 0 & W_{5,6} & 0 & 0 & 0 & 0 & 0 \\ 0\,0\,0 & 0 & 0 & 0 & W_{6,7} & W_{6,8} & W_{6,9} & 0 & 0 \\ 0\,0\,0 & 0 & 0 & 0 & 0 & 0 & W_{7,9} & 0 & 0 \\ 0\,0\,0 & 0 & 0 & 0 & 0 & 0 & W_{8,9} & 0 & 0 \\ 0\,0\,0 & 0 & 0 & 0 & 0 & 0 & 0 & W_{9,10} & 0 \\ 0\,0\,0 & 0 & 0 & 0 & 0 & 0 & 0 & 0 & 0 \\ 0\,0\,0 & 0 & W_{11,5} & W_{11,6} & W_{11,7} & W_{11,8} & W_{11,9} & W_{11,10} & 0 \end{bmatrix}$$

we observe that by using this form the storage is not complete, in the sence that information related to nodes $C4$ to $C10$ is not stored in the fuzzy rule database. Therefore we can not use a procedure similar to eq. (41) in order to reversly obtain the control inputs. In this case we propose an alternative mechanism, which is described in the following subsection.

5.2.2 Control of the Process Using the FCN

Once the FCN has been trained using experimental data, it is capable to adjust the values of the control nodes in order to drive the real system to a new desired equilibrium point The control mechanism is described bellow. Suppose that the system is in a specific equilibrium point with nodes values given from the next D vector:

$$D = \begin{bmatrix} D_1 & D_2 & D_3 & D_4 & D_5 & D_6 & D_7 & D_8 & D_9 & D_{10} & D_{11} \end{bmatrix}$$

In that specific point the designer-engineer demands to move the real system to a new equilibrium point, which is different from the one that the system already has. Once the designer -engineer determines the desired values of the node or the nodes, the control system must decide the values of the control nodes (C1, C2, C3 and C11) in order to drive the real system to the desired equilibrium point. Suppose, for example, that the designer-engineer determines as desired value of node C10 the value A_{10}^{des}. The error for the value of node C10 is:

$$p_{10} = A_{10}^{des} - \frac{1}{1+e^{-\left(\sum\limits_{i=1, i\neq 10}^{N} A_i^{system} Wij + A_{10}^{system}\right)}}$$

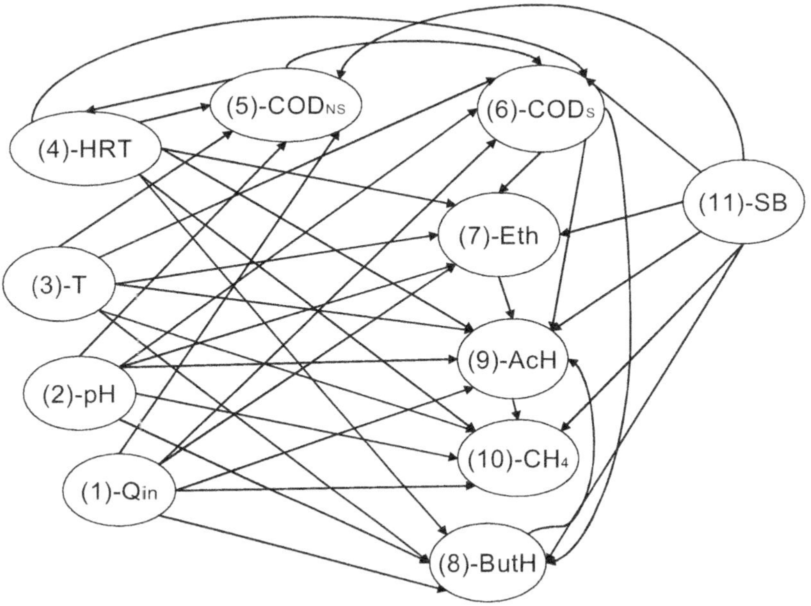

Fig. 23 The FCN designed for the control of the Anaerobic Digestion Process

where A_i^{system} are the values of the variables of the real system corresponding to the respective FCN nodes. These values are taken by using a continous interaction with the real system as depicted in fig. 24. By taking the partial derivative of the above equation in respect to the control node values the following delta rule is derived, which determines the required change of the control nodes' values:

$$A_1^k = A_1^{k-1} + p_{10}(1 - p_{10})W_{1,10}$$

$$A_2^k = A_2^{k-1} + p_{10}(1 - p_{10})W_{2,10}$$

$$A_3^k = A_3^{k-1} + p_{10}(1 - p_{10})W_{3,10}$$

$$A_{11}^k = A_{11}^{k-1} + p_{10}(1 - p_{10})W_{11,10}$$

In a more general form the above equations can be rewritten as follows:

$$p_j = A_j^{desired} - \frac{1}{1 + e^{-\left(\sum_{i=1,i\neq j}^{N} A_i^{system} Wij + A_j^{system}\right)}} \tag{42}$$

$$A_i^{k,control} = A_i^{k-1,control} + p_j(1 - p_j)W_{ij} \tag{43}$$

where the 'control' superscript is used to indicate that the updating is performed only to the control nodes.

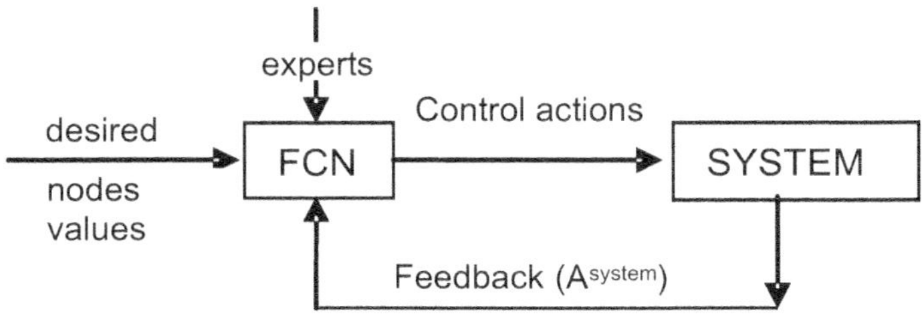

Fig. 24 FCN in close interaction with the system

By using the fuzzy rule database of the already trained FCN we calculate the values of the interconnections related to the new control nodes' values. We repetitively apply eqs 42 and 43 in order to minimize the error. Once the error reaches zero (actually becomes sufficiently small), the FCN control mechanism is sending the new control nodes' values to the physical system. When the real system is triggered by the control values it returns feedback from the measurable nodes values. In case the feedback value of node C10 is not the desired one, this means that the FCN is facing an operational condition not encountered during its training stage. It can enrich its knowledge by using the mechanism described in the previous section. First, eqs 26 and 36 are repetitively executed in order to adjust the FCN weights, which in turn reflect the new operational knowledge for the FCN. Next, the fuzzy rule database is updated according to the procedure described in the previous section so that it incorporates the new acquired knowledge. A pictorial representation of the above procedure is given in fig. 25.

The method presented here presents some advantages in comparison to traditional control methods. The most important innovation of the FCN control mechanism is that, it does not require any mathematical knowledge for the description of the real system. Actually FCNs combine the knowledge of experts and the operational knowledge (data) derived from the operation of the system. The experts' knowledge is mostly used to construct the cognitive graph and probably to give initial weight sets. Experimental data from the real system are used to enrich the knowledge of the FCN on the system's operating conditions. Moreover, during its set points control actions, the FCN adapts its knowledge. Therefore the proposed control mechanism is an adaptive control scheme.

5.2.3 Results of Anaerobic Control

To test the proposed methodology we used data obtained from the operation of the experimental anaerobic digestion unit of the laboratory of wastewater management and treatment technologies of DUTH. For various representative values of the control nodes, the resulting steady state values of the other nodes were measured. The steady state values are the values measured at reasonable time intervals (usually hours) after the application of the new control values to the test unit. The 11 node values obtained this way form a vector representing a real operational condition of

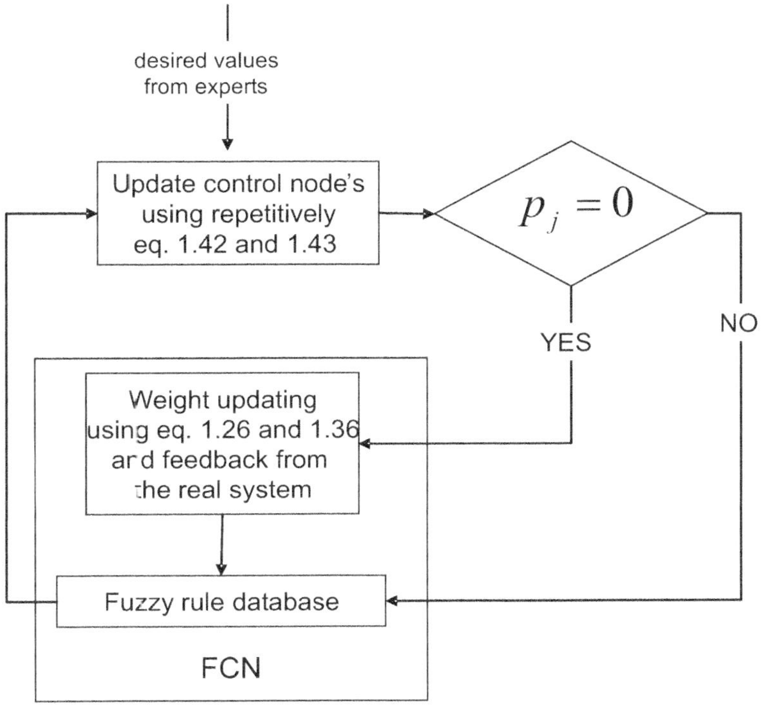

Fig. 25 Control structure in order to achieve the desired equilibrium point defined from the experts

the system and can be used to train or test the FCN. A number of 142 such data vectors was experimentally obtained. 98 of these data vectors are selected to initially train the FCN using the procedure described in section II. The rest 44 data vectors are used to test the generalization ability of the trained FCN.

Fig. 26 shows 28 of the 44 test data values of the three control nodes Q_in, T and pH. These values were selected randomly. However, they are arranged and displayed in respect to the time (in days) they were measured. Fig. 27 shows the production of CH_4 (node 10) when the above control node values are imposed on the real system and on the FCN alone respectively.

The dotted spots represent the measured CH_4 values, while the solid line connects the values of CH_4, which are estimated by the FCN. It can be observed that the estimation error is relatively small having a mean value of 7.4%. This error is expected to be much lower when more and more densely collected training data are used. However, Fig. 27 clearly demonstrates that the FCN can provide with relatively accurate results even for operational conditions it has not been taught about.

The above procedure is testing only the approximation and generalization ability of the FCN. To test the proposed control mechanism introduced in the previous

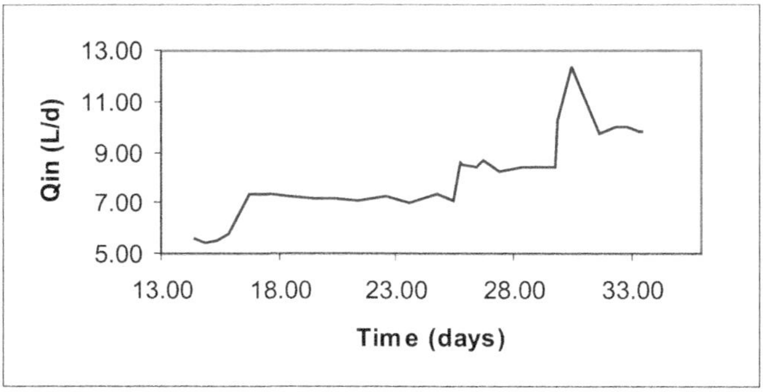

A

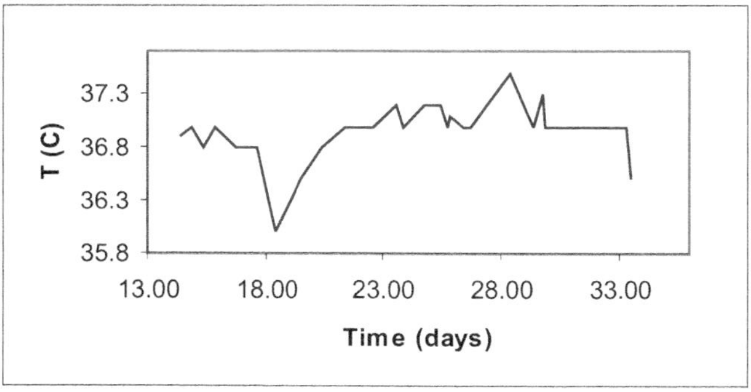

B

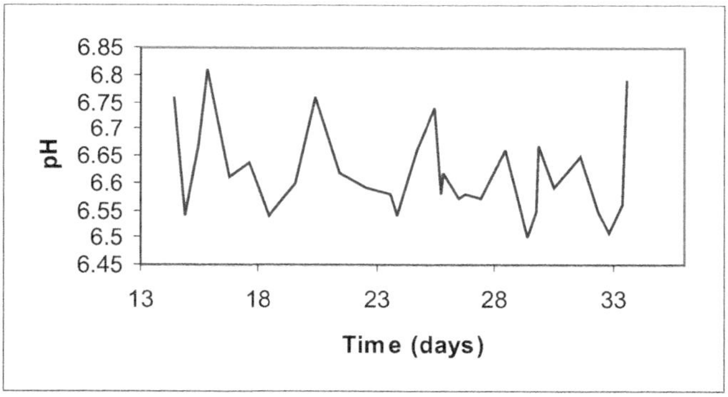

C

Fig. 26 A part of the experimental data used to test FCN a) Qin: inflow to the UASB reactor, b) T: reactor temperature, c) pH: reactor pH

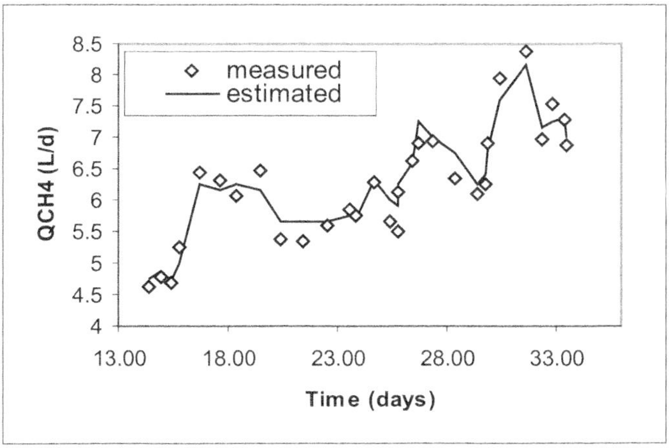

Fig. 27 A comparison between estimated and measured QCH4 values for the experimental anaerobic digestion process

section we perform the following experiment. Suppose we want to regulate Q_in, T and pH in order to change the value of Q_{CH_4}. Let us select the time instant 33,52d, where according to our measurements Q_{CH_4} is equal to value 6,87158. At that specific time the values of control nodes are:

Node 1(Q_{in}): 9,85, node 2(pH): 6,79, node 3(T): 36,5 and node 11(SB): 0,5.

Suppose now, that we require raising Q_{CH_4} to 8,072414 (this value is one of the already measured experimental values, not used for training) in a relatively small time interval. The latter implies that control node 11 can not actually affect the process and therefore we should rely only in changes on the other three control nodes. The control mechanism must regulate the values of control nodes (Q_in, T and pH) in order to reach the desired value of Q_{CH_4}. By applying eq. 42, with the desired value of Q_{CH_4}, we calculate p_{10}, which is equal to 1,200834. For node 11 (SB) there is no need to change its value because we assumed that it does not change its value. Applying p_{10} to eq. 43 (with i=1,2,3 and i $\neq$ 11) the FCN regulates, according to the procedure described in the previous section, the values of nodes 1, 2 and 3 to the values:

Node 1(Q_{in}): 11,79, node 2(pH): 6,60, node 3(T): 36,7 and node 11(SB): 0,5.

It has to be noted here that the experimental data for nodes 1, 2, 3 and 11 associated with Q_{CH_4} = 8,072414 are:

Node 1(Q_{in}): 11,7835, node 2(pH): 6,5912, node 3(T): 36,727 and node 11(SB): 0,5.

Therefore, the control procedure provides with realistic estimations for the desired values of the control nodes According to the proposed control mechanism, these values have to be applied on the real system, which in turn will provide with feedback value for Q_{CH_4} and other measured node values. If Q_{CH_4} does not meet the desired goal the control mechanism proceeds in updating the knowledge of the FCN.

6 Conclusions and Discussion

This chapter presented an alternative use of Fuzzy Cognitive Maps. Traditionally FCMs have been used to model the behavior of complex systems appearing in socio-economical and political problems. The formation of the cognitive graph is mainly based on experts' opinion regarding the interacting concepts of the system. One characteristic of the operation of an FCM is its diversity regarding its steady state behavior, being likely to appear equilibrium, limit cycle or even chaotic behavior. This diversity is however undesirable if FCMs are to be used to model and control traditional engineering systems. In this case the equilibrium after an initial perturbation is a desirable property. To this end, the authors proposed the Fuzzy Cognitive Networks (FCN), which constitute an operational extension of Fuzzy Cognitive Maps (FCM).

The initial graph of the FCN also depends on experts' opinion. However, the interconnecting weights are estimated based on historical data from the operation of the physical system. FCNs are equiped with weight estimation algorithms guaranteeing that they always reach equilibrium points during their operation. Moreover, they are in continuous interaction with the system they describe and may be used to control it. It was shown that FCNs equipped with continuous differentiable sigmoid functions having contractive or at least non expansive properties can always converge to unique equilibrium points, provided that some condition on the weight set is met. For the estimation of the weights corresponding to a new equilibrium point an adaptive estimation algorithm is developed and proofs for its stability and convergence properties are given. The algorithm takes into account the convergence conditions and incorporates them in parameter projection schemes. Fuzzy Cognitive Network (FCN) working in close interaction with the physical system it describes, stores information related to different operational points using fuzzy *meta rules* of the form "If weights then equilibrium point". This information can be used in determining subsequent control actions. Two alternative control paradigms were presented, one in simulations of a traditional benchmark control problem and the other for the control of a pilot unaerobic digestion unit, both of them demonstrating the very good performance of the proposed control approaches.

One interesting extension of the work presented here would be to devise algorithms which adapt not only the weights of the FCN but also the parameters of the sigmoids. This would enhance the flexibility of the FCN in capturing a large variety of operational points of the physical system it describes. Another point that needs also special consideration is its storage mechanism. As it is currently designed, the FCN requires a fuzzy rule database for storing previously acquired knowledge from the operation of the system. This data base may become quite large in demanding applications. Therefore, techniques that will reduce the storage requirements or replace the fuzzy rule database by another more efficient representation is a topic of future research.

References

1. Kosko, B.: Fuzzy Cognitive Maps. International Journal of Man-Machine Studies, 65–75 (1986)
2. Axelrod, R.: Structure of Decision. In: The Cognitive Maps of Political Elites, Princeton University Press, New Jersey (1976)
3. Stylios, C., Groumpos, P.: Fuzzy Cognitive Maps in Modelling Supervisory Control Systems. Journal of Intelligent and Fuzzy Systems 8, 83–98 (2000)
4. Stylios, C., Groumpos, P.: A soft computing approach for modelling the supervisor of manufacturing systems. Journal of Intelligent and Robotics Systems 26(34), 389–403 (1999)
5. Stylios, C., Groumpos, P., Georgopoulos, V.: A Fuzzy Cognitive Maps approach to process control systems. Journal of Intelligent and Robotics Systems 26(34), 389–403 (1999)
6. Schneider, M., Shnaider, E., Kandel, A., Chew, G.: Constructing fuzzy cognitive maps. In: International Joint Conference of the Fourth IEEE International Conference on Fuzzy Systems and The Second International Fuzzy Engineering Symposium, vol. 4(1), pp. 2281–2288 (1995)
7. Kosko, B.: Differential Hebbian Learning. In: Proceedings American Institute of Physics, Neural Networks for Computing, pp. 277–282 (1986)
8. Craiger, P., Coovert, M.D.: Modeling dynamic social and psychological processes with fuzzy cognitive maps. In: IEEE World Congress on Computational Intelligence and Fuzzy Systems, vol. 3, pp. 1873–1877 (1994)
9. Tsadiras, A., Kouskouvelis, I.: Using Fuzzy Cognitive Maps as a Decision Support System for Political Decisions: The Case of Turkey's Integration into the European Union. In: Bozanis, P., Houstis, E.N. (eds.) PCI 2005. LNCS, vol. 3746, pp. 371–381. Springer, Heidelberg (2005)
10. Koulouriotis, D.E., Diakoulakis, I.E., Emiris, D.M.: A fuzzy cognitive map-based stock market model: synthesis, analysis and experimental results. In: 10th IEEE International Conference on Fuzzy Systems, pp. 465–468 (2001)
11. Carvalho, J.P., Tome, J.A.B.: Qualitative modelling of an economic system using rule-based fuzzy cognitive maps. In: IEEE International Conference on Fuzzy Systems, vol. 2, pp. 659–664 (2004)
12. Xirogiannis, G., Glykas, M.: Fuzzy Cognitive Maps in Business Analysis and Performance Driven Change. IEEE Transactions on Engineering Management 51(3), 334–351 (2004)
13. Glykas, M., Xirogiannis, G.: A soft knowledge modeling approach for geographically dispersed financial organizations. Soft Computing 9(8), 579–593 (2005)
14. Xirogiannis, G., Glykas, M.: Intelligent Modeling of e-Business Maturity. Expert Systems with Applications 32(2), 687–702 (2007)
15. Xirogiannis, G., Chytas, P., Glykas, M., Valiris, G.: Intelligent impact assessment of HRM to the shareholder value. Expert Systems with Applications 35(4), 2017–2031 (2008)
16. Kottas, T., Boutalis, Y., Devedzic, G., Mertzios, B.: A new method for reaching equilibrium points in Fuzzy Cognitive Maps. In: Proceedings of 2nd International IEEE Conference of Intelligent Systems, Varna Burgaria, pp. 53–60 (2004)

17. Georgopoulos, V., Malandraki, G., Stylios, C.: A fuzzy cognitive map approach to differential diagnosis of specific language impairment. Artificial Intelligence in Medicine 29(3), 261–278 (2003)
18. Zhang, W., Chen, S., Bezdek, J.: Pool2: A Generic System for Cognitive Map Development and Decision Analysis. IEEE Transactions on Systems, Man, and Cybernetics 19(1), 31–39 (1989)
19. Satur, R., Liu, Z.-Q.: A contextual fuzzy cognitive map framework for geographic information systems. IEEE Transactions on Fuzzy Systems 7(5), 481–494 (1999)
20. Liu, Z.-Q., Satur, R.: Contextual fuzzy cognitive map for decision support in geographic information systems. IEEE Transactions on Fuzzy Systems 7(5), 495–507 (1999)
21. Satur, R., Liu, Z.-Q.: Contextual fuzzy cognitive maps for geographic information systems. In: IEEE International Conference on Fuzzy Systems, vol. 2, pp. 1165–1169 (1999)
22. Carvalho, J.P., Carola, M., Tome, J.A.B.: Using Rule-based Fuzzy Cognitive Maps to Model Dynamic Cell Behavior in Voronoi Based Cellular Automata. In: IEEE International Conference on Fuzzy Systems, pp. 1687–1694 (2006)
23. Papakostas, G., Boutalis, Y., Koulouriotis, D., Mertzios, B.: Fuzzy Cognitive Maps for Pattern Recognition Applications. International Journal at Pattern Recognition and Artificial Intelligence (2008) (in press)
24. Papakostas, G., Boutalis, Y., Koulouriotis, D., Mertzios, B.: A First Study of Pattern Classification using Fuzzy Cognitive Maps. In: International Conference on Systems, Signals and Image Processing - INSSIP 2006, pp. 369–374 (2006)
25. Stach, W., Kurgan, L.A., Pedrycz, W.: Numerical and Linguistic Prediction of Time Series With the Use of Fuzzy Cognitive Maps. IEEE Transactions on Fuzzy Systems 16(1), 61–72 (2008)
26. Silva, P.C.: Fuzzy cognitive maps over possible worlds. In: International Joint Conference of the Fourth IEEE International Conference on Fuzzy Systems and The Second International Fuzzy Engineering Symposium, vol. 2, pp. 555–560 (1995)
27. Dickerson, J.A., Kosko, B.: Virtual worlds as fuzzy cognitive maps. In: Virtual Reality Annual International Symposium, pp. 471–477 (1993)
28. Koulouriotis, D.E., Diakoulakis, I.E., Emiris, D.M.: Anamorphosis of fuzzy cognitive maps for operation in ambiguous and multi-stimulus real world environments. In: 10th IEEE International Conference on Fuzzy Systems, pp. 1156–1159 (2001)
29. Parenthoen, M., Reignier, P., Tisseau, J.: Put fuzzy cognitive maps to work in virtual worlds. In: 10th IEEE International Conference on Fuzzy Systems, vol. 1, pp. 252–255 (2001)
30. Xin, J., Dickerson, J.E., Dickerson, J.A.: Fuzzy feature extraction and visualization for intrusion detection. In: 12th IEEE International Conference on Fuzzy Systems, pp. 1249–1254 (2003)
31. Zhang, W., Chen, S., Wang, W., King, R.: A Cognitive Map Based Approach to the Coordination of Distributed Cooperative Agents. IEEE Transactions on Systems, Man, and Cybernetics 22(1), 103–114 (1992)
32. Hagiwara, M.: Extended fuzzy cognitive maps. In: IEEE International Conference on Fuzzy Systems, pp. 795–801 (1992)
33. Zhang, J.Y., Liu, Z.-Q., Zhou, S.: Quotient FCMs-a decomposition theory for fuzzy cognitive maps. IEEE Transactions on Fuzzy Systems 11(5), 593–604 (2003)
34. Zhang, B.Y., Liu, Z.-Q.: Quotient fuzzy cognitive maps. In: 10th IEEE International Conference on Fuzzy Systems, vol. 1, pp. 180–183 (2001)
35. Miao, Y., Liu, Z., Siew, C., Miao, C.: Dynamical Cogntive Network-an Extension of Fuzzy Cognitive Map. IEEE transactions on Fuzzy Systems 9(5), 760–770 (2001)

36. Zhang, J., Liu, Z.-Q., Zhou, S.: Dynamic Domination in Fuzzy Causal Networks. IEEE Transactions on Fuzzy Systems 14(1), 42–57 (2006)
37. Miao, Y., Liu, Z.-Q.: On causal inference in fuzzy cognitive maps. IEEE Transactions on Fuzzy Systems 8(1), 107–119 (2000)
38. Zhang, J.Y., Liu, Z.-Q.: Dynamic domination for fuzzy cognitive maps. In: IEEE International Conference on Fuzzy Systems, vol. 1, pp. 145–149 (2002)
39. Liu, Z.-Q., Miao, Y.: Fuzzy cognitive map and its causal inferences. In: IEEE International Conference on Fuzzy Systems, vol. 3, pp. 1540–1545 (1999)
40. Zhou, S., Liu, Z.-Q., Zhang, J.Y.: Fuzzy causal networks: general model, inference, and convergence. IEEE Transactions on Fuzzy Systems 14(3), 412–420 (2006)
41. Smarandache, F.: An Introduction to Neutrosophy, Neutrosophic Logic, Neutrosophic Set, and Neutrosophic Probability and Statistics. In: Proceedings of the First International Conference on Neutrosophy, Neutrosophic Logic, Neutrosophic Set, Neutrosophic Probability and Statistics, University of New Mexico - Gallup, (1-3), pp. 5–22 (2001)
42. Kandasamy, V., Smarandache, F.: Fuzzy Cognitive Maps and Neutrosophic Cognitive Maps. In: ProQuest Information & Learning, University of Microfilm International (2003)
43. Kottas, T.L., Boutalis, Y.S., Christodoulou, M.A.: Fuzzy Cognitive Networks: A General Framework. Inteligent Desicion Technologies 1(4), 183–196 (2007)
44. Huerga, A.: A Balanced Differential Learning algorithm in Fuzzy Cognitive Maps. In: Proc. of the Sixteenth Intern.Workshop on Qualitative Reasoning (2002) (poster)
45. Papageorgiou, E., Groumpos, P.: A weight adaptation method for Fuzzy Cognitive Maps to a process control problem. In: Budak, M., et al. (eds.) ICCS 2004. LNCS, vol. 3037, pp. 515–522. Springer, Heidelberg (2004)
46. Papageorgiou, E., Stylios, C., Groumpos, P.: Active Hebbian Learning Algorithm to train Fuzzy Cognitive Maps. International Journal of Approximate Reasoning 37(3), 219–247 (2004)
47. Aguilar, J.: Adaptive Random Fuzzy Cognitive Maps. In: Garijio, F.J., Riquelme, J.C., Toro, M. (eds.) IBERAMIA 2002. LNCS (LNAI), vol. 2527, pp. 402–410. Springer, Heidelberg (2002)
48. Stach, W., Kurgan, L.A., Pedrycz, W.: Data-driven nonlinear hebbian learning method for fuzzy cognitive maps. In: 2008 World Congress on Computational Intelligence, WCCI 2008 (2008)
49. Koulouriotis, D., Diakoulakis, I., Emiris, D.: Learning Fuzzy Cognitive Maps using evolution strategies: A novel schema for modeling a simulating high-level behavior. In: Proceedings of IEEE Congress on Evolutionary Computation, vol. 1, pp. 364–371 (2001)
50. Papageorgiou, E., Parsopoulos, K., Stylios, C., Groumpos, P., Vrahatis, M.: Fuzzy Cognitive Maps Learning Using Particle Swarm Optimization. International Journal of Intelligent Information Systems 25(1), 95–121 (2005)
51. Khan, M., Khor, S., Chong, A.: Fuzzy cognitive maps with genetic algorithm for goal-oriented decision support. Int. J. Uncertainty, Fuzziness and Knowledge-based Systems 12, 31–42 (2004)
52. Stach, W., Kurgan, L., Pedrycz, W., Reformat, M.: Evolutionary Development of Fuzzy Cognitive Maps. In: 14th IEEE International Conference on Fuzzy Systems, pp. 619–624 (2005)
53. Stach, W., Kurgan, L., Pedrycz, W., Reformat, M.: Genetic learning of fuzzy cognitive maps. Fuzzy sets and systems 153(3), 371–401 (2005)
54. Dickerson, J., Kosko, B.: Virtual worlds as Fuzzy Cognitive Maps. Presence 3(2), 173–189 (2006)
55. Kosko, B.: Fuzzy Engineering. Prentice Hall, Englewood Cliffs (1997)

56. Tsadiras, A.: Comparing the inference capabilities of binary, trivalent and sigmoid fuzzy cognitive maps. Information Science 178, 3880–3894 (2008)
57. Boutalis, Y., Kottas, T., Christodoulou, M.: On the Existence and Uniqueness of Solutions for the Concept Values in Fuzzy Cognitive Maps. In: Proceedings of 47th IEEE Conference on Decision and Control - CDC 2008, Cancun, Mexico, December 9-11, pp. 98–104 (2008)
58. Boutalis, Y., Kottas, T., Christodoulou, M.: Adaptive Estimation of Fuzzy Cognitive Maps With Proven Stability and Parameter Convergence. IEEE Transactions on Fuzzy Systems (2009), doi:10.1109TFUZZ, 2017519
59. Kottas, T., Boutalis, Y., Christodoulou, M.: A new method for weight updating in Fuzzy cognitive Maps using system Feedback. In: 2nd International Conference on Informatics in Control, Automation and Robotics (ICINCO 2005), Barcelona, Spain, pp. 202–209 (2005)
60. Kottas, T.L., Boutalis, Y.S., Karlis, A.D.: A New Maximum Power Point Tracker for PV Arrays Using Fuzzy Controller in Close Cooperation with Fuzzy Cognitive Networks. IEEE Transactions on Energy Conversion 21(3), 793–803 (2006)
61. Kottas, T., Boutalis, Y., Diamantis, V., Kosmidou, O., Aivasidis, A.: A Fuzzy Cognitive Network Based Control Scheme for an Anaerobic Digestion Process. In: 14th Mediterranean Conference on Control and Applications, Ancona, Italy. Session TMS Process Control, vol. 1 (2006)
62. Rudin, W.: Principles of Mathematical Analysis, pp. 220–221. McGraw-Hill Inc., New York (1964)
63. Ioannou, P., Fidan, B.: Adaptive Control Tutorial. Society for Industrial and Applied Mathematics (SIAM), Philadelphia (2006)
64. Karlis, A.D., Kottas, T.L., Boutalis, Y.S.: A novel maximum power point tracking method for PV systems using fuzzy cognitive networks (FCN). Electric Power System Research 77(3-4), 315–327 (2007)
65. Kranas, A., Dugundji, J.: Fixed Point Theory. Springer, New York (2003)
66. Kottas, T., Boutalis, Y., Christodoulou, M.: Bilinear Adaptive Parameter Estimation in Fuzzy Cognitive Networks. In: Alippi, C., Polycarpou, M., Panayiotou, C., Ellinas, G. (eds.) ICANN 2009. LNCS, vol. 5769, pp. 875–884. Springer, Heidelberg (2009)
67. Passino, K.M., Yurkovich, S.: Fuzzy Control. Addison-Wesley Longman, Amsterdam (1998)

Modeling of Operative Risk Using Fuzzy Expert Systems

Santiago Medina Hurtado

Abstract. This study is focusing on the modeling of operative loss exposure of the Allowances and Retirement Funds with the use of a Fuzzy Expert System, which evaluates the environmental and managerial factors, this allow obtain a qualification about the possibility that the company incurs in operative losses. The system can be very useful either when the quantitative information is limited due to the discrete character of the risk events or where the information about the risk factors are associated to expert's knowledge, for these reasons the modeling using formal statistical tools is difficult. Too we propose, a simple methodology to complete the knowledge matrix of the Fuzzy Expert System which combines the scoring method with the expert knowledge that allows obtain the rules of system in a simple and quick way. This extraction method saves time in the modeling of complex systems where many variable interact and where there are restrictions with the expert interactions. These systems allow having a structured vision of the sources of operational risk and where the manager should concentrate efforts to diminish the exposure. The Fuzzy Expert System can help to complement the operative risk analysis carried out with quantitative methods as Extreme value theory or Montecarlo simulation.

1 Introduction

The growing necessity to give appropriate solution to socio-economic, administrative and financial problems, starting from human perceptions and that as such we don't have enough information to apply conventional mathematical models, forcing us to search alternative models that allow obtain conclusions starting from linguistic variables. The Fuzzy Logic appears like a tool that allows us to make this transformation and provide a different answer to many real world problems where the information is incomplete, vague and subjective.

Santiago Medina Hurtado
Minas Faculty - National University of Colombia
e-mail: smedina@unal.edu.co

M. Glykas (Ed.): Fuzzy Cognitive Maps, STUDFUZZ 247, pp. 135–159, 2010.
springerlink.com © Springer-Verlag Berlin Heidelberg 2010

Every day companies and institutions must to face different kind of risks, according their activities or environment. The financial literature classifies the risks in three big groups which are not independent, the market risks, the credit risks and the operative risk (Basel II 2004). However the risks are more wide than these classification (for example: regulatory risk, tributary risk, politics risk, social risk, legal risk, etc). On the other hand, the operative risk includes a great quantity of risk factors whose modeling is difficult for two aspects, first the information relative with the factor is in many cases of qualitative type and second the scarce quantity of data due to discreet character of the events that make it difficult to modeling the risk using formal statistical tools.

Most of the methodologies that address the risks have a statistical foundation and they are centered in the modeling of market and credit risks, however, for the modeling of operative risk the statistical approaches are centered in simulation processes and extreme value theory, this one fix a distribution to the tail lost. This would not have bigger difficulty if the quantitative information is enough but this is not this way. For this reason in the last years, the interest of experts and investigators has been directed toward the development of methodologies that treat the subjective characteristics. actually we are applying methodologies like Bayesian Analysis, Fuzzy Inference Systems and the Fuzzy Expert Systems to treat problems with incomplete information, in order to guide make decisions problems.

The first time that the financial community spoke of "Operative risk" was probably in 1995 when of the English Baring Bank PLC, crashed, due to the transactions carried out by an operator which consumed the capital stock of the company. This event reveled to the financial market the exist risks that could not be classified as market or credit risks and could still effect the results considerably. In the context of the financial institutions the operative risk refers to the range of possible losses associated to risks of the environment, by human fault, processes or technology.

The operative risk is a wide concept. In some cases it is difficult to make a clear distinction among the operative risk and the normal risk that institutions face in their daily operations, in some cases it is difficult to distinguish the limits. For example if a customer fails to pay their loan, it can either be due to the normal credit risk that institution face or it can also be due to the human error that occurred when the analyst studied the client's solvency. The administration program of the risk operative should clearly define that be included as operative risk to diminish the grade of ambiguity with other risk types.

Many managers of financial institutions believe that the operative risk losses could be more significant than the market risks or the credit risks losses (although this isn't clearly demonstrated). The importance of risk measure and control were considered when the Norte American banks divided and qualified its risk into 50% in credit risks, 15% for the market and liquidity risks, and 35% operative risks (Cruz 2002).

The first step toward a stricter administration of risk to financial institutions were the agreements signed at the end of 1988 by the central banks group of industrialized countries (Basel I 1988) . These agreements have led the proposal of risk management that has evolved from credit risks, to the market risk and now the operative risks. In this sense they demand the implementation of administration models with the goal of diminishing the probabilities of loss and to have an adequate technical capital to cover the potentials loss.

The BASLE is the main international agreement that regulate and propose methods that are still in evolution process to treat the risks of financial institutions, however any type of business are exposed to risks. For this reason schemes referencing to risks management have been proposed like COSO, ISO, Australian Standard, PMI, etc.

The operative risk is subdivided in two main components, first the operative risk due to internal causes and second the strategic risk due to external causes, this is indicated in the Figure 1.

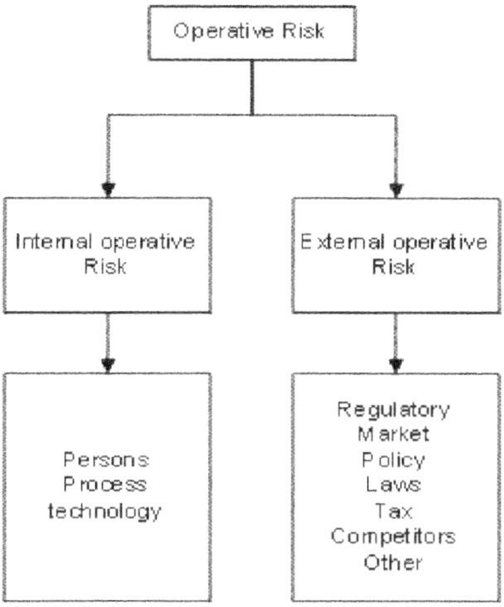

Fig. 1 Operative Risk classifications

The internal operative risks are associated to people, processes or technology used in the normal business course, anyone of these factors can experience some fails and produce financial loss. The external operative risks arise of environment factors as new competitors, political and regulatory environments, climatic factors, technology, but they also can arises of strategic plans; for example a new business lines. These factors are outside of company control. A proportion of this loss can be estimated and they should be covered, the other however can't be estimated. In both cases forecast the impact and the frequency loss is a difficult task to achieve.

In the operative risks analysis is important to quantify the potential loss expected with the purpose of covering them. The potential loss is the product of the probability that the event happens and its associated cost (severity), however, both quantities are difficult to quantify because these events don't occur frequently, that is, they are discreet events that are difficult to register and are time requires. For this reason is difficult to apply, in some cases, statistical analysis to make an objective valuation of the operative risk (Vose 1996, Cruz 2002, Moscadelli 2004, Reiss 2001, Cornalba 2004, Chernobai 2006, Chavez 2006 et all). However the difficulty in the valuation doesn't imply that the risk should be ignored, but rather, we should look for methods that allow us to model this reality.

The impact (severity) of a financial loss can be divided in two categories, a the expected, and the unexpected. In turn the unexpected loss can be classified as "severe" or "catastrophic". The expected losses should be covered with a proportion of the profits, while the severe or catastrophic losses will be covered with a proportion of the company's economic capital, insurance contracts or derivatives (Hull 2008).

The lack of historical data doesn't allow the application of statistical models, however there are people with knowledge that understand and management its processes . In this sense the qualitative information is very important because it allows, starting from a systematic process, to identify the factors that expose the company or business line to lost operational with the purpose of treating these sources of risk in an opportune way.

Actually many researches apply Fuzzy Logic (FL), to treat problems where the information is subjective and vague. The potential use of FL is extendeds to the entire field of the social sciences (Kaufman 1990, Kulkarni 2001, Medina 2008, Glykas et al. 2004, Xirogiannis et al. 2008). In the case of the operative risk, the FL allows to the analysis of the risk with a smaller cost and a structured vision of the operational risk to be carried out when the quantitative information is limited. This doesn't mean that the Fuzzy Logic can replace the statistical methods that measures of the operational risks, however it can provide a rigorous theoretical scheme for the treatment of many problems and to help make decisions. FL is an administrative tools that allows measure the state of a process and help us to makes better management.

This article is organized in the following way. The numeral 2 describes the operation of a FIS. In this numeral a method to extract the base of knowledge in an easy and quick way is proposed; this is the critical part of the FIS. The numeral 3 formally defines the Fuzzy Expert Systems FES. The numeral 4 expose an application of modeling of operative risk. The last numeral presents the summations.

2 Fuzzy Logic Systems

The Classic Logic isn't appropriate when we try to describe facts that are not completely true or completely false, because it completely excludes the possibilities among these two values. The Fuzzy Logic on the other hand, allows us to use relative concepts of reality, defining the grades of membership and following the reasoning patterns similar to the human thought (Kosko, 1995).

The Fuzzy Logic formalized by Zadeh in 1965 is based on the theory of the Fuzzy Set. It indicates the grade of membership of an element to the group determined by its membership function that takes values in the interval [0, 1] (Jang, 1997, Kulkarni, 2001, Kasabov, 1998 and Kosko, 1995, Buckley 2002). This way, while in the framework of classic logic, for example, the utility of a company take only two values, it is low wiht grade zero value (0) or it is high wiht one value (1), for the Fuzzy Logic is possible also all intermediate conditions of utility like "very low", "relatively high", "middle", "lightly low", etc. whit values in [0, 1].

The extreme values assumed by the classic logic are only a particular case inside the universe of the Fuzzy Logic, because it allows being relatively imprecise to represent problems and to arrive to adequate solutions (Kosko, 1995). The

Fuzzy Logic opens the possibilities, by giving solution to problems that are expressed from the human perspective and for this simple condition, they can't have an unique solution from the "false" or "true", since it can take intermediate conditions to give satisfactory solutions to the problems.

The Mamdani Fuzzy Inference Systems FIS, (Mamdani, 1977, 1981) was the first system proven in a practical way as universal approximator of functions. Later in 1992 Kosko and Wang formally settled that any relationship among input and output variables can be approximated by means of FIS, built in linguistic terms with a high grade of accuracy (universal approximator).

To build a FIS the steps to follow are indicated in the Figure 2.

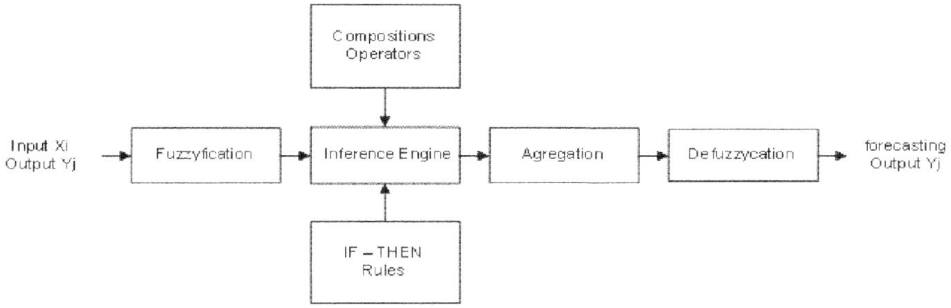

Fig. 2 Steps to building a FIS

Next, we explain the steps to build a FIS:

Fuzzyfication. The first step to build a FIS is to define the inputs and outputs of the system (linguistic variables), their linguistic terms (fuzzy sets) and their membership functions. The "linguistic variable" can take ambiguous or inexact values, for example the linguistic variable "Profitability" can take the linguistic terms "lower", "middle" and "high" and have a semantic meaning for the annalist. This terms are represented by means of fuzzy sets. The Figure 3 present three fuzzy sets lower, middle and high (linguistic terms) for the variable Operative Margin.

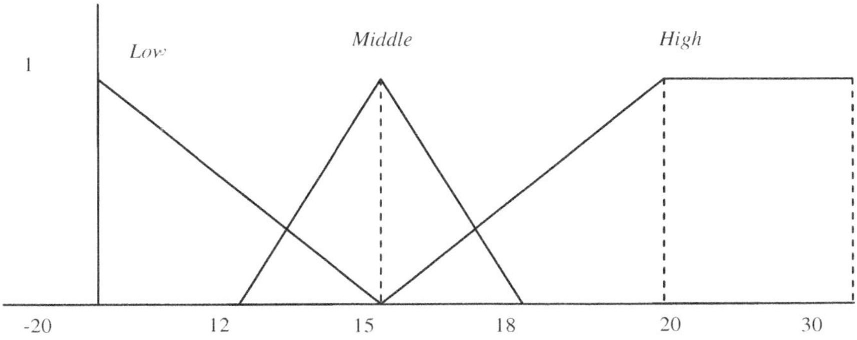

Fig. 3 Fuzzy sets assign to Operative Margin

Formally, to fuzzy Set is defined by:

Let, X the universe of discourse
x is an element anyone of X and
A is a collection of elements $x \in X$ and they have a defined characteristic ($A \subset X$). A is a fuzzy set in X and define the group of orderly couples

$$A = [(x, \mu_A(x)) / x \in X] \quad (1)$$

Where μ_A (x) is a membership function of fuzzy set A.

The membership function assigns to each element of X a real number among [0,1], denominated "grade of membership" of the element to the fuzzy set A. The types of membership functions commonly utilized are: Triangular, Trapezoidal, Gaussiana, Sigmoidal and Generalized of Bell. They are chosen to reach an appropriate correspondence among the inputs and output spaces of a system.

Construction of the rule base: to build a FIS we should define the fuzzy rules If-then: These rules specify the relationship among input and output variables of the system. The fuzzy relationships determine the grade of association between elements of two or more sets.

The IF – THEN Rules type Mandani has the shape:

"IF X1 is A1 and X2 is A2 and.......and Xk is Ak THEN Y is B"

Where A1, A2,...., Ak, B are linguistic terms defined by mean of fuzzy sets to each linguistic variable in the universe of discourse X1,X2,.....,Xk and Y. "Xi is Ai and..." is called the antecedent of the rule or premise and "Then Y is B" it is called the consequent or conclusion. The previous rule, defines a fuzzy relationship in the dimensional space k+1, characterized by a function of membership $\mu_{Ak \to B}$ (X1,X2,....., Xk,Y) $\in$ [0; 1].

The set of fuzzy rules (also call "base of knowledge") are obtained by knowledge experts through interviews, questionnaires or panel technical, however, in many occasions the access to experts isn't available but a numeric database of input and output variables is available. This database allows obtain the set of fuzzy rules applying some extraction algorithms. [Hammell and Sudkamp 1994, Wang et al. 2001, Wang and Mendel 1992, Tsang and Yeung 1997, Wang and Hong 1999, Männle e 1999, Herrera and Verdegay 1995, Chin et al. 2003, Espinosa and Vandewalle 2000].

To represent causal relationships among input and output variables of systems can be a very useful use of fuzzy cognitive maps which give us a complete representation of the situation and their relationships (Kosko 1994, Xirogiannis and Glykas 2004, Xirogiannis and Glykas 2007). For example suppose that an analyst tries to measure the effect of the slowness system in the level of operative risk. As a first step some causes and effects are identified and bound to the event of "slowness of the system." A slow system is caused by the reduction of the satisfaction of the employees, which redounds in a reduction of procedural transactions; this satisfaction commits to other events that increase the operative risk of the business just as it is shown in the Figure 4 (Cruz 2002). Later this representation should be expressed in form of IF-THEN fuzzy rules. The Figure 4 also represent the effect of factors like electric risk, regulatory policy, socio-political risks over the energy spot prices and the decision of load deal for a trader company of electrical energy.

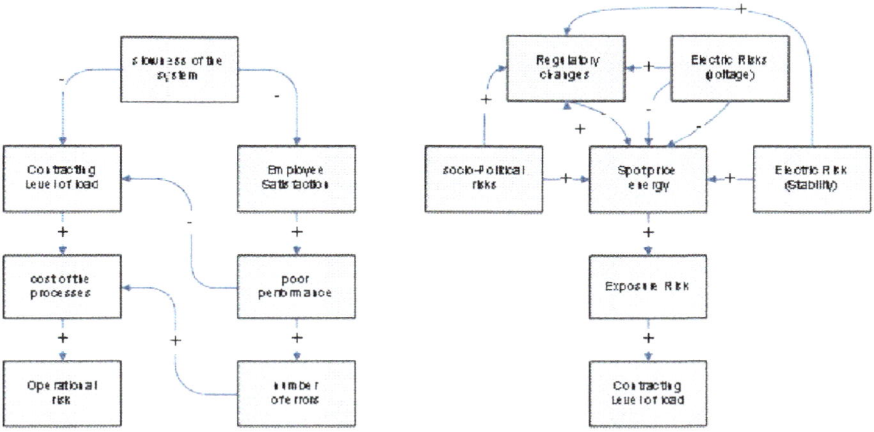

Fig. 4 Examples of Fuzzy cognitive maps

The fuzzy rules can be represented by means of matrix arrangements like in the example indicated in the figure 5. The inputs are located in the columns or row of the matrix and show the linguistic labels of their fuzzy set. The levels of the output variable (Low, Middle or High organizational Administration) are located in the central part of the matrix. This matrix represent the fuzzy rules in a concise form.

ORGANIZATIONAL MANAGEMENT				
Strategic managements	Risk Managements	Exposure to process factor		
		Low	Middle	High
Inadequate	wrong	HIGH	HIGH	HIGH
Adequate	wrong	MIDDLE	MIDDLE	MIDDLE
Inadequate	moderately good	MIDDLE	HIGH	HIGH
Adequate	moderately good	LOW	MIDDLE	MIDDLE
Inadequate	Good	MIDDLE	MIDDLE	HIGH
Adequate	Good	LOW	LOW	MIDDLE

Fig. 5 Matrix arrangements for Organizational management

The matrix should be completed following the procedures like panel sessions, interviews or surveys, etc. In this article we suggest a method to capture the expert opinion by means of scoring method which allows us to obtain the rules quickly, as it is indicated next:

- The first step is ponder each input Xi, according to importance or effect that each Xi has on output Yj.

  Let P_i weights assign to each input. i=1,.,n. $P_i \in [0,1]$ and $\sum_{i=1}^{n} P_i = 1$

  Where n = number of input variables to the FIS.

  To higher weight (Pi), more relative importance of the input variable (is bigger the effect over the output variable)

  Example: If we are evaluating the variable: ***"Exposition to External Factors"***, we can use three input variables and these are qualified with the following weight according to their importance by the expert's opinion:

Other Entities dependence	(24%)
Legal norms Knowledge	(29%)
Regulatory stability	(47%)

- Ponder the importance of each fuzzy set (linguistic label) assigned to each input "Xi". Then "Cij" is the weight assigned to each fuzzy set of the input Xi, and indicates the effect on the output Yj.

  Where:

  i=1,…,n n is number of input variables to the FIS.
  j=1,…,m m is each fuzzy set defined to input i.

  And $\sum_{j=1}^{m} C_{ij} = 1 \qquad \forall i = 1,..,n$

  Example: the input "*Regulatory stability*" has assigned three fuzzy set (three levels: very low, low and middle). The expert should assign weights to each level according to importance or effect that this one has over the output variable:

Very low	weight (50%)
Low	weight (39%)
Middle	weight (11%)

- The expert must qualify any scenarios of the knowledge matrix, that is, the experts should define a set of reference rules, that allows to later, validate the FIS output. In general extreme and middle scenarios are evaluated easier for qualifying when the system is very big. The rules are expressed by:

  If X1 is A1 and X2 is A2 and...... and Xk is Ak Then Y is B

Example:
"IF *Regulatory stability* is Very Low And *Legal norms Knowledge* is Low And *Other Entities dependence* is middle Then *Exposition to External Factors* is High"

- The score for each cell of matrix knowledge is calculated by means of the relationship:

$$K_{ij} = \sum_{i=1}^{n} C_{ij}.P_i \qquad \text{For all j=1,...,m} \tag{2}$$

Where: i=1,2, n number of input variables.
j=1,..m number of fuzzy set assign to each input variables.

- The next step is correlating all scores in each cell of the score matrix with the fuzzy sets assign to output variable Yj. This procedure is carried out by obtaining the maximum and minimum value of the scores matrix, be these K_{max} and K_{min} respectively. K_{max} and K_{min} determine the scores range that can take the output Yj, therefore, the fuzzy sets must be assigned to this range. The score on the intersection of fuzzy sets determine the score from which is assigned a fuzzy set to each cell of the matrix (Compared the score of each cell with the score on the intersection).

  The Figure 6 shows an example of three triangular fuzzy sets assigned to output Yj. These are distributed on the range K_{max} and K_{min}. The scores on the intersections are L1 and L2, they determine the points from which for a specific score, a fuzzy set should be assigned. For example to scores smaller than L1 assigned *Low*, between L1 and L2 *medium* and *high* on other cases (bigger than L2).

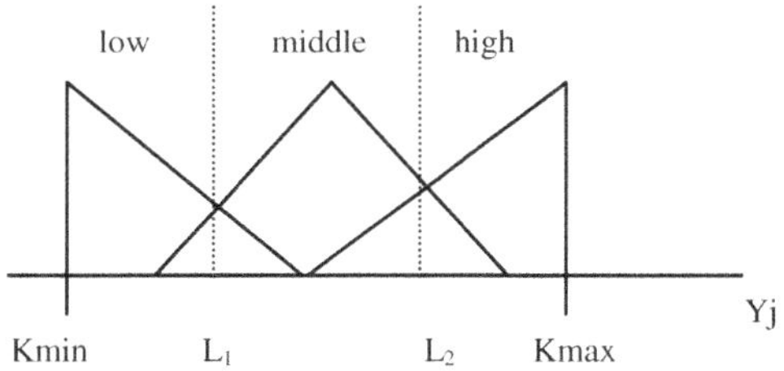

Fig. 6 Triangular fuzzy set to output Yj

The system response can be modified by changing the position of the limits L_1 and L_2 such that the system response is adjusted to the referenced fuzzy rules qualified by the experts previously. This process permits us to obtain the fuzzy rules matrix.

- Then, a validation process should begin wiht the fuzzy rules matrix suggested on the previous step to adjust the system. All cells are evaluated, first it needs to be verified that the reference rules are obtained and then it needs to be verified that all other rules are consistent, otherwise, the limits L1 and L2 must be moved to adjust the system response.

Composition operations. The basic operations carried out with fuzzy sets are Union, Intersection, Complementation, the Cartesian product and the Cartesian Coproducts. These operations are carried out by means of binary operators, classified as T-norms (for intersection operations) or S-norms (for operations of union). (Kaufman 1990, Kaufman 1982, Jang and Mizutani 1997, Kulkarni 2001, Kasabov 1998).

The fuzzy rules define a fuzzy relationship in the space dimensional k+1, characterized by a membership function $\mu_{Ak \to B}$ $(X_1, X_2, \ldots, X_k, Y) \in [0; 1]$, the basic operations with fuzzy sets are utilized like implication relationships and used to derive membership functions of fuzzy set n-dimensional.

Inference Mechanisms (Approximate Reasoning): The Approximate Reasoning is an inference procedure to derive conclusions from fuzzy rules type IF-THEN and values that take the inputs Xi of systems, applying composition relationships Max-Min or Max-product. These permits to infer a fuzzy value B´ if we have fuzzy inputs into k-dimensional space (A'_k) and we have defined the implications relationships:

$$R : A_k \to B, \qquad \text{then}$$

$$B' = A'_k \circ (A_k \to B) \tag{3}$$

For example if the system is defined by the following two fuzzy rules (the rules defined fuzzy relationship):

Rule 1: IF x is A1 AND y is B1 Then z is C1, or
Rule 2: IF x is A2 AND y is B2 Then z is C2

Figure 7 represents the approximate reasoning. The systems response C´ is obtained from the inputs: "x" is A´ and "y" is B´ (A´ and B´ can be fuzzy Sets) and the previous fuzzy rules. Each rule is expressed by:

$$R1 = (A1xB1) \to C1 \text{ and } R2 = (A2xB2) \to C2. \tag{4}$$

The composition max-min can be used to infer $\mu_{C'}(z)$. The composition operator "o" is distributed on the union of two fuzzy relations by:

$$C' = (A' \times B')^{\circ}(R1 \cup R2) \tag{5}$$
$$C' = \{(A' \times B')^{\circ}R1\} \cup \{(A' \times B')^{\circ}R2\}$$

$$C' = C1' \cup C2' \tag{6}$$

Where C1´ y C2´ are fuzzy sets derivate of fuzzy rules 1 and 2 respectively. This result can be extended for n-rules.

Aggregation. In this step the outputs of each rule (C_1', C_2',...,C_n' activation strength of each rule) are combined to obtain only one fuzzy set C´. The inputs of the aggregation process are the truncated membership functions obtained by approximate reasoning for each n-rule. The aggregation method is commutative; it doesn't care the order in which the output of each rule is added.

This process defines a method to find C´= ($C_1' \cup C_2' \cup \ldots\ldots \cup C_n'$). Where C_1', C_2',...,C_n' are fuzzy sets inferred by rules 1,2,..., n. and C´ is a fuzzy set or output system with membership function $\mu_{C'}(z)$.. The *Maximum* operator is the aggregation operator that is most utilized. The output C´ is defined by:

$$C' = [(z, \mu_{C'}(z) / z \in Z] \tag{7}$$

Where Z is the discourse universe of output variable and their membership function is given by:

$$\mu_{C'}(z) = \text{Max}\,(C_1', C_2', \ldots, C_n') \tag{8}$$

The fuzzy set C´ = C1´ U C2´ aggregate the truncated fuzzy set of each rule. See Figure 7.

Defuzzyfication: In this last step, is obtained a crisp value (K) from the fuzzy set C´ that provides the solution to the output system (see Figure 7) and to permit to make decisions. The concretion methods mostly used are gravity (center of area), Bisector, Middle of maxima, More small of maximum and bigger of maxima (Jang and Mizutani 1997, Kasavov 1998).

The center of area is the concretion method most utilized, because it is similar to calculate the expected value of a distributions probability. The crisp value Zc´, is obtained by integrating the output membership function $\mu_{C'}$ by the means of the following relationship:

$$Z_{C'} = \frac{\int_z \mu_{C'}(z) z \, dz}{\int_z \mu_{C'}(z) dz} \tag{9}$$

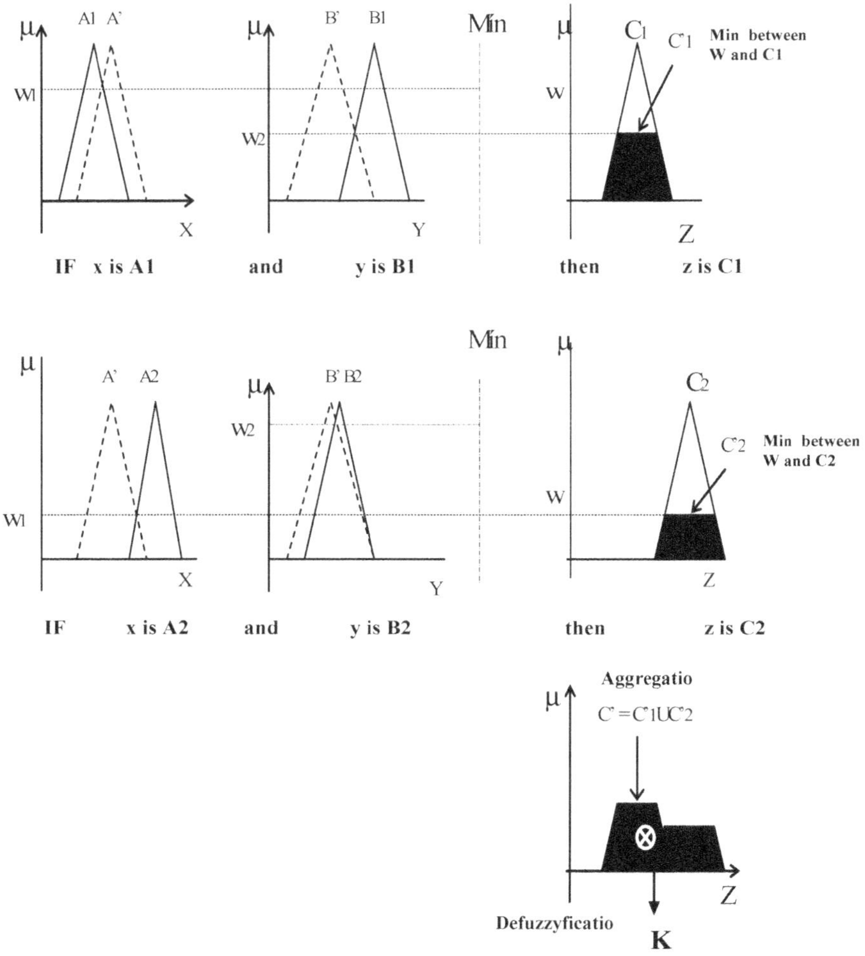

Fig. 7 Approximate reasoning

3 Fuzzy Expert Systems (FES)

The interrelation of FIS allows the configuration of an FES and can be used to model complex decision making systems. A FES is a system that uses Fuzzy Logic instead of Boolean Logic to deduce conclusions from input variables. FES is different from conventional expert systems which are applied to domains well defined, where the knowledge phenomenon is wide however, problems where inaccuracy, ambiguity and bad definition exists, the development of a conventional expert system will be more difficult. The treatment of uncertainty is an important aspect in the evolution of expert systems.

In the domain of a expert system there are different sources of uncertainty and this implies that there are not exact solutions, there are approximate solutions (making decisions about management and economics, psychological, medicine etc.). For example, if a problem can be solved in different ways, the expert system should be able to recognize the possible solutions, whose answers should be overlapped; therefore the solution is also inaccurate and uncertain. The expert system should be able to accept questions that incorporate uncertainty and provide an interpretation or reasoning appropriate to this uncertainty.

The following characteristics of decisioning problems are inherent to the process of human reasoning:

- The learning process or acquisition of knowledge is imprecise
- The knowledge of the expert contains uncertainty.
- The process of the expert's reasoning is imprecise.
- The knowledge represented in natural languages introduces uncertainty.
- The information isn't totally defined.

The previous characteristics limited the use of conventional expert systems. The incorporation of fuzzy logic in the expert systems allows their potential use in many areas of knowledge due to the treatment of characteristics associated to learning processes and human reasoning.

Let $x_i = (x_1 , x_2 , x_3 ,\ldots.., x_n)$ the inputs to FES, these will feed j-Fuzzy Inferences Systems (FIS_{1j}), that is, each FISj is integrated by n+1-tupla ($x_1, x_2, x_3,\ldots.., x_n.,y_j$) and they have a fuzzy sets by output variable (Y_j), which allow to obtain a crisp values $Z(C_j^`)$ after apply a defuzzyfication method. See Figure 8.

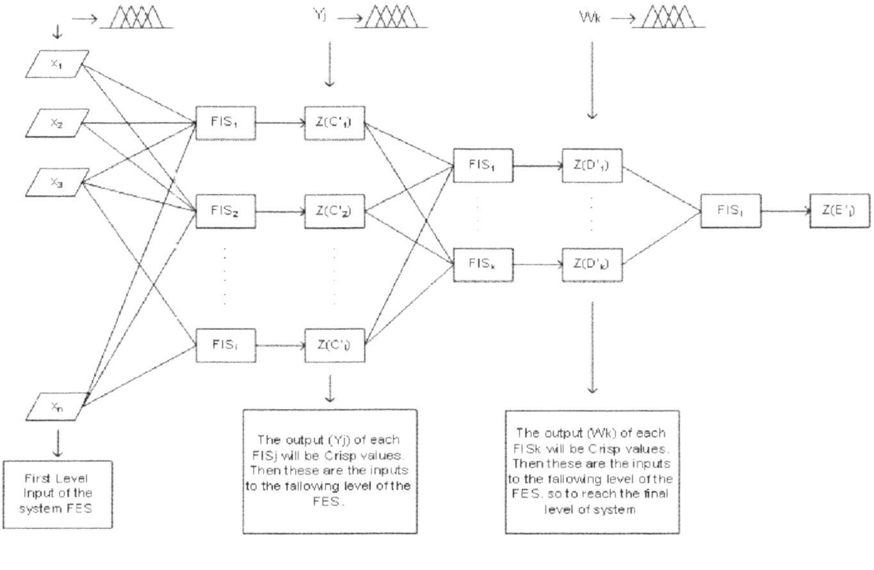

Fig. 8 FES structure

The output Y_j will be the input for the following level of the FES with crisp values $Z(C_j`)$, that is, the variable $Y_j$ will feed again k-new $FIS_k$, each one will be formed for j+1-tuplas $[Z(C_1`), Z(C_2`),\ldots,Z(C_j`), W_k]$, and they have a fuzzy sets by output (W_k), which allow obtained crisp values $Z(D_k`)$ after apply a defuzzyfication method. Repeating this process successively can reach the output $Z(Et`)$ of the fuzzy expert systems. (Medina 2007)

A formal expression for this process is shown below:

Level	Fuzzy Set	Crisp Value (Defuzzyfication)
	Y_1	$Z(C_1)=f_1(x_1 , x_2 , x_3 ,\ldots, x_n)$
Level 1	Y_2	$Z(C_2)=f_2(x_1 , x_2 , x_3 ,\ldots, x_n)$
	....	....
	Y_j	$Z(C_j)=f_j(x_1 , x_2 , x_3 ,\ldots, x_n)$
	W_1	$Z(D_1)= f'_1 (Z(C_1), Z(C_2),\ldots,Z(C_j))$
Level 2	W_2	$Z(D_2)= f'_2 (Z(C_1), Z(C_2),\ldots,Z(C_j))$
	....	....
	W_k	$Z(D_k)= f'_k (Z(C_1), Z(C_2),\ldots,Z(C_j))$
Level 3		$Z(E_1)= f''_1 (Z(D_1), Z(D_2),\ldots,Z(D_k))$
Outputs		$Z(E_2)= f''_2 (Z(D_1), Z(D_2),\ldots,Z(D_k))$
		....
		$Z(E_t)= f''_t (Z(D_1), Z(D_2),\ldots,Z(D_k))$

The output $Z(E_t)$ can be written as:

$$Z(E_t)= f''_t (Z(D_1), Z(D_2),\ldots,Z(D_k))$$

$$Z(E_t)= f''_t \{f'_1 [Z(C_1), Z(C_2),\ldots,Z(C_j)],\ldots, f'_k [Z(C_1), Z(C_2),\ldots,Z(C_j)]\}$$

$$Z(E_t)= f''_t \{f'_1 [f_1(x_1,..,x_n),\ldots, f_j(x_1,..,x_n)],\ldots, f'_k [f_1(x_1,..,x_n),\ldots, f_j(x_1,..,x_n)]\} \quad (10)$$

The functions $f(\cdot)$ are FIS that transform the initial input $(x_1 , x_2 , x_3 ,\ldots, x_n)$. The equation 7.10 is a highly not lineal function.

FES also allows the incorporation of information that come from models of the financial economy (such as a valuation model of companies) or forecasting models (such as a neuronal network or ARIMA model). These structures combine the entire information that comes from the expert knowledge and the mathematical or statistical models that are necessary in the analysis of complex problems. Figure 9.

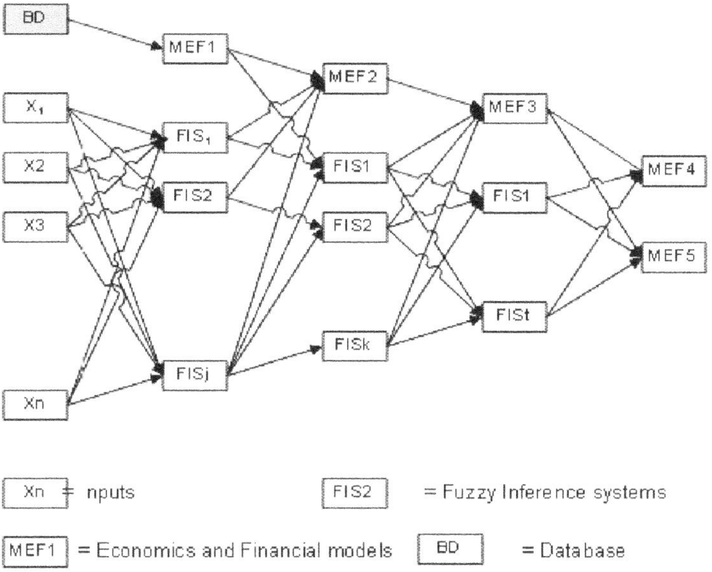

Fig. 9 FES that incorporate financial economic models

The following section shows an application of the operational risk analysis.

4 Modeling Operational Risk Factors

In this paragraph apply the methods described above to the operational risk modeling of a financial institution associated with the pension Funds and layoffs

4.1 Fuzzyfication

The first stage that should be carried out is the fuzzyfication process of system. We should identify and define clearly the input and output variables, their relationships, the number and type of fuzzy sets assigned to all variables. This generally is done by the entity's executive staff at all levels. After carrying out the consultation regarding the factors that expose the entity to operative loss or factors that can reduce their exposure, we proceeded to analyze each ideas expressed. The result of the identification and grouping of variables is shown in Figure 10.

The first level is composed by the input variables of the systems that will be qualified by experts when the systems is operating. These are:

- Regulatory stability
- knowledge of regulations
- Dependency of other entities
- Knowledge Management
- Management of human talent
- Problems of the technological platform

- Planning Design and Development
- Standardization of processes
- Supply Management
- Information security
- Risk Management
- Strategic Management

The variables of the second to fifth correspond to clusters of the inputs variables which have a clear meaning to the experts. These higher level variables define the number of FIS that contains the expert system; in this case seven FIS should be completed.

- Exposure to External Factors
- Exposure to Human factor
- Exposure to process factor
- Risk Management
- Organizational Management
- Exposure to internal factors
- Exposure to Operational Loss

This model allows the evaluation of the level of exposition to operative loss to the company. Each variable in each level was clearly defined for experts to avoid difficulties with the interpretation. This model can be integrated with the other process of risks management.

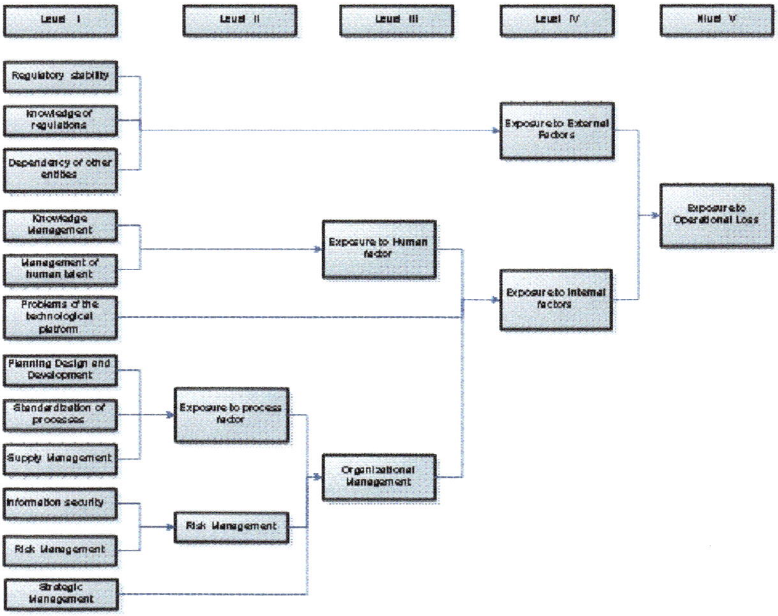

Fig. 10 FES for Operative Risk measurement

Then the fuzzy sets must be assign to each input and output variables of the system. These are defined by a range [0,1]. For example the "Exposure to Operational Loss" variable, has assigned three fuzzy sets wiht linguistic label of low, medium and high.

The Figure 11 shows triangular fuzzy sets used for the FES. All fuzzy sets are defined in the universe of discourse [0,1]. The Figure 12, shows all fuzzy variables and their fuzzy sets defined by the fuzzy Expert System.

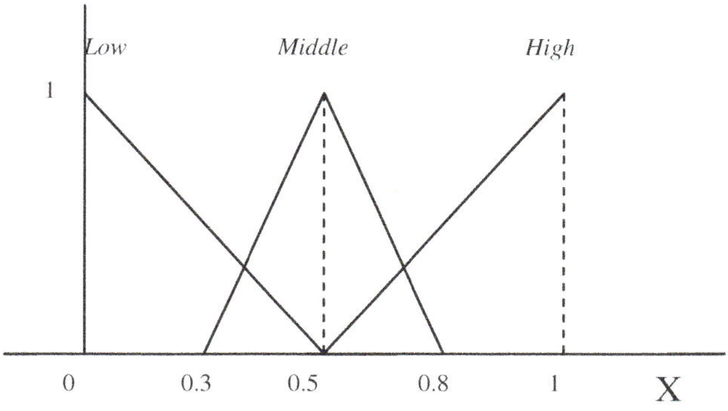

Fig. 11 Triangular Fuzzy sets

INPUT AND OUTPUT VARIABLE	LINGUISTIC LABELS		
Regulatory stability	very low	low	middle
Knowledge of regulations	very low	low	middle
Dependency of other entities	low	middle	high
Knowledge Management	inadequate	adequate	-
Management of human talent	inadequate	adequate	-
technological platform Problems	low	moderate	high
Planning Design and Development	inadequate	adequate	-
Standardization of processes	low	middle	high
Supply Management	inadequate	adequate	-
Information security	regular	good	-
Risk Administrations	regular	good	-
Strategic Management	inadequate	adequate	-
Exposure to External Factors	low	middle	high
Exposure to Human factor	low	middle	high
Exposure to process factor	low	middle	high
Risk Management	wrong	regular	good
Organizational Management	low	middle	high
Exposure to internal factors	low	middle	high
Exposure to Operational Loss.	low	middle	high

Fig. 12 Fuzzy variables and fuzzy sets of systems

4.2 Knowledge Matrix

In this stage we should defined the fuzzy rules that allows the prediction of the behavior of the systems (output variable) with different levels of input variables. It basically corresponds to the definition of the underlying relationships among the input and output variables which are expressed in IF-THEN statements.

The group of statements IF-THEN, conforms the set of fuzzy rules and are the explicit representation of the experts knowledge about the specific process which we wants to evaluate, therefore, the participation of experts in the building of the knowledge matrix is an important issue for both the validity and acceptability of the model.

The fuzzy rules take the form:

"IF $X1$ *is* $A1$ and $X2$ *is* $A2$ and.......and Xk *is* Ak THEN Y *is* B"

From the operative point of view, the process begins with the identification and definition of input and output variables from expert opinion. Then the knowledge matrix should be completed which incorporates the information from the input variables, fuzzy sets and the output variable. The matrixes define the relationships between variables and are completed by using the algorithm presented in Section 2.

Seven FIS should be the developer to measure the "operational loss exposure". For example, to complete the FIS related with "*Organizational Exposure Factor*" the experts defined the six rules of reference shown in the following matrix (Figure 13). The missing fields must be completed with the proposed procedure:

ORGANIZATIONAL MANAGEMENT				
Strategic managements	**Risk Managements**	**Exposure to process factor**		
		Low	**Middle**	**High**
Inadequate	wrong	**HIGH**	**HIGH**	**HIGH**
Adequate	wrong	-	-	-
Inadequate	moderately good	-	-	-
Adequate	moderately good	-	-	-
Inadequate	Good	-	-	-
Adequate	Good	**LOW**	**LOW**	**MIDDLE**

Fig. 13 Organizational management matrix knowledge

Could be thought that it's an easy job to complete the system matrixes from the experts directly, however when the number of input variables as well as the number of fuzzy sets is very large (modeling a complex problem with many chained FIS), the number of cells in the matrix increases exponentially, therefore, it takes a long time for the experts analysis to complete the matrixes.

To complete the scoring matrix (see figure 14), we used the previously proposed procedure and calculate the scores for each cell from weights given to each input variable and the weight given to the fuzzy sets of each variable. For example, the scores in the first cells are calculated as follows (see :

$$0.4549 = 0.32x0.81 + 0.29x0.5 + 0.39x0.13$$
$$0.5212 = 0.32x0.81 + 0.29x0.5 + 0.39x0.30$$
$$0.2565 = 0.32x0.19 + 0.29x0.5 + 0.39x0.13$$
$$0.3228 = 0.32x0.19 + 0.29x0.5 + 0.39x0.30$$

Output Variable				Label Linguistic					
ORGANIZATIONAL MANAGEMENT				LOW	MIDDLE	HIGH			
	Strategic managements 0,3200		Risk Management 0,2900	0,3900 Exposure to Baja 0,130	Media 0,3	Alta 0,6			
0,8	Inadequate	0,5	wrong	0,4549	0,5212	0,6265	HIGH	HIGH	HIGH
0,2	Adequate	0,5	wrong	0,2565	0,3228	0,4281	LOW	MIDDLE	MIDDLE
0,8	Inadequate	0,4	moderately good	0,4259	0,4922	0,5975	MIDDLE	HIGH	HIGH
0,2	Adequate	0,4	moderately good	0,2275	0,2938	0,3991	LOW	MIDDLE	MIDDLE
0,8	Inadequate	0,1	Good	0,3389	0,4052	0,5105	MIDDLE	MIDDLE	HIGH
0,2	Adequate	0,1	Good	0,1405	0,2068	0,3121	LOW	LOW	MIDDLE

Fig. 14 Scoring matrix for organizational management exposition

The next step is to relate the scores for each cell in the matrix with the fuzzy sets assigned to the output variable of the FIS. To make this we must first determine K_{max} and K_{min} and then assign the fuzzy sets defined for the output variable in this range.

In our case $K_{max} =$ 0,59750 and K_{min} = 0,14050. We have assign three fuzzy sets to the output variable ***"Exposition to organizational factor"*** (low, medium, high), then we must define the position of these fuzzy sets into the interval [0,14050 ; 0,59750]. We can make a symmetric partition of interval, in this case we must only determine the amplitude of each interval dividing the rank by three, that is,

$$\text{Rank} = 0{,}5975 - 0{,}1405 = 0{,}4570$$
$$\text{Amplitude} = 0{,}4570/3 = 0{,}15233$$

Then, the limits L_1 and L_2 are determined by adding these amplitude starting from K_{min}. For our case L_1 = 0,29283 and L_2= 0,44517, this is shown in Figure 15.

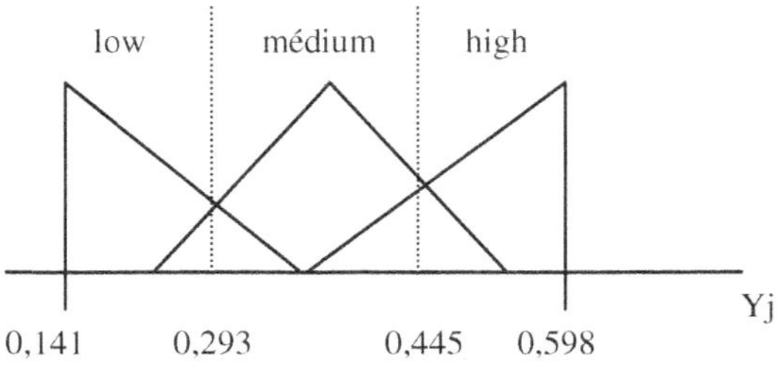

Fig. 15 Fuzzy sets for Exposition to organizational factors

The system response can be modified by varying the position of limits L1 and L2 in a way that the assignment of fuzzy sets is in accordance with all rules described by experts. Through this procedure will be determined all the rules of system.

4.3 FES Validation Process

The inference system must be validated through tests, that is to say, we must give values to the input variables of FES and then we must compare the results of system with the expert opinions. The results must be consistent.

The following validations can be carried out:

- Validation each knowledge matrix (each FIS) that compose the expert system. The fuzzy surfaces allows the validation of the relationships among input and output variables and their consistency. An example of a fuzzy surface is shown in the Figure 16. If the behaviors are not consistent we must review the rules, suggest the use of another fuzzy set to the input or output variables or move the fuzzy sets in the universe of discourse of each variable.
- Validation of the shape and position in the universe of discourse of the fuzzy sets defined to the input and output variables in order to improve the system response.
- Platform Validation, that is to identifying the programming problems that make the system not respond adequately. This process is done through the use of reference cases for which the analyst knows the answer. If the system does not reproduce these predefined answers it must be revised.

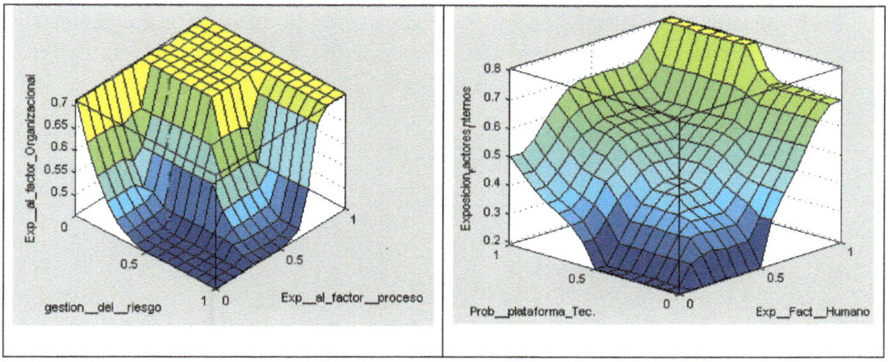

Fig. 16 Fuzzy Surface

4.4 *Platform for Measurement Exposure to Operating Losses*

The proposed model was programmed with the fuzzy logic functions available in MATLAB. The platform developed to measure the exposure of the operational risk as a result of measuring the factors identified from experts, this in turn gives us the evaluation of higher level variables of system, using color coding to give warnings about the state of these variables, as shown in Figure 17.

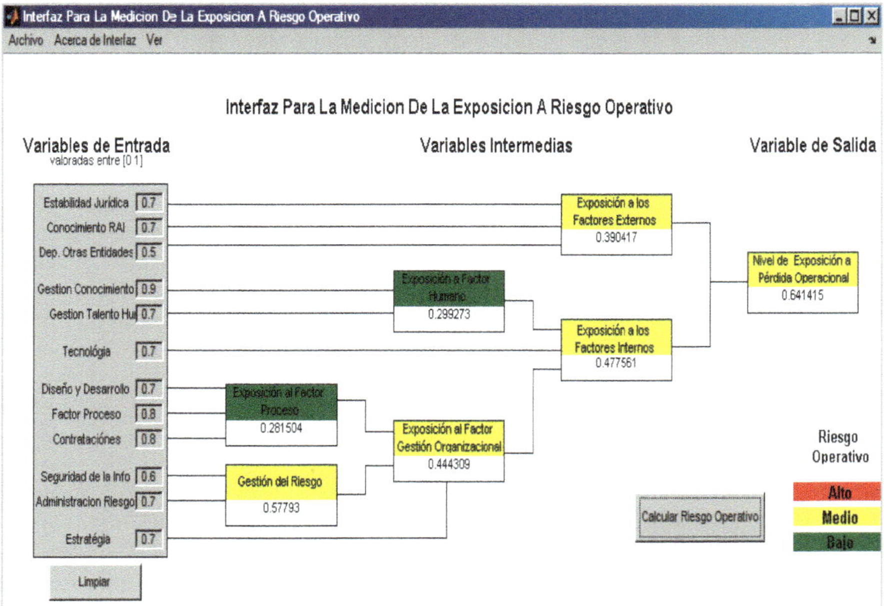

Fig. 17 FES Platform for measure the exposition to operative risk

With the model developed executive staff was asked to qualify the defined aspects related with the exposition of the operative risk and characterized by the inputs previously identified. The survey yielded the following results (figure 18):

Input Variables	1	2	3	4	5	6	prom.
Regulatory stability	0.7	0.3	0.5	0.3	0.5	0.5	0.47
Knowledge of regulations	0.7	0.5	0.5	0.3	0.2	0.7	0.48
Dependency of other entities	0.5	0.5	0.5	0.3	0.4	0.6	0.47
Knowledge Management	0.9	0.95	0.7	0.65	0.2	0.9	0.72
Management of human talent	0.7	0.95	0.8	0.3	0.2	0.9	0.64
technological platform Problems	0.7	0.5	0.7	0.6	0.4	0.5	0.57
Planning Design and Development	0.7	0.9	0.8	0.8	0.4	0.7	0.72
Standardization of processes	0.8	0.95	0.7	0.7	0.4	0.8	0.73
Supply Management	0.8	0.8	0.9	0.8	0.2	0.8	0.72
Information security	0.6	0.99	0.6	0.6	0.3	0.9	0.67
Risk Administrations	0.7	0.99	0.8	0.7	0.3	0.9	0.73
Strategic Management	0.7	0.95	0.9	0.7	0.1	0.9	0.71

Fig. 18 Input Qualification by Executive staff

5 Conclusions

The previous model can be applied to measure the executive staff's perception about factors that expose the company to operative loss such as human factors, technologic factors, and processes factors, etc. That is, FES allows us to evaluate the activities of corporate structure in all aspects. FES allows the desegregations of detailed factors associated with activities of processes or sub processes in such a way that lets the manager improve the process of management control.

Qualitative variables associated with operative risk can be identifying by expert opinions that manage and understand their process in all ranges of organizational structure. This information is modeling by FES. Therefore, the development of support systems to make decisions have more acceptances of the people, since the developed models are participative and intuitive. (The experts are building).

All companies are exposed to risks, some of these are easy to avoid and measure, for other however, their quantification is more difficult; for this reason, the appropriate management should be search new methodologies to improve the management toward competitive results. FES presented in this article, will allow identifying and measuring factors that generate a bigger or lower exposition to operational risk and prioritize the necessary resources to improve the management. The manager can develop strategic activities that allow getting accepted levels of operative loss.

With the purpose of lowers the possibility of operative loss, the companies can improve the risk management systems to promote the continuous improvement that allow to detect and to prevent the potential risks appropriately, also can be used insurance policy to cover the operative losses that can be materialized, the third alternative is transfer the risk through outsourcing agreement (transfer activities that bigger vulnerability and don't have a high added value), lastly the company can avoid carry out risks activities and that they don't affect the normal operation. An appropriate combination of previous alternatives can be employees with the purpose of minimize the exposition levels to operative losses and guarantee the survival and growth of company.

Future works that can be explored is the relationship between the statistical models used to measure operative value at risk VaR (POT Peak over threshold or Monte Carlo Simulation) and the FES model as proposed in this article. ¿How these models can be integrated? ¿how arrive to economic losses starting from the FES system?, is yet a question to be solved.

Other subject that can be explored is the relationship between the Net Bayesian and a FES, both models capture the expert's knowledge by inputs but the Net Bayesian obtain as output a probability and the FES obtain a possibility, which is the relationship between both in a real problem, could be investigated.

References

Basel Committee on Banking Supervision (BASEL I). International convergence of capital measurement and capital standards. Julio de (1988)

Basel Committee on Banking Supervision (BASEL II). Convergencia internacional de medidas y normas de capital. Junio de, Basilea – Suiza. p. 228 (2004)

Basel Committee on Banking Supervision (BASEL II).Sound practices for the management and supervisión of operational risk. Febrero de, Basilea – Suiza. p. 12 (2003)

Buckley, J., Eslami, E.: An introduction to Fuzzy Logic and Fuzzy set. Physica Verlag, Heidelberg (2002)

Chavez-Demoulin, V., Embrechts, P.y., Neslehová, J.: Quantitative models for operational risk: Extremes, dependence and aggregation. En: Journal of baking & Finance, 2635–2658 (2006)

Chernobai, A., Rachev, S.: Applying robust methods to operacional risk modeling. En: The Journal of operational risk 1(1) (2006)

Chin, K.S., Pun, K.F., Lau, H.: Development of a knowledge-based self-assessment system for measuring organisational performance. Expert Systems with Applications 24, 443–455 (2003)

Cornalba, C., Giudici, P.: Statistical models for operational risk management. En: Physica, 166–172 (2004)

Cruz, G., Marcelo: Modeling, Measuring and heading Operational Risk, 330 Pág. John Wiley & Sons, Ltd., Estados Unidos (2002)

Cruz, G., Marcelo: Operational Risk Modelling and Analys: Theory and practice, 360 Pág. Incisive Media Investments Limited, London (2004) (Book Risk)

Espinosa, J., Vandewalle, J.: Constructing Fuzzy Models with Linguistic Integrity from Numerical Data-AFRELI Algorithm. IEEE Transactions On Fuzzy Systems 8(5) (October 2000)

Glykas, M., Xirogiannis, G.: A soft knowledge modeling approach for geographically dispersed financial organizations. Soft Computing 9(8), 579–593 (2004)

Hammell, R., Sudkamp, T.: Interpolation, completion, and learning fuzzy rules. IEEE transaction on systems. Man and cybernetics 24(2), 332–342 (1994)

Herrera, F., Lozano, M., Verdegay, J.L.: Generating Fuzzy Rules from Examples using Genetic Algorithms. In: Bouchon-Meunier, B., Yager, R.R., Zadeh, L.A. (eds.) IPMU 1994. LNCS, vol. 945. Springer, Heidelberg (1995)

John, H.C.: Options, Futures, and Other Derivatives, 7th edn. McGraw Hill, New York (2008)

International Organization for Standardization (ISO). Draft: Risk management – Principles and guidelines on implementation (ISO 31000), p. 18 (2008)

Jang, J., Mizutani, E., Sun, C.: Neuro-Fuzzy and Soft Computing: A Computional Approach to Learning and Machine Intelligence, p. 600. Prentice Hall, Estados Unidos (1997)

Rodriguez, J., José, E., MaríN, M., Luis, J.: El nuevo acuerdo de Basilea y la gestión del riesgo operacional. En: Universia Business Review – actualidad económica. Tercer trimestre del, 54–67 (2005)

Kasabov, N.: Foundations of Neural Networks. In: Fuzzy Systems and Knowledge Engineering. The MIT Press, Estados Unidos (1998)

Kaufman, A., Gil Aluja, J. (1990). Las matemáticas del azar y la incertidumbre. Elementos básicos para su aplicación en economía. Editorial Ramon Areces. Madrid (1990)

Kaufman, A.: Introducción a la teoría de los subconjuntos borrosos. Elementos teóricos de Base. CECSA (1982)

Kosko, B.: Pensamiento Borroso. La nueva ciencia de la lógica borrosa. ed. Critica. Barcelona (1994)

Kulkarni, A.: Computer Vision and Fuzzy-Neuronal Systems. Prentice Hall, Englewood Cliffs (2001)

Mamdani, E., Gaines, B.: Fuzzy Reasoning and its applications. Publ. Academic Press, London (1981)

Mamdani, E.H.: Application on Fuzzy Logic to approximate reasoning using linguistic synthesis. IEEE Transactions on computers C26. Dic., 1182–1191 (1977)

Männle, M.: Identifying Rule-Based TSK Fuzzy Models. In: Proceedings of EUFIT, Aachen, Germany, September 1999, p. 286 (1999)

Medina, S.: Diseño de un sistema de soporte para la negociación de contratos de energía eléctrica. Tesis de doctorado. Universidad Complutense de Madrid. Madrid enero (2007)

Medina, S., Jaramillo, J.: Riesgo Financieros y Económicos editorial Universidad de Medellín- Medellín -Colombia, pp. 49-80 (2009)

Moscadelli, M.: The modelling of operational risk: experience with the analysis of the data collected by the Basel Committee. Julio de, p. 72 (2004)

Reiss, R.Y., Thomas, M.: Statistical analysis of extreme values, 2nd edn. Birkhauser, Basel (2001)

The Committee of Sponsoring Organizations of the Treadway Commission (Coso). Administración de riesgos corporativos – Marco integrado. Diciembre de, USA. p. 135 (2005)

Trillas, E.: Conjuntos Borrosos. Ed. Vicens vives (1980)

Tsang, E.C.C., Yeung, D.S.: Modelling Fuzzy Production Rules with Fuzzy Expert Networks. Expert Systems With Applications 13(3), 169–178 (1997)

Vose, D.: Quantitative Risk Analysis: A Guide to Monte Carlo Simulation Modelling, 328 Pág. John Wiley & Sons, Ltd., Estados Unidos (1996)

Wang, L.: Fuzzy systems are universal approximators. In: Proc. of Int. Conf. on Fuzzy Engineering, pp. 471–496 (1992)

Wang, L.X., Mendel, J.M.: Generating Fuzzy Rules by learning from examples. IEEE Trans. Systems Man Cybernet. 22, 1414–1427 (1992)

Wang, X.Z., Wang, Y.D., Xu, X.F., Ling, W.D., Yeung, D.S.: A New approach to Fuzzy Rule Generating: fuzzy extension matrix. Fuzzy Set and Systems 123, 291–306 (2001)

Xizhao, W., Jiarong, H.: Learning Optimization in simplifying fuzzy rules. Fuzzy Set and Systems 106, 349–356 (1999)

Xirogiannis, G., Chytas, P., Glykas, M., Valiris, G.: Intelligent impact assessment of HRM to the shareholder value. Expert Systems with Applications 35(4), 2017–2031 (2008)

Xirogiannis, G., Glykas, M.: Intelligent Modeling of e-Business Maturity. Expert Systems with Applications 32/2, 687–702 (2007)

Xirogiannis, G., Glykas, M.: Fuzzy Cognitive Maps in Business Analysis and Performance Driven Change. Journal of IEEE Transactions in Engineering Management 13(17) (2004)

Zadeh, L.A.: Probability Measures of Fuzzy events. J. Math. Anal. Appl. 23, 421–427 (1968)

Fuzzy Cognitive Maps in Banking Business Process Performance Measurement

George Xirogiannis, Michael Glykas, and Christos Staikouras

Abstract. This paper addresses the problem of designing an "intelligent" decision support methodology tool to act as a back end to financial planning. The methodology tool proposes a novel approach to supplementing typical financial strategy formulation projects by utilizing the fuzzy causal characteristics of Fuzzy Cognitive Maps (FCMs) to generate a hierarchical and dynamic network of interconnected profit and loss (P&L) concepts. By using FCMs, the mechanism simulates the efficiency of complex hierarchical financial models with imprecise relationships and external stimuli while quantifying the impact of strategic changes to the overall P&L status. Generic maps that supplement the decision making process demonstrate a roadmap for integrating hierarchical FCMs into the P&L model of typical financial sector enterprises. Preliminary experiments indicate that *ex ante* reasoning of the impact of strategic changes (actual or hypothetical) to the status of financial performance can be effective and realistic, without employing detailed P&L numerical calculations.

Keywords: Fuzzy cognitive maps, financial performance management, strategy planning, banking, profit and loss accounts.

George Xirogiannis
University of Piraeus, Department of Informatics, 80,
Karaoli & Dimitriou St. 185 34 Piraeus, Greece
Tel.: +30-210-4142000; Fax.: +30-210-4142328
e-mail: `georgex@unipi.gr`

Michael Glykas
University of Aegean, Department of Financial and Management Engineering,
31, Fostini Street, Chios, 82 100, Greece
Tel.: +30- 22710–35400; Fax.: +30- 22710-34499
e-mail: `mglikas@aegean.gr`

Christos Staikouras
Athens University of Economics and Business, Department of Accounting and Finance,
76, Patision Street, Athens, 10 434, Greece
Tel.: +30-210-8203459; Fax.: +30-210-8228816
e-mail: `cstaik@aueb.gr`

M. Glykas (Ed.): Fuzzy Cognitive Maps, STUDFUZZ 247, pp. 161–200, 2010.
springerlink.com © Springer-Verlag Berlin Heidelberg 2010

1 Introduction

Knowledge has been defined by Western philosophy [52] as "justified true belief". However as long as there is a chance that this belief is mistaken and as long as there is evolution of technologies, theories, practice and behaviors, this definition invites individuals and groups to develop decision mechanisms to utilize effectively what they know [47]. This continuous process of creation of new insights and beliefs is what fuels the entire paradigm of decision management and even constitutes the fundamental rationale for the existence of an enterprise. Some argue that instead of merely solving problems, organizations create and define problems, develop and apply new decision practices to solve problems and then develop further new knowledge through the action of problem solving. In this view an enterprise creates continuously new decision practices through action and interaction, not acting simply as an information-processing machine [47, 55].

Today, there is an increasing demand for a strategic-level qualitative assessment of the financial capabilities that can be assembled and analyzed rapidly at low cost and without significant intrusion into the subject accounting controllers. Typical financial plans consist mainly of quantitative estimates of key financial indicators. It is the view of this paper that a qualitative supplement of typical financial plans can contribute significantly to strategic planning at financial sector enterprises (and vise versa). The benefits of such an exercise are quite straightforward, for instance, identification of strengths and weaknesses, establishment of a strategic rationale for action, reference point for measuring future financial progress, etc.

This paper proposes a novel supplement to strategic-level financial planning at the banking sector based on fuzzy cognitive maps (FCMs). This decision aid mechanism proposes a new approach to supplementing the financial status analysis and financial objectives composition phases of typical financial strategy formulation projects, by supporting cognitive modeling of profit and loss (P&L) analysis and "intelligent" reasoning of the anticipated impact of strategic change initiatives to the financial status of a typical financial sector enterprise. The proposed mechanism utilizes the fuzzy causal characteristics of FCMs as a new modeling technique to develop a causal representation of dynamic financial principles in order to generate a hierarchical network of interconnected financial performance indicators. The proposed mechanism aims at simulating the efficiency of complex hierarchical financial models with imprecise relationships while quantifying the impact of strategic changes to the overall P&L status. The FCM hierarchies also integrate stimuli of the external environment (e.g. sector characteristics, market opportunities, national economy, etc), which may drive the strategic positioning of a typical bank but their effect to the P&L performance may not be quantified precisely. Finally, this paper proposes an updated FCM algorithm to model effectively the hierarchical and distributed nature of all such financial concepts.

This application of FCMs in modeling the P&L characteristics at the financial sector is considered to be novel. Moreover, it is the belief of this paper that the fuzzy reasoning capabilities enhance considerably the usefulness of the proposed mechanism while reducing the effort of identifying precise P&L measurements during a strategy formulation exercise. However, the FCM approach does not pose as a substitute of traditional accounting nor does it offer an alternative to numerical P&L calculation. It merely presents a holistic framework for strategic-level financial planning based on scenario building and *ex ante* impact assessment.

The proposed model has both theoretical and practical benefits. Given the demand for effective financial positioning of strategic change initiatives, such a succinct mechanism of conveying the essential dynamics of P&L planning is believed to be useful for anyone contemplating or undertaking a strategy formulation exercise at the banking sector. Primarily, the proposed model targets the principle beneficiaries and stakeholders of strategy formulation projects (bank administration, client facing division managers, internal auditors, planning and budgeting managers, etc) assisting them to reason effectively about the status of financial maturity metrics, given the (actual or hypothetical) implementation of strategic changes. Nevertheless, the explanatory nature of the mechanism can prove to be useful in a wider educational setting.

This paper consists of five sections. Section 2 presents a short literature overview, Section 2 presents an overview of the FCM based system, while Section 3 discusses the new approach to P&L modeling based on FCMs. Section 4 presents a sample set of P&L maps and section 5 discusses the preliminary experiments with the FCM based system. Finally, Section 6 concludes this paper and briefly discuses future research activities.

2 Literature Overview

2.1 Financial Performance at the Banking Sector

2.1.1 Multidimensional Financial Transactions

Typical banks at the end of the day are institutions that accept deposits and provide loans. On the microeconomic level, they represent the primary source of credit. In market economies, they serve the key purposes of providing financial intermediation and transaction services. They raise funds primarily by issuing checkable deposits (deposits on which checks can be written), savings deposits (deposits that are payable on demand but do not allow their owner to write checks), and time deposits (deposits with fixed terms to maturity). They often use these funds to make commercial, consumer, and mortgage loans.

Measuring bank performance is not always a trivial task due to the multidimensional and intangible nature of financial sector products and the lack of explicit prices for some of the output. Not so long ago the main objective of typical banks was growth and, consequently, the size of their balance sheet total. But, modern economies are based on production and consumption of increasingly differentiated products and services. In the case of the banking sector, this increased variety leads to the fragmentation and changing nature of the banking services. Banks that systematically manage for shareholder value at a strategic level stand the best chance of competing successfully in the new economy. Many banks are keenly interested in earning maximum profits in order to provide the highest possible returns to their shareholders and secure additional funds to support strategic (long-term) growth. They realise that higher profits may enhance the confidence among depositors and investors, making it easier to raise capital in the future.

2.1.2 Multidimensional Financial Performance

The Basle Committee has identified financial performance to be one of the six broad information categories, which should be addressed in clear terms and appropriate level of detail to achieve adequate financial transparency. The other categories are financial position (e.g. capital, solvency, liquidity, etc); risk management strategies and practices; risk exposures (e.g. credit, market, liquidity, operational and legal risks); accounting policies; basic business, and corporate government information.

Evaluating the banks' overall performance and monitoring their financial status is important to depositors, owners, potential investors, managers, and regulators. Currently, financial ratios are often used to measure the overall financial soundness of a bank and the quality of its financial management at the tactical level. It is convenient to analyse the results of a depository institution's operations using a series of performance dimensions like liquidity, credit risk exposure, financial leverage, efficiency or productivity, profitability etc. It is the view of this paper that ratios should be grouped according to the performance dimension, which they are most closely connected to. The performance measures employed in the literature are generally based on accounting profitability since many banks lack market data. Ricketts and Stover in [51] use ratios classified into seven major categories, namely liquidity, loan volume, loan quality, capital adequacy, efficiency, revenue sources, and profitability. Similarly, the performance ratios used by [9] standardize the portfolio composition (see also [14], [24], [41]), the capital composition like equity capital, total assets, etc (see also [5], [6], [7], [41], [44], [58]), the operating efficiency, the prices of services and finally the bank's profitability.

Along with increasing emphasis on asset / liability management, it becomes more important to embody the interactions between the various performance measures. Research [1] defines the performance of a new bank as an index of its profitability, pricing of its services (average loans and deposits rates) and its loan market share in the trade sector. While high loan rates (service fees) and low deposit rates could contribute to short-term profitability, the critical growth of a new bank's market share in a given trade sector can be affected negatively. Banks can also 'bundle' their services by adopting pricing strategies, which offer low loan rates and low deposit rates or high loan rates and high deposit rates with the same impact on profitability. Thus, the relationships between rates, market share, and profitability must be considered simultaneously.

Research [3, 7, 18] characterize the asset quality as the ratios of net loan losses over average loans, loan loss provisions over average loans and loan loss reserves over loans. Similarly, the research in [14] proposes a total of 26 measures of bank performance, like profitability, revenue, expenses, composition of assets, liabilities, loans, etc. However, the two most frequently used performance measures are the effective rate charged on loans and the average rate paid on time deposits. Kaufman [25] uses as activity measures the ratio of loans to total assets and the ratio of time to total deposits. Research [12] proposes that the enterprises should be measured by six financial performance categories, namely liquidity, growth, owner's earnings, management profit performance, leverage, and capital investment.

Meinster and Elyasiani [42] also propose sixteen measures grouped into four categories, namely asset management measures (e.g. asset structure, liquidity and portfolio risk, etc), liability and capital management measures (e.g. sources of fund management), pricing measures (e.g. service charges on deposit accounts, interest paid on deposits, interest charged on loans, etc) and expense / profitability measures (e.g. management performance, total operating expenses over total assets, return on equity, dividend payout ratio, loan losses over total loans, etc). Finally, the research in [43] considers the effect of a number of financial ratios that measure asset management (e.g. lending and investment), liability management (e.g. funding), productivity and efficiency, etc, on bank performance.

2.1.3 Multidimensional Profitability

Financial institutions usually focus on profits; hence we can define performance in terms of economic capacity as measured by a host of financial indicators. The performance measures employed in the literature are generally based on accounting costs or profitability. However, market-based measures have been also used, for example price-to-earnings ratios, the stock beta and alpha, and Tobin's q ratios. The research in [40] argues that some indication of the price of products and services is not a always a good performance measure. Banking is a multi-product industry and cross subsidization among products and services often occurs. Prices can only be used if costs are directly associated with these prices and are explicitly accounted for as explanatory variables. Individual prices of products can often be misleading. The use of profit measures should eliminate many of these potential problems. Profitability measures, where all product profits and losses are consolidated into one figure, are generally viewed as more suitable because they by-pass the problem of cross-subsidization.

Profitability can be used as a summary index of performance as argued by [19]. Adequacy of earnings is needed to provide shareholders a sufficient return, to generate sufficient cash flows in order to cover borrowers demand and to provide for future needs through the development of capital. Gilbert [16] has identified that profit rates are the only performance measures that do not impose major measurement problems.

Other common scale factors are total assets, total deposits, total equity capital (net worth), total loans, total revenue (operating income), total expenses, and number of employees. Usually, comparisons of profitability are made using accounting return on assets (ROA), and accounting return on equity (ROE). The profitability variable equals the total profits before tax (or net income) of all banks in a country divided by the total assets (or total equity) of all banks in that country. ROA reflects management ability to utilize the bank's resources in order to generate net income. Analysts looking to compare profitability (while ignoring differences in equity capital ratios) generally focus on ROA, while those wishing to focus on returns to shareholders focus on ROE. For profitability comparison between countries, net income, assets, and book value of equity are aggregated to the state level by summing across all banks operating in a given country. As an even larger portion of banks' activities moves off the balance sheet, bank management is becoming more sensitive to the shortcomings of conventional measures of operating performance like ROA and ROE. Many banks replace such accounting-based measures with economic performance measures like risk-adjusted return on capital (RAROC) and economic value added (EVA) [57]. These measures quantify better the fact that banks put their capital at risk, seeking capital returns high enough to reward the shareholders.

2.2 Accounting Performance and Valuation

To assess the financial performance of a bank (at least as far as P&L analysis is concerned), it is the view of this paper that it is essential to have a comprehensive breakdown of income and expenses incurred. This information is necessary to assess the quality of earnings, to identify the causes for profitability fluctuation over the years, and to compare the bank's financial performance with that of the competition. The income statement usually includes items for interest income / expenses, fees, commissions, other non-interest income, operating expenses, charge for credit losses, any extraordinary items, tax expenses, net income, etc. Key figures and ratios should include the return on average equity, return on average assets, net interest margin, and cost-to-income ratio.

2.2.1 Return in Equity

Return on equity (ROE) measures profitability from the shareholders' perspective. This enables the evaluation of the source and magnitude of profits relative to selected risks taken. Generally speaking, it is defined as the net income divided by the average equity or by the period-ending figure. The ROE model simply relates ROE to return on assets (ROA) and financial leverage [equity multiplier (EM)], and then decomposes ROA into its contributing elements.

ROA is defined as net income divided by average or total assets. Four pieces of accounting information are required to start ROE analysis: net income, total operating income, average assets and average equity. The first two pieces of information are flow variables that come from a bank's income statement, while the last two are stock variables that come from the balance sheets. The second stage of ROE decomposition shows ROA to be derived from the bank's profit margin (PM) and asset utilization (AU), where PM equals net income divided by total revenue and AU equals total revenue divided by total assets. The components of the third stage of the ROE decomposition framework usually are analyzed with respect to a bank's total revenue or total assets. The objective of this stage of the analysis is to identify symptoms of good and bad performance by pinpointing trends and significant peer-group differences.

2.2.2 Profit and Loss Analysis

A bank's income statement indicates the amount of revenue received and expenses incurred over a specific period of time. To understand the financial status of a bank, one may need to look at the bank's income statement, the description of the sources of income and the expenses that affect the bank's profitability.

Operating income is the income that comes from a bank's ongoing operations. Most of a bank's operating income is generated by interest on its assets, particularly loans. The principal source of bank revenue is the **interest income** generated by a bank's earning assets, mainly its loans, securities, any interest-bearing deposits and any miscellaneous assets generating revenue (including any income earned by subsidiaries of the bank or rental income from property that it owns). The relative importance of these income items fluctuates from year to year with shifts in interest rates and loan demand, though loan income is nearly always the dominant revenue source. The difference between a bank's total interest income and total interest expense is called net interest income. For a traditional bank engaged in funding loans with deposits, net interest income represents the "bread and butter" of the business.

Non-interest income is a mixture of heterogeneous components that differ in terms of their relative importance (i.e. their share in banks' non-interest income). The most important non-interest income source is **fee and commission income**. Fees and service charges on deposits typically generate the bulk of non-interest income. Fees and commissions can in turn be divided into various sub-components, such as net commissions on lending activities, on mutual funds and asset under management, on capital markets, and on network activities and other services. Recently, bankers have targeted fee income as a key source of future revenues. By selling services other than loans more aggressively (such as security brokerage, insurance, and trust services), bankers have identified a promising channel for boosting the bottom line of their income statements, for diversifying their income sources, and for insulating their banks more adequately from fluctuations in interest rates. Fees and commissions are the most stable component of non-interest income. Therefore, further development of this source of revenue, could attribute to a further reduction in the global volatility of banks' income.

The net result (profit or loss) of financial operations (e.g. net result of transactions in securities which are not held as financial assets, value adjustments and value re-adjustments on such securities, net result on exchange activities, net result on other buying and selling operations involving financial instruments, etc) constitutes the second component of non-interest income in terms of its relative importance. Trading activities, although often highly profitable, are risky because they make it easy for financial institutions and their employees to make significant bets both easily and quickly. Securities gains (or losses) arise when a bank sells securities from its investment portfolio at prices above (or below) their cost to the bank. Securities gains are generally viewed as an unpredictable and unstable source of income because forecasting interest rates and whether the bank can sell securities for a profit or loss is difficult. A particular problem for managing trading activities is the principal-agent problem. Given the ability to place large bets, a trader (the agent), whether he/she trades in bond markets, in foreign exchange markets, or in financial derivatives, has an incentive to take on excessive risks.

Operating expenses are the expenses incurred in conducting the bank's ongoing operations. It is composed primarily of personnel expenses, which include salaries and fringe benefits paid to bank employees, occupancy expense from rent and depreciation on equipment and premises, and other operating expenses, including utilities, legal fees, marketing costs, office supplies and repair costs. Reducing this burden will improve profitability.

Another expense item that banks can deduct from current income is known as the **provisions** for potential loan losses. This provision account is really a non-cash expense, created by a simple bookkeeping entry. Its purpose is to shelter a portion of the bank's current earnings before taxes to help prepare for bad loans. The annual loan loss provision is deducted from current revenues before taxes are applied to earnings. When a bank has a bad debt or anticipates that a loan might become a bad debt in the future, it can write up the loss as a current expense in its income statement under the "provision for loan losses" heading. Provisions for loan losses are directly related to loan loss reserves.

2.3 FCMs as a Modeling Technique

Fuzzy Cognitive Maps is a modeling methodology for complex decision systems, which originated from the combination of Fuzzy Logic [61] and Neural Networks. An FCM describes the behavior of a system in terms of concepts; each concept represents an entity, a state, a variable, or a characteristic of the system [10].

Kosko in [29] defined a concept C_i that constitutes causal relationships in FCM as

$$C_i = (Q_i \cup \sim Q_i) \cap M_i$$

where Q_i is a quantity fuzzy set and $\sim Q_i$ is a dis-quantity fuzzy set. $\sim Q_i$ is the negation of Q_i. Each Q_i and $\sim Q_i$ partitions the whole set C_i. Double negation $\sim \sim Q_i$ equals to Q_i, implying that $\sim Q_i$ corresponds to Q_i^c, the complement of Q_i. However, negation does not mean antonym. Therefore, if a dis-quantity fuzzy set $\sim Q_i$ does not correspond to the complement of Q_i, we will call it as anti-quantity fuzzy set to clarify the subtle meaning in the dis-quantity fuzzy set, as proposed by [26]. M_i is a modifier fuzzy set that modifies Q_i or $\sim Q_i$ concretely. The modifier fuzzy set fuzzily intersects the fuzzy union of a quantity fuzzy set and a dis-quantity fuzzy set.

Kosko in [29] also formally defined the positive and negative fuzzy causal relationships (or fuzzy causality) as follows.

Definition 1. C_i causes C_j iff $(Q_i \cap M_i) \subset (Q_j \cap M_j)$ and $(\sim Q_i \cap M_i) \subset (\sim Q_j \cap M_j)$.

Definition 2. C_i causally decreases C_j iff $(Q_i \cap M_i) \subset (\sim Q_j \cap M_j)$ and $(\sim Q_i \cap M_i) \subset (Q_j \cap M_j)$.

Here "$\subset$" stands for fuzzy set inclusion (logical implication).

A more insightful and practical definition of FCMs follows. FCM nodes are named by concepts forming the set of concepts $C=\{C_1, C_2, \ldots, C_n\}$. Arcs (C_j, C_i) are oriented and represent causal links between concepts; that is how concept C_j causes concept C_i. Arcs are elements of the set $A=\{(C_j, C_i)_{ji}\}\subset CxC$. Weights of arcs are associated with a weight value matrix W_{nxn}, where each element of the matrix $w_{ji}\in[-1,..,1]\subset R$ such that if $(C_j, C_i)\notin A$ then $w_{ji}=0$ else excitation (respectively inhibition) causal link from concept C_j to concept C_i gives $w_{ji}>0$ (respectively $w_{ji}<0$). The proposed methodology framework assumes that $[-1,..,1]$ is a fuzzy bipolar interval, bipolarity being used as a means of representing a positive or negative relationship between two concepts.

In practice, the graphical illustration of an FCM is a signed graph with feedback, consisting of nodes and weighted interconnections (e.g. $\xrightarrow{\text{Weight}}$). Signed and weighted arcs (elements of the set A) connect various nodes (elements of the set C) representing the causal relationships that exist among concepts. This graphical representation (e.g. Figure 1) illustrates different aspects in the behavior of the system, showing its dynamics [29] and allowing systematic causal propagation (e.g. forward and backward chaining). Positive or negative sign and fuzzy weights model the expert knowledge of the causal relationships [31]. Concept C_j causally increases C_i if the weight value $w_{ji}>0$ and causally decreases C_i if $w_{ji}<0$. When $w_{ji}=0$, concept C_j has no causal effect on C_i. The sign of w_{ji} indicates whether the relationship between concepts is positive ($C_j \xrightarrow{W_{j,i}} C_i$) or negative ($C_j \xrightarrow{W_{j,i}} \sim C_i$), while the value of w_{ji} indicates how strongly concept C_j influences concept C_i. The forward or backward direction of causality indicates whether concept C_j causes concept C_i or vice versa.

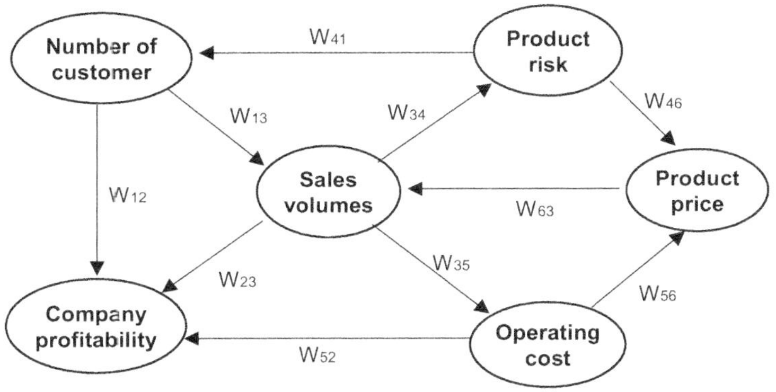

Fig. 1 Simple FCM

Simple variations of FCMs mostly used in business decision-making applications may take trivalent weight values **[-1,0,1]**. This paper allows FMCs to utilize fuzzy word weights like strong, medium, or weak, each of these words being a fuzzy set to provide complicated FCMs. In contrast, [32] adopted only a simple relative weight representation in the interval **[-1,..,1]**. To this extend, research [32] offered reduced functionality since it did not allow fuzzy weight definitions.

Generally speaking FCM concept activations take their value in an activation value set $V = \{0,1\}$ or $\{-1,0,1\}$ if in crisp mode or $[-\delta,1]$ with $\delta=0$ or 1 if in fuzzy mode. The proposed methodology framework assumes fuzzy mode with $\delta=1$. At step $t \in N$, each concept C_j is associated with an inner activation value $a_j^t \in V$, and an external activation value $e_{a_j}^t \in R$. FCM is a dynamic system. Initialization is $a_j^0=0$. The dynamic obeys a general recurrent relation $a^{t+1} = f(g(e_a^t, W^T a^t))$, $\forall t \geq 0$, involving weight matrix product with inner activation, fuzzy logical operators (**g**) between this result and external forced activation and finally normalization (**f**). However, this paper assumes no external activation (hence no fuzzy logical operators), resulting to the following typical formula for calculating the values of concepts of FCM:

$$a_i^{t+1} = f\Big(\sum_{j=1, j\neq i}^{n} w_{ji} a_j^t\Big) \tag{1}$$

where a_i^{t+1} is the value of concept C_i at step $t+1$, a_j^t the value of the interconnected concept C_j at step t, w_{ji} is the weighted arc from C_j to C_i and $f: R \rightarrow V$ is a threshold function, which normalizes activations. Two threshold functions are usually used. The unipolar sigmoid function where $\lambda>0$ determines the steepness of the continuous function $f(x) = \dfrac{1}{1+e^{-\lambda x}}$. When concepts can be negative ($\delta<0$), function $f(x)=\tanh(x)$ is used.

To understand better the analogy between the sign of the weight and the positive/negative relationship, it may be necessary to revisit the characteristics of fuzzy relation [25, 36]. A fuzzy relation from a set A to a set B or (A,B) represents its degree of membership in the unit interval $[0,1]$. Generally speaking, sets A and B can be fuzzy sets. The corresponding fuzzy membership function is $\mu f: AxB \rightarrow [0,1]$. Therefore, $\mu f(x,y)$ is interpreted as the "strength" of the fuzzy membership of the fuzzy relation (x,y) where $x \in A$ and $y \in B$. Then this fuzzy relation concept can be denoted equivalently as $x \xrightarrow{\mu f} y$ and applied to interpret the causality value of FCM, since w_{ji} (the causality value of the arc from nodes C_j to C_i) is interpreted as the degree of fuzzy relationship between two nodes C_j and C_i. Hence, w_{ji} in FCMs is the fuzzy membership value $\mu f(C_j, C_i)$ and can be denoted as $C_j \xrightarrow{w_{j,i}} C_i$.

However, we understand that the fuzzy relation (weight) between concept nodes is more general than the original fuzzy relation concept. This is because it can include negative (-) fuzzy relations. Fuzzy relations mean fuzzy causality; causality can have a negative sign. In FCMs, the negative fuzzy relation (or causality) between two concept nodes is the degree of a relation with a "negation" of a concept node. For example, if the negation of a concept node C_i is noted as $\sim C_i$, then $\mu f(C_j, C_i)=-0.6$ means that $\mu f(C_j, \sim C_i)=0.6$. Conversely, $\mu f(C_j, C_i)=0.6$ means that $\mu f(C_j, \sim C_i)=-0.6$.

FCMs help to predict the evolution of the system (simulation of behavior) and can be equipped with capacities of hebbian learning [28], [30]. FCMs are used to represent and to model the knowledge on the examining system. Existing knowledge of the behavior of the system is stored in the structure of nodes and interconnections of the map. The fundamental difference between FCMs and a Neural Networks is in the fact that all the nodes of the FCM graph have a strong semantic defined by the modeling of the concept whereas the nor input / nor output nodes of the neural network have a weak semantic, only defined by mathematical relations.

2.4 Applications of Fuzzy Cognitive Maps

Over the last years, a variety of FCMs have been used for capturing - representing knowledge and intelligent information in engineering applications, for instance, GIS [39] and fault detection (e.g. [45], [48]). FCMs have been used in modeling the supervision of distributed systems [56]. They have also been used in operations research [8], web data mining [20, 36], as a back end to computer-based models and medical diagnosis (e.g. [15]).

Several research reports applying basic concepts of FCMs have also been presented in the field of business [59] and other social sciences. Research in [2] and [49] have used FCM for representing tacit knowledge in political and social analysis. FCMs have been successfully applied to various fields such as decision making in complex war games [27], strategic planning [11], [50], information retrieval [21] and distributed decision process modeling [62]. Research like [34] has successfully applied FCMs to infer rich implications from stock market analysis results. Research like [35] also suggested a new concept of fuzzy causal relations found in FCMs and applied it to analyze and predict stock market trends. The inference power of FCMs has also been adopted to analyze the competition between two companies, which are assumed to use differential games mechanisms to set up their own strategic planning [37]. FCMs have been integrated with case-based reasoning technique to build organizational memory in the field of knowledge management [46]. Recent research adopted FCMs to support the core activities of highly technical functions like urban design [60]. Summarizing, FCMs can contribute to the construction of more intelligent systems, since the more intelligent a system becomes, the more symbolic and fuzzy representations it utilizes.

In addition, a few modifications have been proposed. For example, the research in [54] proposed new forms of combined matrices for FCMs, the research in [17] extended FCMs by permitting non-linear and time delay on the arcs, the research in [53] presented a method for automatically constructing FCMs. More recently, [39] has carried extensive research on FCMs investigating inference properties of FCMs, proposed contextual FCMs based on the object-oriented paradigm of decision support and applied contextual FCMs to geographical information systems [38].

2.5 Updated FCM Algorithm

This paper extends the basic FCM algorithm (as discussed in section 2.3 and also used by [32]), by proposing the following new FCM algorithm:

$$a_i^{t+1} = f(k_1 a_i^t + k_2 * \sum_{j=1, j\neq i}^{n} w_{ji} a_j^t) \tag{2}$$

This paper assumes that coefficients k_1 and k_2 can be fuzzy sets.

Coefficient k_1 represents the proportion of the contribution of the value of the concept a_i at time t in the computation of the value of a_i at time $t+1$. In practice, this is equivalent to assume that $w_{ii}=k_1$. The incorporation of this coefficient results in smoother variation of concept values during the iterations of the FCM algorithm. Coefficient k_2 expresses the "influence" of the interconnected concepts in the configuration of the value of the concept a_i at time $t+1$. It is the proposal of this paper that such a coefficient should be used to align indirectly causal relationships (essentially, the value of concept C_i) with the influence of concept C_j as follows:

- If the set of identified performance concepts C_j, $j\neq i$, is incomplete (e.g. incomplete maps, missing concepts, etc), then the estimation of the value of concept C_i may prove imprecise. In this case coefficient k_2 may indicate the sufficiency of the set of concepts C_j $j\neq i$, in the calculation of the value of the concept C_i.
- If the information necessary to approximate the input values of concepts C_j, $j\neq i$, is incomplete (e.g. incomplete estimation of bad loans), then the estimation of the value of concept C_i may also prove imprecise. In this case coefficient k_2 may indicate the completeness of information utilized in the approximation of the input values of concepts C_j during the calculation of the value of the concept C_i.

Ideally, coefficient $\mathbf{k_2}$ could break down into two separate coefficients (say $\mathbf{k_2 = x * k_2^x + y * k_2^y}$), where $\mathbf{k_2^x}$ aligns indirectly the value of concept $\mathbf{c_i}$ with the completeness of the set of concepts $\mathbf{c_j}$ (e.g. completeness of P&L performance indicators), while $\mathbf{k_2^y}$ aligns indirectly the value of concept $\mathbf{c_i}$ with the completeness of available information for concepts $\mathbf{c_j}$ within the enterprise (e.g lack of information for certain P&L indicators). Parameters $\mathbf{x}$, $\mathbf{y}$ could present the relative importance of $\mathbf{k_2^x}$ and $\mathbf{k_2^y}$ in mixed interconnection problems (e.g. incomplete set of P&L indicators with partial accounting results). However, preliminary experiments showed that this separation imposed unnecessary initialization overheads without increasing significantly the accuracy of the FCM algorithm.

In contrast to the basic FCM algorithm adopted by most relevant research practices the updated one may suit better the financial domain because:

- Coefficient $\mathbf{k_2}$ "normalizes" the FCM calculations based on incomplete information sets. The updated algorithm supports better the qualitative and trend-based financial planning, providing less conservative results. This normalization proves important at business domains mainly because:
  - ➢ it relaxes the need for extra calculations of error margins as a result of incomplete background information,
  - ➢ it provides reasonable decision modeling approximations without requiring extensive background financial analysis (e.g. complete estimation of bad loans, identification of all financial concept links, etc).
- Coefficient $\mathbf{k_1}$ results in smoother variation of concept values during the iterations of the algorithm.

3 FCM–Based Decision Modeling

3.1 FCM as a Supplement to Financial Planning

A typical financial strategy methodology consists of a series of phases for re-designing the financial aspects of an enterprise:

- **Phase 1:** Current status analysis, best practices & benchmarking
- **Phase 2:** Financial strategy vision & positioning
- **Phase 3:** Financial objectives synthesis (including selection of financial performance indicators)
- **Phase 4:** Action planning

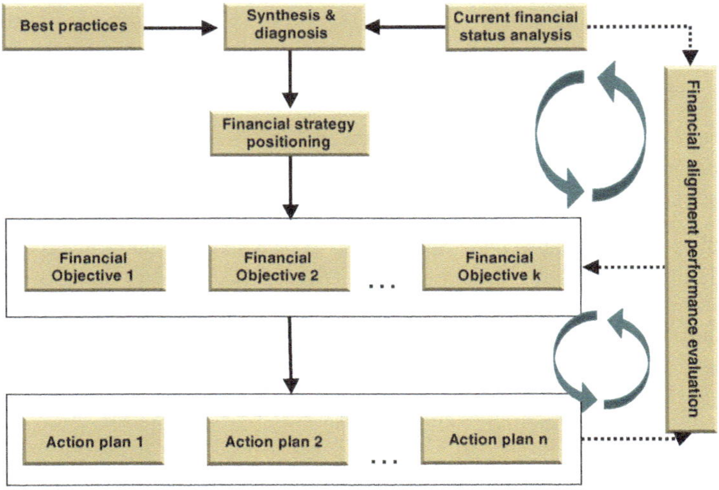

Fig. 2 Overview of financial planning

The proposed mechanism focuses on supplementing a typical financial strategy methodology by providing a holistic evaluation framework based on P&L performance indicators augmented by external environment stimuli. In practice the mechanism supplements the recurring feedback loop between the current financial status, the future strategic objectives and the action plans for improving profitability (the action plan, in turn, affects the future financial status). This mechanism creates a strategy-level utilization of P&L model based on a qualitative and trend-based technique. The proposed mechanism actually generates two financial assessment flows:

- Flow A: "Current financial status analysis → Financial objectives" to estimate the gap between existing ("as-is") and future ("to-be") profitability and support the establishment of objectives which should bridge this gap.
- Flow B: "Action plans → Financial objectives" to estimate the impact of strategic change actions to the evolution of financial status, assess anticipated financial maturity and align financial objectives to meet any potential deviations.

During the third phase of this typical financial strategy formulation exercise, the top management of the bank sets the overall financial performance targets (measured by associated metrics). These targets are exemplified further to action plan performance targets (measured by tactical financial metrics) and then to operational financial performance targets (measured by operational financial metrics). All such targets and metrics present inherent relationships. In practice, overall financial strategy metrics must cascade to tactical financial metrics to allow the middle management to comprehend inherent relations among the different managerial levels of the bank. Similarly, tactical financial metrics must propagate up the overall financial metrics. While P&L inheritance is usually clear, external stimuli with no apparent relationships to financial indicators are not always well defined.

This paper proposes the utilization of such indicators (Figure 3) to develop the FCMs and reason about the impact of strategic positioning changes to the desired ("to-be") financial models. The proposed mechanism utilizes FCMs to interpret:

- financial metrics (P&L indicators, external stimuli, etc) as concepts (graphically represented as nodes),
- decision weights as relationship weights (graphically represented as arrowhead lines),
- decision variables as financial concept values,
- hierarchical decomposition (top-down decomposition) of financial metrics to constituent sub-metrics as a hierarchy of FCMs. This interpretation allows the stakeholders to reason about lower level FCMs first (constituent financial metrics indicators) before they reason about higher-level metrics (affected metrics).

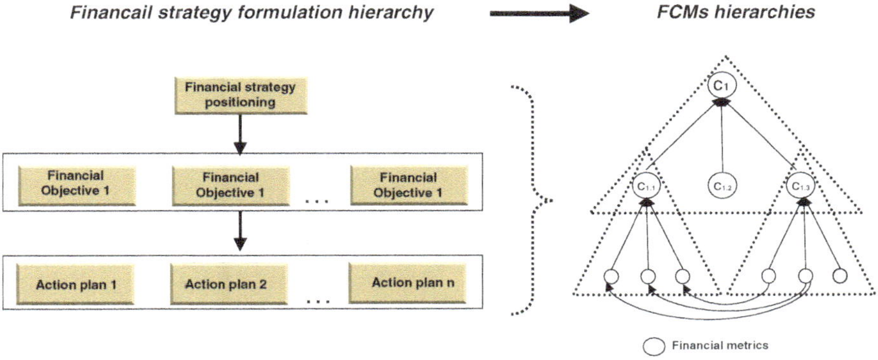

Fig. 3 Inherent relationships between financial strategy and FCM hierarchies

The proposed mechanism supports reasoning about the overall or partial financial strategy implementation using indicators from the P&L philosophy. In contrast to [32], the proposed mechanism builds on hierarchical metrics interrelationships identified and utilized by the financial strategy formulation methodology. The proposed approach does not perform or guide the implementation of any stage of the strategy formulation methodology. Also, the approach does not perform or guide the estimation of the absolute value of any of the financial metrics and/or the overall P&L performance. It only allows the stakeholders to reason about the qualitative state of financial maturity metrics using fuzzy linguistic variables like high–neutral-margin, high-neutral-low impact of loans volume to income, etc.

3.2 Assigning Fuzzy Linguistic Variables to FCM Weights and Concepts

In order to define weight value of the association relationships in an adaptive and dynamic manner, the following methodology is proposed. Managers are asked to describe the interconnection *influence* of concepts using linguistic notions. *Influence* of one concept over another, is interpreted as a linguistic variable in the interval **[-1,1]**. Its term set

T(influence) = {negatively very-very high, negatively very high, negatively high, negatively medium, negatively low, negatively very low, negatively very-very low, zero, positively very-very low, positively very low, positively low, positively medium, positively high, positively very high, positively very-very high}.

This paper proposes a semantic rule **M** to be defined at this point. The above-mentioned terms are characterized by the fuzzy sets whose membership functions μ are shown in Figure 4.

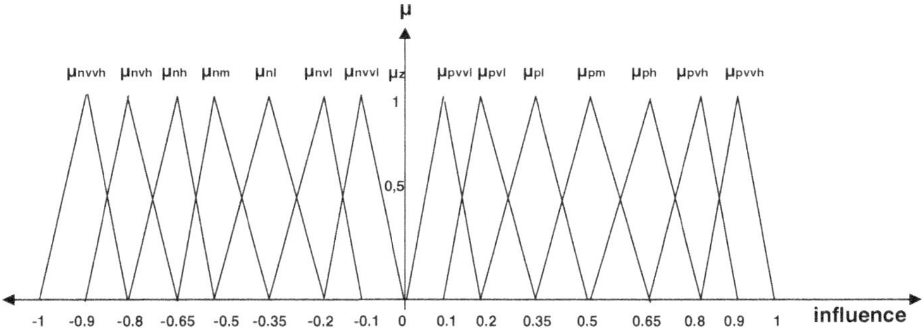

Fig. 4 Membership functions of linguistic variable influence

- **M(zero)=** the fuzzy set for "an influence close to 0" with membership function μ_z
- **M(positively very-very low)=** the fuzzy set for "an influence close to 10%" with membership function μ_{pvvl}
- **M(positively very low)=** the fuzzy set for "an influence close to 20%" with membership function μ_{pvl}
- **M(positively low)=** the fuzzy set for "an influence close to 35%" with membership function μ_{pl}
- **M(positively medium)=** the fuzzy set for "an influence close to 50%" with membership function μ_{pm}

- **M(positively high)=** the fuzzy set for "an influence close to 65%" with membership function μ_{ph}
- **M(positively very high)=** the fuzzy set for "an influence close to 80%" with membership function μ_{pvh}
- **M(positively very-very high)=** the fuzzy set for "an influence close to 90%" with membership function μ_{pvvh}
- Similarly for negative values

The membership functions are not of the same size since it is desirable to have finer distinction between grades in the lower and higher end of the influence scale. Linguistics are integrated using a sum combination method and then the defuzzification method of center of gravity (CoG) produces the weight value in the interval **[-1,1]**. This approach has the advantage that experts do not have to assign numerical causality weights but to describe the degree of causality among concepts. The same semantic rule and term set can be used to define the coefficients k_1 and k_2.

A similar methodology can be used to assign values to concepts. The financial managers are also asked to describe the *measurement* of each concept using once again linguistic notions. *Measurement* of a concept is also interpreted as a linguistic variable with values in the interval **[-1,1]**. Its term set **T(Measurement)** = **T(Influence)**. A new semantic rule M_2 *(*analogous to **M***)* is also defined and these terms are characterized by the fuzzy sets whose membership functions μ_2 are analogous to membership functions μ.

In practice, stakeholders may provide their expert estimates based on past P&L analysis, statistical analysis of trends and opportunities, best practices, benchmarking, brainstorming with client facing division managers, etc. The exact definition method of the expert linguistic variables is of little concern to this research approach. The definition of fuzzy membership functions allows rich interpretation of such variables.

As an example, consider again Figure 1. Assume that three experts propose different linguistic weights for the same interconnection w_{ij} from concept C_i = "Product Risk" to concept C_j = "Product Price" as follows: (a) positively very low (b) positively high (c) positively very high.

- The linguistic variable "positively very low" (see Figure 5, triangle A) with a membership function μ_{pl} has a $CoG_A = 0.216$
- The linguistic variable "positively strong" (see Figure 5, triangle B) with a membership function μ_{ph} has a $CoG_B = 0.65$
- The linguistic variable "positively very strong" (see Figure 5, triangle C) with a membership function μ_{pvh} has a $CoG_C = 0.783$
- The defuzzified linguistic variables produce a weight $w_{ij} = \dfrac{GoC_A + GoC_B + GoC_C}{3} = 0.549$.

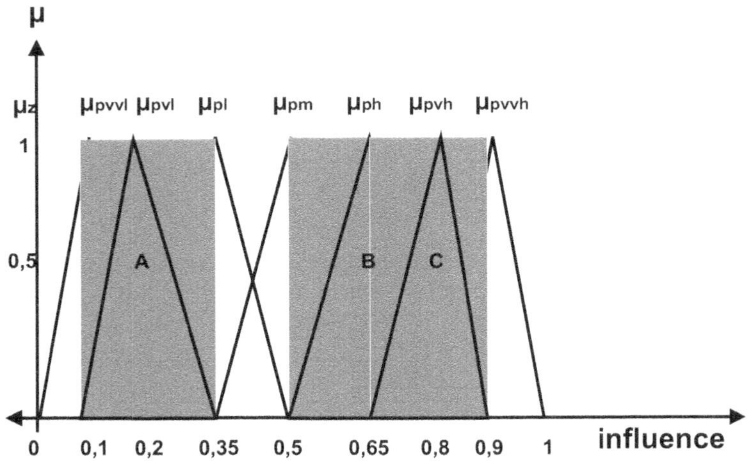

Fig. 5 Linguistic assignments

The fuzzy definitions in cognitive maps at such a business domain are considered to be contemporary. In contrast to other research practices (sections 2.3 and 2.4) the proposed utilization of fuzzy variables:

- Departs from the crisp and/or probabilistic definition of weight and concept values. In contrast, other research practices support only crisp definitions (e.g. −1, 0 or 1) asking for substantial information analysis which may not be available during ex-ante decision making.
- Departs from accurate arithmetic definitions of concept / weight values. It requires term set estimations (e.g. very high, high, etc) which couple closely to the human representation of expert knowledge.
- Caters for independent weight value definition. In contrast, other research practices support relative weight definitions (that is $\sum_{j=1}^{n} w_{ji} = 1$), reducing the flexibility of expert input. Similarly for concept value definitions.
- Caters for qualitative definitions to align with the soft modeling characteristics of cognitive maps. It also aligns with the trend-based decision support usually required for strategic level decisions. The interpretation of fuzzy linguistic variables to arithmetic values is transparent to the user of the tool. In contrast, the quantitative approach offered by relevant research practices usually asks for substantial information analysis which may not be available during ex-ante decision making.

3.3 *Financial Modelling Process*

Consider once again Figure 2. The following basic steps demonstrate the utilization of FCMs in ex-ante financial modelling and planning, that is modelling information flow "Action plans → Financial objectives".

- Step 1 - Financial planning: Assume that a bank has already developed a strategic-level financial plan following the typical structure presented in Figure 2.
- Step 2 – Skeleton FCMs: Utilize the technique discussed in section 3.1 to interpret strategic objectives and actions to skeleton maps. Skeleton FCMs present concepts and links with no value assignments, in practice, generic P&L interconnections.
- Step 3 – Weight value assignment: Bank experts are asked to provide linguistic weight variables for the map developed during the second step. Different weight value assignments generate different business cases for the same skeleton map.
- Step 4 – Quantify potential changes: Stakeholders assign fuzzy linguistic input values to P&L concepts, to quantify potential actions.
- Step 5 – Simulation: The FCM model simulates the impact of strategic changes (actions) to the financial objectives.

Similar steps demonstrate flow "Current financial status analysis → Financial objectives".

The following generic example demonstrates financial modelling using FCMs.

- Step 1 - Financial planning: Assume that a bank has already developed a simple strategic-level financial plan. Assume that an objective asks for increased profitability. Also, assume that the financial plan proposes the increase of sales volume as the sole action to achieve this objective.
- Step 2 – Skeleton FCMs: Utilization of the technique discussed in section 3.1 to interpret strategic objectives and actions to skeleton maps.
- Step 3 – Weight value assignment: Experts provide linguistic weight variables which are interpreted to weight values (Figure 6, LHS).
- Step 4 – Quantify strategic changes: Stakeholders assign input values to concept "sales volume" (Figure 6, RHS).
- Step 5 – Simulation: The FCM model simulates the impact of sales volume changes to the financial objectives and outputs the estimated effect.

Figure 6 depicts a graphical example with no feedback loops followed by sample numerical calculations using formula (2). This example assumes that $k_1=k_2=1$ and $\lambda=5$ as the steepness of the normalization function. Setting the input variable of "sales volume" to "positively medium" (defuzzified to 0,5 or 50%) triggers the FCM formula (1st case). A zero external concept value indicates that the concept remains neutral, waiting for causal relationships to modify its current value. A generic interpretation of the first case indicates that a "positively medium" increase in sales volume increases the income 88% and the profitability by 95%.

In contrast, a "positively very low increase" in "sales volume" (defuzzified to 0,2 or 20%) increases the income and the profitability by 59% and 89% respectively (2nd scenario).

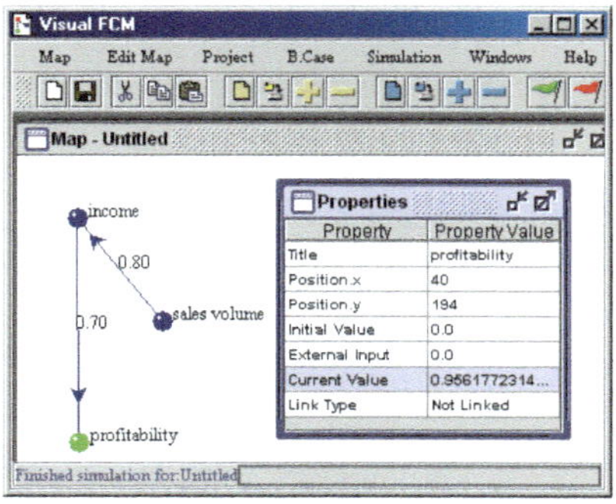

Concept	External input value (t=0)		Current Value	
	Change case A	Change case B	Change case A	Change case B
Sales volume	0.5	0,2	0.5	0,2
Income	0	0	0,8807	0,5986
Profitability	0	0	0,9561	0,8904

Fig. 6 Sample FCM calculations with no feedback loop

Figure 7 presents a typical example of a feedback loop. Similarly to Figure 6, changing the external input value of "sales volume" triggers the FCM formula. However, the feedback loop dictates that calculations stop only when an equilibrium state for all affected concepts has been reached, modifying all input values accordingly.

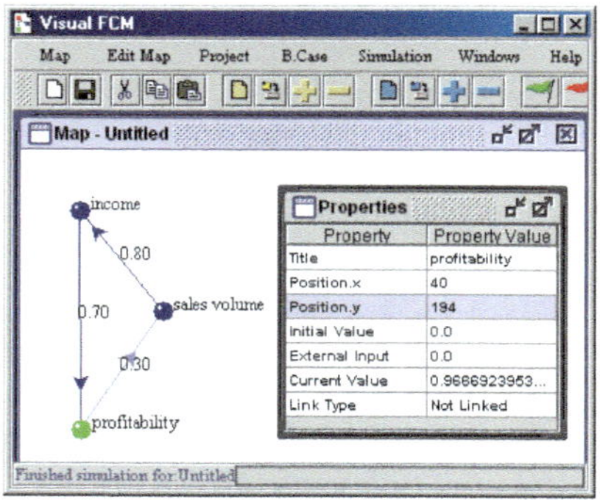

Concept	Initial input (t=0)		Current Value	
	Change case A	Change case B	Change case A	Change case B
Sales volume	0.5	0,2	0,81	0,5128
Income	0	0	0.9623	0,8860
Profitability	0	0	0,9666	0,9569

Fig. 7 Sample FCM calculations with feedback loop

4 FCM Hierarchies

4.1 FCMs Overview

This research team uses the Quanta application tool, a robust visual implementation of FCMs. The implementation of Quanta has been funded by the

ESPRIT E.U. programme. The current implementation of the proposed methodology tool encodes generic maps that can supplement the maturity modeling by storing concepts under different map categories (), namely:

- Business category: all concepts relating to core financial activities.
- Social category: all personnel related financial concepts and external stimuli concepts.
- Technical category: all concepts relating to infrastructure and technology related expenses.
- Integrated category: all top-most concepts (e.g. a concept C_i with no backward causality such that $\forall j : w_{ji}=0$), or concepts which may fall under more than one main categories.

The dynamic nature of the approach allows easy reconfiguration. Further P&L indicators may be added, while concepts may be decomposed further to comply with specialized analyses of the bank. This categorization is compatible with the P&L view of the bank to allow greater flexibility in modeling dispersed financial flows. The hierarchical decomposition of concepts generates a set of dynamically interconnected hierarchical maps (Figure 13).

Currently, the mechanism integrates more than 250 concepts, forming a hierarchy of more than 10 maps. The dynamic interface of the mechanism lets its user to utilize a sub-set of these concepts by setting the value of the redundant ones and/or the value of their weights to zero. Concepts and weight values have been obtained as follows:

- Strategic-level financial plans of two typical (though major) E.U. commercial banks have been selected and analyzed.
- The interpretation of financial plans into FCM hierarchies using the technique presented in section 3.1 has generated the skeleton maps.
- FCM hierarchies comply with the typical P&L reports of most financial sector enterprises.
- Financial sector and bank experts have provided weight values for skeleton FCMs.

It should be noted that as a working hypothesis for the FCM-based financial modelling, this paper adopted the operational and financial characteristics of typical / average European Union (EU) commercial banks, currently the majority of financial sector enterprises. Therefore, experts assigned values under the assumption that the market for intermediated finance was characterised by relationship rather than arm's length lending. Also, experts considered that fact that both banks operated in a bank-oriented (rather than a market-oriented) environment, in which banks predominated as financial intermediaries by collecting savings (through deposits) and providing the bulk of external funding to the non-financial sector.

4.2 Map Linking and Decomposition

A typical example of the interconnection mechanism in now commented briefly. Figure 8 presents a sample map hierarchy.

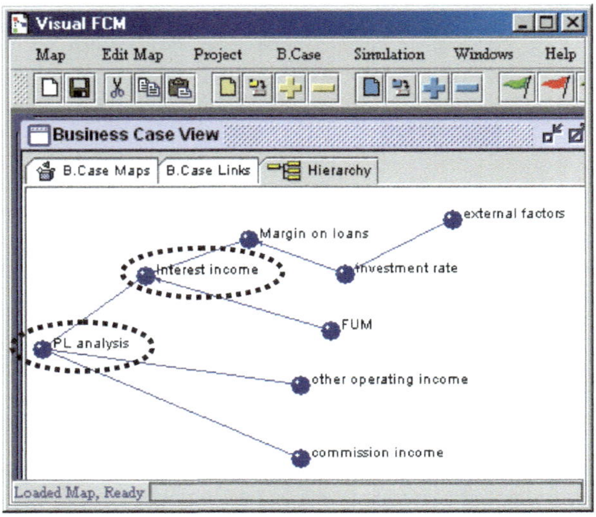

Fig. 8 Sample map hierarchy

Consider now maps "PL analysis" and "Interest income" (to be presented fully in Figure 21 and Figure 15 respectively). Linking a concept, which is defined into two maps, generates a hierarchy. Figure 9 presents the system interface for the generation of the hierarchical relationship between maps "PL analysis" and "Interest income". Map "PL analysis" decomposes further concept "net interest income" by using this concept as the link to map "Interest income".

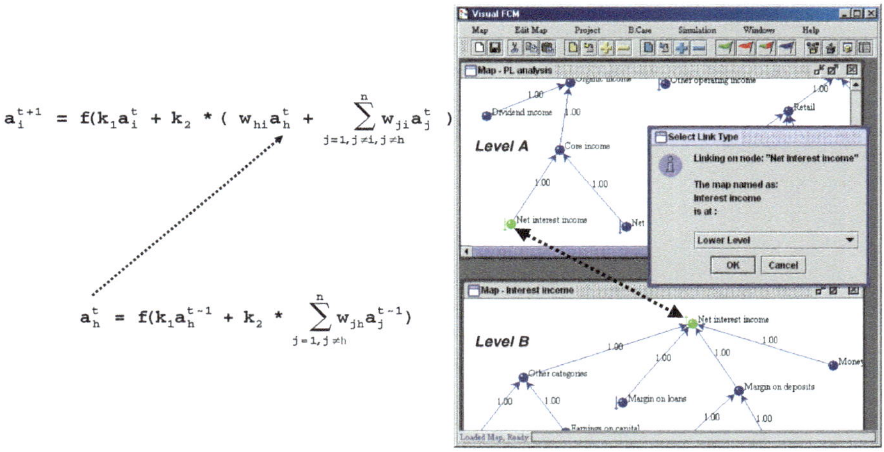

Fig. 9 Interconnection mechanism

Figure 9 also presents how algorithm (2) is decomposed to integrate hierarchical links. The proposed system can portray the financial model following either a holistic or a scalable approach. This is analogous to seeing the bank either as a single, "big bang" event or as an ongoing activity of setting successive financial targets to selected banking operations. The proposed mechanism can accommodate both approaches. Essentially, the implementation can decompose financial concepts to their constituent parts (sub concepts) on demand and let the user reason about lower level hierarchies of FCM before it passes values to the higher-level hierarchies. The proposed mechanism also allows the user to specify the degree of FCM decomposition during the map traversal (Figure 10). Instead of waiting for a lower level FCM to traverse its nodes and pass its value to higher level map hierarchies, the user may assign directly an external value to nodes which link hierarchies. In practice, the simulation is carried out as if there are no links with other maps.

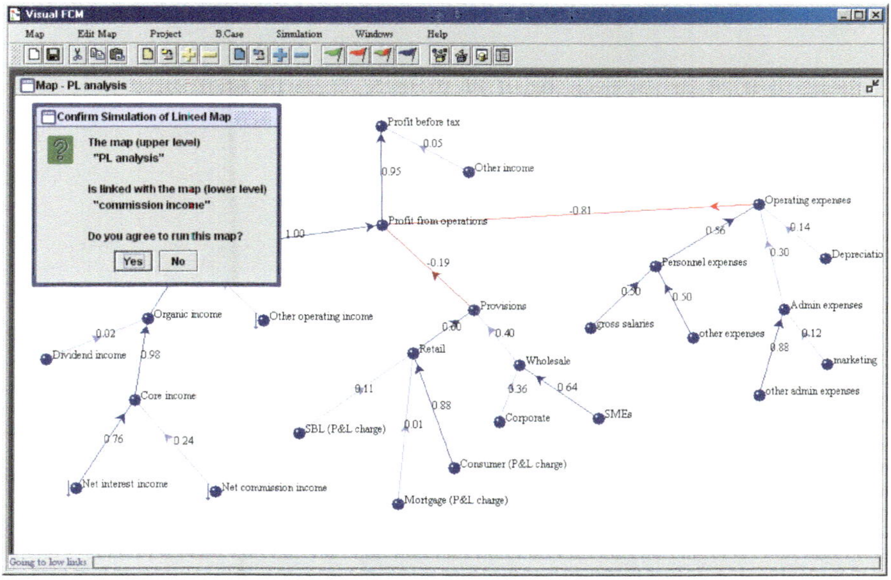

Fig. 10 User-defined map decomposition

The following table summarizes the available variations of the proposed FCM algorithm, which encode dynamic map decomposition and user-defined decomposition bound.

Map decomposition	*FCM algorithm hierarchical calculations*	
	Higher level hierarchy	*Lower level hierarchy*
Unrestricted	$a_i^{t+1} = f(k_1 a_i^t + k_2 * (w_{hi} a_h^t + \sum_{j=1, j \neq i, j \neq h}^{n} w_{ji} a_j^t))$	$a_h^t = f(k_1 a_h^{t-1} + k_2 * \sum_{j=1, j \neq h}^{n} w_{jh} a_j^{t-1})$
User-defined, no external value assigned	$a_i^{t+1} = f(k_1 a_i^t + k_2 * (w_{hi} a_h^t + \sum_{j=1, j \neq i, j \neq h}^{n} w_{ji} a_j^t))$	$a_h^t = f(k_1 a_h^{t-1})$
User-defined, external value assigned	$a_i^{t+1} = f(k_1 a_i^t + k_2 * (w_{hi} a_h^t + \sum_{j=1, j \neq i, j \neq h}^{n} w_{ji} a_j^t))$	$a_h^t \in [-1,..,1]$

Also the current implementation allows:

- easy customization of the function $\mathbf{f}$ and easy re-configuration of the formula $\mathbf{A}_i^{t+1}$ to adapt to the specific characteristics of individual enterprises,
- generation of scenarios for the same skeleton FCM,
- automatic loop simulation until a user-defined equilibrium point has been reached. Alternatively, step-by-step simulation (with graphical output of partial results) is also available to provide a justification for the partial results.

The proposed framework exemplifies further financial performance by decomposing maps into their consistent concepts. The following sections exhibit sample (though typical) skeleton maps for all categories, which provide relevance and research interest to this paper.

4.3 Map Categories

4.3.1 Business Category

This paper proposes five basic maps each consisting of generic financial metrics as follows:

- Margin on loans: Figure 14 summarizes relations between loan margins and loan volumes, considering various categories of retail and wholesale products. Concepts denoted as "↓" expand further to lower level hierarchies, while "↑" denotes bottom-up causal propagation.
- Net interest income: Figure 15 summarizes concept relationships between margin on deposits, margin on loans, earnings on capital and bond and money market mismatch (includes customer repos, swaps, and money market funds).
- Funds under management: Figure 16 summarizes concept relationships between liquid funds (e.g. repos, saving deposits, time deposits, etc), mutual funds and other funds (e.g. portfolio management, bond sales, other investment products, etc).
- Other Operating income: Figure 17 summarizes concept relationships between income from FX, bonds and derivatives, and equities.
- Commission income: Figure 18 summarizes concept relationships between commissions on lending activities, fees on investment banking and capital markets, commissions on mutual funds and other fees (e.g. platform fees, bank charges, FX transactions, etc).

4.3.2 Social Category

The mechanism proposes two basic maps each consisting of generic financial metrics as follows:

- External factors: Figure 19 summarizes concept relationships between external stimuli that may affect the P&L performance of a bank like national economy, credit expansion, stock exchange index, interest rates, etc.
- Investment rate and marketing: Figure 20 summarizes the impact of investment rates and marketing operations to the sales volumes of the bank.

4.3.3 Integrated Category

The mechanism proposes one map consisting of generic financial performance metrics as follows:

- Profits before taxes: Figure 21 summarizes concept relationships between the major components of P&L analysis, like total operating income and its components, other income (income from associates, extraordinary income, etc), operating expenses, provisions etc.

5 Preliminary Experiments

5.1 The Nature of the Experiments

Experiments were conducted by utilizing metrics and financial planning scenarios from the same two commercial banks discussed in section 4.1. Several financial planning scenarios have been generated to test the modeling capabilities of the proposed tool.

The experiments were executed as follows:

- The authors generated the set of skeleton maps following the technique presented in section 3.1. These maps are compatible with the P&L view of a typical bank.
- The tool utilized the same skeleton maps during experiments (e.g. Figure 14 to Figure 21). Financial sector experts provided two different sets of linguistic variables for the causal weights. Each set generated one business case (essentially a skeleton map with weight values) for each bank.
- The authors assigned linguistic values to $\mathbf{k}_1$ and $\mathbf{k}_2$. The experts validated these values.
- Bank stakeholders chose random (though typical) impact assessment cases (sets of objectives and actions) and provided input values to the relevant actions (financial planning scenarios). Upper level nodes measure the performance of the objectives while input nodes measure the value of the action.

- Given impact assessment cases (sets of objectives and actions) and the input values to the relevant actions (financial planning scenarios), the bank experts and stakeholders also provided their independent expert estimates (using similar linguistic variables) of the impact of the change actions.
- FCM impact estimates were benchmarked against expert estimates. This paper presents the results of one typical example for each bank.

The majority of the financial concepts of these examples cascade to several constituent metrics. This allows the mechanism to express its reasoning capabilities by traversing complicated concept interrelations spreading over different maps and hierarchies. As an example, the FCM mechanism was asked to support the following decision problem: "is a certain change in interest rates and administration expenses and consumer charge, and FUM, etc, going to have a certain change in the operating income". The first experiment involves a bank with limited retail market share seeking to penetrate the customer base of its competitors. The second experiment involves a bank with an established retail market presence seeking to enhance further its customer base. The following table presents sample actions / objectives, the associated nodes and their input / desired values.

Change action	*Associated node*	*Input value – Bank A*	*Input value – Bank B*
Increase FUM	FUM	0,65	0,9
Increase deposits	deposits	0,8	0,65
Increase consumer P&L charge	consumer P&L charge	0,1	0,35
Increase SBL margin	SBL margin	0,5	0,68
Increase other administration expenses	other administration expenses	0,68	0,8
Increase interest rates	interest rates	0,58	0,68
etc	etc	etc	etc
Objective	*Associated node*	*Desired value – Bank A*	*Desired value – Bank B*
Increase operating income	operating income	> positively high ⇒ > 0,65	> positively high ⇒ > 0,65
Increase profit from operations	profit from operations	> positively very low ⇒ > 0,21	> positively very low ⇒ > 0,21

For both cases, the Quanta tool iterated a subset of approximately 120 concepts spread over 10 sample hierarchical maps in order to calculate their equilibrium values. Both cases run on a typical business PC with a 2.4GHz Pentium processor and 512MB RAM. As far as the number of iterations is concerned, lower level maps iterated 8 times on average. The average number of iterations increased to 20 for middle and upper level maps. Only the top-most map increased the average number of iterations to approximately 90 (depending on the initial concept and weight values) due to the volume of map links. In practice, the actual process time was negligible on such a typical PC.

Figure 11 compares two decision values for a set of nodes as estimated by the FCM mechanism and the team of experts respectively for the first bank.

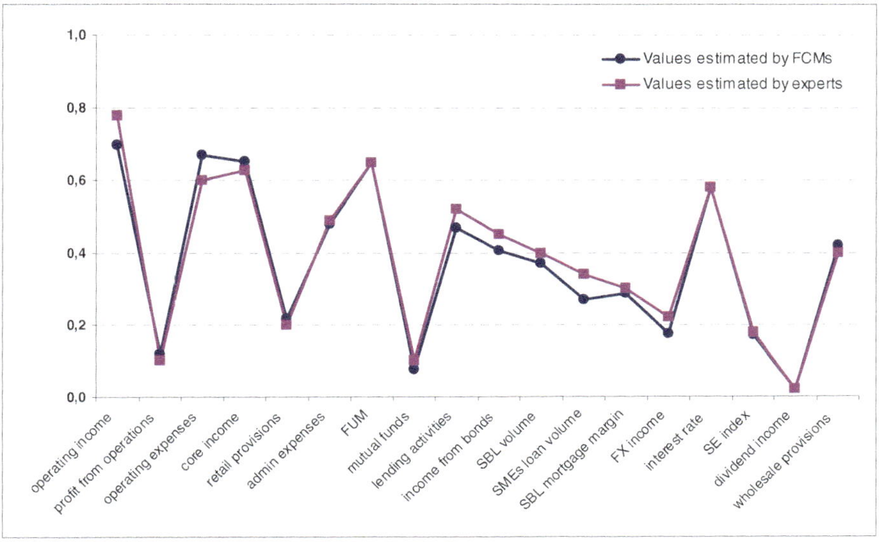

Fig. 11 Financial performance – Bank A

Given the input values and Figure 11, the following table presents the estimated impact of the change actions. These sample calculations indicate that if the actions are implemented as planned then the impact to the first objective will exceed the expectations. On the other hand, the impact to the second objective will not meet the expectations. Figure 11 also presents the values of other affected nodes for comparison purposes.

Objective	*Associated node*	*Desired value*	*FCM estimation*	*Expert estimation*
Increase operating income	operating income	> positively high ⇒ > 0,65	0,698	0,780
Increase profit from operations	profit from operations	> positively very low ⇒ > 0,21	0,120	0,100

Similarly, Figure 12 compares decision values as estimated by the FCM mechanism and the team of experts respectively for the second bank.

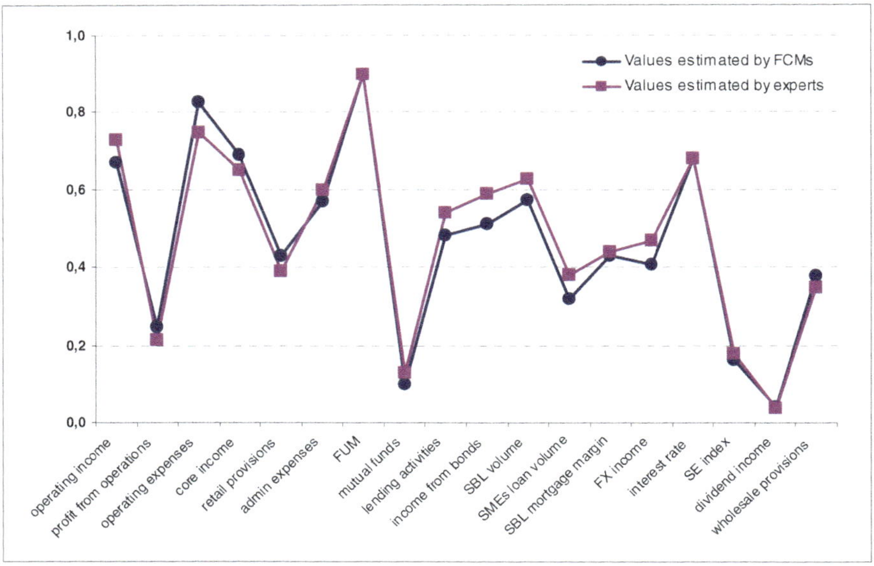

Fig. 12 Financial performance – Bank B

Given the input values and Figure 12, the following table presents the estimated impact of the change actions. These sample calculations indicate that if the actions are implemented as planned then the impact to the first objective will marginally fail the expectations. The impact to the second objective will meet the expectations. Figure 12 also presents the values of other affected nodes for comparison purposes.

Objective	*Associated node*	*Desired value*	*FCM estimation*	*Expert estimation*
Increase operating income	operating income	> positively high ⇒ > 0,65	0,644	0,730
Increase profit from operations	profit from operations	> positively very low ⇒ > 0,21	0,246	0,216

5.2 Discussion

5.2.1 Theoretical and Practical Value

Various aspects of the proposed modeling mechanism are now commented on. As far as the theoretical value is concerned, this mechanism extends previous research attempts by:

- allowing fuzzy definitions in the cognitive maps,
- introducing a specific interpretation mechanism of linguistic variables to fuzzy sets

- proposing an updated FCM algorithm to suit better the financial performance domains
- introducing the notion of linked performance hierarchies, as discussed in section 4.2
- concentrating on the actual strategic formulation activity and its impact on the financial model,
- allowing dynamic map decomposition and reconfiguration, as discussed in section 4.2.

As far as the practical value of the proposed mechanism is concerned:

- The additions to classical strategic-level financial planning at the banking sector are worth mentioning. For example:
  - It proposes a qualitative approach to strategic-level financial planning, which minimizes the utilization of crisp arithmetic values while offering adequate trend-based decision support.
  - The fuzzy qualitative approach bypasses the problem of detailed numerical and/or statistical calculations. Statistical calculations support decision-making under the restrictions of error margins, which usually complicate strategic-level decisions.
  - Fuzzy linguistic variables increase the understanding of concept / weight importance without compromising significantly their arithmetic interpretation. Therefore, the proposed trend-based decision support may suit better strategic-level exercises, which ask for less numerical analyses and more qualitative answers.
  - In contrast to relevant research in the business domain, the proposed mechanism builds on hierarchical metrics interrelationships identified and utilized by the financial strategy formulation methodology (see section 3.1). While the proposed approach does not perform or guide the implementation of any stage of the strategy formulation methodology, it supports the stakeholders to reason about the qualitative state of financial maturity metrics using fuzzy variables and imprecise relationships.
- When compared to the expert estimates, the mechanism does not provide fundamentally different "diagnosis". On the contrary, it provides reasonably good approximations of the impact of strategic change activities to the financial model.
- In comparison to the expert estimates, the proposed mechanism tends to under-estimate slightly the impact of strategic changes to financial concepts (metrics), which have several constituent sub-concepts or concepts, which have several hierarchical dependencies. This conservatism, however, does not reduce the effectiveness of the proposed mechanism. It simply indicates that when several complex financial performance factors are involved, it may be safer to assume a conservative financial improvement scenario.
- The justification of the "diagnosis" (essentially the P&L decomposition) proved extremely helpful in comprehending the sequence of complex concept interactions (essentially the P&L roadmap).

- The concept-based approach did not restrict the interpretation of the estimated financial status. The fuzzy interpretation of concept and weight values served as indications rather than precise arithmetic calculations.
- The hierarchical (or partial) traversal of financial metrics improved the distributed monitoring of strategic change activities throughout different hierarchical levels of the bank and stipulated targeted communication of the associated financial status (e.g. partial financial status of partial bank operations).
- The realism of financial status estimation depended on the number of concepts and weights, as well as on the estimation of their fuzzy characteristics (weights values, input values, coefficients, etc). However, the complexity and the length of the concept domain did not discourage the maintenance of the mechanism. Irrelevant and/or unnecessary maps could be isolated on demand to reduce the reasoning effort.

5.2.2 Added Value

Having established the theoretical and practical value of the proposed mechanism, it is useful to discuss also the added value of incorporating such a mechanism into financial planning exercises. It is the belief of this paper that the resulting tool provides real value to the principle beneficiaries and stakeholders of financial planning projects. For example:

- This decision aid mechanism proposes a new approach to supplementing financial planning. It provides a qualitative though "intelligent" support during the financial status analysis and financial objectives composition phases of typical financial strategy formulation projects. It utilizes cognitive modeling and offers a strategy-level utilization of P&L in order to shift focus from quantitative analysis to strategy-level impact assessment of the financial characteristics of a financial sector enterprise.
- The main purpose of this approach is to drive strategic change activities for continuous improvement rather than limit itself to qualitative simulations.
- The mechanism eases significantly the complexity of deriving expert decisions concerning the financial planning. Informal experiments indicated that the time required by experts to estimate manually the extensive impact of major strategic changes to realistic financial models could impose considerable overheads. On the other hand the elapsed time for automated estimations using FCM decision support can be insignificant, once the map hierarchies have been set up.
- To extend further this syllogism, realistic financial strategy formulation projects should involve continuous argument of strategic change options (e.g. application of best practices, alternative scenarios, alternative customer focus, etc) until an equilibrium solution has been agreed by all stakeholders. Informal discussions with the principle beneficiaries and stakeholders of the two financial planning projects revealed that the proposed FCM decision support can reduce significantly the estimation overheads of financial

maturity, letting the stakeholders focus on the actual planning exercise while exploring in depth all alternatives and controlling effectively major strategic change initiatives.

- The proposed mechanism can also assist the post financial planning evaluation of the enterprise on a regular basis. FCMs may serve as a back end to performance scorecards (e.g. [4], [23], [22]) to provide holistic strategic performance evaluation and management. However a detailed analysis of this extension falls out of the scope of this paper.

5.2.3 Preliminary Usability Evaluation

Senior managers of the two major financial sector enterprises have evaluated the usability of the proposed tool and have identified a number of benefits that can be achieved by the utilization of the proposed FCM tool as a methodology framework for financial planning. Detailed presentation of the usability evaluation results fall out of the scope of the paper. However, a summary of major business benefits (as identified by senior managers) is provided to improve the autonomy of this paper:

- Shared Goals
  - Concept-driven financial simulation pulls individuals together by providing a shared direction and determination of strategic change.
  - Shared financial planning and performance measurement enables business units to realize how they fit into the overall business model of the bank and what their actual contribution is.
  - Senior management receives valuable inputs from the business units (or the individual employees) who really comprehend the weaknesses of the current strategic model as well as the opportunities for financial performance change.
- Shared Culture
  - All business units at the bank feel that their individual contribution is taken under consideration and provide valuable input to the whole change process.
  - All business units and individuals at the bank feel confident and optimistic; they realize that they will be the ultimate beneficiaries of the financial planning exercise.
  - The information sharing culture supports the bank's competitive strategy and provides the energy to sustain this by exploiting fully the group and the individual potential.
- Shared Learning
  - The bank realizes a high return from its commitment to its human resources.
  - There is a constant stream of improvement within the bank.
  - The entire bank becomes increasingly receptive to strategic changes, since the financial benefit can be easily demonstrated to individual business units.

- Shared Information
  - All business units and individuals have the necessary information needed to set clearly their financial objectives and priorities.
  - Senior management can control effectively all aspects of the strategic re-design process.
  - The bank reacts rapidly to threats and opportunities.
  - It reinforces trust and respect throughout the bank.

Summarizing, experimental results showed that FCM-based *ex ante* reasoning of the impact of strategic changes (actual or hypothetical) to the status of financial performance can be effective and realistic, without employing detailed P&L numerical calculations. This is considered to be a major contribution of the proposed methodology tool to actual strategic change exercises.

6 Conclusion

This paper presented a supplement to the financial strategy formulation methodology based on FCMs. This decision aid mechanism proposes a new approach to supplementing the current financial status analysis and financial objectives composition phases of typical financial strategy formulation projects, by supporting fuzzy cognitive modeling of profit and loss (P&L) analysis and "intelligent" reasoning of the anticipated impact of strategic change initiatives to the financial status of a typical bank. The mechanism utilizes the fuzzy causal characteristics of FCMs as a new modeling technique to develop a causal representation of dynamic financial principles in order to generate a hierarchical network of linked financial performance indicators.

This paper discussed the FCM approach in putting realistic and measurable objectives in financial planning projects and presented sample maps with causal relationships. Preliminary experiments indicate that the mechanism does not provide fundamentally different estimates than expert decisions. Moreover, the decomposition of financial metrics into their constituent parts supported reasoning of the financial performance roadmap. The main purpose of the mechanism is to drive strategic change activities rather than limit itself to qualitative simulations. Moreover, the proposed mechanism should not be seen as a "one-off" decision aid. It should be a means for setting a course for continuous improvement [33].

Future research will focus on conducting further real life experiments to test and promote the usability of the tool, but also to identify potential pitfalls. Furthermore, future research will focus on the automatic determination of appropriate fuzzy sets (e.g. utilizing pattern recognition, mass assignments, empirical data, etc) for the representation of linguistic variables to suit each particular project domain. Finally, further research will focus on implementing backward map traversal, a form of adbuctive reasoning [13]. This feature offers the functionality of determining the condition(s) C_{ij} that should hold in order to infer the desired C_j in the causal relationship $C_{ij} \xrightarrow{w_{jk}} C_k$. Incorporating performance integrity constraints reduces the search space and eliminates combinatory search explosion. Backward reasoning has been tested extensively in other applications and its integration in the proposed methodology framework may prove beneficiary.

Appendix

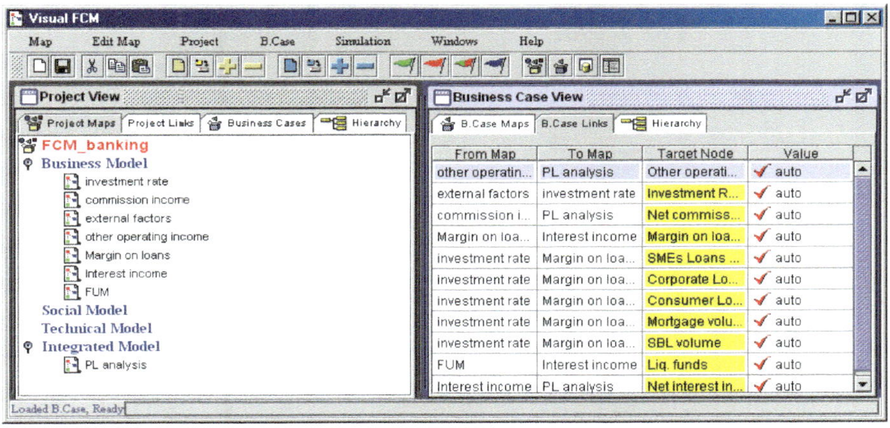

Fig. 13 Sample maps and map links

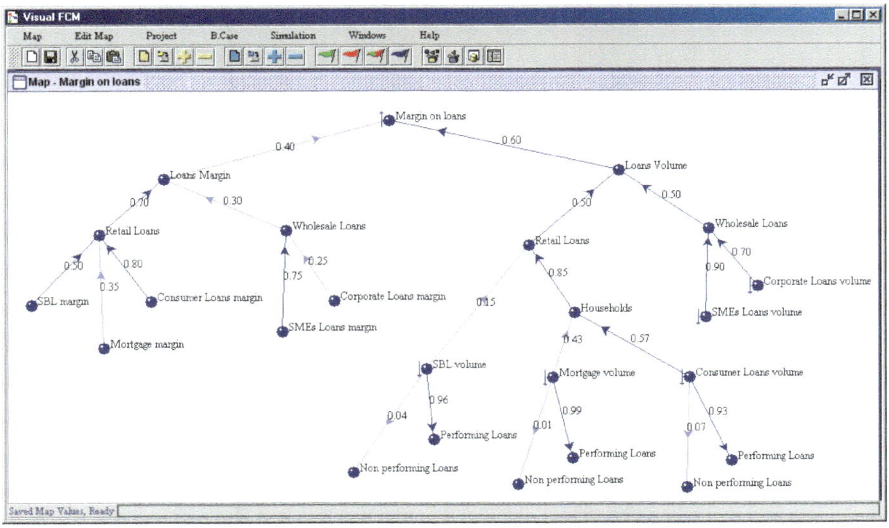

Fig. 14 Margin on loans concepts map

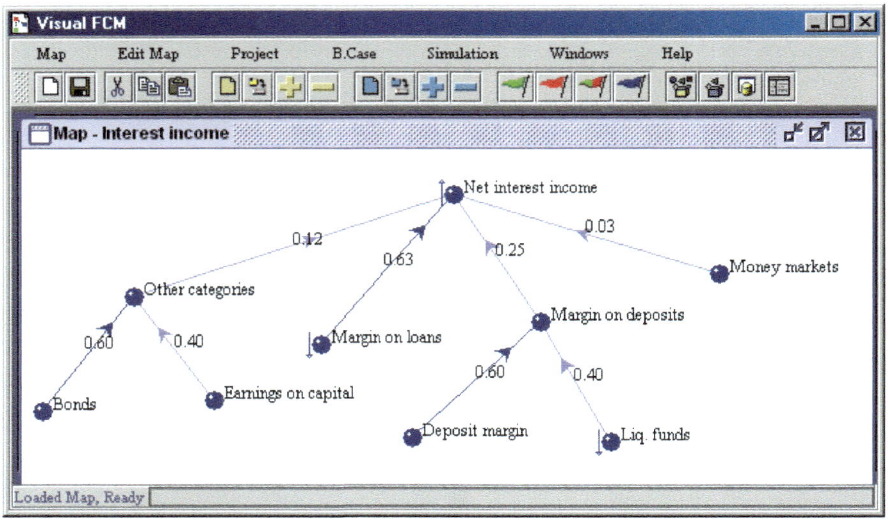

Fig. 15 Interest income concepts map

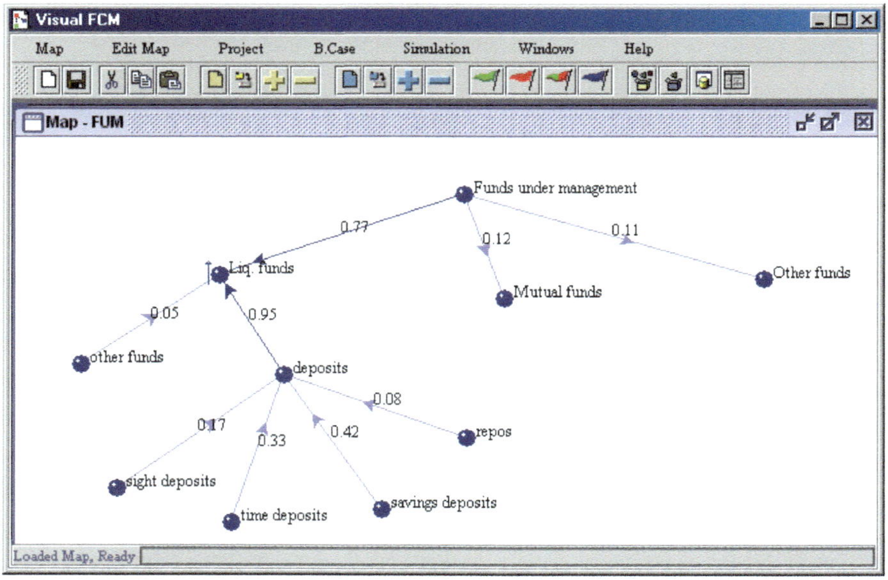

Fig. 16 Funds under management concepts map

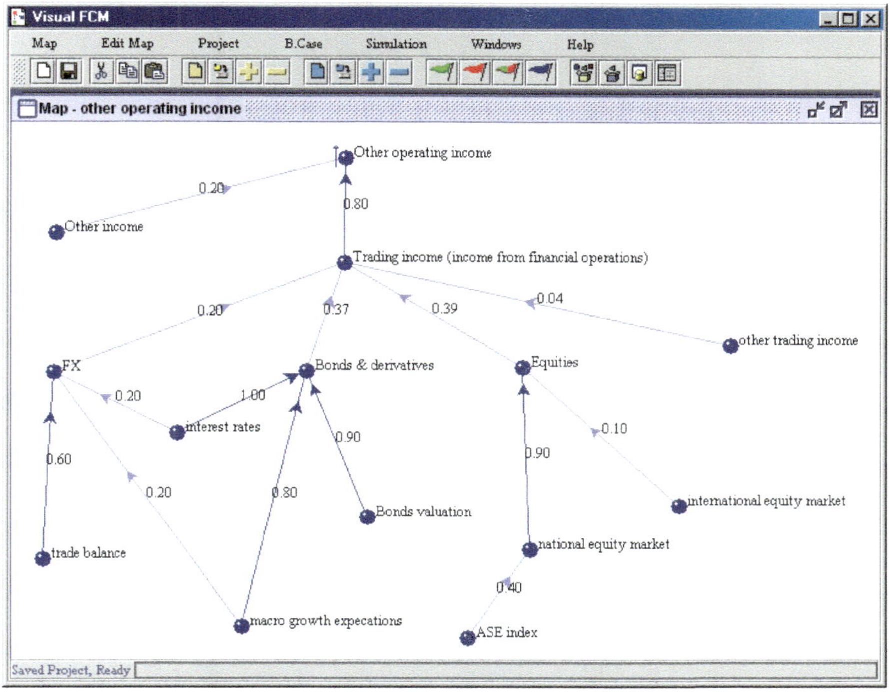

Fig. 17 Other operating income concepts map

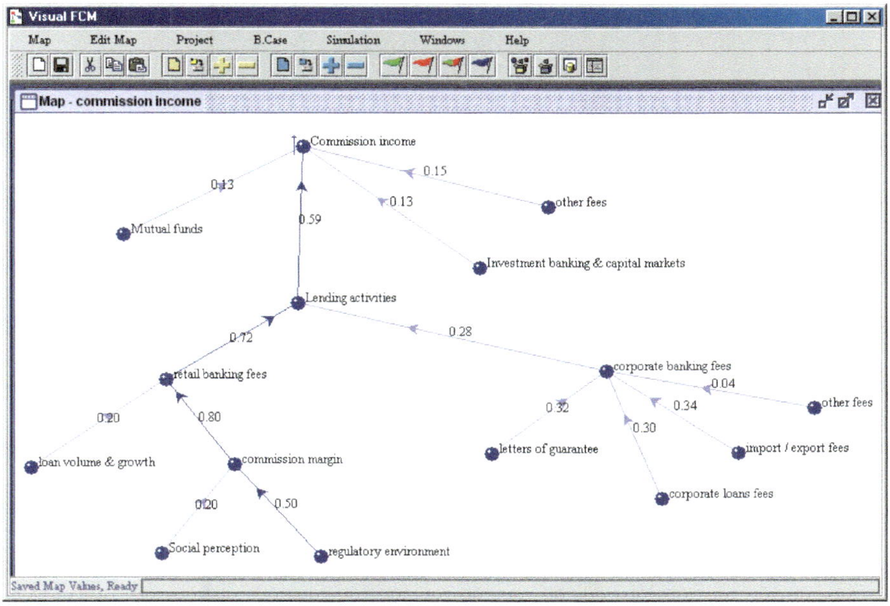

Fig. 18 Commissions income concepts map

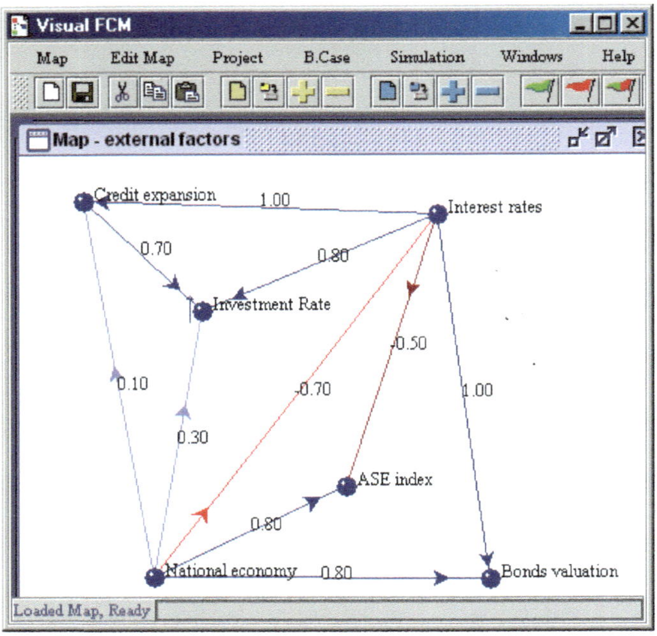

Fig. 19 External factors concepts map

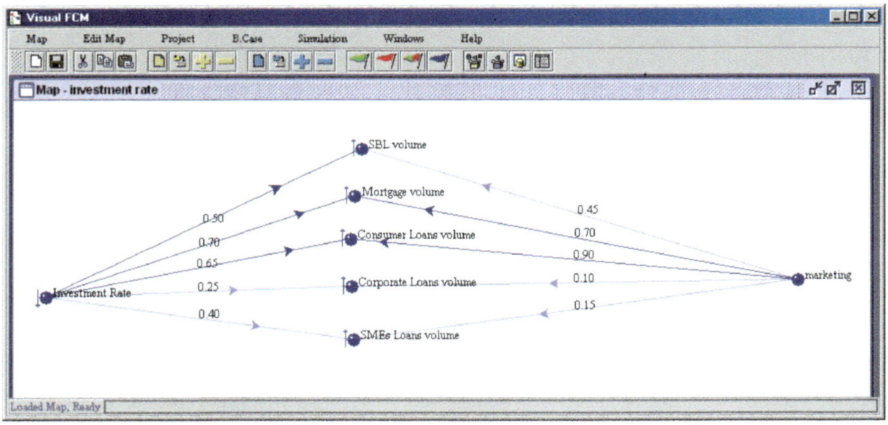

Fig. 20 Investment rate and marketing concepts map

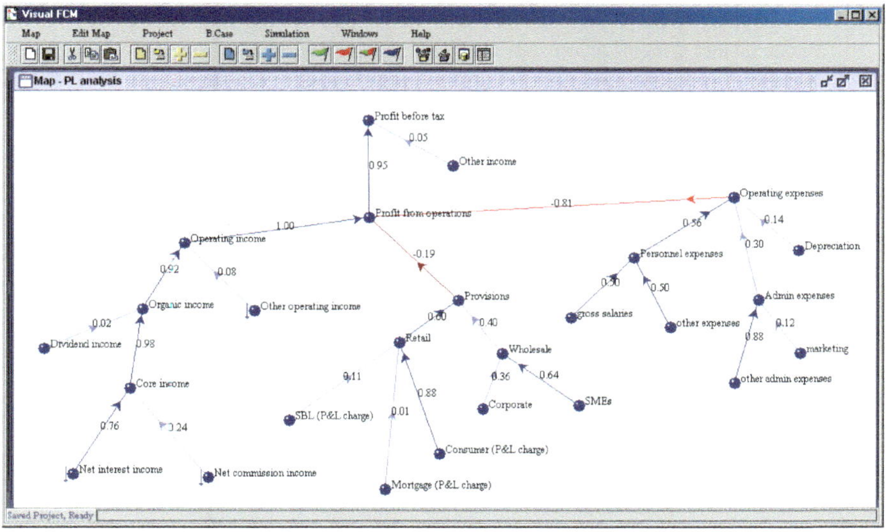

Fig. 21 Profit concepts map

Bibliography

[1] Arshadi, N., Lawrence, E.: An empirical investigation of new bank performance. Journal of Banking and Finance 11, 33–48 (1987)

[2] Axelrod, R.: Structure of decision: The cognitive maps of political elites. Princeton University Press, Princeton (1976)

[3] Boorman, J.T.: The prospects for minority-owned commercial banks: a comparative performance analysis. Journal of Bank Research 4(4), 263–279 (1974)

[4] Bourne, M., Mills, J., Wilcox, M., Neely, A., Platts, K.: Designing, implementing and updating performance measurement systems. International Journal of Operations and Production Management 20, 754–771 (2000)

[5] Boyd, J., Runkle, D.: Size and performance of banking firms: Testing the predictions of theory. Journal of Monetary Economics 31, 47–67 (1993)

[6] Brimmer, A.: The black banks: An assessment of performance and prospects. Journal of Finance 26, 379–405 (1971)

[7] Cole, R.T.: Financial performance of small banks. Federal Reserve Bulletin 67(6), 480–485 (1981)

[8] Craiger, J.P., Goodman, D.F., Weiss, R.J., Butler, A.: Modeling Organizational Behavior with Fuzzy Cognitive Maps. Journal of Computational Intelligence and Organisations 1, 120–123 (1996)

[9] Curry, T.J., Rose, J.T.: Bank holding company presence and banking market performance. Journal of Bank Research 14(4), 259–265 (1984)

[10] Dickerson, J.A., Kosko, B.: Virtual Worlds as Fuzzy Cognitive Maps. In: Kosko, B. (ed.) Fuzzy engineering, pp. 125–141. Prentice-Hall, Upper Saddle River (1997)

[11] Diffenbach, J.: Influense diagrams for complex strategic issues. Strategic Management Journal 3, 133–146 (1982)

[12] Elliott, J.W.: Control, size, growth, and financial performance in the firm. Journal of Financial and Quantitative Analysis 7(1), 1309–1320 (1972)
[13] Flach, P.A., Kakas, A.: Abduction and Induction in AI: Report of the IJCAI97 workshop. Logic Journal of the Interest Group on Pure and Applied Logic 6(4), 651–656 (1998)
[14] Fraser, D., Rose, P.: Bank entry and bank performance. Journal of Finance 27, 65–78 (1972)
[15] Georgopoulos, V., Malandraki, G., Stylios, C.: A fuzzy cognitive map approach to differential diagnosis of specific language impairment. Journal of Artificial Intelligence in Medicine 679, 1–18 (2002)
[16] Gilbert, R.A.: Bank market structure and competition. Journal of Money, Credit, and Banking 16(4), 617–660 (1984)
[17] Hagiwara, M.: Extended fuzzy cognitive maps. Presented at Proceedings of the 1st IEEE International Conference on Fuzzy Systems, New York (1992)
[18] Haslem, J.: A statistical estimation of commercial bank profitability. Journal of Business 42, 22–35 (1969)
[19] Heggestad, A.: Market structure, risk and profitability in commercial banking. Journal of Finance 32(4), 1207–1216 (1977)
[20] Hong, T., Han, I.: Knowledge based data mining of news information of the internet using cognitive maps and neural networks. Journal of Expert Systems with Applications 23(1), 1–8 (2002)
[21] Johnson, R.J., Briggs, R.O.: A model of cognitive information retrieval for ill-structured managerial problems and its benefits for knowledge acquisition. In: 27th Annual Hawaii International Conference on System Sciences, Hawaii (1994) (presented)
[22] Kaplan, R.S., Norton, D.P.: Leading change with the balanced scorecard. Financial Executive 17, 64–66 (2001)
[23] Kaplan, R.S., Norton, D.P.: Using the balanced scorecard as a strategic management system. Harvard Business Review, 75–85 (January/February 1996)
[24] Kaufman, G.: Bank market structure and performance: the evidence from Iowa. Southern Economic Journal 32, 429–439 (1965)
[25] Kaufmann, A.: Introduction to the theory of of fuzzy subsets. In: Fundamental theoretical elements. Academic Press, New York (1975)
[26] Kim, H.S., Lee, K.C.: Fuzzy implications of fuzzy cognitive map with emphasis on fuzzy causal relationship and fuzzy partially causal relationship. Fuzzy Sets and Systems 97, 303–313 (1998)
[27] Klein, J.C., Cooper, D.F.: Cognitive maps of decision makers in a complex game. Journal of Operation Research Society 33, 63–71 (1982)
[28] Kosko, B.: Differential hebbian learning. Presented at AIP Conference Proceedings, vol. 151 (1986)
[29] Kosko, B.: Fuzzy Cognitive Maps. Journal of Man-Machine Studies 24, 65–75 (1986)
[30] Kosko, B.: Hidden patterns in combined and adaptive knowledge networks. International Journal of Approximate Reasoning 2, 377–393 (1998)
[31] Kosko, B.: Neural networks and fuzzy systems. Prentice Hall, Englewood Cliffs (1991)
[32] Kwahk, K.Y., Kim, Y.G.: Supporting business process redesign using cognitive maps. Decision Support Systems 25(2), 155–178 (1999)
[33] Langbert, M., Friedman, H.: Continuous improvement in the history of human resource management. Journal of Management Decision 40(8), 782–787 (2002)

[34] Lee, K.C., Kim, H.S.: A fuzzy cognitive map-based bi-directional inference mechanism: An application to stock investment analysis. Journal of Intelligent Systems in Accounting Finance & Management 6(1), 41–57 (1997)
[35] Lee, K.C., Kim, H.S.: Fuzzy implications of fuzzy cognitive map with emphasis on fuzzy causal relationship and fuzzy partially causal relationship. Journal of Fuzzy Sets and Systems 3, 303–313 (1998)
[36] Lee, K.C., Kim, J.S., Chung, N.H., Kwon, S.J.: Fuzzy Cognitive Map approach to web-mining inference amplification. Journal of Expert Systems with Applications 22, 197–211 (2002)
[37] Lee, K.C., Kwon, O.B.: A strategic planning simulation based on cognitive map knowledge and differential game. Journal of Simulation 7(5), 316–327 (1998)
[38] Liu, Z.Q.: Fuzzy cognitive maps: Analysis and extension. Springer, Tokyo (2000)
[39] Liu, Z.Q., Satur, R.: Contexual fuzzy cognitive map for decision support in geographical information systems. Journal of IEEE Transactions on Fuzzy Systems 7, 495–507 (1999)
[40] Lloyd-Williams, M., Molyneux, P., Thornton, J.: Market structure and performance in Spanish banking. Journal of Banking and Finance 18, 433–443 (1994)
[41] Mayne, L.S.: Management policies of bank holding companies and bank performance. Journal of Bank Research 7(1), 37–48 (1976)
[42] Meinster, D., Elyasiani, E.: The performance of foreign owned, minority owned, and holding company owned banks in the U.S. Journal of Banking and Finance 12, 293–313 (1988)
[43] Miller, A., Noulas, A.: Portfolio mix and large-bank profitability in the USA. Journal of Applied Economics 29, 505–512 (1997)
[44] Mingo, J.: Capital management and profitability of prospective holding company banks. Journal of Financial and Quantitative Analysis 10, 191–203 (1975)
[45] Ndouse, T.D., Okuda, T.: Computational intelligence for distributed fault management in networks using fuzzy cognitive maps. Presented at IEEE International Conference Communication Converging Technology Tomorrow's Application, New York (1996)
[46] Noh, J.B., Lee Lee, K.C., Kim, J.K., Lee, J.K., Kim, S.H.: A case- based reasoning approach to cognitive map driven -driven tacit knowledge management. Journal of Expert Systems with Applications 19, 249–259 (2000)
[47] Nonaka, I., Takeushi, H.: The knowledge creating company. Oxford University Press, Oxford (1994)
[48] Pelaez, C.E., Bowles, J.B.: Applying fuzzy cognitive maps knowledge representation to failure modes effect analysis. Presented at IEEE Annual Reliability Maintainability Symposium. New York (1995)
[49] Perusich, K.: Fuzzy cognitive maps for political analysis. Presented at International Symposium on Technology and Society. Technical Expertise and Public Decisions, New York (1996)
[50] Ramaprasad, A., Poon, E.: A computerised interactive technique for mapping influense diagrams (MIND). Strategic Management Journal 6(4), 377–392 (1985)
[51] Ricketts, D., Stover, R.: An examination of commercial bank financial ratios. Journal of Bank Research 9(2), 121–124 (1978)
[52] Russel, B.: Wisdom of the West. Crescent Books, New York (1989)
[53] Schneider, M., Schnaider, E., Kandel, A., Chew, G.: Constructing fuzzy cognitive maps. Presented at IEEE Conference on Fuzzy Systems, New York (1995)

[54] Silva, P.C.: New forms of combinated matrices of fuzzy cognitive maps. Presented at IEEE International Conference on Neural Networks, New York (1995)

[55] Spencer, J.C., Grant, R.M.: Knowledge and the form: overview. Strateigc management journal 17(1), 256–267 (1996)

[56] Stylios, C.D., Georgopoulos, V.C., Groumpos, P.P.: Introducing the Theory of Fuzzy Cognitive Maps in Distributed Systems. Presented at 12 th IEEE International Symposium on Intelligent Control, Istanbul, Turkey (1997)

[57] Uyemura, D., Kantor, C., Pettit, J.: EVA for banks: value creation, risk management, and profitability measurement. Journal of Applied Corporate Finance 9(2), 156–175 (1996)

[58] Wall, L.: Why are some banks more profitable than others? Journal of Bank Research 15(4), 240–256 (1985)

[59] Xirogiannis, G., Glykas, M.: Fuzzy cognitive maps in business analysis and performance-driven change. IEEE Transactions on Engineering Management 13(17), 111–136 (2004)

[60] Xirogiannis, G., Stefanou, J., Glykas, M.: A fuzzy cognitive map approach to support urban design. Journal of Expert Systems with Applications 26(2), 257–268 (2004)

[61] Zadeh, L.A.: Fuzzy Sets. Journal of Information and Control 8, 338–353 (1965)

[62] Zhang, W.R., Wang, W., King, R.S.: A-pool: An agent-oriented open system for distributed decision process modeling. Journal of Organisational Computing 4(2), 127–154 (1994)

Fuzzy Cognitive Maps-Based IT Projects Risks Scenarios

Jose L. Salmeron

Abstract. Firms have spent billions of dollars in IT projects. Therefore, IT risk management is a critical issue. According to this context, the applied efforts to look for the correct IT implementation should be accompanied by mechanisms for managing the implementation risks. The goal is to reduce the risk of implementation failure. This paper analyzes IT projects implementation risks and the relationships between using an innovative soft computing technique called Fuzzy Cognitive Map. Through this proposal, it is possible to observe which the most relevant risks are, and, above all, which have a greater impact on the IT projects. Finally, three what-if analyses are done.

1 Introduction

Most real-life decision-making processes are dynamic. Critical decisions in business areas as manufacturing, sales, marketing, finance, and other domains require multiple and interrelated time-constrained decisions within strongly uncertain and so complex environments.

Information Technology (IT) projects have certain features that make them slightly different from other engineering projects. These include increased complexity and higher rates of project failure. Improving general knowledge about IT projects and IT risks specially can make it easier to implement and to increase the success rates.

This paper is structured in six sections. In the second section, IT projects risks fundamentals are presented. Next, in the third section, the author explains a decision support technique called Fuzzy Cognitive Maps (FCM). In the fourth section an experimental analysis about IT projects risks is done. In the fifth section, FCM-based what-if scenarios are designed. Then, in the last section the conclusions are exposed.

Jose L. Salmeron
School of Engineering, University Pablo de Olavide, Ctra. de Utrera, km. 1, 41013 Seville, Spain

M. Glykas (Ed.): Fuzzy Cognitive Maps, STUDFUZZ 247, pp. 201–215, 2010.
springerlink.com © Springer-Verlag Berlin Heidelberg 2010

2 IT Projects Risks

IT can enhance database access, analytical power, and the communication capacity of decision makers. It is based on the premise that more and better quality information will result in reduced uncertainty and complexity in decision-making (Ritchie and Brindley, 2001). As noted, most approaches to decision making often revolve around the assumptions that alternative courses of action (future scenarios) can be established, the outcome of choosing an alternative is known, or is at least calculable.

Several issues measure IT projects success: to be deployed on time, to budget, and meeting the specifications. Literature confirms that a few of those projects were finished on time, and within the estimated budget (Saleh and Alshawi, 2005).

It seems clear that the criteria mentioned (delivered on time, to budget and meeting the specification) are not enough to guarantee the success of an IT project avoiding their risks. An IT project could be implemented on time, within cost forecasts and according the specifications, but if it is not useful to the users, not liked by the sponsors and does not increase the efficiency of the company, that IT project is not a success. Therefore, there is a need to identify what the critical risks influencing IT projects success are and the relationships between them.

In order to begin to assess the risks involved within major IT projects it is necessary first to understand them.

Large IT projects are inherently risky undertakings. Yet the literature (Griffiths, 1995) suggests that risks factors are strongly underplayed and underassessed. Management challenges are compounded when technological innovation integral to a project is a factor. This is specifically the case in large (IT) projects. Indeed a range of studies show that the IT component adds a different dimension of risk which all too often can tip the balance towards project failure, rather than towards project success.

IT risk management is a critical process in any effective IT project (Cuellar and Gallivan, 2006; Rodriguez-Repiso et al., 2007). Risk management in IT ones is needed according to the high risk and cost of those projects (Hakim and Hakim, 2010). It causes huge loss for companies. In fact, risks are intrinsic to any IT project and risk taking is a necessary component of any process of decision-making (Cadle and Yeate, 2001). Poor risk management often leads IT projects to failure.

However, before we can develop meaningful risk management strategies, however, we must identify these risks and the relationships between them. Risk is the occurrence of an event that has consequences for, or impacts on, IT projects (Kliem and Ludin, 2000). The risk may halt from the nature of the work, from the resources available, from the contractual relationship, or from social issues with impact over the project (Cadle and Yeate, 2001).

So, if it is not possible to eradicate risks altogether; it must certainly be possible to manage IT projects in a way that identify the risks and plans methods of mitigate them if they occur (Cadle and Yeate, 2001). Risk identification, valuation and assessment are therefore the fundamental basis to support the entire risk management process. This paper proposes the use of Fuzzy Cognitive Maps for risks analysis in IT projects.

3 Fuzzy Cognitive Maps

This paper applies a soft computing technique called Fuzzy Cognitive Map (FCM) in IT projects risks management. The FCM process is an innovative and flexible technique (Bueno and Salmeron, 2008) for modelling human knowledge. In addition, FCM provide excellent mechanisms to develop forecasting exercises, specially what-if analysis.

Cognitive maps (Axelrod, 1976) and, after, Fuzzy Cognitive Maps (Kosko, 1986), have emerged as tools for representing and studying the behaviour of systems and people. Cognitive maps are a set of nodes linked by edges. The nodes represent concepts relevant to a given domain. The causal links between these concepts are represented by the edges, which are oriented to show the direction of influence. The other attribute of an edge is its sign, which can be positive (a promoting effect) or negative (an inhibitory effect).

The main goal of building a cognitive map (or FCM) around a problem is to be able to predict the outcome by letting the relevant issues interact with one another.

FCMs were proposed as an extension of cognitive maps. The FCM technique was proposed by Kosko (1986) to describe a cognitive map model with two significant characteristics. The first one, causal relationships between nodes have different intensities represented by fuzzy numbers. A fuzzy number is a quantity whose value is uncertain, rather than exact. It can be thought of as a function whose domain is usually the interval between 0 and 1 (or -1 and 1), including both (Xirogiannis and Glykas, 2007). Each numerical value in the interval is the grade of membership to the set, where 0 represents the smallest possible grade, and 1 is the largest possible grade.

The second one, the system is dynamic, that is, it evolves with time. It involves feedback, where the effect of change in a concept node may affect other concept nodes, which in turn can affect the node initiating the change.

After an inference process, the FCM reaches either one of two states following a number of iterations. It settles down to a fixed pattern of node values, the so-called hidden pattern or fixed-point attractor. Alternatively, it keeps cycling between several fixed states, known as a limit cycle. Using a continuous transformation function, a third possibility known as a chaotic attractor exists. This occurs when, instead of stabilizing, the FCM continues to produce different results (known as state-vector values) for each cycle. Feedback plays a prominent role in FCMs by propagating causal influences in complicated pathways.

FCMs have been applied in such different fields as medicine (Georgopoulos et al., 2003; Papageorgiou et al., 2006a), computer science (Kim and Lee, 1998; Konar and Chakraborty, 2005; Osei-Bryson, 2004; Stach et al., 2005), IT projects success (Rodriguez-Repiso et al., 2007), and other domains (Kang et al., 2004; Lee and Han, 2000; Rai and Kim, 2002; Xirogiannis et al., 2008). As yet, there have been few attempts at applying FCMs as real-world tools for supporting decisions (Salmeron, 2009) within real business environments. In addition, FCMs have been used in simulation and prediction projects (Fu, 1991).

Moreover, FCM provide an intuitive, yet precise way of expressing concepts and reasoning about them at their natural level of abstraction. By transforming decision modelling into causal graphs, decision makers with no technical background can understand all of the components in a given situation. In addition, with a FCM, it is possible to identify and consider the most relevant factor that seems to affect the expected target variable.

3.1 Fuzzy Cognitive Map Fundamentals

The FCM nodes (x_i) would represent such concepts as costs, sales, tool selection, investment, or marketing strategy, to name a few. Figure 1 shows an example of an FCM model.

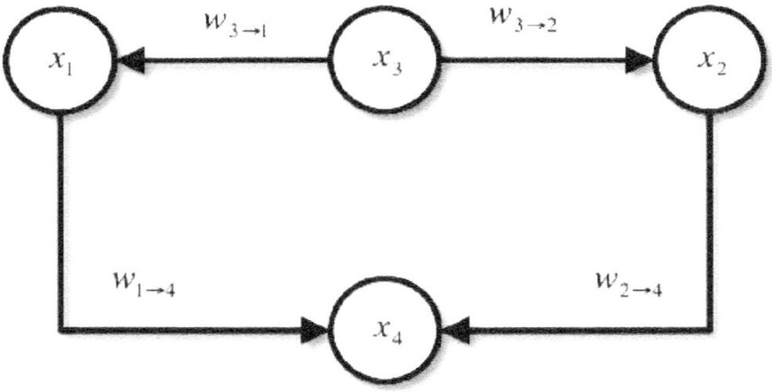

Fig. 1 Fuzzy Cognitive Map example

The relationships between nodes are represented by directed edges. An edge linking two nodes models the influence of the causal variable on the effect variable (Xirogiannis and Glykas, 2004). Since FCMs are hybrid methods (Xirogiannis and Glykas, 2008) mixing fuzzy logic (Zadeh, 1965, Bellman and Zadeh, 1970) and neural networks (Kosko, 1992), each cause is assessed by its intensity $w_{i\to j} \in [0,1]$, where i is the pre-synaptic (causal) node and j the post-synaptic (effect node) one.

An adjacency matrix A represents the Figure 1 FCM nodes connectivity. FCMs measure the intensity of the causal relation between two factors and if no causal relation exists it is denoted by 0 in the adjacency matrix.

$$A = \begin{pmatrix} 0 & 0 & 0 & w_{1\to 4} \\ 0 & 0 & 0 & w_{2\to 4} \\ w_{3\to 1} & w_{3\to 2} & 0 & 0 \\ 0 & 0 & 0 & 0 \end{pmatrix}$$

The high level FCM building process is summarized in Fig. 2.

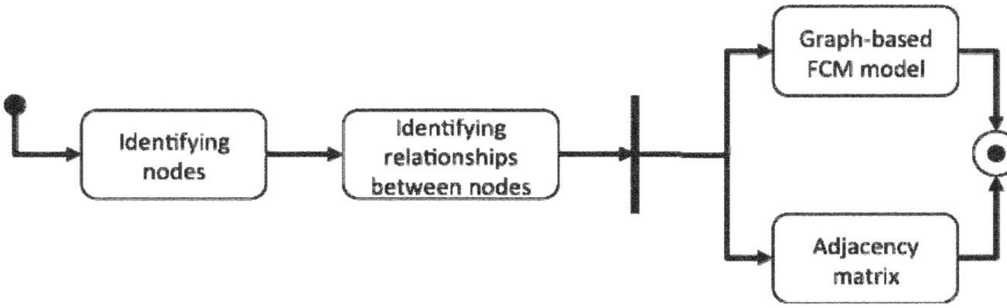

Fig. 2 FCM building process

FCMs are dynamical systems (Xirogiannis et al., 2008) involving feedback, where the effect of change in a node may affect other nodes, which in turn can affect the node initiating the change. The analysis begins with the design of the initial vector state ($\vec{C}_0$), which represents the initial value of each variable or concept (node). The initial vector state with n nodes is denoted as

$$\vec{C}_0 = \left(C_0^{[1]} \quad C_0^{[2]} \quad \ldots \quad C_0^{[n]}\right)$$

where $C_0^{[i]}$ is the initial value of the i concept.

The new values of the nodes are computed in an iterative vector-matrix multiplication process with an activation function, which is used to map monotonically the node value into a normalized range [0,1]. The sigmoid is the most used function (Bueno and Salmeron, 2009) when the concept (node) value maps in the range [0,1]. The vector state $\vec{C}_{(t+1)}$ at the instant $t+1$ would be

$$\begin{aligned}
\vec{C}_{(t+1)} &= S\left[\vec{C}_t \times A\right] \\
&= S\left[\overrightarrow{C_t A}\right] \\
&= S\left[\left(C_t A^{[1]} \quad C_t A^{[2]} \quad \ldots \quad C_t A^{[n]}\right)\right] \\
&= \left(S\left(C_t A^{[1]}\right) \quad S\left(C_t A^{[2]}\right) \quad \ldots \quad S\left(C_t A^{[n]}\right)\right) \\
&= \left(C_{(t+1)}^{[1]} \quad C_{(t+1)}^{[2]} \quad \ldots \quad C_{(t+1)}^{[n]}\right)
\end{aligned}$$

where $\vec{C}_t$ is the vector state at the t instant, $C_t^{[i]}$ is the value of the i concept at the t instant, $S[x]$ is the sigmoid function and A the adjacency matrix. The state is changing along the process.

The component i of the vector state $\vec{C}_{(t+1)}$ at the instant $t+1$ would be

$$C_{(t+1)}^{[i]} = \frac{1}{\left(1 + e^{-1 \times C_t A^{[i]}}\right)}$$

where λ is the constant for function slope (degree of fuzzification). The FCM designer has to specify the lambda value. For large values of lambda (e.g. $\lambda \geq 10$) the sigmoid approximates a discrete function that maps its results to interval (0,1), for smaller values of lambda (e.g. $\lambda = 1$) the sigmoid approximates a linear function, while values of lambda closer to 5 provides a good degree of fuzzification in [0,1] interval (Bueno and Salmeron, 2009; Grant and Osei-Bryson, 2005).

FCM inference process finishes when the stability is reached. The final vector state shows the effect of the change in the value of each node in the FCM. After the inference process, the FCM reaches either one of three states following a number of iterations. It settles down to a fixed pattern of node values, the so-called hidden pattern or fixed-point attractor. Alternatively, the state could to keep cycling between several fixed states, known as a limit cycle. With a continuous function, a third possibility would be a chaotic attractor. This occurs when, instead of stabilizing, the FCM continues to produce different results (state vector values) for each cycle. In this case, the technique is not able to offer a useful outcome for risks analysis.

3.2 Experts' Consensus in FCM

Various methodologies could be used in order to reach a consensus among the experts in FCM (Bryson et al., 1997; Bueno and Salmeron, 2008). Delphi is a well-known methodology used to structure the experts' communication process to reach a consensus regarding a complex problem (Dalkey and Helmer, 1963; Linstone and Turoff, 1975). One of the main features of the Delphi study is when the experts receive feedback reports; they have the opportunity of changing their own opinion based on this feedback (Dalkey and Helmer, 1963).

The Augmented FCM approach (Dickerson and Kosko, 1994; Salmeron, 2009) does not need that experts change slightly their former opinions for consensus. It is needed within Delphi methodology (Cho et al., 2002). The augmented adjacency matrix is built adding the adjacency matrix of each expert (Kosko, 1996).

The Augmented FCM approach combines additively FCM matrices of each expert. The outcoming Augmented includes the union of the causal nodes for all the experts. If an expert's FCM does not include a specific concept then those rows and columns in adjacency matrix are all zero. The resulting augmented matrix is computed by

$$A^{Aug} = \Sigma_{i=1}^{n} A_i$$

where n is the number of experts and A_i is the adjacency FCM matrix for i expert.

4 Experimental Analysis

The experiment is focused on IT projects implementation risks and its relationships. IT projects have certain features that make them different from other projects. These include increased complexity and higher chances of project failures (Rodriguez-Repiso et al., 2007). To increase the chances of these projects to be successful, it is necessary to identify and control the critical risks influencing them. Risk control is needed according to the high risks and costs of IT projects.

According to Kwak and Stoddard (2004), a critical point is that some IT practitioners perceive risk management as extra work. Usually, risk management processes are the first element to be removed from the project tasks when the project schedule slips. For that reason, practitioners' involvement in risk management (e.g.: building FCM models) is so interesting.

4.1 Expert Panel

With the purpose of determining the group of risks in IT projects, as well as the relationships between them, advice was taken from a panel of experts. The number of experts selected is found to be within the recommended range. Experts suggest a range of 10 to 18 to be a fit number for each panel of experts (Okoli and Pawlowski, 2004). Anyway, the optimal number of experts depends of the study itself. In our case, ten participants composed the expert panel.

These experts had formed or form part of an IT implementation team or is working as IT manager, both with more than ten years of experience. They belonged to both the public and private sectors from Europe (6) and USA (4). This team composition guarantees the experts who are finally chosen having thorough profound IT knowledge.

The main selection criterion considered was recognized knowledge in domain topic, absence of conflicts of interest and geographic diversity. All conditions were respected. In addition, respondents were not chosen just because they are easily accessible.

4.2 Building the Model

The experts suggest fifteen ERP projects critical risks. The node description and its relationships are detailed in Table 1 and the graphical model in Fig. 3. The Augmented FCM approach has been used for building consensus.

The FCM graphical model is shown in Figure 4. Note that it is complex to take into account all these concepts at the same time even for experts. IT projects managers gain understanding of the problem in the FCM building process and then they can simulate different initial situations. For those reasons, FCM is a useful tool for IT projects managers.

Table 1 Nodes and outgoing edges

Node source	Description	Weights of outgoing edges
x_1	Wrong IT tools selection	$w_{1\to 2}=0.4$
		$w_{1\to 3}=0.8$
		$w_{1\to 4}=0.4$
		$w_{1\to 6}=0.6$
		$w_{1\to 8}=0.2$
		$w_{1\to 12}=0.7$
		$w_{1\to 13}=0.6$
x_2	IT software releases unstability	$w_{2\to 3}=0.7$
		$w_{2\to 6}=0.5$
		$w_{2\to 7}=0.2$
		$w_{2\to 8}=0.3$
		$w_{2\to 11}=0.6$
		$w_{2\to 12}=0.7$
x_3	Complex IT integration within current infrastructure	$w_{3\to 8}=0.8$
		$w_{3\to 13}=0.3$
x_4	Unrealistic expectations	$w_{4\to 5}=0.6$
		$w_{4\to 6}=0.4$
		$w_{4\to 9}=0.1$
x_5	Low Top management support	No outgoing edges
x_6	Low users' involvement	$w_{6\to 11}=0.2$
		$w_{6\to 13}=0.6$
x_7	Inadequate consulting services	$w_{7\to 1}=0.8$
		$w_{7\to 8}=0.9$
		$w_{7\to 9}=0.6$
		$w_{7\to 11}=0.7$
		$w_{7\to 13}=0.8$
x_8	High costs	$w_{8\to 4}=0.7$
		$w_{8\to 5}=0.8$
x_9	Wrong IT project management	$w_{9\to 3}=0.3$
		$w_{9\to 5}=0.5$

Table 1 (*continued*)

		$w_{9\to 6} = 0.4$
		$w_{9\to 8} = 0.4$
		$w_{9\to 11} = 0.3$
x_{10}	High reliability requirements	$w_{10\to 8} = 0.4$
x_{11}	Wrong legacy systems management	$w_{11\to 3} = 0.7$
x_{12}	IT security issues	$w_{12\to 5} = 0.2$
x_{13}	Low Performance	$w_{13\to 5} = 0.3$

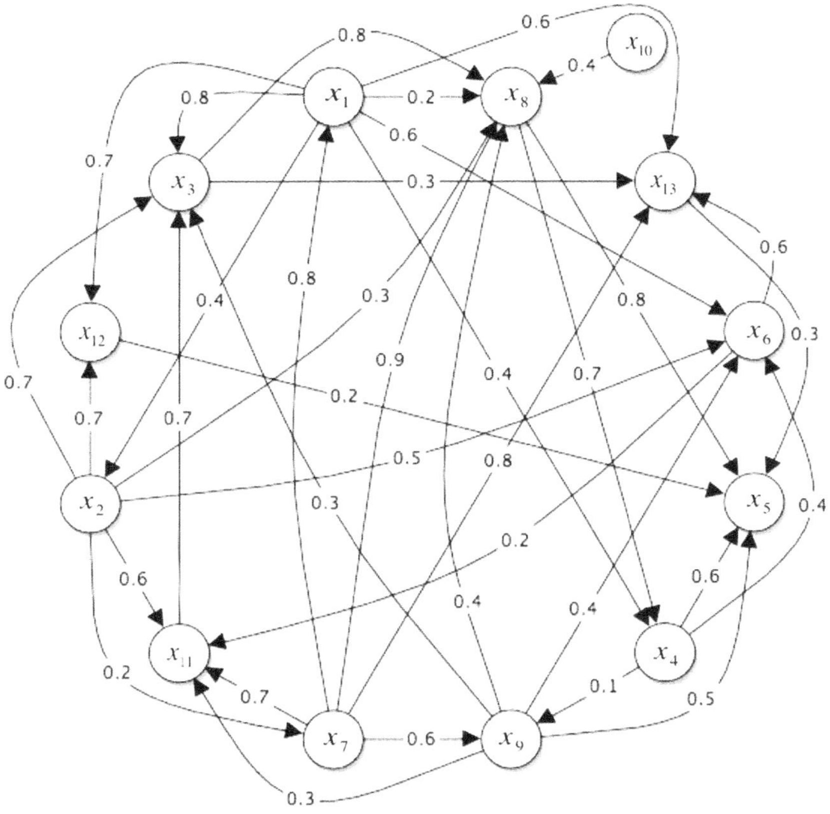

Fig. 3 IT projects risks Fuzzy Cognitive Map

5 FCM-Based Scenarios

It is possible to develop what-if analysis (scenarios) using different initial vector states. With the intention of observing the evolution of several initial scenarios, each analysis begins with the definition of an initial vector ($\vec{C}_0$), which represents a proposed initial situation or scenario. It means that it would start a specific IT project, but one o several risks are known before.

In this experiment the author uses three initial vectors state, each one have some risks activated. Each initial vector state interacts with the adjacency matrix A^{Aug} as detailed before.

5.1 Scenario A

Globalization has led companies to engage in IT projects that are critical to their survival. In order to succeed, companies must finish IT projects on time and within budget, and meet specifications while managing project risks. While large amounts of resources are dedicated to selecting and designing IT projects, it remains of critical importance that projects be adequately managed if they are to achieve their performance objectives (Raymond and Bergeron, 2008; Holland and Light, 1999).

On the other hand, consulting services can be one of the most difficult kinds of services to acquire. The acquirement process required for these services can involve major investments in time, money, and people with no assurance of a successful outcome (Mitchell, 1994). Anyway, one key to a successful IT project implementation is to maintain an effective and smooth consulting process, specially in IT projects as Enterprise Resource Planning systems (Bueno and Salmeron, 2008).

All the risks (nodes) in the initial vector $\vec{C}_0^A$ are not activated at the initial time in the A scenario, but "Wrong IT project management" and "Inadequate consulting services". This scenario is related with an IT project where the external side has been wrong selected.

$$\vec{C}_0^A = \begin{pmatrix} 0 & 0 & 0 & 0 & 0 & 0 & 1 & 0 & 1 & 0 & 0 & 0 & 0 \end{pmatrix} \xrightarrow{n \text{ iterations}}$$

$$\vec{C}_n^A = \begin{pmatrix} .77 & .54 & .98 & .90 & .97 & .96 & .71 & .98 & .66 & 0 & .96 & .92 & .96 \end{pmatrix}$$

Regarding to the results, the stable vector shows as the initial activated risks have a strong influence over the remainder risks. Findings confirm the critical role of project management and external consultants in IT projects. It is coherent with the importance of project management and consulting services in IT projects. According to this, one critical factor to a successful IT implementation is to maintain an effective and smooth consulting process and an effective IT project management.

Higher impacts are got over "Complex IT integration within current infrastructure" (0.98), "Unrealistic expectations" (0.90), "Low top management support" (0.97), "Low users' involvement" (0.96), "High costs" (0.98), "Wrong legacy system management" (0.98) and "Wrong legacy systems management" (0.96), "IT security issue" (0.92) and "Low performance" (0.96).

5.2 Scenario B

All the risks (nodes) in the initial vector $\vec{C}_0^B$ are not activated at the initial time, but "Low top management support" and "Low users' involvement". This scenario is related with an IT project where the internal side is not supporting enough the IT project.

Top management mission in IT projects is to promote the project. Their mission is not focused on the daily activity of development and implementation process, but on supporting the IT system with his/her authority and influence over the rest of the firm. Literature claims that top management support is a critical factor of IT projects success (Bingi et al., 1999; Umble et al., 2003). Without this support, IT may not be implemented optimally and returns on IT investments can be reduced.

User involvement refers to participation in IT projects implementation processes by target users representatives. In a deeply approach, user's involvement is defined (Barki and Hartwick, 1989; Lin and Shao, 2000) as a mental or psychological state of users toward the system and its development process.

It is generally accepted that IT users' involvement is important and necessary, as the lack of their involvement may represent a serious problem for the final system. IT implementation could represent a threat to users' perceptions of control over their own work. User involvement restores or enhances perceived control through participating the whole project plan (Zhang et al., 2005).

$$\vec{C}_0^B = (0 \quad 0 \quad 0 \quad 0 \quad 1 \quad 1 \quad 0 \quad 0 \quad 0 \quad 0 \quad 0 \quad 0 \quad 0) \xrightarrow{n \text{ iterations}}$$

$$\vec{C}_n^B = (0 \quad 0 \quad .24 \quad .45 \quad .79 \quad .34 \quad 0 \quad .42 \quad .07 \quad 0 \quad .15 \quad 0 \quad .62)$$

The final stable vector shows as the initial activated risks have the stronger influence over "Low performance" (0.62). It shows that if top management support and users' involvement are low, the final IT system will get low performance.

Despite the difficulties in explaining the contribution of IT investments to organizational performance, it is a critical issue for companies (Li and Yen, 1999). IT allows companies to obtain, process, store and exchange information. Nevertheless, the presence of IT neither guarantees performance increases.

5.3 Scenario C

All the risks in the initial vector $\vec{C}_0^C$ are not activated at the initial time, but "Complex IT integration within current infrastructure" and "Wrong IT tools selection". This scenario represents an organization with high technical complexity that it fails selecting the IT tools. Effectively integrating the new IT tools in the previous technology infrastructure is not always an easy task. Anyway, it will be more complex if the selected tool is wrong.

$$\vec{C}_0^C = (1 \quad 0 \quad 1 \quad 0 \quad 0 \quad 0 \quad 0 \quad 0 \quad 0 \quad 0 \quad 0 \quad 0 \quad 0) \xrightarrow{n \text{ iterations}}$$

$$\vec{C}_n^C = (.77 \quad .54 \quad .98 \quad .90 \quad .97 \quad .96 \quad .71 \quad .98 \quad .66 \quad 0 \quad .96 \quad .92 \quad .96)$$

The stable vector $\vec{C}_n^C$ shows as the initial activated risks have a strong influence over the whole model, specially on "Low Top management support" (0.97), "Low users' involvement" (0.96), "High costs" (0.98), "Wrong legacy systems management" (0.96) and "Low performance" (0.96). Obviously, more complexity implies more risks as results suggest.

The findings are reasonable, because a more complex integration would generate more costs. The increasing budget decrease the top management support and it would has an impact over the users' involvement. As a result, the system would get a lower performance (Berlin et al., 2009).

6 Conclusions

Companies have spent billions of dollars in IT projects. Therefore, IT risk management is a critical issue. According to this context, IT projects risk management is a relevant issue. To address it and improve scientific background around requires solid methodologies. In this sense, this paper proposes the use of Fuzzy Cognitive Maps for IT projects risk management.

An experimental analysis with three initial scenarios was designed. Each scenario analysis begins with the definition of the initial situation. It means that it would start a specific IT project, but one o several risks are known before. The FCM evolution generates the final situation for each scenario.

6.1 Relevance of the Proposal

The author' proposal offers some advantages in comparison with others similar tools. This paper proposes an innovative and flexible technique called Fuzzy Cognitive Maps to IT risks scenarios. It can be adapted to a wide range of problems, specially in knowledge intensive environments. This flexibility allows this model to be turned into a useful and innovative tool.

Firstly, FCM technique allows the defining of relationships between concepts. Through this characteristic, decisional models that are more reliable for interrelated environments are defined.

Secondly, FCM is able to quantify the influence of the relationships between concepts. Through this attribute, a better support in complex decisions can be reached. Finally, with this FCM model it is possible to develop a what-if analysis with the purpose of describing possible scenarios.

FCM application was used to analyze IT projects risks and interesting findings were extracted. Three risks initial scenarios were simulated and their impact over the model have been detailed. Through this proposal, one can observe which the most relevant risks are, and, above all, which have a greater impact on the IT projects. It allows using FCM as a simulation tool, where the initial risk situation suggests the future problems in IT projects implementation.

6.2 Limitations

However, this research possesses as its main limitation the experts' knowledge dependency. An FCM is based on experts' knowledge. Accumulated experiences are integrated within FCM structure. According to this, experts' selection needs to be extremely solid. Anyway, the opinions of one expert are never completely accurate and always contain some level of subjectivity.

The author highlights the reduced number of experts consulted in spite of it being among the recommended limits.

Consensus is other limitation. The Augmented FCM approach does not need that experts change slightly their former opinions, as others ones (Delphi). The current proposal applies Augmented matrices for reaching consensus, but it is not possible to know if the outlier has the right opinion.

6.3 Future Work

The author considers that the methodology applied in this paper has allowed successfully reaching the desired aim of this research. In this way, the author believes that its application is possible in other equally relevant decision-making process and that occur in the stages of pre-implementation, implementation and, even, maintenance of IT projects.

Moreover, Fuzzy Cognitive Maps are not a close research topic. More research about FCM building, learning and application is needed. Specially, the author is going to make efforts towards building FCM.

Finally, our findings highlight the necessity of continuing with research about this topic. Specifically, we could start future work based on these findings. First, we are interested in analyzing how some factors that influence the technological complexity, such as design, affordance, integrated search functionality and friendly interface, affect the behavioral intention to use an ERP system. Second, we consider that it would be interesting to develop future research to analyze IT projects after completing an implementation.

Acknowledgement

This work has been partly supported by the Spanish Ministry of Science (ECO2009-12853).

References

Axelrod, R.: Structure of decision: The Cognitive Maps of political elites. Princeton University Press, Princeton (1976)

Barki, H., Hartwick, J.: Rethinking the concept of user involvement. MIS Quarterly 13(1), 53–63 (1989)

Bellman, R.E., Zadeh, L.A.: Decision-Making in a Fuzzy Environment. Management Science 17, 141–164 (1970)

Berlin, S., Raz, T., Glezer, C., Zviran, M.: Comparison of estimation methods of cost and duration in IT projects. Information and Software Technology 51(4), 738–748 (2009)
Bingi, P., Sharma, M.K., Godla, J.K.: Critical issues affecting an ERP implementation. Information Systems Management 16, 7–14 (1999)
Bryson, N., Mobolurin, A., Joseph, A.: Generating consensus fuzzy cognitive maps. Intelligent Information Systems 8, 231–235 (1997)
Bueno, S., Salmeron, J.L.: Fuzzy modeling Enterprise Resource Planning tool selection. Computer Standards & Interfaces 30, 137–147 (2008)
Bueno, S., Salmeron, J.L.: Benchmarking main activation functions in fuzzy cognitive maps. Expert Systems with Applications 36(3), 5221–5229 (2009)
Cadle, J., Yeate, D.: Project Management for Information Systems. Financial Times/Prentice-Hall, Harlow (2001)
Cho, H.K., Turoff, M., Hiltz, S.R.: The impacts of Delphi communication structure on small and medium sized asynchronous groups: preliminary results. In: System Sciences-36th Annual Hawaii International Conference, pp. 17–26 (2002)
Cuellar, M.J., Gallivan, M.J.: A framework for ex ante project risk assessment based on absorptive capacity. European Journal of Operational Research 173, 1123–1138 (2006)
Dalkey, N.C., Helmer, O.: An experimental application of the Delphi method to the user of experts. Management Science 9, 458–467 (1963)
Dikerson, J.A., Kosko, B.: Virtual worlds as fuzzy cognitive maps. Presence 3, 173–189 (1994)
Fu, L.: CAUSIM: A rule-based causal simulation system. Simulation 56, 251–256 (1991)
Georgopoulos, V.C., Malandraki, G.A., Stylios, C.D.: A fuzzy cognitive map approach to differential diagnosis of specific language impairment. Artificial Intelligence in Medicine 29, 261–278 (2003)
Grant, D., Osei-Bryson, K.: Using fuzzy cognitive maps to assess MIS organizational change impact. In: 38th Hawaii International Conference on System Sciences (2005)
Griffiths, C.: Assessing the risks in major IT projects. Computer Audit Update 1995(11), 4–10 (1995)
Hakim, A., Hakim, H.: A practical model on controlling the ERP implementation risks. Information Systems 35, 204–214 (2010)
Holland, C.P., Light, B.: A critical success factors model for ERP Implementation. IEEE Software 16, 30–36 (1999)
Kang, I., Lee, S., Choi, J.: Using fuzzy cognitive map for the relationship management in airline service. Expert Systems with Applications 26, 545–555 (2004)
Kim, H.S., Lee, K.C.: Fuzzy implications of fuzzy cognitive map with emphasis on fuzzy causal relationship and fuzzy partially causal relationship. Fuzzy Sets and Systems 97, 303–313 (1998)
Kliem, R., Ludin, I.: Reducing Project Risk. Gower Publishing Limited, Aldershot (2000)
Konar, A., Chakraborty, U.K.: Reasoning and unsupervised learning in a fuzzy cognitive map. Information Sciences 170, 419–441 (2005)
Kosko, B.: Fuzzy Cognitive Maps. International Journal of Man-Machine Studies 24, 65–75 (1986)
Kosko, B.: Neural Networks and Fuzzy Systems. Prentice-Hall, Upper Saddle River (1992)
Kosko, B.: Fuzzy Engineering. Prentice-Hall, Englewood Cliffs (1996)
Lee, S., Han, I.: Fuzzy cognitive map for the design of EDI controls. Information & Management 37, 37–50 (2000)
Kwak, Y.H., Stoddard, J.: Project risk management: lessons learned from software development environment. Technovation 24, 915–920 (2004)

Li, M., Yen, L.R.: Information technology and firm performance: Linking with environmental, strategic and managerial contexts. Information & Management 35(1), 43–51 (1999)
Lin, W.T., Shao, B.B.M.: The relationship between user participation and systems success: a simultaneous approach. Information & Management 37(6), 283–295 (2000)
Linstone, H.A., Turoff, M.: The Delphi Method, Techniques and Applications. Addison-Wesley, London (1975)
Mitchell, V.W.: Problems and risks in the purchasing of consulting services. The Services Industries Journal 14(3), 315–339 (1994)
Okoli, C., Pawlowski, S.: The Delphi method as a research tool: an example, design considerations and applications. Information & Management 42(1), 15–29 (2004)
Osei-Bryson, K.: Generating consistent subjective estimates of the magnitudes of causal relationships in fuzzy cognitive maps. Computers and Operations Research 31, 1165–1175 (2004)
Papageorgiou, E.I., Spyridonos, P.P., Stylios, C.D., Ravazoula, P., Groumpos, P.P., Nikiforidis, G.N.: Advanced soft computing diagnosis method for tumour grading. Artificial Intelligence in Medicine 36, 59–70 (2006a)
Rai, V., Kim, D.: Principal–agent problem: A cognitive map approach. Electronic Commerce Research and Applications 1, 174–192 (2002)
Raymond, L., Bergeron, F.: Project management information systems: An empirical study of their impact on project managers and project success. International Journal of Project Management 26(2), 213–220 (2008)
Ritchie, B., Brindley, C.: The information-risk conundrum. Marketing Intelligence & Planning 19, 29–37 (2001)
Rodriguez-Repiso, L., Setchi, R., Salmeron, J.L.: Modelling IT projects success: Emerging methodologies reviewed. Technovation 27, 582–594 (2007)
Saleh, Y., Alshawi, M.: An alternative model for measuring the success of IS projects: The GPIS model. The Journal of Enterprise Information Management 18(1), 47–63 (2005)
Salmeron, J.L.: Augmented Fuzzy Cognitive Maps for modelling LMS Critical Success Factors. Knowledge-Based Systems 22(4), 275–278 (2009)
Stach, W., Kurgan, L., Pedrycz, W., Reformat, M.: Genetic learning of fuzzy cognitive maps. Fuzzy Sets and Systems 153, 371–401 (2005)
Umble, E.J., Haft, R.R., Umble, M.M.: Enterprise resource planning: Implementation procedures and critical success factors. European Journal of Operational Research 146, 241–257 (2003)
Xirogiannis, G., Chytas, P., Glykas, M., Valiris, G.: Intelligent impact assessment of HRM to the shareholder value. Expert Systems with Applications 35, 2017–2203 (2008)
Xirogiannis, G., Glykas, M.: Fuzzy Cognitive Maps in Business Analysis and Performance Driven Change. IEEE Transactions in Engineering Management 51, 334–351 (2004)
Xirogiannis, G., Glykas, M.: Intelligent Modeling of e-Business Maturity. Expert Systems with Applications 32, 687–702 (2007)
Zadeh, L.A.: Fuzzy sets. Information and Control 8, 338–353 (1965)
Zhang, Z., Leeb, M.K.O., Huang, P., Zhang, L., Huang, X.: A framework of ERP systems implementation success in China: An empirical study. International Journal of Production Economics 98, 56–80 (2005)

Software Reliability Modelling Using Fuzzy Cognitive Maps

Panagiotis Chytas, Michael Glykas, and George Valiris

Abstract. This paper describes a model for software reliability prediction. The output is a 'forward-looking' model, which helps testers to predict and manage software reliability. Despite the availability of various approaches developed in the field of software reliability, there are still issues that require further research in order to succeed in supporting the decision making process and improving software quality. In the present paper we propose the use of Fuzzy Cognitive Maps (FCMs) as an approach to modelling of software reliability. FCMs capture information in the relationships between concepts, are dynamic, express hidden relationships, and are combinable and tunable. Preliminary experiments indicate that the proposed mechanism forms a sound support aid for software reliability modelling.

Keywords: Software Reliability, Fuzzy Cognitive Maps, Modelling, Uncertainty.

1 Introduction

According to (Billinton and Allen 1983) "Reliability is the probability of a device performing its purpose adequately for the period of time intended under the operating conditions encountered". The ANSI IEEE Standard Glossary of Software Engineering Terminology provides the following definitions in reliability engineering:

- An error is "A discrepancy between a computed, observed, or measured value or condition and the true, specified, or theoretically correct value or condition."

Panagiotis Chytas · George Valiris
University of Aegean, Department of Business Administration,
8, Michalon Street, Chios, 82 100, Greece
e-mail: p.chytas@chios.aegean.gr

Michael Glykas
University of Aegean, Department of Financial and Management Engineering,
31, Fostini Street, Chios, 82 100, Greece
e-mail: mglikas@aegean.gr

M. Glykas (Ed.): Fuzzy Cognitive Maps, STUDFUZZ 247, pp. 217–230, 2010.
springerlink.com © Springer-Verlag Berlin Heidelberg 2010

- A failure is "The termination of the ability of a functional unit to perform its required function."
- A fault is "An accidental condition that causes a functional unit to fail to perform its required function"

(Marciniak 1994) defines software reliability as follows: "Software Reliability is the probability of failure-free operation of a program for a specified time under a specified set of operating conditions". Software reliability modelling evolved in the early 1970s by transferring the hardware theory to software: "Because of the relatively advanced state of hardware reliability, it is natural to try to apply this theory to software reliability" (Myers 1976).

Various techniques have been proposed for software reliability modelling in the past (Singpurwalla and Wilson 1994) mainly based on regression models. However, in order to obtain accurate estimates, a huge number of data is required, which is not usually the case. Furthermore, the statistical estimates might not exist at all and this has caused a lot of problems in practice. Many software developers would be more interested in estimating the software reliability as early as possible for their planning purpose and they are hesitant to use software reliability models. According to (Xie 1999), the traditional software reliability models only make use of failure information for the particular software we are interested in that is the particular version of the software system for which the reliability estimate is to be provided.

In spite of the availability of various approaches, there are still issues that require further research in order to succeed in supporting the decision making process and improving software quality. It has been realized that several factors may affect software reliability behaviour. Among these factors are software development methodology, software development environment, software complexity, software development organization and personnel, etc. (Neufelser 1993; Cai 1998). Usually, trade offs occur among these factors in a non-linear way. This imposes several limitations on existing statistical modelling approaches that depend heavily upon the assumptions of interdependence and linearity.(Fenton and Neil 1999), provide a detailed critique of software defect prediction models. They argue that the essential problem is the oversimplification that is associated with the use of simple regression models. To overcome this, (Fenton and Neil 1999)proposed the use of Bayesian Belief Networks (BBN). However, the large amount of data required by BBN hinders the widespread use of this technique and the accuracy of its results (Li and Smidts 2003). Furthermore, BBN do not provide a mechanism to capture the strength (weight) between relationships of one factor on another.

In this paper we propose the use of FCMs in software reliability modelling to perform "what-lf" analysis in order to identify potential problems and to generate appropriate actions for improvement.

The proposed decision aid mechanism supports software reliability modelling based on fuzzy cognitive maps (FCMs). The mechanism utilizes the fuzzy causal characteristics of FCMs as a new modelling technique to develop a causal representation of dynamic software engineering principles in order to generate an interrelated network of interconnected software engineering factors. It is the belief of the authors that the fuzzy reasoning capabilities enhance considerably the usefulness of the proposed mechanism while reducing the effort of identifying precise data during a software reliability modelling exercise. The proposed model has both theoretical and practical benefits. Given the demand for software quality it is believed to be useful for anyone contemplating or undertaking a software reliability modelling exercise in the software sector. Primarily, the proposed model targets the principle beneficiaries of software engineering (software engineers, managers, etc) assisting them to reason effectively about the status of a piece of software Nevertheless, the explanatory nature of the mechanism can prove to be useful in a wider educational setting.

This paper is structured as follows: Section 2 provides an introduction to FCMs and its application in other fields. Section 3 describes an FCM model to be used for software reliability prediction. Section 4 concludes this paper and presents future steps to validate the proposed model.

2 Fuzzy Cognitive Maps

FCMs are fuzzy signed directed graphs with feedback loops, in which the set of objects is modelled by the nodes, and the set of causal relationships is modelled by directed arcs (Figure 1). FCM theory developed recently (Kosko 1986) as an expansion of cognitive maps that had been employed to represent social scientific knowledge (Axelrod 1976), to make decision analysis (Zhang, Chen et al. 1989) and to analyse extend-graph theoretic behaviour(Zhang and Chen 1988). FCMs combine the strengths of cognitive maps with fuzzy logic. By representing human knowledge in a form more representative of natural human language than traditional concept mapping techniques, FCMs ease knowledge engineering and increase knowledge-source concurrence.

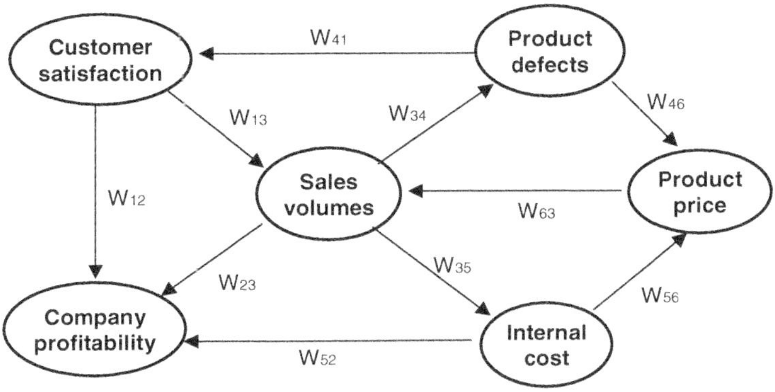

Fig. 1 A Simple FCM

More particularly, the term fuzzy cognitive map (FCM) was coined by (Kosko 1986) to describe a cognitive map model with two significant characteristics:

- Causal relationships between nodes can take on values representing fuzzy set memberships,
- The system is dynamic involving feedback, where the effect of change in a concept node affects other nodes, which in turn can affect the node initiating the change. The presence of feedback adds a temporal aspect to the operation of the FCM.

According to (Cole and Persichitte 2000), concepts in an FCM are not arrayed according to abstractness or centrality of the idea. The centrality of an idea can be naturally determined after the map has been completed. Centrality becomes a function of the number of links to and from a given node and the weight of those links. The abstractness of an idea can be interpreted as a function of its fuzziness. The more abstract an idea, the more fuzzy subsets it contains. Hierarchical conceptual relationships can be embedded within a FCM node. The node then becomes an embedded FCM within the larger framework. The resulting signal strength from the node is a function of the embedded processing. These features offer several advantages to FCMs over traditional mapping methods. FCMs have these specific advantageous characteristics:

✓ FCMs capture more information in the relationships between concepts
✓ FCMs are dynamic
✓ FCMs express hidden relationships
✓ FCMs are combinable
✓ FCMs are tunable

FCM nodes are the concepts that correspond to variables, states, factors and other characteristics incorporated in the model, which describe the behaviours of the system. Directed, signed and weighted arcs, which represent the causal relationships that exist between the concepts, interconnect the FCM concepts (Figure 1). Each concept represents a characteristic, state or variable of the system; concepts stand for events, actions, goals, values and/or trends of the system being modelled as an FCM. Each concept is characterised by a numeric value that represents a quantitative measure of the concept's presence in the model. A high numeric value indicates the strong presence of a concept. The numeric value results from the transformation of the real value of the system's variable, for which this concept stands, to the interval [0,1]. All the values in the graph are fuzzy, so weights of the arcs are described with linguistic values that can be defuzzified and transformed to the interval [-1,1]. Studying this graphical representation, one can conclude which concept influences other concepts and what are the interconnections among them. This representation makes the updating of the structure of the graph easy, as new information becomes available or as more experts are asked. This can be done, for example, by the addition or deletion of an interconnection or a concept.

Between concepts, there are three possible types of causal relationships expressing the type of influence of one concept on another. The weight of an interconnection, Wij, for the arc from concept C_i to concept C_j, can be positive ($W_{ij}>0$), which means that an increase in the value of concept C_i leads to the increase of the value of concept C_j, and a decrease in the value of concept C_i leads to the decrease of the value of concept C_j. Or there is a negative causality ($W_{ij}<0$), which means that an increase of the value of concept C_i leads to the decrease of the value of concept C_j and vice versa. When there is no relationship from concept C_i to concept C_j, then ($W_{ij}=0$).

When the FCM starts to model the system, concepts take their initial values and then the system is simulated. At each step, the value of each concept is determined by the influence of the interconnected concepts on the corresponding weights:

$$a_i^{t+1} = f\left(\sum_{j=1, j\neq i}^{n} w_{ji} a_j^t\right) \tag{1}$$

where a_i^{t+1} is the value of concept C_i at step t+1, a_j^t the value of the interconnected concept C_j at step t, W_{ji} the weighted arc from concept C_j to C_i, and f a threshold function. Three threshold functions have been identified in the literature (Kosko 1998) and are described below:

$$S_i(x_i) = 0, x_i \le 0$$
$$S_i(x_i) = 1, x_i > 0$$

bivalent

$$S_i(x_i) = -1, x_i \le -0.5$$
$$S_i(x_i) = 0, -0.5 < x_i < 0.5$$
$$S_i(x_i) = 1, x_i \ge 0.5$$

trivalent

$$S_i(x_i) = \frac{1}{1+e^{-cx_i}}$$

logistic signal, c = 5

An expert defines the main concepts that represent the model of the system, based on his knowledge and experience on the operation of the system. At first, the expert determines the concepts that best describe the system. He knows which factors are crucial for the modelling of the system and he represents each one by a concept. Moreover, he has observed which elements of the system influence other elements and for the corresponding concepts he determines the positive, negative or zero effect of one concept on the others. He describes each interconnection with a linguistic value that represents the fuzzy degree of causality between concepts. The linguistic weights are transformed into numerical weights using the methodology proposed by (Stylios, Groumpos et al. 1999).

2.1 *Applications of Fuzzy Cognitive Maps*

Over the last 10 years, a variety of FCMs have been used for representing knowledge and artificial intelligence in engineering applications, like geographical information systems (Liu and Satur 1999) and fault detection (Pelaez and Bowles 1995; Ndouse and Okuda 1996). FCMs have been used in modelling the supervision of distributed systems (Stylios, Georgopoulos et al. 1997). FCMs have also been used in operation research (Craiger, Goodman et al. 1996), web data mining (Hong and Han 2002; Lee, Kim et al. 2002), as a back end to computer-based models and medical diagnosis (Georgopoulos 2002).

Several research reports applying basic concepts of FCMs have also been presented in the field of business and other social sciences. (Axelrod 1976) and (Perusich 1996) have used FCM for representing tacit knowledge in political and social analysis. FCMs have been successfully applied to various fields such as decision making in complex war games (Klein and Cooper 1982), strategic planning (Diffenbach 1982; Ramaprasad and Poon 1985), strategic information systems planning (Kardaras and Karakostas 1999), information retrieval (Johnson and Briggs 1994) and distributed decision modelling (Zhang, Wang et al. 1994).

Research like (Lee and Kim 1997) has successfully applied FCMs to infer implications from stock market analysis results. Research like (Lee and Kim 1998) has also suggested a new concept of fuzzy causal relations found in FCMs and has applied it to analyse and predict stock market trends. The inference power of FCMs has also been adopted to analyse the competition between two companies, which have been assumed to use differential games mechanisms to set up their own strategic planning (Lee and Kwon 1998).

FCMs have been integrated with case-based reasoning technique to build organizational memory in the field of knowledge management (Noh, Lee Lee et al. 2000). The research in (Parenthoen, Reignier et al. 2001) proposed the use of FCM as a tool to model emotional behaviour of virtual actors improvising in free interaction within the framework of a "nouvelle vague" scenario, and discussed the problem of de-localizing each agent level to model autonomous agents within a virtual world. Summarizing, FCMs can contribute to the construction of more intelligent systems, since the more intelligent a system becomes, the more symbolic and fuzzy representations it utilizes.

In addition, a few modifications have been proposed. For example, (Silva 1995) has proposed new forms of combined matrices for FCMs, (Hagiwara 1992) has extended FCMs by permitting non-linear and time delay on the arcs, the research in (Schneider, Schnaider et al. 1995) has presented a method for automatically constructing FCMs. More recently, (Liu and Satur 1999) has carried out extensive research on FCMs, investigating inference properties of FCMs and has proposed contextual FCMs based on the object-oriented paradigm of decision support, having applied contextual FCMs to geographical information systems (Liu 2000).

3 An FCM-Based Model for Software Reliability Prediction

According to (Neil and Fenton 1996), "project managers make decisions about quality and cost using best guesses; it seems that will always be the case and the best that researchers can do is a) recognise the fact and b) improve the 'guessing' process". Furthermore, FCMs combine the strengths of cognitive maps with fuzzy logic. By representing human knowledge in a form more representative of natural human language than traditional concept mapping techniques, FCMs ease knowledge engineering and increase knowledge-source concurrence.

Based on the two directions suggested by (Neil and Fenton 1996) as well as the combined strengths provided by FCMs we propose the use of FCMs for software reliability modelling. Our approach to software reliability modelling involves the following three characteristics:

a) Representation of cause-effect relationships useful for viewing the overall structure of the concerning decision problem. The role of FCMs is to identify factors that seem to be relevant to software reliability modelling.
b) Integration of human expertise with FCMs can not only improve the performance of software reliability modelling, but also widen applicability to various software reliability-modelling problems.
c) Simple matrix calculations, which allow the software reliability modelling process to be performed more easily and quickly compared to that of conventional models.

How can a FCM be used in our software reliability modelling process? To answer this question, we will address the following two kinds of possible ways of applying FCMs to software reliability modelling.

i. Identify Potential Factors for Software Reliability Modelling

In this kind of FCM usage the aim is to identify those factors, which have no effect on the software reliability decision outcome. By analysing the connectivity between factors depicted in a FCM, one can easily determine whether or not data about a particular factor are relevant to a software failure. For illustration, consider Figure 8.2, where causal links between six factors are depicted. Suppose that one is concerned with the state of F because it will affect a decision to be made. Figure 8.2 shows us that both A and D influence F directly, and C has an indirect influence on F through D. On the other hand, neither B nor E influences F indirectly or directly. They (B and E) are therefore "irrelevant" in light of outcome F. Determination like this is frequently difficult to make by direct inquiry or observation, while it becomes quite clear when using FCMs purposely.

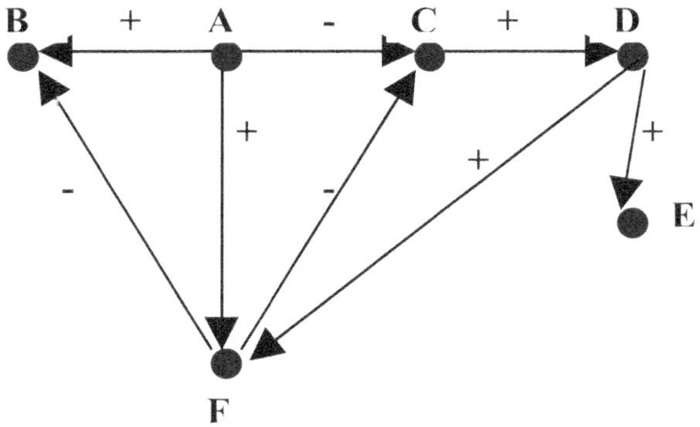

Fig. 2 A Cognitive Map

Furthermore, consider the following. Let us assume that one finds that the value of F is relatively high, while the values of A and D are relatively low. The FCM suggests, however, that low A and low D would lead to a low F. Thus one knows that this is not the expected case. The reason may be that other factors, which remain unknown, might be affecting F in this particular situation, those that were not originally perceived to be relevant. Effort should then be focused on finding such factors affecting F so that the performance of an FCM-based model may not degrade sharply under turbulent situations.

ii. Perform "What-If" Analysis

FCMs have been proposed to represent causal reasoning by using numeric processing. They have been also used to represent arbitrarily complex models of interrelated concepts. FCMs are capable of abstracting non-situated behaviours from a number of such situated learning experiences. FCM can be organized into a matrix, which is capable of not only evolving with time but also producing outputs for some specific inputs, and therefore allowing "what-if" analysis.

In order to create an FC bnM-based model for software reliability we have to go through a series of stages that involve: (1) inputs to be provided, (2) and outcomes to be generated. Software reliability domain experts and/or professionals are people with specific expertise that contribute towards providing the knowledge for the FCM-based model under construction. The following table indicates the stages for the construction of an FCM-based model together with the inputs, outcomes and benefits of each stage:

Table 1 The Inputs and Outcomes derived from the Stages of the model

STAGE	**INPUT**	**OUTCOME**
1) Node Specification	Meeting Preparation, Participant Identification	List of Nodes (Concepts)
2) Node Relationship Table Production	Finalised Nodes and filled table from participants	Final FCM with no weights
3) Weight Definition and FCM Prototype	First weights from participants in the FCM	Final FCM with weights are send to participants
4) Manual Assessment and Final FCM Production	Corrections to weights are received from participants	Final FCM

Before proceeding to each of the aforementioned stages a kick-off meeting takes place between the domain experts and the researchers. The aim of this meeting is for all participants to contribute towards:

- Establishing the FCM team
- Clarifying the objectives of the team
- Identifying the context of the FCM under construction
- Selecting the reference material to be used for the construction of theFCM
- Anticipating possible user benefits
- Preparing further actions for the participants

1) Node Specification

In order to produce the nodes of the FCM under construction the participants focus on the actual definition of the FCM by defining the nodes that are of relevance to the particular context-FCM. This exercise is expected to be quite laborious at the beginning with the researchers taking an active role in interviewing and guiding the domain experts in expressing their beliefs and knowledge in the form of nodes and their interrelationships.

Concepts could represent either software reliability metrics or decision variables (factors) in the particular domain. At the end of this workshop a list of nodes will be produced. At this point it must be also explained the definition of a "Relationship" to the participants in order for them to produce the node association table where cause nodes are related to effect nodes by associating "+" or "-" signs.

2) Node Relationship Table Production

During the meeting, a table that integrates the participants' beliefs should be produced. If there is a cause or effect relationship between 2 nodes, a '+' or '-' sign is marked between these 2 nodes. If there is no cause or effect relationship a '0' should indicate that. The corresponding FCM contains the positive and negative causal relationships of the table only without weights yet.

3) Weight Definition and FCM Prototype

In order to define weight value of the association rules in an adaptive and dynamic manner, the following methodology is proposed. Experts are asked to describe the interconnection influence of concepts using linguistic notions. Influence of one concept over another, is interpreted as a linguistic variable in the interval [-1,1]. Its term set **T(influence)** is:

T(influence) = {negatively very-very high, negatively very high, negatively high, negatively medium, negatively low, negatively very low, negatively very-very low, zero, positively very-very low, positively very low, positively low, positively medium, positively high, positively very high, positively very-very high}.

We propose a semantic rule M to be defined at this point. The above-mentioned terms are characterized by the fuzzy sets whose membership functions μ are shown in Figure 3.

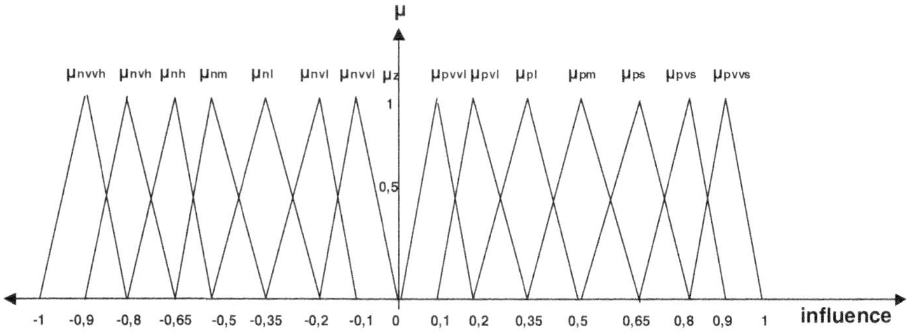

Fig. 3 Membership functions of linguistic variable influence

- M(negatively very-very high)= the fuzzy set for "an influence close to -90%" with membership function μ_{nvvh}
- M(negatively very high)= the fuzzy set for "an influence close to 80%" with membership function μ_{nvh}
- M(negatively high)= the fuzzy set for "an influence close to 65%" with membership function μ_{nh}
- M(negatively medium)= the fuzzy set for "an influence close to -50%" with membership function μ_{nm}
- M(negatively low)= the fuzzy set for "an influence close to -35%" with membership function μ_{nl}
- M(negatively very low)= the fuzzy set for "an influence close to -20%" with membership function μ_{nvl}
- M(negatively very-very low)= the fuzzy set for "an influence close to -10%" with membership function μ_{nvvl}
- M(zero)= the fuzzy set for "an influence close to 0" with membership function μ_z
- M(positively very-very low)= the fuzzy set for "an influence close to 10%" with membership function μ_{pvvl}
- M(positively very low)= the fuzzy set for "an influence close to 20%" with membership function μ_{pvl}
- M(positively low)= the fuzzy set for "an influence close to 35%" with membership function μ_{pl}
- M(positively medium)= the fuzzy set for "an influence close to 50%" with membership function μ_{pm}

- M(positively high)= the fuzzy set for "an influence close to 65%" with membership function μ_{ph}
- M(positively very high)= the fuzzy set for "an influence close to 80%" with membership function μ_{pvh}
- M(positively very-very high)= the fuzzy set for "an influence close to 90%" with membership function μ_{pvvh}

The membership functions are not of the same size since it is desirable to have finer distinction between grades in the lower and higher end of the influence scale. As an example, three experts propose different linguistic weights for the interconnection W_{ij} from concept C_i to concept C_j: (a) positively high (b) positively very high (c) positively very-very high. The three suggested linguistics are integrated using a sum combination method and then the defuzzification method of centre of gravity (CoG) is used to produce a weight W_{ij}=0,73 in the interval [-1,1]. This approach has the advantage that experts do not have to assign numerical causality weights but to describe the degree of causality among concepts.

4) Manual Assessment and Final FCM Production
Over this workshop the initial values of the concept-node values have to be set and the system is simulated so that its functionality and performance is assessed. A similar methodology to the one described in stage 3 can be used to assign values to concepts. The experts are also asked to describe the measurement of each concept using once again linguistic notions. Measurement of a concept is also interpreted as a linguistic variable with values in the interval [-1,1]. Its term set **T(Measurement)** = **T(Influence)**. A new semantic rule M2 (analogous to M) is also defined and these terms are characterized by the fuzzy sets whose membership functions μ_2 are analogous to membership functions μ. Several initial values and simulation scenarios are tried according to scenario hypotheses from the experts.

3.1 Preliminary Experiments

Two preliminary experiments were conducted by utilizing factors from actual software engineering exercises in two major Greek software enterprises. For each experiment, a team of experts was engaged to provide linguistic variables for the causal weights and the concept values to let the FCM algorithm reason about the impact of potential change initiatives.

As far as the theoretical value is concerned, the FCM mechanism extends previous research attempts in software reliability modeling by (a) allowing fuzzy definitions in the cognitive maps, (b) introducing a specific interpretation mechanism of linguistic variables to fuzzy sets, and (c) introducing the notion of interconnected factors that affect software reliability behavior. As far as the practical value is concerned, preliminary experimental results indicate that (a) when compared to the expert estimates, the mechanism provides reasonably good approximations, and (b) it allows performing "what-if" analysis in order to identify potential problems and to generate appropriate actions for improvement.

4 Conclusions

This paper proposed a methodology for software reliability modelling based on a Fuzzy Cognitive Map (FCM) – driven approach. The proposed methodology utilises the fuzzy causal characteristics of FCMs as a new modelling technique to develop a causal representation of software reliability factors and indicators. The proposed methodology should not be regarded only as an effective software reliability modelling support tool. Its main purpose is to set a course for continuous improvement in software quality.

Future research will focus on conducting further experiments to test and promote the usability of the methodology, but also to identify potential pitfalls. Furthermore, future research will focus on the automatic determination of appropriate fuzzy sets (e.g. utilizing pattern recognition, mass assignments, empirical data, etc) for the representation of linguistic variables to suit each particular software engineering stage. Finally, further research will focus on implementing backward map traversal, a form of adbuctive reasoning (Flach and Kakas 1998). This feature offers the functionality of determining the condition(s) C_{ij} that should hold in order to infer the desired C_j in the causal relationship $C_{ij} \xrightarrow{w_{jk}} C_k$. Incorporating performance integrity constraints reduces the search space and eliminates combinatory search explosion. Backward reasoning has been tested extensively in other applications and its integration in the proposed methodology framework may prove beneficiary.

References

Axelrod, R.: Structure of decision: The cognitive maps of political elites. Princeton University Press, Princeton (1976)

Billinton, R., Allen, R.: Reliability Evaluation of Engineering Systems: Concepts and Techniques. Pitman Books, Ltd., London (1983)

Cai, K.Y.: Software Defect and Operational Profile Modelling. Kluwer Academic Publishers, Dordretch (1998)

Cole, J.R., Persichitte, K.A.: Fuzzy Cognitive Mapping: Applications in Education. International Journal of Intelligent Systems 15, 1–25 (2000)

Craiger, J.P., Goodman, D.F., et al.: Modeling Organizational Behavior with Fuzzy Cognitive Maps. Journal of Computational Intelligence and Organisations 1, 120–123 (1996)

Diffenbach, J.: Influense diagrams for complex strategic issues. Strategic Management Journal 3, 133–146 (1982)

Fenton, N., Neil, M.: A Critique of Software Defect Prediction Research. IEEE Transactions on Software Engineering 25(5) (1999)

Flach, P.A., Kakas, A.: Abduction and Induction in AI: Report of the IJCAI97 workshop. Logic Journal of the Interest Group on Pure and Applied Logic 6(4), 651–656 (1998)

Georgopoulos, V., Malandraki, G., et al.: A fuzzy cognitive map approach to differential diagnosis of specific language impairment. Journal of Artificial Intelligence in Medicine 679, 1–18 (2002)

Hagiwara, M.: Extended fuzzy cognitive maps. In: 1st IEEE International Conference on Fuzzy Systems, New York (1992)

Hong, T., Han, I.: Knowledge based data mining of news information of the internet using cognitive maps and neural networks. Journal of Expert Systems with Applications 23(1), 1–8 (2002)

Johnson, R.J., Briggs, R.O.: A model of cognitive information retrieval for ill-structured managerial problems and its benefits for knowledge acquisition. In: 27th Annual Hawaii International Conference on System Sciencesi, Hawai (1994)

Kardaras, D., Karakostas, B.: The use of fuzzy cognitive maps to simulate the information systems strategic planning process. Journal of Information and Software Technology 41(1), 197–210 (1999)

Klein, J.C., Cooper, D.F.: Cognitive maps of decision makers in a complex game. Journal of Operation Research Society 33, 63–71 (1982)

Kosko, B.: Fuzzy Cognitive Maps. International Journal of Man-Machine Studies 24, 65–75 (1986)

Kosko, B.: Hidden patterns in combined and adaptive knowledge networks. International Journal of Approximate Reasoning 2, 377–393 (1998)

Lee, K.C., Kim, H.S.: A fuzzy cognitive map-based bi-directional inference mechanism: An application to stock investment analysis. Journal of Intelligent Systems in Accounting Finance & Management 6(1), 41–57 (1997)

Lee, K.C., Kim, H.S.: Fuzzy implications of fuzzy cognitive map with emphasis on fuzzy causal relationship and fuzzy partially causal relationship. Journal of Fuzzy Sets and Systems 3, 303–313 (1998)

Lee, K.C., Kim, J.S., et al.: Fuzzy Cognitive Map approach to web-mining inference amplification. Journal of Expert Systems with Applications 22, 197–211 (2002)

Lee, K.C., Kwon, O.B.: A strategic planning simulation based on cognitive map knowledge and differential game. Journal of Simulation 7(5), 316–327 (1998)

Li, M., Smidts, C.S.: A Ranking of Software Engineering Measures Based on Expert Opinion. IEEE Transactions On Software Engineering 29(9) (2003)

Liu, Z.Q.: Fuzzy cognitive maps: Analysis and extension. Springer, Heidelberg (2000)

Liu, Z.Q., Satur, R.: Contexual fuzzy cognitive map for decision support in geographical information systems. Journal of IEEE Transactions on Fuzzy Systems 7, 495–507 (1999)

Marciniak, J.J.: Encyclopedia of Software Engineering. John Wiley and Sons, Inc., New York (1994)

Myers, G.J.: Software Reliability:Principles and Practices. Wiley, New York (1976)

Ndouse, T.D., Okuda, T.: Computational intelligence for distributed fault management in networks using fuzzy cognitive maps. In: IEEE International Conference Communication Converging Technology Tomorrow's Application, New York (1996)

Neil, M., Fenton, N.: Predicting Software Quality using Bayesian Belief Networks. In: 21st Annual Software engineering Workshop, NASA/Goddard Space Flight Centre (1996)

Neufelser, A.M.: Ensuring Software Reliability. Marcel Dekker, New York (1993)

Noh, J.B., Lee Lee, K.C., et al.: A case- based reasoning approach to cognitive map driven - driven tacit knowledge management. Journal of Expert Systems with Applications 19, 249–259 (2000)

Parenthoen, M., Reignier, P., et al.: Put Fuzzy Cognitive Mapsto Work in Virtual Worlds. In: Fuzz-IEEE 2001. IEEE, Australia (2001)

Pelaez, C.E., Bowles, J.B.: Applying fuzzy cognitive maps knowledge representation to failure modes effect analysis. In: IEEE Annual Reliability Maintainability Symposium, New York (1995)

Perusich, K.: Fuzzy cognitive maps for political analysis. In: International Symposium on Technology and Society, Technical Expertise and Public Decisions, New York (1996)

Ramaprasad, A., Poon, E.: A computerised interactive technique for mapping influense diagrams (MIND). Strategic Management Journal 6(4), 377–392 (1985)

Schneider, M., Schnaider, E., et al.: Constructing fuzzy cognitive maps. In: IEEE Conference on Fuzzy Systems, New York (1995)

Silva, P.C.: New forms of combinated matrices of fuzzy cognitive maps. In: IEEE International Conference on Neural Networks, New York (1995)

Singpurwalla, N., Wilson, S.: Software reliability modeling. International Statistical Review 62, 289–317 (1994)

Stylios, C.D., Georgopoulos, V.C., et al.: Introducing the Theory of Fuzzy Cognitive Maps in Distributed Systems. In: 12th IEEE International Symposium on Intelligent Control, Istanbul, Turkey (1997)

Stylios, C.D., Groumpos, P.P., et al.: A fuzzy cognitive map approach to process control systems. J. Adv. Comput. Intell. 3, 1–9 (1999)

Xie, M., Hong, G.Y., Wohlin, C.: Software reliability prediction incorporating information from a similar project. The Journal of Systems and Software 49, 43–48 (1999)

Zhang, W.R., Chen, S.S.: A logical architecture for cognitive maps. In: Second IEEE International Conference on Neural Networks, San Diego, CA (1988)

Zhang, W.R., Chen, S.S., et al.: Pool 2: a generic System for cognitive map development and decision analysis. IEEE Transactions on Systems, Man and Cybernetics 19, 31–39 (1989)

Zhang, W.R., Wang, W., et al.: A-pool: An agent-oriented open system for distributed decision process modelling. Journal of Organisational Computing 4(2), 127–154 (1994)

Fuzzy Cognitive Networks for Maximum Power Point Tracking in Photovoltaic Arrays

Thodoris L. Kottas, Athanasios D. Karlis, and Yiannis S. Boutalis

Abstract. The studies on the photovoltaic (PV) power generation are extensively increasing, since it is considered as an essentially inexhaustible and broadly available energy resource. However, the output power of the photovoltaic modules depends on solar radiation and temperature of the solar cells. Therefore, to maximize the efficiency of the renewable energy system, it is necessary to track the maximum power point of the PV array and make the array operate near it. Maximum power operation is a challenging problem, since it requires that the system load is capable of using all power available from the PV system at all times. The I-V characteristic of the load must intersect the focus of maximum power points on the I-V characteristics of the PV array for varying insolation and temperature levels. Fuzzy Cognitive Networks (FCN) have been proposed as an operational extension of Fuzzy Cognitive Maps (FCM), which work in continuous interaction with the system they describe and may be used to control it. In this chapter FCN is used to construct a maximum power point tracker (MPPT), which may operate in cooperation with a fuzzy MPPT controller. The proposed scheme outperforms other existing MPPT schemes of the literature giving very good maximum power operation of any PV array under different conditions such as changing insolation and temperature. Moreover it has the ability to adapt to different changes that might happen during the life cycle of the PV module, such as a destroyed cell of the PV array.

1 Introduction

It is well established that energy production and use based on consumption of fossil fuels can have deleterious environmental and human health impacts, including the potential for global warming of the earth through changes in the atmosphere's

Thodoris L. Kottas · Yiannis S. Boutalis
Automatic Control Systems Lab., Dept. of Electrical & Computer Engineering,
Democritus University of Thrace, V. Sofias 12, 67100, Xanthi, Greece
e-mail: tkottas@ee.duth.gr, ybout@ee.duth.gr

Athanassios D. Karlis
Electrical Machines Lab., Dept. of Electrical & Computer Engineering,
Democritus University of Thrace, V. Sofias 12, 67100, Xanthi, Greece
e-mail: akarlis@ee.duth.gr

M. Glykas (Ed.): Fuzzy Cognitive Maps, STUDFUZZ 247, pp. 231–257, 2010.
springerlink.com © Springer-Verlag Berlin Heidelberg 2010

concentration of carbon dioxide. Worldwide the conventional energy sources are rapidly depleting, while Population growth, increased expectations and means, and scientific and technological developments have dramatically increased the global demand for energy in its various forms. What this all implies is that the world is in the initial stages of an inevitable transition to a new energy system that, over time, will be less dependent on traditional uses of fossil fuels and increasingly dependent on renewable energy resources. In particular, solar photovoltaic (PV) energy is attracting a lot of attention, since it is clean, pollution-free, and inexhaustible. Applications of PV systems include water pumping, domestic and street lighting, electric vehicles, hybrid systems, military and space applications, refrigeration and vaccine storage, power plants, etc., all in either stand-alone or grid-connected configurations. A PV array is by nature a non-linear power source, which under constant uniform irradiance has a current–voltage (I–V) characteristic like that shown in Fig. 1. There is a unique point on the curve, called the maximum power point (MPP), at which the array operates with maximum efficiency and produces maximum output power. As it is well known, the MPP of a PV power generation system depends on array temperature, solar insolation, shading conditions and PV cells ageing, so it is necessary to constantly track the MPP of the solar array. A switch-mode power converter, called a maximum power point tracker (MPPT), can be used to maintain the PV array's operating point at the MPP. The MPPT does this by controlling the PV array's voltage or current independently of those of the load. If properly controlled by an MPPT algorithm, the MPPT can locate and track the MPP of the PV array. However, the location of the MPP in the I–V plane is not known a priori. It must be located, either through model calculations or by a search algorithm. Figure 2 shows a family of PV I–V curves under increasing irradiance, but at constant temperature, and Fig. 3 shows I–V curves at the same irradiance values, but at various temperatures. Needless to say there is a change in the array voltage at which the MPP occurs.

For years, research has focused on various MPP control algorithms to draw the maximum power of the solar array. These techniques include look-up table methods using Neural Networks [1]-[3] perturbation and observation (P&O) methods [4] - [7] and computational methods [8]-[10].

Other techniques were developed according to a linear approximation between the maximum output power and the optimal operating current [11], [12]. Another algorithm [13] measures and compares the output of two modules to track the MPP. Most recent applications of MPPT are based on the extremum value theorem [14]-[17]. According to the theorem the extremum, which is maximum or minimum, occurs at the critical point. If a function $y = f(x)$ is continuous on a closed interval, it has the critical points at all points x_0, where $f'(x_0) = 0$. Using the centered differentiation authors in [18] manage to reduce the oscillation around the MPP in steady state by controlling active perturbation.

One of the computational methods which have demonstrated fine performances under different environmental operating conditions is the fuzzy based MPPT technique [19] - [25]. The fuzzy controller uses dP/dI and its variations Δ(dP/dI) as the inputs and computes MPPT converter duty cycle. An on-line search algorithm that

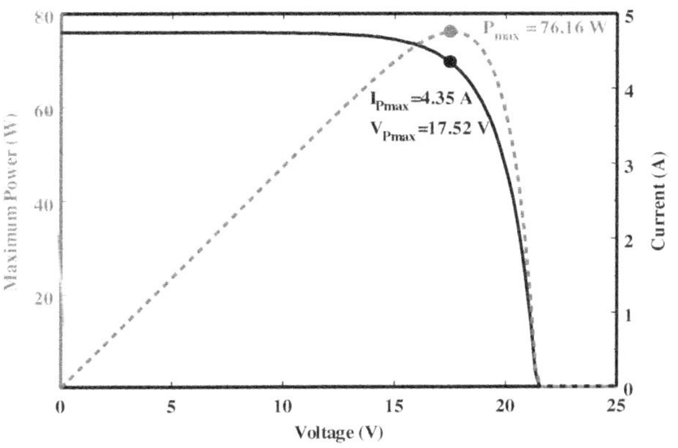

Fig. 1 PV array I-V and P-V Characteristics

does not require the measurement of temperature and solar irradiation level is proposed in reference [26]. Other researchers analyzed and compared the various MPPT techniques [8], [27], [28]. Beside that, in [24] a simple DSP-based MPPT algorithm is proposed. Finally, in [29] efforts have been made to model the dynamic behavior of a PV system in order to study its interaction with the pertinent MPPT system, while in [30] MPPT assessment and testing methods were presented in order to identify the accuracy, error and efficiency of the MPPTs.

This chapter presents an alternative MPPT method, which is based on the Fuzzy Cognitive Network Framework (FCN) for the representation of unknown systems. FCNs constitute an extension of the well-known Fuzzy Cognitive Maps (FCM) [31]. Traditionally FCMs have been used to model the behavior of complex systems appearing in socio-economical and political problems [32]-[42]. The formation of the cognitive graph is mainly based on experts' opinion regarding the interacting concepts of the system. One characteristic of the operation of an FCM is its diversity regarding its steady state behavior, being likely to appear equilibrium, limit cycle or even chaotic behavior. This diversity is however undesirable if FCMs are to be used to model and control traditional engineering systems. In this case the equilibrium after an initial perturbation is a desirable property. To this end, the authors proposed the Fuzzy Cognitive Networks (FCN), which constitute an operational extension of Fuzzy Cognitive Maps (FCM). Their main characteristic is that they work adaptively in close interaction with the system they describe receiving feedback from it. They capture steady state conditions of the system and associate them with equilibrium points of the FCN, which are also associated with specific weight sets of the FCN. Therefore, FCN assume and guarantee that they can always converge to unique equilibrium points associated with the steady state conditions of the system they describe. In the sequel they store the acquired knowledge in a fuzzy *if-then* rules database for future use, thus eliminating the future adaptation effort. FCN appeared gradually as operational extensions of

FCM [43-45]. The complete framework is presented in [45], but the conditions that guarantee their convergence to unique equilibrium points are given in [46], [47]. Based on these conditions, new improved weight adaptation algorithms have been recently proposed in [47], [48].

In this chapter a FCN is appropriately designed to solve the MPPT problem of a PV system. The nodes of the FCN represent essential operational (Voltage, Current, Insolation, Temperature) and control (Current) variables of the PV system. The node interconnection weights are determined using data, which are constructed so that they cover the operation of a PV system under a wide range of different climatic conditions. Once the FCN is trained it can be mounted on any PV system. Moreover, during the operation of the PV array the FCN weights are continuously updated based on data from the encountered operating conditions. The FCN is also combined with a Fuzzy Controller [19] to form a novel control system for the MPPT problem of a PV system. The performance of the method, as well as its adaptation abilities under changing environmental conditions or physical changes (ageing etc) of the PV cells is tested using climatic data for a specific PV system of the market, which reaches its MPP for various operational conditions, such as changing insolation and temperature and seasonal variations with great accuracy. The work presented here is based on the work of the authors presented in [49-51].

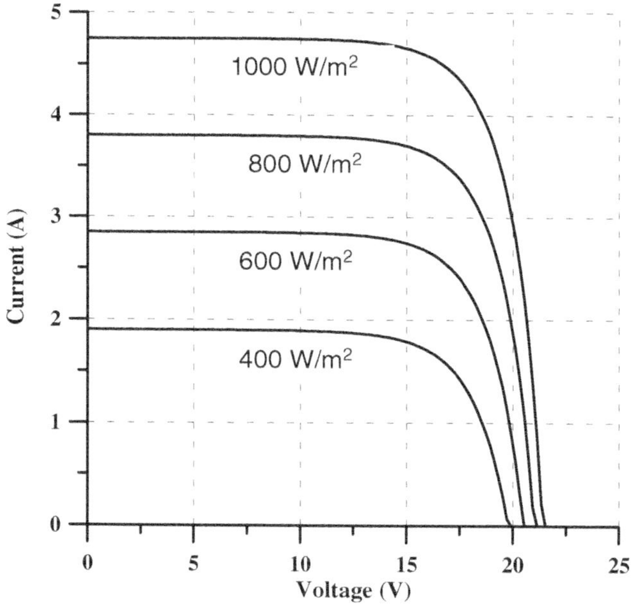

Fig. 2 PV array I-V Characteristics at various insolation levels

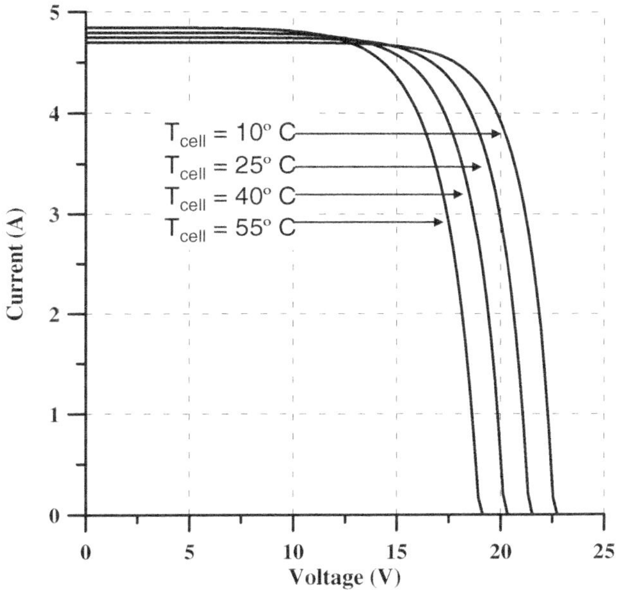

Fig. 3 PV array I-V Characteristics at various Cell Temperatures

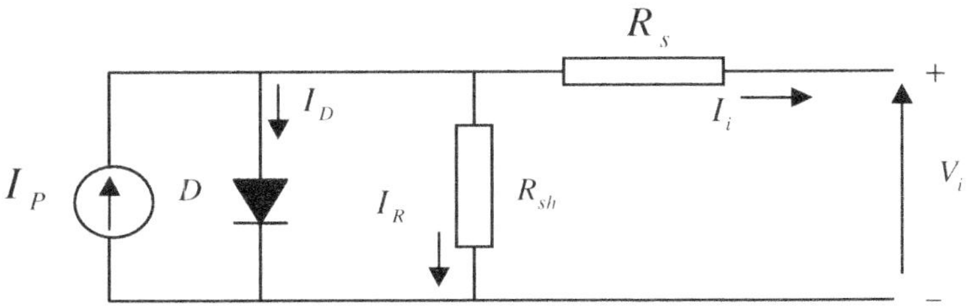

Fig. 4 Equivalent circuit of a solar cell

The chapter is organized as follows. In Section 2 mathematical relations between the essential variables of a PV system are presented. These relations are necessary for simulating its operation under different insolation and temperature levels. In Section 3 the FCN essentials and the MPPT method which is based on FCNs and a Fuzzy Controller is analyzed. In Section 4 simulated results from the presented method are given. Section 5 analyzes the way by which the presented method can adapt its knowledge in order to follow the physical changes that might happen to the PV system. Finally, section 6 concludes the work.

2 Simulation of the PV System

Using the equivalent circuit of a solar cell (Fig. 4) and the pertinent equations [24] the non-linear I-V characteristics of a solar array, consisting of N_s solar cells in series and N_p solar cells in parallel, are extracted, neglecting the series resistance:

$$I_i = N_p I_{ph} - N_p I_{rs}(e^{\frac{qV_i}{kTAN_s}} - 1) - \frac{V_i}{R_{sh} N_s} \tag{1}$$

where I_i is the PV array output current (A); V_i is the PV array output voltage (V); q is the charge of an electron; k is Boltzmann's constant in J/K; A is the p–n junction ideality factor; T is the cell temperature (K); and I_{rs} is the cell reverse saturation current. The factor A in (1) determines the cell deviation from the ideal p–n junction characteristics.

The photocurrent I_{ph} depends on the solar radiation and the cell temperature as stated in the following equation:

$$I_{ph} = (I_{scr} + k_i(T - T_r))\frac{S}{100} \tag{2}$$

where I_{scr} is the PV array short circuit current at reference temperature and radiation (A); k_i is the short circuit current temperature coefficient (A/K) and S is the solar radiation (mW/cm^2).

The reverse saturation current I_{rs} varies with temperature according to the following equation:

$$I_{rs} = I_{rr}\left(\frac{T}{T_r}\right)^3 e^{\frac{1.115}{k'A}\left(\frac{1}{T_r} - \frac{1}{T}\right)} \tag{3}$$

where T_r is the cell reference temperature, I_{rr} is the reverse saturation current at T_r, k' is the Boltzmann's constant in eV/K and the band-gap energy of the semiconductor used in the cell is equal to 1.115.

Finally, the next equation was used in the computer simulations to obtain the open circuit voltage of the PV array:

$$V_{oc} = N_s \frac{AkT}{q} \ln\left(\frac{I_{ph} + I_{rs}}{I_{rs}}\right) \tag{4}$$

From (2), (3) and (4) we get:

$$I_{rr} = \frac{(I_{scr} + k_i(T - T_r))\frac{S}{100}}{e^{\frac{V_{oc}q}{AkTN_s}} - 1}\left[\left(\frac{T_r}{T}\right)^3 e^{-\frac{1.115}{k'A}\left(\frac{1}{T_r} - \frac{1}{T}\right)}\right] \tag{5}$$

and from (1):

$$R_{sh} = \frac{V_{oc}}{-N_s N_p I_{rs}(e^{\frac{qV_{oc}}{kTAN_s}} - 1)} \tag{6}$$

The required data for identifying the maximum operating point at any insolation level and temperature are the following:

- k_i ,
- Open Circuit Voltage V_{oc}
- Short Circuit Current I_{scr}
- Maximum Power Voltage V_{mp}
- Maximum Power Current I_{mp}

(for initial conditions $T_r = 25°C$, $S = 100mW/cm^2$), all given by the PV array manufacturer.

3 MPPT Approach by Using Fuzzy Cognitive Networks and Fuzzy Logic Controller

In this section we will present the FCN+Fuzzy Logic Controller approaches [51] designed to represent the operation of a PV system. The control objective is to track and extract maximum power from the PV arrays for a given solar insolation level. The maximum power corresponding to the optimum operating point is determined for a different solar insolation level.

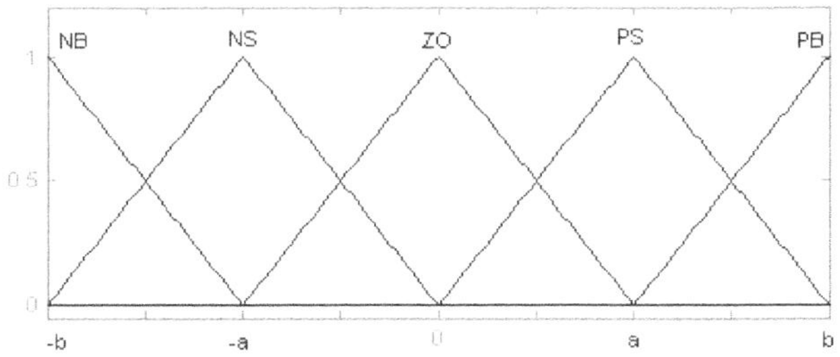

Fig. 5 Membership function for input and output

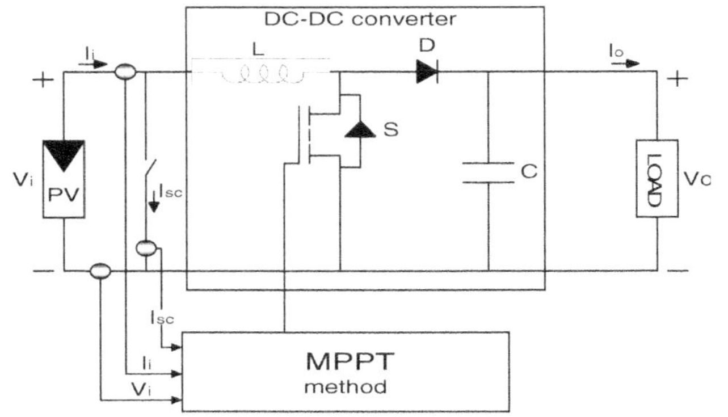

Fig. 6 Step-up boost converter for MPPT

3.1 Fuzzy Logic Controller [19]

In [19] the authors focused on single input- single output plant in which control is determined on the basis of satisfaction of two criteria relating to two input variables of the presented controller, namely error (*E*) and change of error (*CE*), at a sampling instant k.

The variable *E* and *CE* are expressed as follows:

$$E(k) = \frac{P_{pv}(k)-P_{pv}(k-1)}{I_{pv}(k)-I_{pv}(k-1)}, \tag{7}$$

$$CE(k) = E(k) - E(k-1), \tag{8}$$

where $P_{pv}(k)$ and $I_{pv}(k)$ are the power and current of the PV array, respectively. Therefore $E(k)$ is zero at the MPP of a PV array. In Fig. 5 the fuzzy sets of input $E(k)$, input $CE(k)$ and output dD is presented. Values, a and b are based on the range of values of the numerical variables. Output dD represents the change of the on/off duty ratio of the switch S of a step-up boost converter similar to the one shown in Fig. 6. The characteristics of the step-up boost converter are given in Appendix A.

Table 1 shows the rule table of the fuzzy controller where all the entries of the matrix are fuzzy sets of error $E(k)$ change of error $CE(k)$, and change of duty ratio dD to the converter. In the case of fuzzy control, the control rule must be designed in order that input variable $E(k)$ has to always be zero.

Table 1 Fuzzy rule table

E \ CE	NB	NS	ZO	PS	PB
NB	ZO	ZO	NB	NB	NB
NS	ZO	ZO	NS	NS	NS
ZO	NS	ZO	ZO	ZO	PS
PS	PS	PS	PS	ZO	ZO
PB	PB	PB	PB	ZO	ZO

The Fuzzy system is of Mamdani type using the min conjunction operator and max-min synthesis for the inference procedure. For the defuzzification the Center of Area (COA) and the Max Criterion Method (MCM) is used [52].

Authors choose to use the Fuzzy Controller method as MPPT instead of the simple P&O method [53], because by doing so there is a reduction not only in the time required to track the MPP but also in the fluctuation of power, as it is clearly presented in Figs 7 and 8.

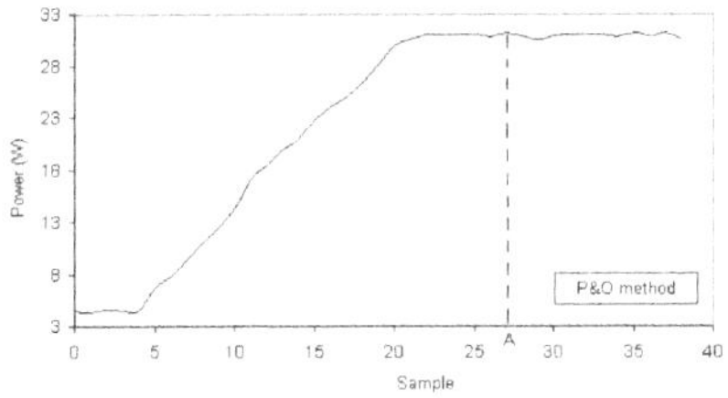

Fig. 7 Sample vs Power of PV array using P&O method
A. The P&O method reaches the MPP for the first time after the 27th iteration.

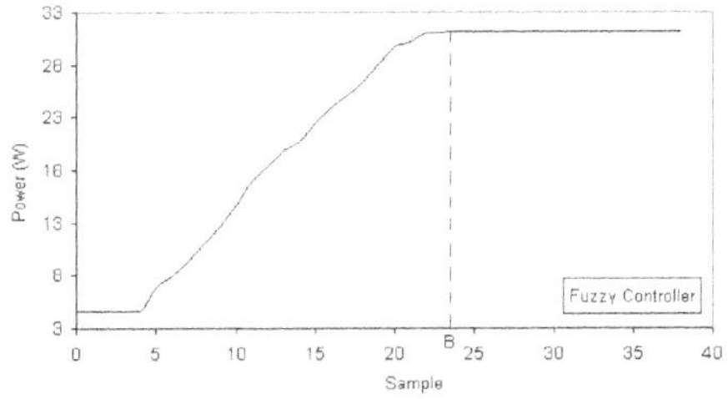

Fig. 8 Sample vs Power of PV array using Fuzzy Controller method
B. The Fuzzy Controller's method reaches MPP after the 24th iteration.

3.2 The Cognitive Graph Designed for the PV Project

The graph shown in Fig. 9 shows a FCN representation of a PV system, for a MPPT use. Using standard FCM representation the nodes take values in the interval [0, 1] and represent physical quantities of the PV system which are scaled to this interval. The weight interconnections take values in the interval [-1, 1]. The graph have 6 nodes, where nodes C1, C2 and C6 are steady value nodes and nodes C3, C4, C5 could be control nodes. Steady value nodes could also be termed input nodes. They are nodes that influence the other ones but they are not influenced by the other nodes. In this approach, node C4 is the control node whose value is used to regulate the current of the system. The regulation of the current of the system means that a different power is now the output power of the PV. Control nodes are the nodes the values of which will be used to the real system as control actions. Node C4 is used to calculate the optimum current needed to regulate the output power of the PV in the maximum point.

The weight of the interconnection between node C_i and node C_j denoted by W_{ij}, could be positive (W_{ij} >0) for positive causality or negative (W_{ij} <0) for negative causality or there is no relationship between node C_i, and node C_j, thus $W_{ij} = 0$. The nodes of the graph are related to the following physical quantities of the PV system.

Node C1 represents the irradiation with range in the interval [0→(0 mW/cm^2), 1→(100 mW/cm^2)].

Node C2 represents the temperature which also must be in the interval [0→(-30°C), 1→(70°C)].

Node C3 represents the optimum voltage of the PV system for the climatologic data obtained at the specific point of time, which also must be in the interval [0→(0 Volts), 1→($V_{\max}$)].

Node C4 represents the optimum current of the PV system for the climatologic data obtained at the specific point of time, which also must be in the interval [0→(0 Amps), 1→($I_{\max}$)].

Node C5 expresses the optimum output power of the PV system for the climatologic data obtained at the specific point of time, which also must be in the interval [0→(0 Watts), 1→($W_{\max}$)].

Node C6 is an artificial design node the value of which is used to regulate the equilibrium point in the nodes C3, C4 and C5. The value of C6 is steady and equals 1.

The value of each node is influenced by the values of the connected nodes with the corresponding causal weights and by its previous value. So, the value A_j for each node C_j is calculated by the following rule [54]:

$$A_j^{s,FCM} = f\left(\sum_{i=1, i \neq j}^{N} A_i^{s-1,FCM} W_{ij} + A_j^{s-1,FCM} \right), \tag{9}$$

And for the steady state nodes the correction equation is:

$$A_j^{s,FCM} = A_j^{system}, \tag{10}$$

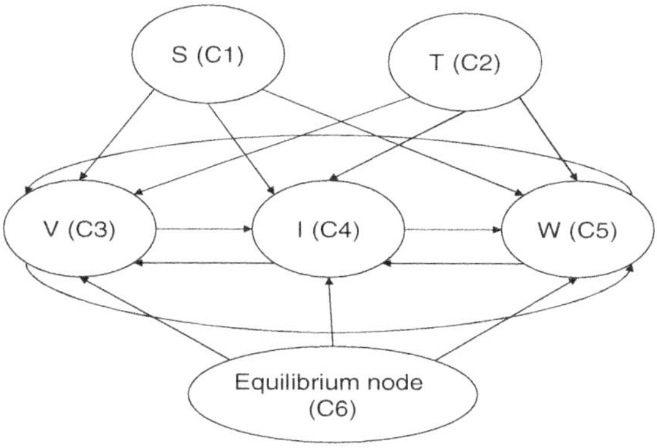

Fig. 9 An FCN designed for the photovoltaic project

where A_j^{system} is the node's value, derived from the physical system, A_j^s, is the value of node C_j at step s, A_i^{s-1} is the value of node C_i, at step s-1, A_j^{s-1} is the value of node C_j at step s-1, and W_{ij} is the weight of the interconnection between nodes C_i and C_j f is a squashing function: $f = \frac{1}{1+e^{-cx}}$. By using c = 1 we convert the nodes values in the range [0, 1].

3.3 The FCN Approach for the PV Project

As it has already been mentioned in the introduction, Fuzzy Cognitive Networks (FCN), constitute an operational extension of Fuzzy Cognitive Maps. They rely on experts' knowledge for the description of node interdependencies and the construction of the graph. Experts could also provide some initial estimation of the weights, but this is optional. FCNs may use these values only as a starting point or may not use them at all. The operation of FCNs is tightly connected with the operation of the physical system providing control values and taking feedback from the system. Moreover, during its initial training or its subsequent interaction with the physical system, the FCN keeps track of its previous equilibrium points by means of a collection of fuzzy if-then rules. Using these characteristics, the FCN becomes a dynamic control system. In this chapter we use the FCN in close cooperation with a fuzzy MPPT controller and with a PV system as shown in Fig. 10. The FCN is first off-line trained by appropriately constructed data and then it is connected to any PV system to get feedback and send control values to regulate its output. Once the FCN is trained its knowledge can be updated and the FCN acts as an adaptive controller of the PV system. The off-line training and the subsequent operation are described bellow.

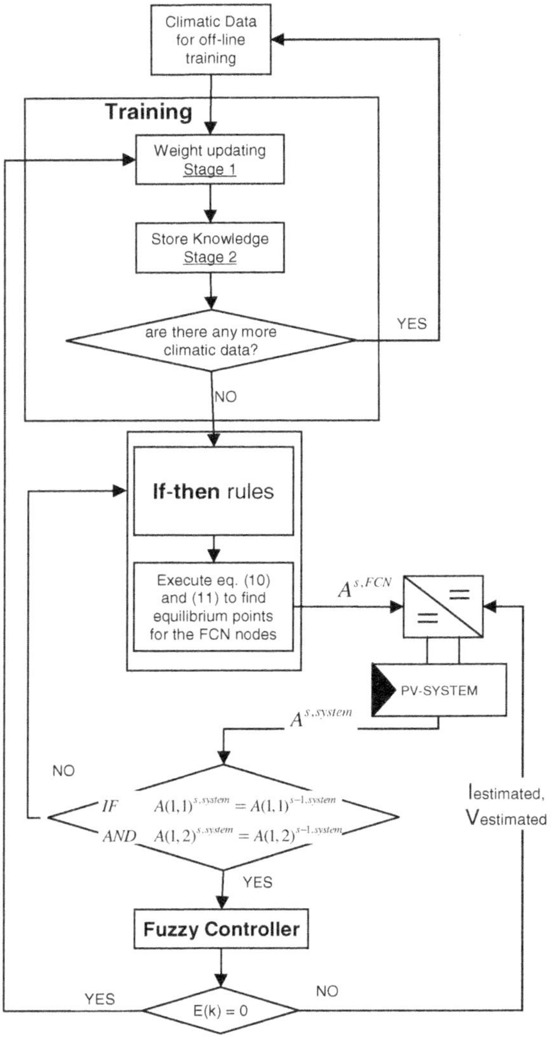

Fig. 10 Simplified flowchart of the proposed method

1) Initial off-line training of the FCN

The off-line training is being performed in an incremental manner. This means that for each training data vector which contains PV value variables corresponding to different operation conditions, the FCN weights are updated to comply with the data vector. Moreover, this new acquired knowledge is been stored in a fuzzy-rule data base. We can divide the training into two cooperating stages.

2) Stage 1 – Weight updating using new data

This stage is concerned with the method of updating the interconnections weights of FCN taking into account training data. Since the training is being performed incrementally, during stage 1, only one data vector is used. The FCN converges to its new weights values after a number of iterations. In each training iteration the FCN uses the updated weights to reach new equilibrium node values by means of (9) and (10). These values are compared to the given training values and the error is given for the new updating iteration. The weight updating is used by the following extended delta rule [43]:

$$p = A_j^{system} - \frac{1}{1+e^{-\left(\sum_{i=1,i\neq j}^{N} A_i^{s,FCN} W_{ij} + A_j^{s,FCN}\right)}}, \tag{11}$$

$$W_{ij}^{k} = W_{ij}^{k-1} + R_{ij} * (ap(1-p)A_i^{s,FCN}, \tag{12}$$

where p is the error, k is the number of iteration, a is the learning rate (usually $a = 0.1$) and R_{ij} is a calibration variable, which prevents the FCN node and weight values from being driven in their saturation point. R_{ij} can be computed by the following formula [43]:

$$R_{ij} = \eta \frac{\sum_{i=1}^{i=n} |W_{ij}|}{|W_{ij}|} \text{ if Wij} \neq 0 \text{ and } R_{ij} = 0 \text{ if Wij} = 0 \tag{13}$$

where constant value η is used to drive values R_{ij} in the range [0, 1]. In most practical situations $\eta = 0.1$.

It has to be noted that (11), (12) and (13) have been proposed so that the FCN reaches equilibrium points corresponding to new PV system desired values. However, they do not provide any guarantee that such equilibrium will be reached by the FCN. In [46], [47], the weight conditions that guarantee that the FCN converges to unique equilibrium points are derived. Based on these conditions new weight updating laws were derived in [47], [48], which guarantee that these weight conditions are always met. However, these weight updating laws are not employed here because equations (11) – (13) were initially used in [49 – 51] and were found sufficient for the problem at hand.

3) Stage 2 – Storage of the new knowledge in a fuzzy rule database

The procedure described in the previous stage modifies our knowledge about the system by continuously modifying the weight interconnections and consequently the node values. After the weight updating is taking place, the FCN reaches a new equilibrium point using (9) and (10). Since a new training vector might produce different weights and different equilibrium point we have to keep track of the current knowledge (weights and equilibrium points) to be used after the training phase. We do that by producing fuzzy if-then rules according to the following procedure [46].

Suppose, for example, that the FCN after being trained by a data vector converges to the following weight matrix:

$$W = \begin{bmatrix} 0 & 0 & W13 & W14 & W15 & 0 \\ 0 & 0 & W23 & W24 & W25 & 0 \\ 0 & 0 & 0 & W34 & W35 & 0 \\ 0 & 0 & W43 & 0 & W45 & 0 \\ 0 & 0 & W53 & W54 & 0 & 0 \\ 0 & 0 & W63 & W64 & W65 & 0 \end{bmatrix}$$

and concludes to an equilibrium point, which is:

$$A = \begin{bmatrix} A1 & A2 & A3 & A4 & A5 & A6 \end{bmatrix}$$

Suppose also that for a new training data vector it concludes to a new equilibrium point:

$$B = \begin{bmatrix} B1 & B2 & B3 & B4 & B5 & B6 \end{bmatrix}$$

with weight matrix:

$$K = \begin{bmatrix} 0 & 0 & K13 & K14 & K15 & 0 \\ 0 & 0 & K23 & K24 & K25 & 0 \\ 0 & 0 & 0 & K34 & K35 & 0 \\ 0 & 0 & K43 & 0 & K45 & 0 \\ 0 & 0 & K53 & K54 & 0 & 0 \\ 0 & 0 & K63 & K64 & K65 & 0 \end{bmatrix}$$

The fuzzy rule database, which is obtained using the information from the two previous equilibrium points, is depicted in Figs 11 and 12 and is resolved as follows: two rules related to the above two different equilibrium situations:

<u>Rule 1</u>
if node 1 is mf1 and node 2 is mf1 and node 3 is mf1 and
node 4 is mf1 and node 5 is mf1 and node 6 is mf1
then w13 is mf1 and w14 is mf1 and w15 is mf1 and
w23 is mf1 and w24 is mf1 and w25 is mf1 and
w34 is mf1 and w35 is mf1 and w43 is mf1 and
w45 is mf1 and w53 is mf1 and w54 is mf1 and
w63 is mf1 and w64 is mf1 and w65 is mf1

<u>Rule 2</u>
if node 1 is mf1 and node 2 is mf2 and node 3 is mf2 and
node 4 is mf2 and node 5 is mf2 and node 6 is mf2
then w13 is mf2 and w14 is mf2 and w15 is mf2 and
w23 is mf2 and w24 is mf2 and w25 is mf2 and

w34 is mf2 and w35 is mf2 and w43 is mf2 and
w45 is mf2 and w53 is mf2 and w54 is mf2 and
w63 is mf2 and w64 is mf2 and w65 is mf2

The number and shape of the fuzzy membership functions of the variables of both sides of the rules are gradually modified as new desired equilibrium points appear to the system during its training. To add a new triangular membership function in the fuzzy description of a variable, the new value of the variable must differ from one already encountered value more than a specified threshold. The threshold comes usually as a compromise between the maximum number of allowable rules and the detail in fuzzy representation of each variable. Once the new knowledge has been stored using the above procedure we run again stage 1 using a new operational point.

A simple illustration of the creation or not of new triangular membership function is shown in Fig. 13. A new triangular mf2 (fig 13.b.) appears together with triangular mf1 (fig 13.a.) due to the fact that distance c between values a and b is larger from the selected threshold value. In Fig 13.c. the value of c is below the threshold value and therefore there is no creation of new triangular, in order to store the knowledge that comes from the new equilibrium point of the real system. Value a represents an equilibrium point of the real system in terms of the variable, illustrated on the triangulars, while value b represents a new equilibrium point. We need to decide whether it is necessary or not to save the knowledge derived from value b, as a specific triangular, or is so close to value a and hence it is adequate to cover value b.

The threshold that will be determined is clearly an objective target of the designer engineer who will analyze the system and will design the targets of the control system and therefore the precision on which the system will work. High threshold implies the creation of a small number of membership functions. The stored knowledge is "poor" but the fuzzy rules are few and so is the storage space and the time required by the fuzzy inference procedure. On the other hand, if the threshold value is very low then the knowledge is extremely detailed, the required reaction time is big and a significant amount of storage space is required.

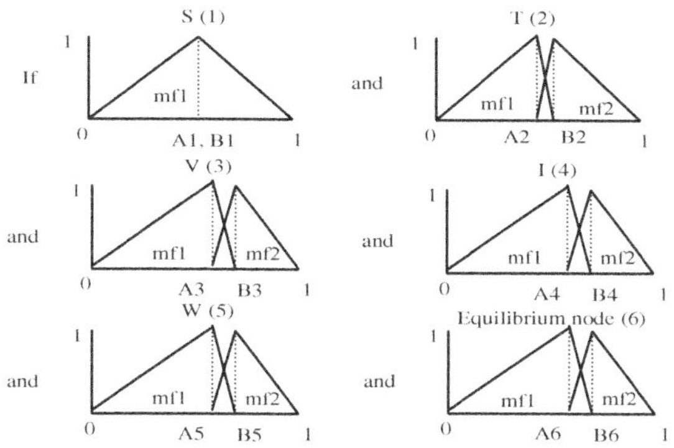

Fig. 11 Left hand side (if part)

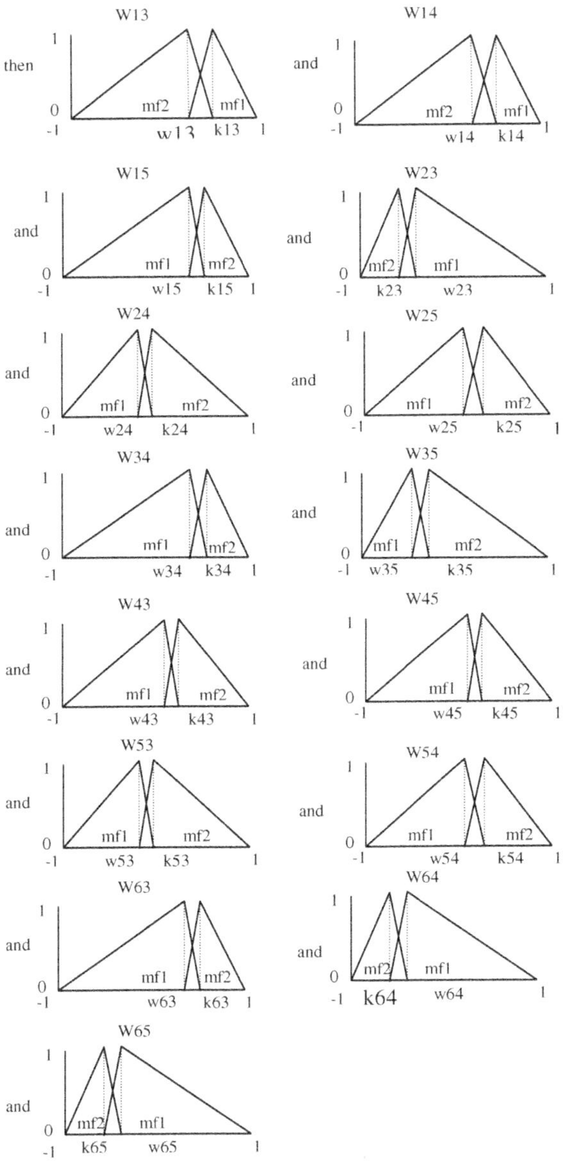

Fig. 12 Right hand side (Then part)

If the issue of the significant amount space required can be resolved, then it can be said that a convenient dynamic system is available for solving and resembling control systems that can be used in any complex system for which the initial knowledge is minor.

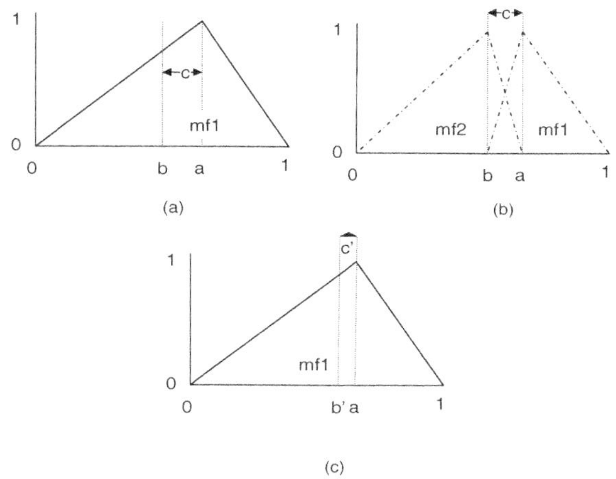

Fig. 13 Creation of new triangular membership function: (a) present triangular, (b) c>specified threshold (creation of new triangular), (c) c<specified threshold (non creation of new triangular)

4) Control of a PV system using the trained FCN and the Fuzzy Controller
Once the FCN is off-line trained it can be connected to the PV system according to Fig. 10. The FCN receives feedback from the Fuzzy Controller and from the PV array also. When the error of the Fuzzy Controller becomes zero, this means that the duty ratio of the switch S of the boost converter is set to the proper value, so that the PV array is in its maximum power point. This new maximum power point gives a new equilibrium point to the FCN. The new equilibrium point is used to train further the FCN. If there is a change in the values of temperature and insolation before the Fuzzy Controller drives the duty ratio of the switch S to the proper value corresponding to the MPP then the FCN interferes to the procedure and sends a new proper value for MPP voltage for the new insolation and temperature.

4 Results

Needless to say that irradiation and temperature play the most significant role on the maximum power that is drawn from a PV module. In order to measure these two quantities a pyranometer and a thermocouple is often used, although the output from these two measuring devices is not always the most adequate information to identify the operating point yielding the maximum power, which is of course a drawback of this methodology. The short circuit current from the PV array gives the most adequate information of the effective insolation and temperature using equations (1) to (6).

We construct training data for the FCN using the following procedure:

We use some typical climatic data. These data are chosen to be:

Irradiation (S-node 1). We select values in the range $0\, mW/cm^2$ to $100\, mW/cm^2$ using a step of $2\, mW/cm^2$.

Temperature (T-node 2). We select values in the range $-30\,°C$ to $70\,°C$ using a step of $2\,°C$.

By using all the possible combination of these data and by using the simulation of the photovoltaic array we calculate the values of the optimum voltage (node 3), current (node 4), and output power (node 5) from equations (1) to (4). Using these node values for nodes 1 to 5 we update the weights of the FCN according to stage 1 of the training procedure and for the equilibrium point derived for any possible combination we store the knowledge according to stage 2.

The possible combinations of the climatic data are 2601 and the FCN creates 51 triangular fuzzy numbers for nodes 1 and 2, 47 for node 3, 89 for node 4, 64 for node 5, 13 for node 6. Also, 867 fuzzy if-then rules are created to store the knowledge. The number of rules appears to be large because they account for all possible combinations of climatic data, even for those which are unlikely ever to occur. It could be significantly reduced if we excluded this kind of combinations.

When we connect the FCN system to the PV-array the PV-array sends the values of nodes 1 and 2 and through the fuzzy rule database, we decide which weights values are appropriate to express the values of nodes 3, 4 and 5. Executing equations (8), (9) and using the weights derived above we calculate the new equilibrium point which expresses the values of the optimum current, voltage and output power of the PV-array for the climatic data obtained at the specific point of time. In the next step the FCN sends the values of the control nodes, to the PV-array controller thus determining the optimum current, which reflects to optimum maximum output power for the climatic data obtained at the specific point of time.

In order to evaluate the effectiveness of the presented method we used the trained FCN for controlling the operation of the BP270L PV array, the parameters of which are given in Appendix B. Fig. 14 presents a comparison between (A) the theoretical (computed by (1), (2), (3) and (4)) MPP, (B) the calculated by the proposed FCN and Fuzzy Controller method, (C) the well known P&O method and (D) the calculated by the Fuzzy Controller [19]. It is clearly depicted that by using the FCN + Fuzzy Controller MPPT [51] system we have a significant energy gain. Actually, the combined method needs only 5 iterations in order to reach the new MPP, while the Fuzzy Controller method alone needs 12 iterations in order to reach the same MPP, and the P&O method needs 11 iterations to reach the Maximum Power Point. Each iteration corresponds to one second, following the same sampling procedure with [19]. It should be pointed out that in Figure 14 the fluctuations around the MPP are due to the P&O method (C) and not to the Fuzzy Controller (D), which after it reaches the MPP it stays there.

In Fig. 15 a similar comparison among the performance of the four methods but in different temperature and insolation conditions than those of Fig. 14 is given. We can see that the new MPP was exactly the point that the FCN instantly returned to the DC/DC converter for the new insolation and temperature levels.

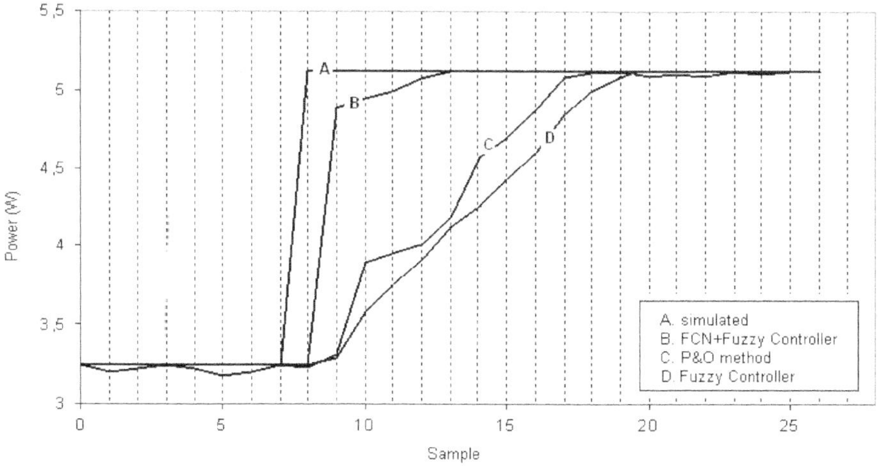

Fig. 14 Sample vs Power of PV-array,
A. theoretical, B. FCN+Fuzzy Logic method, C. P&O method, D. Fuzzy controller.

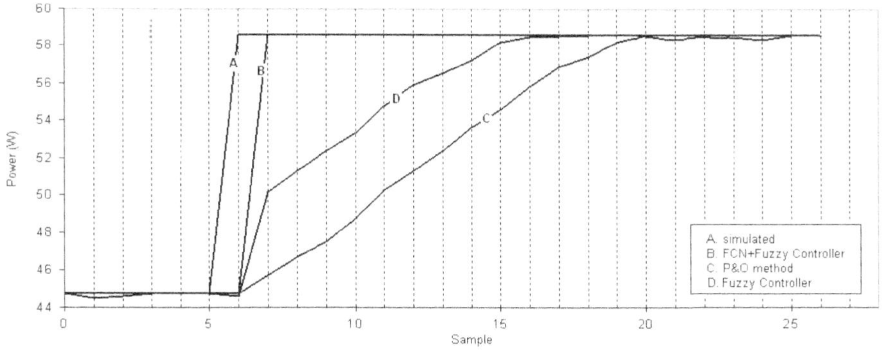

Fig. 15 Sample vs Power of PV-array,
A. theoretical, B. FCN+Fuzzy Logic method, C. P&O method, D. Fuzzy controller

That's why the proposed method (B) does not need to use the Fuzzy Controller in order to reach the MPP. In this case by using the proposed method we reach the MPP after only one iteration, while by using only the Fuzzy controller the system reached the MPP after 20 iterations, while P&O method reaches the MPP after 16 iterations.

Finally, in order to estimate the energy gain of the presented method (FCN + Fuzzy Controller MPPT) in comparison to the method which uses Fuzzy Controller only authors performed the following experiment. Using data from the year 2002 ran both methods to give the maximum power points and calculated the energy acquired from the PV array. Both methods are compared to the optimal one (theoretical MPP values, computed by using (1) to (4) and the results are shown in Table 2. It can be observed that the presented method outperforms method C (Fuzzy controller only). Actually, when the off-line trained FCN is used the proposed method provides with only a 0.0022% less energy production than the optimal (theoretical) case.

Table 2

MPPT Method	kWh (P_X)	error (%) $(P_A-P_X)/P_A$
A. Theoretical total energy within year 2002	P_A=70.389	
B. Off-line trained FCN + Fuzzy Controller	P_B=70.38745	0.0022
C. Fuzzy Controller	P_C=65.96857	6.28
D. Only on-line trained FCN + Fuzzy Controller	P_D=68.995	1.98

5 Adaptation Abilities of the Presented Method

In this Section the advantage of the presented method in respect to its adaptation performance is presented [49] and its contribution to the optimal performance of the PV array is demonstrated. Our aim is to ascertain how the method can adapt to physical changes taken place on the PV array under study. We recall from the previous section, that the test PV array is the BP270L PV model, which has 36 cells in series connection. In Figs 14 and 15 we saw that the presented method tracks the MPP of the PV array, through the DC/DC converter faster than any other method examined. The advantage of the presented method is that it can adapt to changes that happen during the circle of life of the PV array, so that it can always be a fast tracker of the MPP. The way by which the presented method operates is dynamically changing, by taking feedback from the Fuzzy Controller which tracks the MPP of the PV array. When the characteristics of the PV array change, the Fuzzy Controller will detect the MPP, which however is different than the one that has already been encountered under the same environmental conditions. Therefore, the knowledge already stored in the FCN is updated so that it incorporates the new information and discards the past and now erroneous one.

After the initial off-line training, the FCN simulates in part the solid line appearing in Fig 16, which represents the MPP line of the PV array for $T_r = 25°C$ and various insolation levels. Whenever the insolation changes, the FCN responds, giving the proper control rules, which regulate the current of the DC/DC converter. Suppose that the two of the 36 cells of the PV array are now destroyed or somehow obscured. The knowledge stored in the FCN corresponds to the previous solid line, which however is no longer correct. The correct line representing the current condition of the PV array is shown in Fig. 16 by the dotted line which is figured out using equations (1-4). The FCN has to change its knowledge so that it switches from the solid line to the dotted one.

Suppose that insolation and temperature are $S = 60mW/cm^2$ and $T_r = 25°C$ respectively. The knowledge stored in the FCN responds by giving $I_{MPP} = 2.4721\ A$ and $V_{MPP} = 16.9\ V$, denoted by point A on Fig. 16. But due to the two destroyed cells the proper FCN response should be $I_{MPP} = 2.4777\ A$ and $V_{MPP} = 15.9\ V$, denoted by point B (Fig 16). The past knowledge for this insolation and temperature is already stored in the FCN through triangular membership function mf76 for Voltage and through mf53 for Current. When the Fuzzy Controller detects the MPP, which now is far more different than the response of the FCN, the FCN will change its knowledge by permuting the triangular memberships function mf76 with the new-mf76. Due to the way by which the storage of fuzzy membership function and fuzzy rules takes place, the above permutation affects also the membership functions corresponding to neighbouring to A operating points.

Therefore, based on only one different operational point, the knowledge stored in FCN is not represented by the solid line anymore. Instead, it has been replaced by the knowledge represented by the dashed line of Fig. 16. When new operational points appear to the "spoiled" PV array, they will further change the knowledge of the FCN, so that it gradually reaches the one depicted by the dotted line of Fig.16.

In Fig. 17 we can see the changes that might happen to the MPP line for the same situation, as above. The MPP line (solid line) in Fig. 17 represents the MPPs for 100mW/cm^2 insolation and for different temperature values. The changes in the FCN are located into new fuzzy values for the current and the voltage. These two fuzzy changes are responsible for delaying the FCN MPP line (dashed line) in respect to the real one (dotted line).

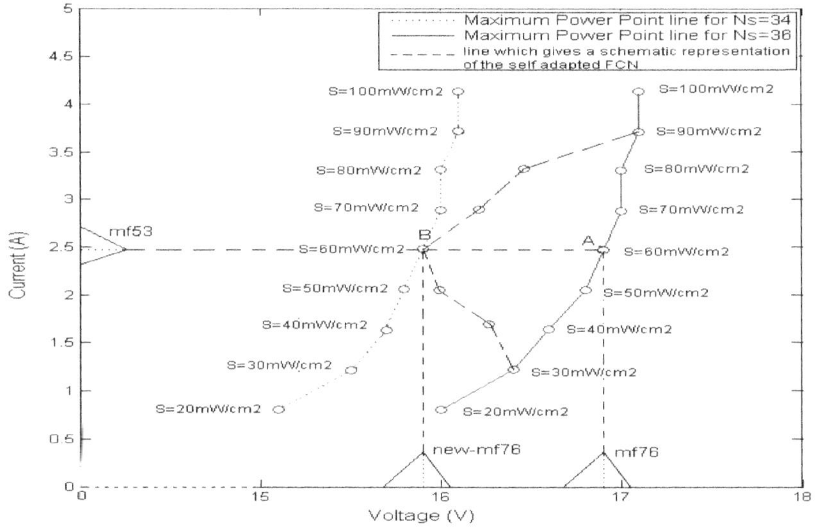

Fig. 16 MPP lines corresponding to the temperature of 25°C and different insolation values for the BP270L PV model

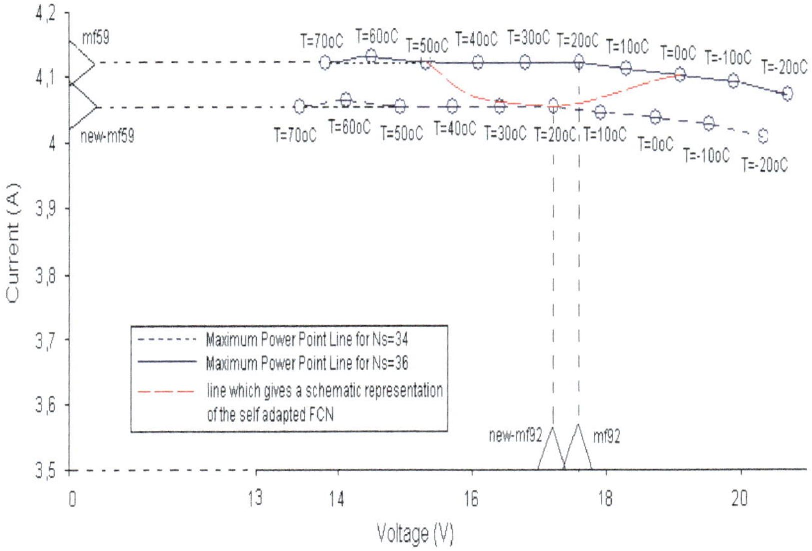

Fig. 17 MPP lines corresponding to the insolation of 100mW/cm^2 and different temperature values for the BP270L PV model

The adaptation abilities of the proposed scheme can be equally used when the initial training data set of the FCN was not accurate due to erroneous use of manufacturers' PV data. That is, if our method is initially trained based on data that are different than the real one, during its operation it will be adapted to meet the specifications of the real system.

6 Conclusions

The use of Fuzzy Cognitive Maps (FCM) in an appealing engineering problem was considered in this chapter. Fuzzy Cognitive Networks (FCN), which are extensions of FCM, were actually used, because FCN are designed to operate in continuous interaction with the system they describe and are able to control it. Using FCN, a new method for Maximum Power Point Tracking (MPPT) and its adaptive abilities were presented. The method combines a fuzzy MPPT with an appropriately designed Fuzzy Cognitive Network (FCN) to speed-up the procedure of reaching the accurate MPP of a PV array under changing environmental conditions. The graph of the FCN is designed using experts' opinion regarding the interacting concepts of the PV system but the weights of the interconnections are adapted according to PV system's operation data. The method presents very good results in Energy production compared to the theoretical expected one of a commercially available PV array. The methodology can be applied on any PV array of the market. Due to the existence of the Fuzzy Cognitive Network the method could track and adapt to any physical variations of the PV array through time.

Therefore, the method is guaranteed to present its very good performance independently of these variations.

Appendix A

The DC/DC boost converter has been simulated according to characteristics described below:

- array: BP270L PV array
- DC/DC converter input voltage (V_i): 13.7 to 24.7 V
- DC/DC converter output voltage (V_o): 48 V
- switching frequency (fs): 33kHz

Below, are shown the basic equations necessary for the DC/DC boost converter design [54]:

$$\frac{V_O}{V_i} = \frac{1}{1-D}, \frac{I_i}{I_O} = \frac{1}{1-D} \text{ where } D = \frac{t_{on}}{T} \text{ and } T = \frac{1}{fs}$$

Appendix B

PV SYSTEM DATA:

- $k_i = 2.8\ mA/{}^\circ C$
- Open Circuit Voltage $V_{oc} = 21.4\ V$,
- Short Circuit Current $I_{scr} = 4.48\ A$,
- Maximum Power Voltage $V_{mp} = 17.1\ V$,
- Maximum Power Current $I_{mp} = 4.15\ A$

References

[1] Hiyama, T., Kouzuma, S., Imakubo, T.: Identification of Optimal Operating Point of PV Modules using Neural Network for Real Time Maximum Power Tracking Control. IEEE Trans. Energy Convers 10(2), 360–367 (1995)
[2] Ro, K., Rahman, S.: Two-Loop Controller for Maximizing Performance of a Grid-Connected Photovoltaic-Fuel Cell Hybrid Power Plant. IEEE Trans. Energy Convers 13(3), 276–281 (1998)
[3] Veerachary, M., Senjyu, T., Uezato, K.: Neural-Network-Based Maximum-Power-Point Tracking of Coupled-Inductor Interleaved-Boost-Converter-Supplied PV System Using Fuzzy Controller. IEEE Transactions on Ind. Electronics 50(4), 749–758 (2003)
[4] Miyatake, M., Kouno, T., Nakano, M.: A Simple Maximum Power Tracking Control Employing Fibonacci Search Algorithm for Power Conditioners of Photovoltaic Generators. In: 10th International Power Electronics and Motion Control Conference (EPE-PEMC 2002) Cavtat & Dubrovnik, Cavtat & Dubrovnik, Croatia (2002)
[5] Bahgat, A.B.G., Helwa, N.H., Ahmad, G.E., El Shenawy, E.T.: Maximum power point tracking controller for PV systems using neural networks. Renewable Energy 30, 1257–1268 (2005)
[6] Salameh, Z., Taylor, D.: Step-up Maximum Power Point Tracker for Photovoltaic Arrays. Solar Energy 44, 57–61 (1990)
[7] Koutroulis, E., Kalaitzakis, K., Voulgaris, N.C.: Development of a Microcontroller-Based, Photovoltaic Maximum Power Point Tracking Control System. IEEE Trans. on Power Electronics 16(1), 46–54 (2001)
[8] Masoum, M.A.S., Dehbonei, H., Fuchs, E.F.: Theoretical and Experimental Analyses of Photovoltaic Systems With Voltage- and Current-Based Maximum Power-Point Tracking. IEEE Trans. Energy Convers 17(4), 514–522 (2002)
[9] Kasa, N., Iida, T., Chen, L.: Flyback inverter controlled by sensorless current MPPT for photovoltaic power system. IEEE Trans.Ind.Electron. 52(4), 1145–1152 (2005)
[10] Kim, I., Kim, M., Youn, M.: New Maximum power point tracker using sliding-mode observer for estimation of solar array current in the grid-connected photovoltaic system. IEEE Trans. Ind. Electron. 53(4), 1027–1035 (2006)
[11] Noguchi, T., Togashi, S., Nakamoto, R.: Short-current pulse-based maximum-power-point tracking method for multiple photovoltaic and converter module system. IEEE Trans. Ind. Electron. 49(1), 217–223 (2002)
[12] Noguchi, M., Takayoshi, I.: A control method to charge series connected ultraelectric double-layer capasitors suitable for photovoltaic generation systems combining MPPT control method. IEEE Trans. Ind. Electron. 54(1), 374–383 (2007)
[13] Park, J., Ahn, J., Cho, B., Yu, G.: Dual-module-based maximum-power-point tracking control of photovoltaic systems. IEEE Trans. Ind. Electron. 53(4), 1036–1047 (2006)
[14] Koizumi, H., Mizuno, T., Kaito, T., Noda, Y., Goshima, N., Kawasaki, M., Nagasaka, K., Kurokawa, K.: A novel microcontroller for grid-connected photovoltaic systems. IEEE Trans. Ind. Electron. 53(6), 1889–1897 (2006)
[15] Xiao, W., Lind, M., Dunford, W., Capel, A.: Real time identification of optimal operating points in photovoltaic power systems. IEEE Trans. Ind. Electron. 53(4), 1017–1026 (2006)

[16] Yeong-Chau, K., Tsorng-Juu, L., Jiann-Fuh, C.: Novel maximum-power-point-tracking controller for photovoltaic energy conversion system. IEEE Trans. Ind. Electron. 48(3), 594–601 (2001)
[17] Chihchiang, H., Jongrong, L., Chihming, S.: Implementation of a DSP-controlled photovoltaic system with peak-power tracking. IEEE Trans. Ind. Electron. 45(1), 963–973 (1998)
[18] Xiao, W., Dunford, W., Palmer, P., Capel, A.: Application of centered differentiation and Steepest descent to Maximum Power Point Tracking. IEEE Trans. Ind. Electron. 54(5), 2539–2549 (2007)
[19] Won, C.Y , Kim, D.H., Kim, S.C., Kim, W.S., Kim, H.S.: A New Maximum Power Point Tracker of Photovoltaic Arrays using Fuzzy Controller. In: 25th Annual IEEE Power Electronics Specialists Conference, PESC 1994, June 20-25, vol. 1, pp. 396–403 (1994)
[20] Hilloowala, R.M., Sharaf, A.M.: A rule-based fuzzy logic controller for a PWM inverter in photo-voltaic energy conversion scheme. In: Proceedings in IEEE Industrial Application Society, Annual Meeting, pp. 762–769 (1992)
[21] Senjyu, T., Uezato, K.: Maximum power point tracker using fuzzy control for photovoltaic arrays. In: Proceedings in IEEE International Conference in Industrial Technology, pp. 143–147 (1994)
[22] Simoes, M.G., Franceschetti, N.N., Friedhofer, M.: A Fuzzy Logic based photovoltaic peak power tracking control. In: Proceedings in IEEE International Symposium in Industrial Electronics, pp. 1429–1432 (1998)
[23] Simoes, M.G., Franceschetti, N.N.: Fuzzy Optimization Based Control of a Solar Array", Electric Power Applications. IEE Proceedings 146(5), 552–558 (1999)
[24] Patcharaprakiti, N., Premrudeepreechacharn, S.: Maximum power point tracking using fuzzy logic control for grid-connected photovoltaic system. In: IEEE Power Engineering Society, Winter Meeting, pp. 372–377 (2002)
[25] Wilamowski, B.M., Li, X.: Fuzzy system based maximum power point tracking for PV system. In: Proceeding of the 28th Annual Conference in IEEE Industrial Electronics Society, pp. 3280–3284 (2002)
[26] Altas, I.H., Sharaf, A.M.: A Novel On-Line MPP Search Algorithm for PV Arrays. IEEE Transactions on Energy Conversion 11(4), 748–754 (1996)
[27] Hua, C., Shen, C.: Study of Maximum Power Tracking Techniques and Control of DC/DC Converters for Photovoltaic Power System. In: 29th Annual IEEE Power Electronics Specialists Conference (1998)
[28] Hua, C., Lin, J., Shen, C.: Implementation of a DSP-controller photovoltaic system with peak power tracking. IEEE Trans. Ind. Electron. 45(1), 99–107 (1998)
[29] Matsukawa, H., Koshiishi, K., Koizumi, H., Kurokawa, K., Hamada, M., Bo, L.: Dynamic evaluation of maximum power point tracking operation with PV array simulator. Solar Energy Materials & Solar Cells 75, 537–546 (2003)
[30] Jantsch, M., et al.: Measurement of PV Maximum Power Point Tracking Performance. In: Proceedings of the 14th European Photovoltaic Solar Energy Conference and Exhibition, Barcelona, Spain, 30 June-4 July (1997)
[31] Kosko, B.: Fuzzy Cognitive Maps. International Journal of Man-Machine Studies, 65–75 (January 1986)
[32] Craiger, P., Coovert, M.D.: Modeling dynamic social and psychological processes withfuzzy cognitive maps. In: IEEE World Congress on Computational Intelligence and Fuzzy Systems, vol. 3, pp. 1873–1877 (1994)

[33] Tsadiras, A., Kouskouvelis, I.: Using Fuzzy Cognitive Maps as a Decision Support System for Political Decisions: The Case of Turkeys Integration into the European Union. In: Bozanis, P., Houstis, E.N. (eds.) PCI 2005. LNCS, vol. 3746, pp. 371–381. Springer, Heidelberg (2005)
[34] Koulouriotis, D.E., Diakoulakis, I.E., Emiris, D.M.: A fuzzy cognitive map-based stock market model: synthesis, analysis and experimental results. In: 10th IEEE International Conference on Fuzzy Systems, pp. 465–468 (2001)
[35] Carvalho, J.P., Tome, J.A.B.: Qualitative modelling of an economic system using rulebased fuzzy cognitive maps. In: IEEE International Conference on Fuzzy Systems, vol. 2, pp. 659–664 (2004)
[36] Xirogiannis, G., Glykas, M.: Fuzzy Cognitive Maps in Business Analysis and Performance Driven Change. IEEE Transactions on Engineering Management 51(3), 334–351 (2004)
[37] Glykas, M., Xirogiannis, G.: A soft knowledge modeling approach for geographically dispersed financial organizations. Soft Computing 9(8), 579–593 (2005)
[38] Xirogiannis, G., Glykas, M.: IntelligentModeling of e-BusinessMaturity. Expert Systems with Applications 32(2), 687–702 (2007)
[39] Xirogiannis, G., Chytas, P., Glykas, M., Valiris, G.: Intelligent impact assessment of HRM to the shareholder value. Expert Systems with Applications 35(4), 2017–2031 (2008)
[40] Kottas, T., Boutalis, Y., Devedzic, G., Mertzios, B.: A new method for reaching equilibrium points in Fuzzy Cognitive Maps. In: Proceedings of 2nd International IEEE Conference of Intelligent Systems, Varna Burgaria, pp. 53–60 (2004)
[41] Georgopoulos, V., Malandraki, G., Stylios, C.: A fuzzy cognitive map approach to differential diagnosis of specific language impairment. Artificial Intelligence in Medicine 29(3), 261–278 (2003)
[42] Zhang, W., Chen, S., Bezdek, J.: Pool2: A Generic System for CognitiveMap Development and Decision Analysis. IEEE Transactions on Systems, Man, and Cybernetics 19(1), 31–39 (1989)
[43] Kottas, T., Boutalis, Y., Christodoulou, M.: A new method for weight updating in Fuzzy cognitive Maps using system Feedback. In: 2nd International Conference on Informatics in Control, Automation and Robotics (ICINCO 2005), Barcelona, Spain, September 13-17, pp. 202–209 (2005)
[44] Boutalis, Y., Kottas, T., Mertzios, B., Christodoulou, M.: A Fuzzy Rule Based Approach for storing the Knowledge Acquired from Dynamical FCMs. In: 5th International Conference on Technology and Automation (ICTA 2005), Thessalonica, Greece, October 15-16, pp. 119–124 (2005)
[45] Kottas, T.L., Boutalis, Y.S., Christodoulou, M.A.: Fuzzy Cognitive network: A general Framework. Intelligent Decision Technologies 1, 1–14 (2007)
[46] Boutalis, Y., Kottas, T., Christodoulou, M.: On the Existence and Uniqueness of Solutions for the Concept Values in Fuzzy Cognitive Maps. In: Proceedings of 47th IEEE Conference on Decision and Control – CDC 2008, Cancun, Mexico, December 9-11, pp. 98–104 (2008)
[47] Boutalis, Y., Kottas, T., Christodoulou, M.: Adaptive Estimation of Fuzzy Cognitive Maps with proven Stability and Parameter Convergence. IEEE Trans. on Fuzzy Systems (2009), doi:10.1109/TFUZZ, 2017519
[48] Kottas, T., Boutalis, Y., Christodoulou, M.: Bilinear Adaptive Parameter Estimation in Fuzzy Cognitive Networks. In: Alippi, C., Polycarpou, M., Panayiotou, C., Ellinas, G. (eds.) ICANN 2009. LNCS, vol. 5769, pp. 875–884. Springer, Heidelberg (2009)

[49] Boutalis, Y.S., Karlis, A.D., Kottas, T.L.: Fuzzy Cognitive Networks + Fuzzy Controller as a self adapting control system for Tracking Maximum Power Point of a PV-Array. In: Proceedings in IEEE International Conference on Industrial Electronics, pp. 4355–4360 (2006)
[50] Karlis, A.D., Kottas, T.L., Boutalis, Y.S.: A novel maximum power point tracking method for PV systems using fuzzy cognitive networks (FCN). Electric Power Systems Research 77(3-4), 315–327 (2007)
[51] Kottas, T.L., Boutalis, Y.S., Karlis, A.D.: New Maximum Power Point Tracker for PV Arrays Using Fuzzy Controller in Close Cooperation with Fuzzy Cognitive Networks. IEEE Transactions on Energy Conversion 21(3), 793–803 (2006)
[52] Lee, C.: Fuzzy Logic in Control Systems: Part I and Part II. IEEE Transactions on Systems, Man and Cybernetics 20(2), 404–435 (March/April)
[53] Hua, C., Shen, C.: Comparative Study of Peak Power Tracking Technics for Solar Storage System. In: Proceedings of the 13th Annual Applied Power Electronics Conference and Exposition, vol. 2, pp. 679–685 (1998)
[54] Stylios, C., Groumpos, P.: Fuzzy Cognitive Maps: a model for intelligent supervisory control systems. Computers in Industry 39, 229–238 (1999)
[55] Santos, J., Antunes, F.: Maximum Power Point Tracker for PV Systems, World Climatic and Energy Event, Rio de Janeiro, Brasil, pp. 75–80 (2003)

Fuzzy Cognitive Maps Applied to Computer Vision Tasks

Gonzalo Pajares, María Guijarro, P. Javier Herrera, José J. Ruz, and Jesús M. de la Cruz

Abstract. Computer vision is an emerging area which is demanding solutions for solving different problems. The data to be processed are bi-dimensional (2D) images captured from the tri-dimensional (3D) scene. The objects in 3D are generally composed of related parts that joined form the whole object. Fortunately, the relations in 3D are preserved in 2D. Hence, we can exploit this fact by considering specific and basic elements which are related to other elements in the 2D images. The relations with other elements allow establishing a link among them. Hence, we have the necessary ingredients to build a structure under the Fuzzy Cognitive Maps (FCMs) paradigm. FCMs have been satisfactorily used in several areas of computer vision including: pattern recognition, image change detection or stereo vision matching. In this chapter we establish the general framework of fuzzy cognitive maps in the context of 2D images and describe three applications in the three mentioned areas of computer vision. We also give some details about the performance of this paradigm in these applications.

1 Introduction

Computer vision is an emerging area which is demanding solutions for solving different problems. The basic data are bi-dimensional (2D) images captured from the three-dimensional (3D) real world.

Gonzalo Pajares
Dpto. Ingeniería del Software e Inteligencia Artificial, Facultad de Informática, Universidad Complutense, 28040 Madrid, Spain
e-mail: pajares@fdi.ucm.es

María Guijarro
Centro Superior de Estudios Felipe II, Ingeniería Técnica en informática de Sistemas 28300 Aranjuez, Madrid, Spain
e-mail: mguijarro@cesfelipesegundo.com

P. Javier Herrera · José J. Ruz · Jesús M. de la Cruz
Dpto. Arquitectura Computadores y Automática, Facultad de Informática, Universidad Complutense, 28040 Madrid, Spain
e-mail: pjherrera@pdi.ucm.es, {jjruz,jmcruz}@dacya.ucm.es

M. Glykas (Ed.): Fuzzy Cognitive Maps, STUDFUZZ 247, pp. 259–289, 2010.
springerlink.com © Springer-Verlag Berlin Heidelberg 2010

A look to the 3D world allows observing objects in the scene which are formed by structures or elements grouped together. Fortunately when the objects are mapped from the 3D world to the 2D image they preserve this grouping.

From this observation, an important issue addressed in computer vision tasks, based on image applications, is referred to how sensory devices perceive the objects in the scene, i.e. how the scene analysis problem is addressed. To deal with real-world scenes some criterion for grouping elements or features in the scene is required. In the work of Wang (2005) a list of major grouping principles is exhaustively studied. They are inspired in the Gestalt's principles (Koffka, 1935). The most important, from the point of view of computer vision tasks, are: *proximity*, labeled features that lie close in space tend to group; *similarity*, labeled features with similar properties or attributes tend to group; *connectedness*, labeled features that lie inside the same connected region tend to group.

These principles allow defining a spatial neighborhood in the 2D images which is a projection from the 3D scene. Now the problem is to build some structure where the above principles are to be mapped. Several approaches can be used, the Fuzzy Cognitive Maps (FCMs) is one of them.

Indeed, as explained in the next section, a FCM is a network of concepts where each concept represents some feature in the image. The concepts are joined by links representing some kind of relations among the features.

The application of FCMs for computer vision tasks requires the identification of the features to be represented as concepts; each concept should be characterized by a property, quantified by a specific value. Moreover, the concepts are linked among them and the strength of each link is also characterized by a value, which is determined by the application of the Gestal's principles.

This chapter describes the application of FCMs to three well addressed computer vision tasks, namely: 1) classification, 2) image change detection and 3) stereovision matching. The features used in this chapter are pixels for the tasks 1) and 2) and edge segments for the task 3). Other features (corners, regions) should be possible depending on the application.

Through this chapter we shall be concerned with the definition of the FCM framework for computer vision. Afterwards, we apply this framework for the three applications mentioned above. Much of our attention will be devoted to studying the method for finding concepts and relations between them. Some details about the performance of FCM on these applications are also provided.

2 Fuzzy Cognitive Maps: Topology and Basic Concepts

Fuzzy Cognitive Maps is a well developed modelling methodology for complex systems originating that allow to describe the behaviour of a system in terms of concepts. They have been used as a decision modelling tool under different approaches including those related with enterprises orientations (Xirogiannis et al., 2008; Xirogiannis and Glykas, 2004), e-business (Xirogiannis and Glykas, 2007) or financial applications among others (Xirogiannis and Glykas, 2005). Under its most general approach, each concept represents an entity, a state, a variable or a feature of the system (Tsardias and Margaritis, 1997, 1999; Kosko, 1986, 1992; Miao and Liu, 2000). According to Kosko (1992), FCMs are fuzzy signed directed graphs with feedback. The directed edge e_{ik} from causal concept C_i to concept C_k measures how much C_i causes C_k. The edges e_{ik} take values in the fuzzy causal interval $[-1,+1]$, $e_{ik} = 0$ indicates no causality; $e_{ik} > 0$ indicates causal increase, this means that C_k increases as C_i increases and vice versa, C_k decreases as C_i decreases; $e_{ik} < 0$ indicates causal decrease or negative causality, C_k decreases as C_i increases and C_k increases as C_i decreases.

Given an FCM with a number n of concepts C_i, i.e. $i = 1,\ldots,n$, the value assigned to each concept, called activation level, can be updated iteratively, until convergence, based on the external influences exerted by the other nodes C_k on C_i and its self-influence through C_i. Several approaches have been proposed to map these influences, including those averaging both the internal and the external ones (Xirogiannis et al., 2008; Xirogiannis and Glykas, 2007, 2005, 2004). In this work we have chosen the proposed by Tsardias and Margaritis (1997, 1999) according to equation (1), because it includes a decay factor, explained below, which introduces a mechanism for achieving high stability in the network.

$$A_i(t+1) = f\left(A_i(t), \sum_{k=1}^{n} e_{ki}(t) A_k(t)\right) - d_i A_i(t) \tag{1}$$

The explanation of the terms in the equation (1) is as follows:

1. $A_i(t)$ and $A_k(t)$ are respectively the activation levels of the concepts C_i and C_k at the iteration, t. The sum is extended to all n concepts available. Nevertheless, only the concepts with edge values different from zero exert influences over the concept C_i trying to modify its current activation level $A_i(t)$ towards $A_i(t+1)$.
2. e_{ki} (t) are the fuzzy causalities between concepts, defined as above but considering that they could vary dynamically with the iterations.
3. $d_i \in [0,1]$ is the decay factor of certainty concept C_i. It determines the fraction of the current activation level that will be subtracted from the new one as a result of the concept's natural intention to get stable activation levels. The bigger the decay factor, the stronger the decay mechanism. This factor was introduced in Tsardias and Margaritis (1997, 1999) as a mechanism for introducing a degree of instability, so that those concepts destabilised intentionally but with a high degree of real stability tend towards its stabilized activation level. On the contrary, if the activation level is unstable, the decay mechanism induces a continuous variability on the activation level.
4. f is a non-linear function that determines the activation level of each concept. Some common choices of f are the following:
   a) sigmoid (Haykin, 1994),

$$f(x) = \tanh(x) \tag{2}$$

b) logistic (Haykin, 1994),

$$f(x) = \left(1 - e^{-x}\right)^{-1} \tag{3}$$

c) the function defined in the MYCIN expert system for the aggregation of the certainty factors (Shorliffe, 1976; Tsardias and Margaritis 1998)

$$f(x,y) = \begin{cases} x + y(1-x) & \textit{if} \;\; x, y \geq 0, \\ x + y(1+x) & \textit{if} \;\; x, y < 0, \\ (x+y)/\left(1 - \min\left(|x|, |y|\right)\right) & \text{else} \end{cases} \tag{4}$$

where $|x|, |y| \leq 1$

The variable x in (2) and (3) represents a combination of $A_i(t)$ and $\sum_{k=1}^{n} e_{ki}(t) A_k(t)$, some times as a weighted average as mentioned before. In (4) $x = A_i(t)$ and $y = \sum_{k=1}^{n} e_{ki}(t) A_k(t)$.

The inference mechanism of FCMs can be summarized as follows: first, the FCMs are initialized. The activation level of each concept is set to a value based on a specific mechanism that must be defined. Because the various concepts are free to interact (Yaman and Polat, 2009), the activation of one concept influences the concepts to which it is connected. This interaction continues until:

1) a fixed-point equilibrium is reached.
2) a limit cycle is reached or equivalently a number of iterations.
3) chaotic behavior is exhibited (Tsardias and Margaritis, 1997).

3 Application of FCMs to Image Classification

This application is based on the work of Pajares et al. (2009). It describes a combined classifier where each pixel in the image must be classified as belonging to a cluster. There are different approaches for classification, two of the most popular are Fuzzy Clustering (*FC*) (Zimmermann, 1991) and the parametric Bayesian (*BP*) one (Duda et al., 2000). These classifiers perform a training phase and then make a decision based on this training. Actually, the tendency for making better decisions is the combination of classifiers (Kuncheva, 2004). Here, in the combination of *FC* and *BP*, is where the FCMs paradigm is applied.

We briefly describe both the training and decision phases which are concerning the simple classifiers. During the decision one, we will identify the pixels as features, which define the causal concepts, their activation levels, the causalities between concepts and the decay factor. The final goal is to classify the pixels as belonging to a cluster. This is carried out by combining the supports provided by the simple classifiers. These supports are membership degrees and probabilities provided by *FC* and *BP* respectively. The combination is made by defining a coefficient which can cope with both supports. This coefficient is part of the causal edges between concepts.

3.1 Training Phase

We start with the observation of a set X of N training samples, i.e., $X = \{\boldsymbol{x}_1, \boldsymbol{x}_2, ..., \boldsymbol{x}_N\} \in \Re^p$, where p is the data dimensionality. Each sample belongs to a known cluster w_j, where the number of clusters is c, i.e., $j = 1, 2, \ldots, c$. Each training sample is described by its vector $\boldsymbol{x}_i$ representing a pixel in the image, where its components are the three spectral RGB values of that pixel at the image location (x,y); hence in this example $p = 3$.

Under the above assumption, we know the distribution of the samples into the c clusters: this fact makes the training process supervised. For combination purposes, we apply two training procedures based on both *FC* and *BP* as described below.

3.1.1 Training through the Fuzzy Clustering

This process receives the input training patterns $\boldsymbol{x}_i$ and computes for each one, at the iteration t, its degree of membership μ_i^j in the cluster w_j and also the cluster centres v_j according to equation (5).

$$\mu_i^j(t+1) = \frac{1}{\sum_{r=1}^{c}\left(d_{ij}(t)/d_{ir}(t)\right)^{2/(b-1)}} \tag{5}$$

where $d_{ij}^2 \equiv d^2(\boldsymbol{x}_i, \boldsymbol{v}_j)$ is the squared Euclidean distance. The number $b > 1$ is called the exponential weight (Bezdek, 1981; Duda et al., 2000). The stopping criterion of the iteration process is achieved when $\left\|\mu_i^j(t+1) - \mu_i^j(t)\right\| < \varepsilon \;\; \forall ij$ or a number t_{max} of iterations is reached. Note, that the number of iterations for estimating the membership degrees and the cluster centers is different from the iterations in the *FCM*s, that appear in equation (1).

3.1.2 Training through the Parametric Bayesian Classifier

Following Duda et al. (2000), the Bayesian's classifier makes its decision about any pattern $\boldsymbol{x}$, which is to be classified, according to the following rule,

$$\boldsymbol{x} \in w_j \;\; \text{if} \;\; p(\boldsymbol{x} \mid w_j)P(w_j) > p(\boldsymbol{x} \mid w_h)P(w_h) \quad \forall h \neq j \tag{6}$$

$P(w_j)$ and $P(w_h)$ are the prior probabilities that $\boldsymbol{x} \in w_j$ and $\boldsymbol{x} \in w_h$ respectively; $p(\boldsymbol{x} \mid w_j)$ and $p(\boldsymbol{x} \mid w_h)$ are the likelihoods that $\boldsymbol{x} \in w_j$ and $\boldsymbol{x} \in w_h$ respectively. Because we do not have prior information about the samples $\boldsymbol{x}$, which are to be classified, we assume that $P(w_j) = P(w_h)$. This means that the decision is made based only on the likelihoods, generally modelled as normal (Gaussian) probability density functions, which are obtained for each cluster w_j as follows,

$$p\left(\boldsymbol{x} \mid w_j\right)=\frac{1}{(2\pi)^{p/2}\left|\Sigma_j\right|^{1/2}}\exp\left[-\frac{1}{2}\left(\boldsymbol{x}-\boldsymbol{m}_j\right)^T \Sigma_j^{-1}\left(\boldsymbol{x}-\boldsymbol{m}_j\right)\right] \quad (7)$$

p is as before the data dimensionality, i.e., $p = 3$.

The goal of the training process for this classifier is to estimate both parameters: the mean $\boldsymbol{m}_j$ and the covariance Σ_j parameters, both for each cluster w_j with n_j samples belonging to such cluster. This is carried out through maximum likelihood estimation considering the distribution of the training samples into clusters, as follows,

$$\boldsymbol{m}_j=\frac{1}{n_j}\sum_{k=1}^{n_j}\boldsymbol{x}_k \qquad \Sigma_j=\frac{1}{n_j-1}\sum_{k=1}^{n_j}\left(\boldsymbol{x}_k-\boldsymbol{m}_j\right)\left(\boldsymbol{x}_k-\boldsymbol{m}_j\right)^T \quad (8)$$

where T denotes transpose.

Figure 1(*a*) displays an original training image of a 3D aerial scene captured with an airborne sensor, (*b*) the distribution of the samples in the four clusters according to the representation given in (*c*). In the left column of (*c*) the spectral signature is represented for each cluster centre, computed according to equation (5), i.e. through the *FC* classifier; in the right column, the label assigned to each cluster.

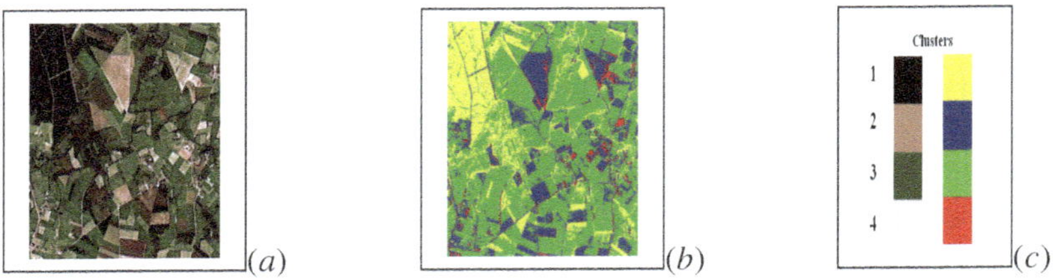

Fig. 1 (*a*) Original training image; (*b*) samples classified in four clusters as displayed in (*c*)

3.2 Decision Phase: Fuzzy Cognitive Maps

After the training, each new incoming sample $\boldsymbol{x}_s \in \Re^p$ must be classified as belonging to a cluster w_j based on the learning and knowledge achieved by the system during the training phase. This sample, like each training sample, represents a pixel at the image location (*x*,*y*) with its corresponding R,G,B spectral components. *FC* computes the membership degrees for $\boldsymbol{x}_s$ to each cluster trough the equation (5) and classifies the pixel according to the following rule: $\boldsymbol{x}_s \in w_j$ if $\mu_s^j > \mu_s^h$ for all $h \neq j$. *BP* computes the probabilities through equation (7); these probabilities can be used for classifying $\boldsymbol{x}_s$ according to the rule in the equation (6). Given an image to be classified, we compute these membership degrees and probabilities for each pixel in the image before the *FCM* structure is built. The membership degrees are used as the initial activation levels for the *FCM* as described in the next section. In what follows, we define the network topology of the *FCM*, the equivalences between this specific *FCM*s derived form the general *FCM* framework, synthesized in equation (1). Afterwards, we combine probabilities and membership degrees for defining the directed dynamical $e_{ik}(t)$ causal edges between concepts and the decay factors, as required by the *FCM*.

3.2.1 Network Topology of the Fuzzy Cognitive Maps

For each cluster w_j, we build a network of concepts, net_j, where the topology of this network is established by the spatial distribution of the pixels in the image to be classified with size $L \times M$. Hence, we have one *FCM* network associated to each cluster. Each concept C_i in the net_j is associated to the pixel location (x,y) in the image, i.e. $i \equiv (x, y)$. Hence, the number of concepts in the net_j is $q = L \times M$ and a concept represents a pixel in the image. For simplicity, instead of using the sample $\boldsymbol{x}_s$ with its R, G, B spectral components at (x,y), we will say that the concept C_i is to be classified as belonging to a cluster w_j. The activation level for the concept C_i in the net_j is initialized with the membership degree μ_i^j provided by *FC* according to the equation (5), but mapped linearly for ranging in $[-1,+1]$ instead of $[0,+1]$. Hence, the activation levels start with the membership degrees associated to the concepts. Through the *FCM*, the activation levels are reinforced or punished iteratively based on the influences exerted by their neighbours. The goal is to make the best decision for classifying the concept C_i, based on more stable activation levels. Because we have one *FCM* for each cluster w_j, both the activation levels and causal edges are identified with the super-index j associated to the cluster w_j, i.e. according to the equation (1), now $A_i(t) \equiv \mu_i^j(t)$ and $e_{ik}^j(t)$ include j, this is applicable to the decay factor, identified as d_i^j. In summary, given an image of size $L \times M$, the goal is to classify the pixel i, equivalently the concept C_i, located at (x,y) as belonging to a cluster w_j The initial activation levels at $t = 0$ are set as $\mu_i^j(0) \equiv \mu_i^j$, computed through the equation (5). Every concept is positively or negatively activated to a certain degree. At the end of the iterative process, the decision about the cluster to which it belongs is made based on maximum activation levels considering all j networks.

Each causal weight $e_{ik}^j(t)$ is defined as a combination of two coefficients representing the mutual influence exerted by the k neighbours concepts C_k, over the concept C_i, they are:

a) a *regularization* coefficient which computes the consistency between the activation levels of the concepts and the supports provided by the classifier *BP* in a given neighbourhood for each net_j;

b) a *contextual* coefficient which computes the consistency between the clustering labels after each decision made by the combined classifier.

Considering the equation (1), the goal is to compute: *a*) the causal weights $e_{ik}^{j}(t)$ and *b*) the decay factor.

3.2.2 Computing the Causal Weights

Taking into account the mapping between a pixel location $i \equiv (x, y)$ and the concept C_i at each net_j, the neighbourhood $\mathbf{N}_i^m$ contains the m pixels (concepts) surrounding i, mapped from the image and representing the m-connected spatial region around the pixel i. A typical value of m used in the literature is 8, which defines a 3×3 region. As can be observed, the index i varies from 1 to q, i.e., this is the number of concepts explored at each net_j. For borders concepts in the image, the neighbourhood only includes the pixels belonging to the image, i.e., m = 3 in the four corners and m = 5 in the remainder borders.

We define the regularization coefficient at the iteration t as follows,

$$r_{ik}^{j}(t) = \begin{cases} 1-\left|\mu_i^j(t) - p_k^j\right| & k \in \mathbf{N}_i^m, \ i \neq k \\ 0 & k \notin \mathbf{N}_i^m, \ i = k \end{cases} \tag{9}$$

where p_k^j is supplied by *BP*, it is the probability that a concept C_k (pixel k) with attributes $\boldsymbol{x}_k$ belongs to the cluster w_j, computed through the equation (7), i.e. $p_k^j \equiv p\left(\boldsymbol{x}_k \mid w_j\right)$. These values are mapped linearly to range between [−1, +1] instead of [0,+1]. From (9) we can see that $r_{ik}^{j}(t)$ ranges between [−1, +1] where the lower/higher limit means minimum/maximum influence respectively. The contextual coefficient for the concept C_i associated to the cluster w_j at the iteration t is defined taking into account the clustering labels l_i and l_k as follows,

$$c_{ik}(t) = \begin{cases} +1 & l_i(t) = l_k(t) & k \in \mathbf{N}_i^m, \ i \neq k \\ -1 & l_i(t) \neq l_k(t) & k \in \mathbf{N}_i^m, \ i \neq k \\ 0 & & k \notin \mathbf{N}_i^m, \ i = k \end{cases} \tag{10}$$

where values of −1 and +1 mean negative and positive influences, respectively.

Labels l_i and l_k could change between iterations and they are obtained as follows: given the concept C_i, we know its activation level, at each iteration t for each net_j as given by the equation (1); we determine that the concept C_i belongs to the cluster w_j if $\mu_i^j(t) > \mu_i^h(t) \ \forall j \neq h$, so we set $l_i(t)$ to the j value which identifies the cluster, j = 1,..., c. The label $l_k(t)$ is set similarly. Thus, this coefficient is independent of the net_j, because it is identical for all networks.

The regularization and contextual coefficients are both combined for computing the causal edges e_{ik}^{j} as an averaged sum, taking into account the signs, as follows,

$$z_{ik}^{j}(t)=\gamma r_{ik}^{j}(t)+(1-\gamma)c_{ik}(t);\quad e_{ik}^{j}=\left[sgn\left(z_{ik}^{j}(t)\right)\right]^{v} z_{ik}^{j}(t);$$

$$sgn\left(z_{ik}^{j}(t)\right)=\begin{cases}-1 & z_{ik}^{j}(t)\le 0\\ +1 & z_{ik}^{j}(t)>0\end{cases} \tag{11}$$

$\gamma\in[0,1]$ represents the trade-off between both coefficients, a typical value for it is 0.8; *sgn* is the *signum function* and v is the number of negative values in the set $B\equiv\left\{W_{ik}^{j}(t),r_{ik}^{j}(t),c_{ik}(t)\right\}$, i.e., given $D\equiv\left\{u\in B/u<0\right\}\subseteq B$, v = card (D). Note that the causal edges vary with the iteration t. The Gestalt's similarity principle is mapped in $r_{ik}^{j}(t)$ through similarities between activation levels and probabilities and also in $c_{ik}(t)$ trough similarity between the labels. The other two Gestalt's principles of proximity and connectedness are embedded in the neighbourhood $\boldsymbol{N}_i^m$.

3.2.3 Computing the Decay Factor

We define the decay factor based on the assumption that high stability in the network states implies that the activation level for the concept C_i in the network net_j would be to lose some of its activation with such purpose. We build an accumulator of cells of size $q=L\times M$, where each cell i is associated to the concept C_i. Each cell i contains the number of times h_i^j, that the concept C_i has changed significantly its activation level in the net_j. Initially, all h_i^j values are set to zero and then $h_i^j=h_i^j+1$ if $\left|\mu_i^j(t+1)-\mu_i^j(t)\right|>\varepsilon$. The stability of the node i is measured as the fraction of changes accumulated by the cell i compared with the changes in its neighbourhood $k\in\boldsymbol{N}_i^m$ and the number of iterations t. The decay factor is computed as follows,

$$d_i^j=\begin{cases}0 & h_i^j=0 \text{ and } h_k^j=0\\ \dfrac{h_i^j}{\left(\bar{h}_k^j+h_i^j\right)t} & \text{otherwise}\end{cases} \tag{12}$$

where h_i^j is defined above and $\bar{h}_k^j$ is the average value accumulated by the concepts $k\in\boldsymbol{N}_i^m$. As one can see, from equation (12), if $h_i^j=0$ and $h_k^j=0$, the decay factor takes the null value, this means that no changes occur in the activation levels of the concepts, i.e., high stability is achieved; if the fraction of changes is small, the stability of the node i is also high and the decay term tends towards zero. Even if the fraction is constant the decay term also tends to zero as t increases, this means that perhaps initially some changes can occur and then no more changes are detected, and this is another sign of stability. The decay factor subtracts from the new activation level a fraction; this implies that the activation level could take values less/greater than ‒1 or +1. In these cases, the activation level is set to ‒1 or +1, respectively.

3.2.4 Summary of the Full FCM Process

In section 2, we introduced three criteria for convergence. So, based on them, the iterative process ends if all nodes in the network fulfil the convergence criterion $\left|\mu_i^j(t-1)-\mu_i^j(t)\right| > \varepsilon$ or a number of iterations t_{max}, is reached or a percentage of concepts, greater than a threshold U, change its activation level in a value greater than ε between two consecutive iterations. This last implies that the system is under a chaotic behaviour.

The FCM process is synthesized as follows:

1. *Initialization*: load each concept with its activation level $\mu_i^j(t=0) \equiv \mu_i^j$ through the equation (5); set $\varepsilon = 0.05$, $t_{max} = 50$ and $U = 0.9$. Define nc as the number of concepts from a total of q that change their activation levels at each iteration. The activation mechanism is that defined in equation (1) and the activation function is that defined in equation (4).

2. *FCM process*:

```
t = 0
while t < t_max and nc/q < U
    t = t + 1; nc = 0;
    for each concept C_i
        update μ_i^j(t) according to the equation (1)
        if |μ_i^j(t) − μ_i^j(t−1)| > ε then
            nc = nc + 1
        end if;
    end for;
end while
```

3. *Outputs*: the activation levels $\mu_i^j(t)$ for all concepts updated.

Once the FCMs processes end, each concept C_i has achieved an activation level $\mu_i^j(t)$ that determines de degree of belonging of the pixel i, represented by C_i, to the cluster j. This decision is made according to the following rule $i \in w_j$ if $\mu_i^j(t) > \mu_i^h(t)$, $\forall\, j \neq h$ where j and h identify the clusters and t represents the last iteration.

Figure 2(*a*) displays an original image, which is to be classified and (*b*) the image classified after the FCM process is applied with eighteen iterations.

(*a*) (*b*)

Fig. 2 Original image; (*b*) classification of the pixels in four clusters through the FCM process

4 Application to Image Change Detection

The image change detection is the second problem addressed under the FCM framework in this work. This problem is formulated as follows: given two registered images $I_1(x,y)$, $I_2(x,y)$ of size $L \times M$ of the same area in the scene, taken at different times, the goal is to detect if a pixel, located at (x,y), has changed and the magnitude of the change. The approach proposed here is based on the work of Pajares et al. (2007) and Pajares (2006).

A difference image D, is computed at each pixel location (x,y) by subtracting the corresponding intensity values of the incoming images, i.e. $D(x,y) = I_2(x,y) - I_1(x,y)$. We also formulate two hypotheses H_0 and H_1 that represent no change and change respectively.

Bruzzone and Fernandez (2000) proposed an image change detection technique that estimates the parameters of the mixture distribution $p(D)$ consisting of all pixels in the difference image. The mixture distribution $p(D)$ can be written as,

$$p\left(D(x,y)\right) = p\left(D(x,y)\middle|H_0\right)P\left(H_0\right) + p\left(D(x,y)\middle|H_1\right)P\left(H_1\right) \tag{13}$$

Under this assumption, the probability density functions $p\left(D(x,y)\middle|H_0\right)$, $p\left(D(x,y)\middle|H_1\right)$ and the *a priori* probabilities $P\left(H_0\right)$ and $P\left(H_1\right)$ are estimated by using the Expectation Maximization (*EM*) algorithm (Duda et al., 2000), which is a general approach to maximum-likelihood estimation. It is assumed that both $p\left(D(x,y)\middle|H_0\right)$ and $p\left(D(x,y)\middle|H_1\right)$ can be modeled by Gaussian distributions. So, the parameters to be estimated are the means μ_0, μ_1 and variances σ_0^2, σ_1^2 respectively. The process is iterated until convergence and the initial values of the estimates are determined by simple differencing. Based on these estimates, we can determine if a pixel location can be classified as changed or unchanged. Nevertheless, better decisions are still possible by applying the Gestalt's principles that consider the spatial information existing between a pixel and its neighbors. This improvement is achieved through the FCMs.

Under the FCM framework, we build a network of $q = L \times M$ concepts, where each concept C_i represents a pixel i at the image location (x,y). The activation level of each concept at each iteration $A_i(t)$ determines the magnitude of the change at this location, ranging from $A_i(t) = +1$, maximum degree of change to $A_i(t) = -1$, without change.

The initialization of the activation levels is carried out by exploiting the characteristics of the difference image. We use the initialization strategy, described in Bruzzone and Fernandez (2000), as follows, from the histogram of the difference image $h(D)$ we compute two different thresholds T_0 and T_1 as $T_0 = M_D(1-\alpha)$ and $T_1 = M_D(1+\alpha)$, where M_D is the middle value of $h(D)$, i.e. $M_D = \left[\max\{D\} - \min\{D\}\right]/2$, and $\alpha \in (0,1)$ is set to 0.5 in this process. Now, given a pixel location (x,y) in the difference image, the initial activation level of its associated concept is given by equation (14).

$$A_i(0) = \begin{cases} -1 & if \quad D(x,y) < T_0 \\ +1 & if \quad D(x,y) > T_1 \\ -1 \text{ or } +1 \; (randomly) & \text{otherwise} \end{cases} \tag{14}$$

Once we have initialized the network, the next step consists in the definition of the edges $e_{ik}(t)$ based on the application of the Gestalt's principles. This is carried out by computing the data and contextual consistency coefficients. Afterwards, the decay factor is also derived. This allows updating the initial activation levels according to the equation (1).

4.1 Computation of the Causal Weights: Data and Contextual Consistencies

The data information is mapped through the *a posteriori* probabilities that given a pixel value $D(x,y)$ in the difference image D it is associated to hypothesis $H_k \in \{H_0, H_1\}$, i.e. we compute $P\left(H_k \mid D(x,y)\right)$. This is carried out by applying the Bayes rule,

$$P\left(H_k \mid D(x,y)\right) = \frac{p\left(D(x,y) \mid H_k\right) P(H_k)}{p\left(D(x,y)\right)} \tag{15}$$

where the *density functions* $p\left(D(x,y) \mid H_k\right)$ and the *a priori* probabilities $P\left(H_k\right)$, are estimated through the EM algorithm as described above. The mixture density distribution $p\left(D(x,y)\right)$ is computed trough equation (13). The initialization required by the EM algorithm is based on the procedure described above, under the assumption that $D(x,y)$ is with hypothesis H_0 or H_1 if $D(x,y) < T_0$ or $D(x,y) > T_1$ respectively. As we have not prior information, the *a priori* probabilities are initialized to 0.5.

Now, each pixel (x,y) in the difference image $D(x,y)$ should be associated to the hypothesis that maximizes the posterior conditional probability, i.e.

$$H_s = \arg \max_{H_s \in \{H_0, H_1\}} \{P(H_s | D(x,y))\}$$
$$= \arg \max_{H_s \in \{H_0, H_1\}} \{P(H_s) p(D(x,y) | H_s)\} \tag{16}$$

From (16) we build a *data map* with the same size as the difference image and identical (x,y) locations that those of the pixels in the image difference and concepts in the FCM network. Each concept C_i associated to the pixel location $i \equiv (x,y)$ is loaded with the data information $r(i)$ according to the criterion in equation (16) as follows,

$$a(i) = (-1)^{s+1} P(H_s | D(x,y)); \quad s = \{0,1\} \tag{17}$$

When a pixel i is considered unchanged, it is with the hypothesis H_0 and $a(i) = -P(H_0 | D(x,y))$ with its minimum value being -1. On the contrary, if a pixel has changed significantly it is with the hypothesis H_1 and $a(i) = P(H_1 | D(x,y))$ with its maximum value being +1.

Now, the goal is to map the data consistency between concepts C_i and C_j into the consistency coefficient r_{ij}. Given the concept C_i we consider its m-connected neighborhood, $\mathbf{N}_i^m$, where m takes the typical value of 8.

For each concept C_i, only consistencies can be established between concepts C_j, where the associated pixels $j \in \mathbf{N}_i^m$ and $i \neq j$ otherwise if $k \notin \mathbf{N}_i^m$ it is assumed that there is not consistency between concepts C_i and C_k. This is justified under the hypothesis that only local relations can be established between changed/unchanged pixels, based on the Gestalt's proximity principle. Indeed, given a changed pixel probably its neighbors should be also changed pixels and vice versa for unchanged pixels. Two concepts C_i and C_j where $j \in \mathbf{N}_i^m$ are said consistent if they have similar data information. Otherwise they should be inconsistent.

The data consistency between the concepts C_i and C_j is mapped into the coefficient r_{ij} as follows,

$$r_{ij} = \begin{cases} 1 - |a(i) - a(j)| & j \in \mathbf{N}_i^m \\ 0 & j \notin \mathbf{N}_i^m \end{cases} \tag{18}$$

r_{ij} ranges in [1,+1], where $-1/+1$ represents strong unchanged/changed respectively; the contribution so made may be positive (excitatory causality) or negative (inhibitory causality). Hence, a positive data consistency will contribute towards the network stability, in terms of small variations on its activation levels.

The mapping of the contextual consistencies involves the current activation levels according to the equation (19).

$$c_{ij}(t) = \begin{cases} h_{ij}\left(1 - |A_i(t) - A_j(t)|\right) & j \in \mathbf{N}_i^m \\ 0 & j \notin \mathbf{N}_i^m \end{cases} \tag{19}$$

where h_{ij} is a coefficient that measures the influence exerted by the concept C_i over the concept C_j taking into account the relevance of each concept. The relevance is a magnitude introduced in this approach in order to measure the strength of each concept against changes on its activation level. We build an accumulator of cells of size $q = L \times M$, where the cell i is associated to the concept C_i. Each cell i contains the number of times, h_i, that the concept C_i has changed significantly its activation level. Initially, all h_i are set to zero and then $h_i = h_i + 1$ if $|A_i(t+1) - A_i(t)| > \varepsilon$, where ε is set to 0.05 as in the classification approach; h_{ij} is computed as follows,

$$h_{ij} = \begin{cases} h_j / (h_i + h_j) & (h_i + h_j) \geq 2 \\ 1 & \text{else} \end{cases} \tag{20}$$

The equation (20) measures the fraction of changes accumulated for the concept C_j as compared to the concept C_i; h_{ji} measures the reverse influence of concept C_j over concept C_i. This equation is interpreted as follows, if $h_i < h_j$ the concept C_i has accumulated less number of changes than the concept C_j, i.e. the relevance of C_i is greater than the concept C_j and vice versa. This implies that h_{ij} could be different from h_{ji}. From (19) we can see that c_{ij} varies with the iteration and ranges in [−1,+1] where the lower/higher limit means minimum/maximum contextual consistency respectively. One can see that c_{ij} could be different from c_{ji} (non symmetry) due to the factors h_{ij} or h_{ji}.

Now the goal is to combine appropriately r_{ij} and $c_{ij}(t)$ in order to derive the causal weights required by the equation (1). Making use of the fuzzy set theory, we consider two fuzzy sets, where their elements are pairs of concepts (C_i, C_j) and the degrees of compatibility (membership functions) are given by r_{ij} and $c_{ij}(t)$ respectively. According to the dissertations of Zimmermann (1991) we propose the Hamacher's union operator for the combination because of its performance,

$$e_{ij}(t) = \frac{(\gamma - 1) r_{ij} c_{ij}(t) + r_{ij} + c_{ij}(t)}{1 + \gamma c_{ij}(t) r_{ij}} \tag{21}$$

where $\gamma \geq -1$; by setting $\gamma = -1$ the Hamacher's union operator matches with the Hamacher sum. We have used this value in the proposed approach.

The Gestalt's principle of similarity is considered in both r_{ij} and $c_{ij}(t)$ coefficients through the differences in the data and activation levels respectively. As in the above classification approach the proximity and connectedness principles are embedded in the neighborhood $\mathbf{N}_i^m$.

4.2 Computing the Decay Factor

The decay factor involved in the equation (1) is computed in a similar fashion that the one defined in the equation (12). Indeed, considering the accumulated changes for each concept and its neighbors, we can write the equation (22)

$$d_i = \begin{cases} 0 & h_i = 0 \text{ and } \overline{h}_k = 0 \\ \dfrac{h_i}{\left(\overline{h}_k + h_i\right) t} & \text{otherwise} \end{cases} \quad (22)$$

where h_i is defined in section 4.1 and $\overline{h}_k$ is the average value accumulated by the concepts $k \in N_i^m$. The same discussion with respect the equation (12) can be applied here.

4.3 Summary of the Full FCM Process

The FCM process is synthesized as follows:

1. *Initialization*: load each concept with its activation level $A_i(t = 0)$ through the equation (14); set $\varepsilon = 0.05$, $t_{max} = 50$ and $U = 0.9$, which are the typical values already used in the classification procedure. Define nc as the number of concepts from a total of q, that change their activation levels at each iteration. The activation mechanism and the activation function are defined in (1) and (4) respectively.

2. *FCM process*:
   $t = 0$
   while $t < t_{max}$ *and* $nc/q < U$
   $t = t + 1$; $nc = 0$;
   for each concept C_i
   update $A_i(t)$ according to the equation (1)
   if $|A_i(t) - A_i(t-1)| > \varepsilon$ *then*
   $nc = nc + 1$
   end if;
   end for;
   end while

3. *Outputs*: the activation levels $A_i(t)$ for all concepts updated.

Once the FCMs processes end, each concept C_i has achieved an activation level $A_i(t)$ that determines de degree of change. So, we consider that a pixel has changed if $A_i(t) > \varepsilon$, where its value in the range $[-1, +1]$ determines the degree.

Figure 3(a) and (b) display two original images from an outdoor environment, where the changes between them are to be identified. After the FCM process, the changes detected are displayed in image 2(c), and represented in the gray scale from 0 to 255, so that 255 means no change and 0 maximum degree of change. Intermediate values represent the magnitude of the change.

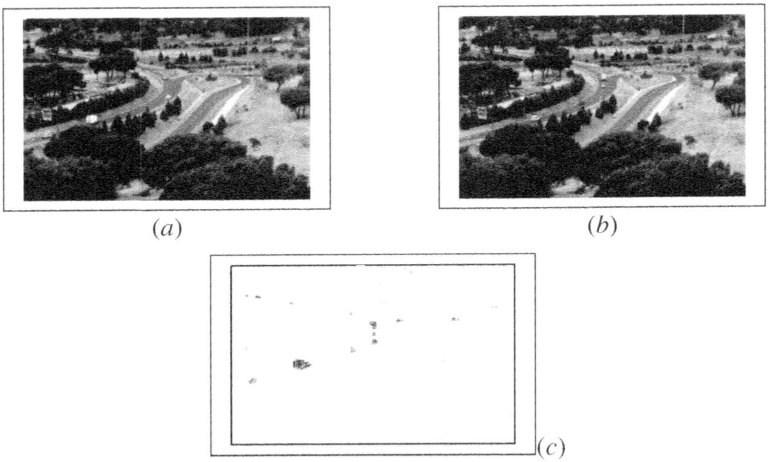

Fig. 3 Outdoor environment; (*a*) and (*b*) two images of the same sequence and scene; (*c*) changes detected with the FCM approach

5 Application to Image Matching in Stereovision

This is the third application addressed in this work. The image matching in stereovision is the process of identifying the corresponding points in two images that are cast by the same physical point in the tri-dimensional space. This can be carried out pixel by pixel or identifying significant features in the images, such as edges, regions or interest points. This is the approach used in this section, where we try to match pairs of edge segments.

Hence, the stereo correspondence problem can be defined in terms of finding pairs of true matches, namely, pairs of edge segments in two images that are generated by the same physical edge segment in space. These true matches generally satisfy some constraints (Scharstein and Szeliski, 2002): 1) *epipolar*, given two segments one in the left image and a second in the right one, if we slide one of them along a horizontal direction, i.e. parallel to the epipolar line, they would intersect (overlap) (figure 4); 2) *similarity*, matched edge segments have similar properties or attributes; 3) *smoothness*, disparity values in a given neighbourhood change smoothly, except at a few depth discontinuities; 4) *ordering*, the relative position among two edge-segments in the left image is preserved in the right one for the corresponding matches; 5) *uniqueness*, each edge-segment in one image should be matched to a unique edge-segment in the other image.

In section 5.1, we define the procedure for extracting edge segments and computing the attributes of each one of them. Afterwards, we build the FCM structure, where each pair of edge segments will define a concept in the FCM. The similarity constraint will allow define the activation levels of the concepts (section 5.2) and the smoothness, ordering and epipolar constraints will allow define the causal edges (section 5.3). The approach described in this chapter is based on the work of Pajares and Cruz (2006).

5.1 Feature and Attribute Extraction

The contour edges in both images are extracted using the Laplacian of Gaussian filter in accordance with the zero-crossing criterion Huertas and Medioni (1986). For each zero-crossing in a given image, its gradient vector (magnitude and direction) Leu and Yau (1991) and Lew et al. (1994) and variance Krotkov (1989) values are computed from the gray levels of a central pixel and its eight immediate neighbors. The edges are obtained by joining adjacent zero-crossings following the algorithm of Tanaka and Kak (1990), where: (1) a margin of deviation of ± 20% in gradient magnitude and of ±45° in gradient direction are allowed; (2) each detected contour is approximated by a series of piecewise linear line segments (Nevatia and Babu, 1980). Finally, for every segment, an average value of the four attributes is obtained from all computed values of its zero-crossings. All average attribute values are normalized in the same range. Now, we have edge segments, which are the features and their attributes.

Each pair of features has two associated 4-dimensional vectors $\boldsymbol{x}_a$ and $\boldsymbol{x}_b$, where the components are the attribute values, and the sub-indices a and b denote features belonging to the left and right images, respectively. A four-dimensional difference measurement vector $\boldsymbol{x}$ is then also obtained from the above $\boldsymbol{x}_a$ and $\boldsymbol{x}_b$ vectors, $\boldsymbol{x} = \boldsymbol{x}_a - \boldsymbol{x}_b = \{x_m, x_d, x_l, x_v\}$. The components of $\boldsymbol{x}$ are the corresponding differences for module and direction gradient, Laplacian and variance values. Only those pairs verifying the following three initial conditions will be processed: 1) their absolute value of the difference in the gradient direction is below a specific threshold, fixed to 25° in this work; 2) their absolute value in the gradient magnitude is also below a fixed threshold, set to 15, and 3) their overlap rate surpasses a certain value, fixed to 0.5 here. The remaining pairs that do not meet such conditions are directly considered as false correspondences. The overlap is a concept introduced in Medioni and Nevatia (1985), two segments u and z overlap if by sliding one of them in a direction parallel to the epipolar line, they would intersect.

Figure 4 clarifies the overlapping concept. Indeed, segment u in the left image overlaps with segment s in the right image, but segment v does not overlap with s. The overlap rate between edge segments (u,z), α_{uz} is defined as the percentage of coincidence, ranging in [0,1], when two segments u and z overlap, and it is computed taking into account the common overlap length l_c defined by c and the two lengths for the involved edge segments l_u and l_z respectively. All lengths are measured in pixels.

$$\alpha_{uz} = 2l_c/(l_u + l_z) \tag{23}$$

Considering the above initial conditions 1) and 3) and the parallel optical axis geometry of the stereovision system, we compute the disparity between two edge-segments (u and z in figure 1) as follows: trace epipolar lines (four) crossing the common overlapping segment (c), for each line compute x_u and x_z, so the disparity is $(x_u - x_z)$. Then the disparity for both edge segments u and z is the averaged disparity for the four pairs of points x_u and x_z.

Figure 4 also illustrates the neighboring N_{ab} concept between edge-segments (a,b). For each edge segment "a" in the left image we define a window $w(a)$ in the right image and, similarly, for each segment "b" in the right image, we define a window $w(b)$ in the left image. Given the pair (a,b), a new pair (g,o) belongs to N_{ab}, if g lies in $w(b)$ and o lies in $w(a)$. It is said that "a segment g lies in $w(b)$" if at least the 30% of the length of the segment "g" is contained in the $w(b)$ window. The shape of this window is a parallelogram; one side is "a", for left to right match, and the other a horizontal vector of length *2.maxd*. The constant *maxd* is said the disparity limit, it is set to 15 pixels in the proposed approach and is used for mapping the smoothness constraint as an interconnection value between concepts.

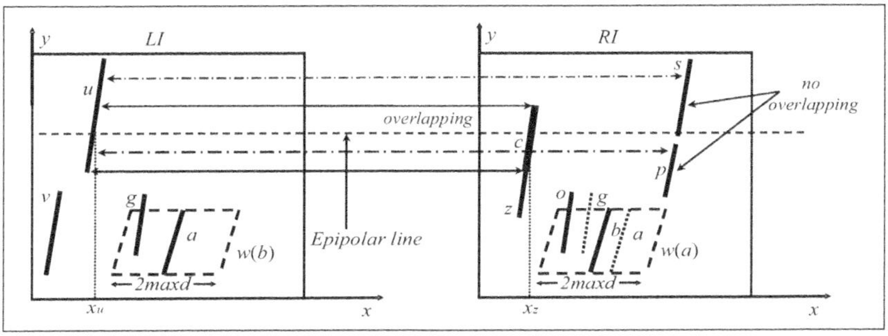

Fig. 4 Overlapping concept, edge-segments interactions and neighborhood for the pair (a, b)

5.2 Causal Concepts and Their Activation Levels

The system receives as inputs a pair of stereo images left (*LI*) and right (*RI*). This pair is processed in order to extract *edge segments* and their attributes; each pair of extracted features (a,b) is to be matched, the features a and b come from *LI* and *RI* respectively. For each pair (a,b) the attribute difference vector $\boldsymbol{x}$ is computed as described above. In this approach, a pair of edge segments (a,b) defines a causal fuzzy concept C_i, where its initial activation level at the iteration $t = 0$ is derived from $\boldsymbol{x}$ as follows,

$$A_i(0) = \left(1 + \|\boldsymbol{x}\|\right)^{-1} \tag{24}$$

where $\|\cdot\|$ is defined as the Euclidean norm, i.e. A_i ranges from 0 to +1 if $\|\boldsymbol{x}\| \to \infty$ (minimum similarity between a and b) or $\|\boldsymbol{x}\| = 0$ (maximum similarity between a and b respectively. This implies the application of the similarity Gestalt's principle. Hence, our FCM structure is built with as many concepts as pairs of edge segments, from *LI* and *RI*, are available.

5.3 Computation of the Causal Weights

Based on the smoothness, ordering and epipolar stereovision matching constraints, we map them for computing the causal weights between concepts. This is carried out in the next three subsections. Afterwards, we describe the *FCM* process, after which the uniqueness constraint is applied.

5.3.1 Mapping the Smoothness Constraint

The smoothness constraint assumes that neighbouring edge segments have similar disparities, except at a few depth discontinuities (Medioni and Nevatia, 1985). This implies the mapping of the proximity and connectedness Gestalt's principles. Generally, when the smoothness constraint is applied, it is assumed there is a bound on the disparity range allowed for any given segment. This limit is denoted as *maxd* and defined before.

Returning to figure 4, given "a" and "g" in $w(b)$ and "b" and "o" in $w(a)$ where "a" matches with "b" and "g" with "o" the differential disparity $|d_{ab} - d_{go}|$, measures how close the disparity between edge segments "a" and "b" denoted as d_{ab} is to the disparity d_{go} between edge segments "g" and "o". The disparity between edge segments is the average of the disparity between the two edge segments along the length they overlap. This differential disparity criterion is used in Ruichek and Postaire (1996), Medioni and Nevatia, (1986), Pajares and Cruz (2004, 2006) among others. We define a compatibility coefficient derived from Nasrabadi and Choo (1992) and Ruichek and Postaire (1996) as follows,

$$c_{(ab)(go)}(D) = \frac{1}{1+\exp\left[\gamma\left(D/m(D)-1\right)\right]} \tag{25}$$

where $D = \left|d_{ab} - d_{go}\right|$, $m(D)$ denotes the average of all values D in the pair of stereo images (LI and LR) under processing. The slope of the compatibility coefficient in (25) is expressed by γ and varies for each pair of stereo images. To determine γ, it is assumed that the probability distribution function of D is Gaussian with average $m(D)$ and standard deviation $\sigma(D)$, i.e. $p(D) = \left[1+\exp\left[\gamma\left(D_{(ab)(go)}/m(D)-1\right)\right]\right]^{-1}$. Under this assumption and following Barnard (1989) and Hattori (1998), to set the possibility value to 0.1 when the value of cumulative distribution function is 0.9, γ value is calculated by $\gamma = \ln 9\left((m(D))/(1.282\sigma(D))\right)$. In this approach, typical values of γ, $m(D)$ and $\sigma(D)$ are about 6, 9 and 2 respectively. So, values of D near 0 should give high values in the compatibility coefficient $c_{(ab)(go)}(\cdot) \approx +1$, but near 25 they give low values, $c_{(ab)(go)}(\cdot) \approx 0$. Note that $c_{(ab)(go)}(\cdot)$ ranges in (0,1). So, a compatibility coefficient of +1 is obtained for a good consistency between nodes (a,b) and (g,o) ($D = 0$) and a compatibility of 0 for a bad consistency ($D >> 0$). This causal weight embedding the smoothness constraint should indicate positive causality for high compatibility coefficient values and vice versa.

5.3.2 Mapping the Ordering Constraint

We define the ordering coefficient $O_{(ab)(go)}$, for the edge-segments according to (26), which measures the relative average position of edge segments "a" and "g" in LI with respect to "b" and "o" in RI, related to the neighboring N_{ab}, it ranges from 0 to 1.

$$O_{(ab)(go)} = -\frac{1}{S}\sum_{S} y_{(ab)(go)}$$
$$\text{where} \quad y_{(ab)(go)} = \left\| R(x_a - x_g) - R(x_b - x_o) \right| - 1 \big| \quad \text{and} \quad R(r) = \begin{cases} 1 & \text{if} \quad r>0 \\ 0 & \text{otherwise} \end{cases} \tag{26}$$

We trace S scanlines (four) along the common overlapping length, each scanline produces a set of four intersection points (a_S and g_S in LI and b_S and o_S in the RI) with the four edge-segments. Hence, the $y_{(ij)(hk)}$ can be computed as in Ruichek and Postaire (1996) considering the above four edge points and it takes 0 and 1 as two discrete values. A value of +1 in the ordering coefficient means that the ordering constraint is preserved. On the contrary, a value of 0 indicates that the ordering constraint is not preserved. The causal weight embedding the ordering constraint should indicate positive causality for a high ordering coefficient value.

5.3.3 Mapping the Epipolar Constraint (Overlapping Concept)

The epipolar constraint is mapped through the overlapping concept described in Medioni and Nevatia (1985), by computing the overlapping coefficient as given in equation (27).

$$\lambda_{(ab)(go)} = 0.5\left(\alpha_{ab} + \alpha_{go}\right) \tag{27}$$

where α is the overlap rate defined in equation (23). Under the epipolar constraint we can assume that correct/incorrect matches should have high/low overlap rates, i.e. the overlapping coefficient should be +1 or 0 respectively and $\lambda_{(ab)(go)}$ for neighborhoods should be high increasing the consistency. The use of the overlapping criterion is justified by the fact that the edge segments are reconstructed by piecewise linear line segments as described in section 5.1. The reasoning for the influence of this coefficient in the causal weight is similar to the previous ones for the compatibility and ordering coefficients.

5.3.4 Fuzzy Criteria for Computing the Causal Weights

Once the smoothness, ordering and epipolar stereovision matching constraints have been mapped, we have available the compatibility, ordering and overlapping coefficients derived from such constraints respectively. Now the goal is to combine the three coefficients for computing the causal weights involved in equation (1). Making use of the fuzzy theory, we can consider these coefficients as membership functions, which can be combined in order to compute the causal weight $e_{(ab)(go)}$.

Hence, in our approach the causal weight is considered as a fuzzy measurement (membership value). Taking into account the dissertations in Zimmermann (1991), a straightforward approach for aggregating fuzzy sets, would be to use the aggregating procedures frequently used in multi-criteria decision theory. They realize the idea of trade-offs between conflicting goals when compensation is allowed, and the resulting trade-offs lie between the most optimistic lower bound and the most pessimistic upper bound, that is, they map between the minimum and the maximum degree of membership of the aggregated sets. Therefore they are called averaging operators.

Following the discussion in Zimmermann (1991) about the criteria for selecting appropriate aggregation operators, we find that adaptability is suitable; this can be achieved by parametrization. Thus *min* and *max* operators cannot be adapted at all. They are acceptable in situations in which they fit, by contrast, there are other operators hat can be adapted to certain contexts by setting their parameters; we have used the Hamacher's union operator. Taking into account that causal weights are considered as fuzzy membership values and making use of the operator's associativity the equation (28) is derived. The parameter τ allows a fitting appropriately

$$\begin{aligned} \rho_{(ab)(go)} &= \frac{(\tau-1)c_{(ab)(go)}O_{(ab)(go)} + c_{(ab)(go)} + O_{(ab)(go)}}{1+\tau c_{(ab)(go)}O_{(ab)(go)}} \\ E_{(ab)(go)} &= \frac{(\tau-1)\rho_{(ab)(go)}\lambda_{(ab)(go)} + \rho_{(ab)(go)} + \lambda_{(ab)(go)}}{1+\tau\rho_{(ab)(go)}\lambda_{(ab)(go)}} \end{aligned} \tag{28}$$

where $\tau \geq -1$, we have found acceptable the behavior of this parameter by setting it to 1. This is because its behavior is a trade-off between maximum and minimum operators Zimmermann (1991). Other values for τ and other parameterized operators (Einstein, Yager, Dubois and Prade among others, Zimmerman) could be used.

As a result of the aggregation's operators, the resulting $E_{(ab)(go)}$ from equation (28) ranges in [0,+1]. So, rescaling this interval to [−1,+1], we can derive the final causal weight $e_{(ab)(go)}$ between features (a,b) and (g,o) as required by the general equation (1). Hence, we obtain,

$$e_{(ab)(go)} = 2E_{(ab)(go)} - 1 \tag{29}$$

So, high coefficient values should give high causal weights i.e. positive causality as expected and vice versa for low coefficient values and negative causality.

5.3.5 Full FCM Process

The FCM process is synthesized as follows; it is similar to the processes defined in sections 3.2.5 and 4.3:

1 Initialization: load each concept with its activation level $A_i(t=0)$ through the equation (24); set $\varepsilon = 0.05$, $t_{max} = 50$ and $U = 0.9$, which are typical values used in the classification procedure, described in the section 3.2.5. Define *nc* as the number of concepts from a total of q concepts representing pairs of edge segments that change their activation levels at each iteration. The activation mechanism is that defined in equation (1), but with null decay factors and the activation function is that defined in equation (2).

2 FCM process:

$t = 0$

while $t < t_{\max}$ *and* $nc/q < U$

  $t = t + 1$; $nc = 0$;

  for each concept C_i

    update $A_i(t)$ according to the equation (1)

    if $|A_i(t) - A_i(t-1)| > \varepsilon$ *then*

      $nc = nc + 1$

    end if; end for; end while

3 Outputs: the activation levels $A_i(t)$ for all concepts updated.

5.3.6 Uniqueness Constraint

The *uniqueness* constraint completes the set of matching constraints used for solving the stereovision matching problem.

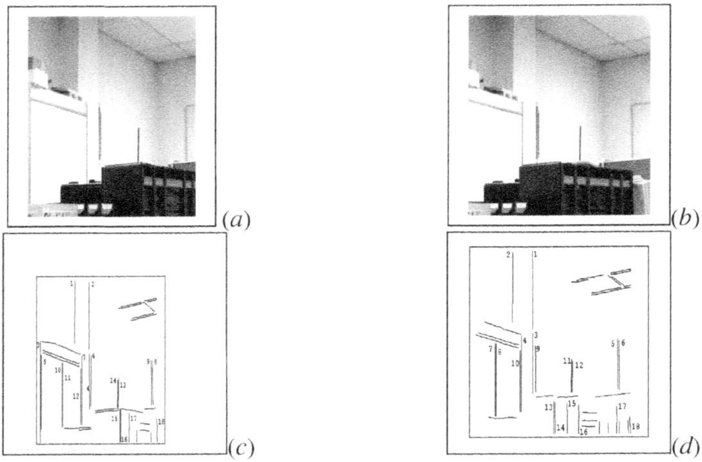

Fig. 5 (*a*) and (*b*) original left and right stereo images; (*c*) and (*d*) labelled edge segments in left and right images

A left edge segment can be assigned to a unique right edge segment (unambiguous pair) or several right edge segments (ambiguous pairs). The decision about whether a match is correct, is made by choosing the greater activation level for the corresponding causal concept (in the unambiguous case there is only one) whenever it surpasses a previous fixed threshold *U* (=0.5), intermediate value for C_i ranging in [0, 1]. A secure true match (*a*,*b*), represented by the causal concept C_i should have at the last iteration an activation level with value the unity.

6 Comparative Analysis and Performance Evaluation

This section contains the corresponding comparative analysis and performances of the proposed methods, as compared to other strategies of similar category. Details about these performances wit respect image classification, image change detection and stereovision matching are provided in sub-sections 6.1 to 6.3 respectively.

6.1 Performance of FCMs in Texture Classification

In texture classification there are two phases involved: training and decision. We used a set of 36 digital aerial images acquired during May, 2006, from the Abadin region of Lugo (Spain). They are multi-spectral images, 512×512 pixels in size. The images were taken during different days from an area with several natural textures. We selected randomly 12 images from the set of 36 available. This set was used for training and the remainder set of 24 images for testing. Each image in the training set is down-sampled by two, obtaining the training samples used in this initial training phase, i.e., the number of training samples was n_0 = $12 \times 256 \times 256 = 786432$.

We assume known the number of clusters, which is fixed to four in our experiments because this is the number of the main textures identified; hence, from the training set we estimate the cluster centres, $\boldsymbol{v}_j$, through *FC*, according to equation (5) and also the mean centres $\boldsymbol{m}_j$, and the covariance matrices Σ_j based on *BP* through equation (8).

Our combined *FCM* (*FM*) method was compared with the base classifiers used for the combination, i.e. *FC* and *BP*. *FM* was also compared with the following classical hybrid strategies described in Kittler et al. (1998): Mean (*ME*), Max (*MA*) and Min (*MI*). Given the node *i*, with supports of belonging to the cluster *j*, μ_i^j and p_i^j provided by *FC* and *BP* respectively, *ME* computes the mean value, i.e., $\left(\mu_i^j + p_i^j\right)/2$; *MA* the maximum value, i.e., $\max\left\{\mu_i^j, p_i^j\right\}$ and *MI* the minimum value, i.e., $\min\left\{\mu_i^j, p_i^j\right\}$. They were studied in terms of reliability (Cabrera, 2006). Given the resulting combined supports that the pixel *i* belongs to each cluster *j*, the final decision is made based on the maximum combined support value for all clusters. Yager (1988) also proposed a multi-criteria decision-making approach based on fuzzy sets aggregation. So, *FM* was also compared against the *Fuzzy Aggregation* (*FA*) defined by the equation (30).

$$\gamma_i^j = 1 - \min\left\{1, \left(\left(1-\mu_i^j\right)^a + \left(1-p_i^j\right)^a\right)^{1/a}\right\} \qquad a \geq 1 \tag{30}$$

This represents the joint support provided by *FC* and *BP*; the decision is made based on the maximum support for all clusters as above. The parameter a is adjusted from the set of 12 images available for training. The experiments were carried out by varying the parameter a from 1 to 8. The minimum error rate was achieved with $a = 4$, which in the end was the value used for this parameter. The decision made according to node i with the supports provided by (30) was based on the maximum value, in keeping with the following rule: $i \in w_j$ if $\gamma_i^j > \gamma_i^h \quad \forall h \neq j$.

To verify the performance for each method we have built a ground truth for each image under the supervision of the expert human criterion. Based on the above assumption that the number of clusters is four, we classify each image pixel with the simple classifiers, obtaining a labelled image with four expected clusters. For each cluster we build a binary image, which is manually touched up until a satisfactory classification is obtained under the human supervision. The pixels belonging to a cluster are labelled as white.

Table 1 shows the percentage of error during the classification for the different classifiers. These percentages are computed as follows. Let I_r an image from the testing set of 24 for testing, (r = 1,…,24); i is the node at the location (x,y) in I_r. An error counter E_r is initially set to zero for each image r and for each classifier. Based on the corresponding decision process, each classifier determines the cluster to which the node i belongs, $i \in w_j$. If the same pixel location on the corresponding ground truth image is black, then the pixel is incorrectly classified and the error counter value is increased by the unity. The error rate of the image I_r is: $e_r = E_r / Z$, where Z is the image size, i.e., 512×512. The average error rate for the testing set is finally computed by: $\overline{e} = \frac{1}{24}\sum_{r=1}^{24} e_r$. The standard deviation is simultaneously computed as: $\overline{\sigma} = \sqrt{\frac{1}{23}\sum_{r=1}^{24}\left(e_r - \overline{e}\right)^2}$.

In table 2, they are displayed as percentages, i.e., $\tilde{e} = 100\overline{e}$ and $\tilde{\sigma} = 100\overline{\sigma}$. The numbers in square brackets in the row *FM* indicate the rounded and averaged number of iterations required for this method.

From results in table 1, it is seen that the best performance is achieved by the proposed *FM* strategy. The best performance for the classical hybrid methods is achieved by *ME* and for the simple methods is *BP*. The best performances are established in terms of the least average percentage of error and the least standard deviation values.

Table 1 Average percentages of error and standard deviations

$\tilde{e}$: average percentage of error $\tilde{\sigma}$: standard deviation of error (%)		$\tilde{e}$	$\tilde{\sigma}$
Iterative and fuzzy hybrid methods	**[iterations]** **FM** (Fuzzy Cognitive Maps)	**[12]** 17.5	0.85
	FA(Fuzzy Aggregation)	20.6	1.55
Classical hybrid methods	**MA** (Maximum)	26.3	2.01
	MI (Minimum)	29.1	2.35
	ME (Mean)	24.2	1.7
Simple methods	**FC** (Fuzzy clustering)	25.9	2.0
	BP(Bayesian Parametric)	24.7	1.8

6.2 Performance of FCMs for Image Change Detection

For comparative purposes, in image change detection, we used four data sets: 1) 40 pairs of real video sequences of outdoor environments of size 1392 *x* 1040; 2) 36 pairs of real video sequences of indoor environments of size 840 *x* 760; 3) 10 pairs of real remote sensing images of size 400 *x* 400; and 4) 30 pairs of synthetic sensing images also of size 400 *x* 400. A full description of this data set can be found in Pajares (2006), where a Hopfield Neural Network (HNN) is applied.

The analysis requires a ground truth for each pair of images tested, useful to assess change-detection errors and successes. So, we prepare a map of changed and unchanged areas as follows. In Rosin and Ioannidis (2003) are evaluated several classical global image thresholding approaches for image change detection. Based on this study, the best performance is achieved with the method described in Kapur et al. (1985) which uses the entropy of the histogram. In Wu et al. (2005) also a new technique based on cumulative histograms is used with acceptable performance. By applying both methods we obtain two binary images. They are combined by using the logical operator "*or*". The resulting image is manually refined and the *ground truth map* obtained.

The HNN method of Pajares (2006) was compared against six existing image change detection strategies including the approach of Bruzzone and Fernández-Prieto (2000) (BRU). The best performances were obtained by HNN followed by BRU. Now, in the present work, we compare the performance of the proposed FCM approach against the iterative HNN and BRU methods. FCM, HNN and BRU use the same initialization process and also a neighborhood region of size 3*x*3. We use a set of seven experiments, a brief description of the experiments is the following: E1: 30 outdoor pairs of images from the same sequence; E2: 10 outdoor pairs of images from different sequences of the same scene; E3: 12 indoor pairs of images from the same sequence without changes in the illumination levels; E4: 12 indoor pairs of images from the same sequence, during the full video capture the illumination levels are on-line changed, i.e. the images have different intensity levels; E5: 12 indoor pairs of images from the same sequence, an image is obtained without changes in the illumination during its capture and the other, as before, by varying on-line the illumination; E6: 10 pairs of remote sensing images of the same scene; E7: 30 pairs of synthetic remote sensing images of the same scene corrupted with Gaussian noise of zero-mean and different variances selected randomly.

The results obtained for each method are compared against the ground truth, based on the *PCC* magnitude described in Rosin and Ioannidis (2003), also used in Pajares (2006): $PCC = (TP + TN)/(TP + FP + TN + FN)$, where TP: number of change pixels correctly detected; FP: number of no-change pixels incorrectly labelled as change; TN: number of no-change pixels correctly detected; FN: number of change pixels incorrectly labelled as no-change.

Table 2 shows the results in terms of the correct classification for the seven experiments. The final result for each experiment is averaged by the number of pairs of images processed. The number of iterations used in our FCM (k_{max}) is set to the number of iterations where HNN gained the convergence for each set of experiments, i.e. E1, E3 = 4, E2, E5 = 8, E4 = 10 and E6, E7 = 5.

Table 2 Averaged PCC scores for each method against the set of experiments

$x10^{-3}$	E1	E2	E3	E4	E5	E6	E7
BRU	921	821	945	653	698	819	615
HNN	987	943	991	789	876	901	847
FCM	944	954	956	823	901	848	844

From results in table 2, one can see that *FCM* improves the performance of *HNN* for experiments E2, E4 and E5 where the number of iterations is higher than the used for the other experiments. This means that the *FCM* approach is suitable for images where the number of iterations is high. The *FCM* outperforms the other approaches when each image in the pair displays high variability due to different illumination conditions as in the experiments E2, E4 and E5. This means that the *FCM* should be applied in image sequences captured under such illumination conditions where it is foreseeable that the number of iterations could become high.

The best performance achieved by the *FCM* approach against *HNN* in those experiments, can be interpreted in the light of the mutual influence between two nodes based on the relevance's values. Indeed, as the number of iterations increases, the relevance of each node achieves higher stability (less number of changes in the activation level). *FCM* and *HNN* achieve both a similar performance for E7 (with noise).

6.3 Performance of FCMs for Stereovision Matching

In stereovision matching, the proposed Fuzzy Cognitive Maps strategy (FCMS) is compared against the Support Simulated Annealing (SANN) in Pajares and Cruz (2004), but also with the method described in Pajares et al. (2000) which is a Relaxation Labeling (RELB) approach and the method described in Pajares et al. (1998), which is an optimization approach based also on the Hopfiel Neural Network (HNNB1).

SANN applies the similarity constraint through Support Vector Machines and RELB and HNNB1 both apply the similarity constraint by computing a matching probability based on the estimation of a probability density function through the Bayes's theory. The matching probabilities are used as the inputs for the relaxation and optimization processes respectively. From these processes, RELB performs an iteration procedure by applying smoothness, ordering and uniqueness constraints. HNNB1 performs the optimization process by mapping the smoothness and uniqueness constraints in an energy function which is to be minimized. From HNNB1, we have implemented a new version HNNB2, by mapping the ordering constraint as an energy function to be minimized and applying the similarity constraint as the 4-dimensional difference null vector $\boldsymbol{x}$. HNNB2 can be considered a very close approach to the described in Ruichek and Postaire (1996), although this work uses edge pixels as features, we have modified the original method in Ruichek and Postaire (1996) to use edge-segments as in SANN.

We have also compared our approach with the Stochastic Stereovision Matching Method (SSVM) in Barnard (1989), also used in Hattori et al. (1998). This method uses the regularization criterion proposed in Poggio et al. (1985), where an energy functional is minimized based on a penalty functional which measures the dissimilarity between corresponding features (similarity constraint) and a stabilizing functional by which the smoothness constraint is imposed. The energy minimization is carried out through the Simulated Annealing algorithm, we have used a value of 50 as in Barnard (1989) for the regularization parameter λ (this works well for images quantized in 8-bit values) and the same neighborhood criteria as that used in this paper. Two differences are considered in this implementation with respect to our implementation: 1) the edge-segments disparities are the outputs obtained in SSVM, which are used to obtain the correspondences and 2) the hierarchical coarse-to-fine control structure with re-heating in Hattori et al. (1998) is not used in our implementation.

Finally we have chosen the Minimum Differential Disparity algorithm (MDDA) in Medioni and Nevatia (1985) for comparative purposes for the following reasons: 1) it is a merit relaxation approach, 2) it applies the commonly used constraints (similarity, smoothness and uniqueness); 3) it uses edge segments as features and the contrast and orientation of the features as attributes; and 4) some concepts of MDDA, such as minimum differential disparity, overlapping concept, disparity limit or average disparity are used in our FCMS approach. Table 3 summarizes the main differences between the eight strategies compared. All methods use edge-segments as features and the same four attributes.

We have selected 82 stereo pairs of realistic stereo images from an indoor environment. All tested images are 512 x 512 pixels in size, with 256 gray levels. Of all the possible combinations of pairs of matches formed by pairs of segments from left and right images only 1701 are considered for testing.

Table 4 displays the percentage of successes for each method (FCMS, SANN, RELB, HNNB1, HNNB2, SSVM, and MDDA) as a function of the number of iterations. These values are averaged over the above full number of stereo pairs considered for matching. Iteration 0 corresponds to the results obtained only by applying the similarity constraint, i.e. before the iteration process is triggered. This is the starting point for the FCM process proposed in this approach. Intermediate results are also displayed for a number of iterations of 15. The final results correspond to a number of iterations of 30, which is on average the number of iterations required for the FCMS for convergence with the 82 pairs of stereo images used for testing.

Table 3 Summary of Stereovision Matching Methods and Constraints

	Stereovision matching constraints				
	Similarity	**Smoothness**	**Ordering**	**Epipolar**	**Uniqueness**
FCMS	Simple difference vector	mapped as coefficients aggregated in the causal weight between concepts	mapped as coefficients aggregated in the causal weight between concepts	mapped as coefficients aggregated in the causal weight between concepts	applied by selecting the highest causal concept values
SANN	Support Vector Machines	mapped as an energy minimized by Simulated Annealing	mapped as an energy minimized by Simulated Annealing	mapped as an energy minimized by Simulated Annealing	applied by selecting the highest state values
RELB	Bayes probability density estimation	Probabilistic relaxation	Probabilistic relaxation	mapped under the overlapping concept	applied by selecting the highest probabilities
HNNB1	Bayes probability density estimation	mapped as an energy minimized by Hopfield	No	mapped under the overlapping concept	mapped as an energy minimized by Hopfield
HNNB2	Euclidean distance without estimation	mapped as an energy minimized by Hopfield	mapped as an energy minimized by Hopfield	mapped under the overlapping concept	mapped as an energy minimized by Hopfield
SSVM	mapped as an energy minimized by regularization	mapped as an energy minimized by regularization	No	implicit application by image registration	No
MDDA	qualitative boolean function	merit function relaxation	No	implicit application by image registration	applied by selecting the highest merits

Table 4 Percentage of successes against the number of iterations

Iteration #	**0**	**15**	**35**
FCMS	69.3	81.1	94.2
SANN	76.1	83.3	92.1
RELB	74.5	82.1	90.2
HNNB1	70.4	81.6	90.2
HNNB2	67.8	69.3	67.8
SSVM	0	39.2	57.4
MDDA	66.1	68.7	69.7

According to values from table 2, the following conclusions may be inferred:

1) *iteration 0*: the best performances are achieved with the methods that apply a learning process, i.e. SANN, RELB and HNNB1. The methods without previous estimation or learning (FCMS, HNNB2 and MDDA) obtain the worst results at this phase.

2) *Iteration process*: As the number of iterations progresses all methods achieve better performances; this means that the mapping of different constraints is positive. The ordering constraint is not decisive: HNNB2 (with ordering) obtains worse results than HNNB1 and HNNP (without ordering). In Pajares and Cruz (2004) is reported that SANN reaches its equilibrium with an average of 65 iterations, with such number of iterations the performance of SANN is comparable to the performance of FCMS with 35 iterations. Therefore, we can conclude that the fuzzy causal approach under FCMs is suitable in terms of performance in stereovision matching, i.e. the causal reasoning and decision process under a fuzzy point of view is consistent with that of the human. This is the reason for the better performance of our FCMS approach against the remainder methods, particularly against the SANN approach. Another important reason comes from equation (1), where the previous causal concept $A_i(t)$ contributes to the updating of the current one $A_i(t+1)$, this implies that the network in the FCMS achieves a rapid stabilization. Finally, SANN only reaches the 96.0 in percentage with 400 iterations, as reported in Barnard (1989). MDDA obtains the worst results.

7 Conclusions

We have applied the Fuzzy Cognitive Maps framework for three computer vision tasks, verifying its performance as compared to other existing methods. Starting from the general definition we can adapt and customize the different parameters and coefficients involved on it for each specific approach. This means that FCMs become an excellent tool when spatial relations among elements in the images can be established based on the Gestalt's principles. In the proposed approaches the elements used were pixels in image classification and change detection and edge segments in the stereovision matching approach. This establishes the base for its application to other different image computer vision problems, where the only requirement is the mapping of the spatial relations or properties.

The main drawback of FCMs is its computational cost due to the iterative process applied. But this problem also appears in the remainder iterative approaches. In the future, this could be improved through parallel implementations.

Another research line will come from the study of the non-linear functions for computing the activation levels in equations (2) to (4), so that the convergence process can be accelerated without loss of performance.

References

Barnard, S.T.: Stochastic stereo matching over scale. Int. J. Comput. Vision 3(1), 17–32 (1989)

Bezdek, J.C.: Pattern Recognition with Fuzzy Objective Function Algorithms. Kluwer, Plenum Press, Dordrecht (1981)

Bruzzone, L., Fernández-Prieto, D.: Automatic Analysis of the difference Image for unsupervised change detection. IEEE Trans. Geoscience Remote Sensing 38(3), 1171–1182 (2000)

Cabrera, J.B.D.: On the impact of fusion strategies on classification errors for large ensambles of classifiers. Pattern Recognition 39, 1963–1978 (2006)

Duda, R.O., Hart, P.E., Stork, D.S.: Pattern Classification. Wiley, Chichester (2000)

Haykin, S.: Neural Networks: A comprehensive Foundation. Mcmillan College Publishing Co., New York (1994)

Hattori, S., Okamoto, A., Hasegawa, H.: Stereo matching by simulated annealing incorporating a diffusion equation. In: Proc. ASPRS 1998 Annual Conference, pp. 1030–1041 (1998)

Huertas, A., Medioni, G.: Detection of Intensity Changes with Subpixel Accuracy Using Laplacian-Gaussian Masks. IEEE Trans. Patt. Anal. Machine Intell. 8(5), 651–664 (1986)

Kapur, J., Sahoo, P., Wong, A.: A new method for gray-level picture thresholding using the entropy of the histogram. Computer Vision Graphics Image Processing 29(3), 273–285 (1985)

Kittler, J., Hatef, M., Duin, R.P.W., Matas, J.: On Combining Classifiers. IEEE Trans. Patt. Anal. Machine Intell. 20(3), 226–239 (1998)

Koffka, K.: Principles of Gestalt Psychology. Harcourt, New York (1935)

Kosko, B.: Fuzzy Cognitive Maps. Int. J. Man Machine Studies 24, 65–75 (1986)

Kosko, B.: Neural Networks and Fuzzy Systems: a dynamical systems approach to machine intelligence. Prentice-Hall, NJ (1992)

Krotkov, E.P.: Active Computer Vision by Cooperative Focus and Stereo. Springer, New York (1989)

Kuncheva, L.I.: Combining Pattern Classifiers: Methods and Algorithms. Wiley, London (2004)

Leu, J.G., Yau, H.L.: Detecting the dislocations in metal crystals from microscopic images. Pattern Recognition 24, 41–56 (1991)

Lew, M.S., Huang, T.S., Wong, K.: Learning and Feature Selection in Stereo Matching. IEEE Trans. Patt. Anal. Machine Intell. 16(9), 869–881 (1994)

Medioni, G., Nevatia, R.: Segment Based Stereo Matching. Computer Vision, Graphics and Image Processing 31, 2–18 (1985)

Miao, Y., Liu, Z.Q.: On Causal Inference in Fuzzy Cognitive Maps. IEEE Trans. Fuzzy Systems 8(1), 107–119 (2000)

Nevatia, R., Babu, K.R.: Linear Feature Extraction and Description. Computer Vision, Graphics, and Image Processing 13, 257–269 (1980)

Pajares, G., Guijarro, M., Herrera, P.J., Ribeiro, A.: Combining Classifiers through Fuzzy Cognitive Maps in natural images. IET Computer Vision 3(3), 112–123 (2009)

Pajares, G., Sánchez-Beato, A., de la Cruz, J.M., Ruz, J.J.: A neural Network Model for image Change Detection Based on Fuzzy Cognitive Maps. In: Martí, J., Benedí, J.M., Mendonça, A.M., Serrat, J. (eds.) IbPRIA 2007. Part I, LNCS, vol. 4477, pp. 595–602. Springer, Heidelberg (2007)

Pajares, G., Cruz. J.M.: Fuzzy Cognitive Maps for stereovision matching. Pattern Recognition 39, 2101–2114 (2006)
Pajares, G.: A Hopfield Neural Network for Image Change Detection. IEEE Trans. Neural Networks 17(5), 1250–1264 (2006)
Pajares, G., Cruz, J.M.: On Combining Support Vector Machines and Simulated Annealing in Stereovision Matching. IEEE Trans. System, Man Cybernetics, Part B 34(4), 1646–1657 (2004)
Pajares, G., Cruz, J.M., López-Orozco, J.A.: Relaxation labeling in stereo image matching. Pattern Recognition 33, 53–68 (2000)
Pajares, G., Cruz, J.M., Aranda, J.: Relaxation by Hopfield network in stereo image matching. Pattern Recognition 31(5), 561–574 (1998)
Poggio, T., Torre, V., Koch, C.: Computational Vision and regularization theory. Nature 317, 314–319 (1985)
Rosin, P.L., Ioannidis, E.: Evaluation of global image thresholding for change detection. Pattern Recognition Letters 24, 2345–2356 (2003)
Ruichek, Y., Postaire, J.G.: A neural network algorithm for 3-D reconstruction from stereo pairs of linear images. Pattern Recognition Letters 17, 387–398 (1996)
Scharstein, D., Szeliski, R.: A Taxonomy and Evaluation of Dense Two-Frame Stereo Correspondence Algorithms. Int. J. Computer Vision 47(1/2/3), 7–42 (2002)
Shorliffe, E.H.: Computer-Based Medical Consultations: MYCIN. Elsevier, NY (1976)
Tanaka, S., Kak, A.C.: A Rule-Based approach to binocular stereopsis. In: Jain, R.C., Jain, A.K. (eds.) Analysis and Interpretation of range images, pp. 33–139. Springer, Berlin (1990)
Tsardias, A.K., Margaritis, K.G.: Cognitive Mapping and Certainty Neuron Fuzzy Cognitive Maps. Information Sciences 101, 109–130 (1997)
Tsardias, A.K., Margaritis, K.G.: An experimental study of the dynamics of the certainty neuron fuzzy cognitive maps. Neurocomputing 24, 95–116 (1999)
Tsardias, A.K., Margaritis, K.G.: The MYCIN certainty factor handling as uniform operator and its use as threshold function in artificial neurons. Fuzzy Sets and Systems 93, 263–274 (1998)
Wang, D.: The time dimension for scene analysis. IEEE Trans. Neural Networks 16(6), 1401–1426 (2005)
Wu, Q.Z., Cheng, H.Y., Jeng, B.S.: Motion detection via change-point detection for cumulative histograms of ratio images. Pattern Recognition Letters 26, 555–563 (2005)
Xirogiannis, G., Chytas, P., Glykas, M., Valiris, G.: Intelligent impact assessment of HRM to the shareholder value. Expert Systems with Applications 35, 2017–2031 (2008)
Xirogiannis, G., Glykas, M.: Intelligent modelling of e-business maturity. Expert Systems with Applications 32, 687–702 (2007)
Xirogiannis, G., Glykas, M.: A soft knowledge modeling approach for geographically dispersed financial organizations. Soft Computing 9, 579–593 (2005)
Xirogiannis, G., Glykas, M.: Fuzzy Cognitive Maps in Business Analysis and Performance-Driven Change. IEEE Trans. Engineering Management 51(3), 334–462 (2004)
Yager, R.R.: On ordered weighted averaging aggregation operators in multicriteria decision making. IEEE Trans. System Man and Cybernetics 18(1), 183–190 (1988)
Yaman, D., Polat, S.: A fuzzy cognitive map approach for effect-based operations: an illustrative case. Information Sciences 179, 382–403 (2009)
Zimmermann, H.J.: Fuzzy Set Theory and its applications. Kluwer Academic Publishers, Dordrecht (1991)

Classifying Patterns Using Fuzzy Cognitive Maps

G.A. Papakostas and D.E. Koulouriotis

Abstract. This chapter is focused on the use of Fuzzy Cognitive Maps (FCMs) in classifying patterns, as alternative to the traditional classifiers such as neural networks or even as collaborators, in achieving better classification capabilities. By defining the classification procedure as the equilibrium point achieved by applying common inference laws, a FCM can simulate a typical classifier that maps a set of inputs to specific output values.

The classification capabilities of the FCM classifiers are studied in several pattern classification problems, while the ability of the FCM to store knowledge about the problem in hand is investigated in conjunction to the nodes' type of activation function and the inference law used. Appropriate experiments are taken place, in order to analyze the behavior of the FCM-based classifiers, in well known benchmark problems.

1 Introduction

Recently, an increased interest about the theory and application of the Fuzzy Cognitive Maps in engineering science is noted. FCMs are characterized by their ability to model the dynamics of complex systems by incorporating the casual relationships between the main concepts that describe the system. They have been used initially to model complex social (Koulouriotis et al. 2003), strategic (Xirogiannis and Glykas 2007) and financial systems (Koulouriotis et al. 2001b, 2005, Koulouriotis 2004, Xirogiannis and Glykas 2004b, Glykas and Xirogiannis 2004, Chytas et al. 2006), where analytical descriptions do not exist or can not be derived.

Fuzzy Cognitive Maps are fuzzy signed directed graphs with feedback. They were proposed by Kosko (Kosko 1986) as a modeling methodology of complex systems, able to describe the casual relationships between the main factors (concepts) that determine the dynamic behavior of a system. The concepts are interconnected through arcs having weights that denote the cause and effect relationship that a concept has on the others.

G.A. Papakostas and D.E. Koulouriotis
Democritus University of Thrace, Department of Production and Management Engineering, Xanthi, Greece

M. Glykas (Ed.): Fuzzy Cognitive Maps, STUDFUZZ 247, pp. 291–306, 2010.
springerlink.com © Springer-Verlag Berlin Heidelberg 2010

At each time, the value of each ith concept A_i is calculated, by summarizing the influences of all the other concepts to this and by squashing the overall impact using a barrier function *f*, according to the following inference rule

$$A_i^{t+1} = f\left(A_i^t + \sum_{i=1, i \neq j} W_{ji} A_j^t \right) \tag{1}$$

where A_i^{t+1} and A_i^t are the values of concept *i* at times *t+1* and *t* respectively, A_j^t the value of concept *j* at time *t*, W_{ji} the weight value of the interconnection with direction from concept *j* to concept *i*, and *f* the barrier function used to restrict the concept value into a specific range, usually in the range [0,1].

In each step, a new state of the concepts is derived according to (1) and after a number of iterations, the FCM may arrive in one of the three states (Kosko 1986) (i) fixed equilibrium point, (ii) limited cycle and (iii) chaotic behavior. When the FCM arrives in a fixed equilibrium point, we can conclude that the FCM has converged and the final state corresponds to the actual system state in which the system concludes, when the initial values of concepts are applied.

Besides the use of FCMs in modeling complex physical systems, their application as part of modern DSS (Decision Support Systems) in decision making, significantly increases through the years (Khan and Quaddus 2004, Xirogiannis et al. 2004a, 2008, Pajares et al. 2006). This motivated the authors (Papakostas et al. 2006, 2008) to investigate the behavior and performance of the FCMs in classifying patterns belonging to pattern recognition problems, under several configurations. The main principles as long as the latest results of this investigation, constitute the subject of the following sections.

2 FCM-Based Classifiers

A crucial part of any modern intelligent system, which learns from its environment and interacts with it, is a pattern recognition process. In general, a pattern recognition process employs four stages: (1) data acquisition (2) data preprocessing (denoising, filtering, etc) (3) feature extraction and finally (4) classification. Among these processing steps, the last one significantly affects the overall performance of the system, comprising the stage where knowledge about the problem in process is stored and takes the decision by classifying the patterns in proper classes. The main functionality of a typical classifier is the mapping of an input set to a specific output one, according to the internal representation of the knowledge stored inside the classifier's structure.

Representative type of commonly used classifiers is the Artificial Neural Networks (ANNs), in which the need of knowledge storing and knowledge based decision have been inspired by the neural networks of a human's brain (Haykin 1999). Due to their knowledge storage capability, neural networks are able to be used for pattern recognition tasks and classification problems, while their ability to repeatedly learn their internal representation makes them very useful to real-time image and signal processing applications. The knowledge in a neural network is stored during the training phase, when its weights take appropriate values, according to the problem being processed. However, since the dynamics of a neural network is still under investigation, one have to make some appropriate trial and error tests in order to decide the optimal architecture of the network to be used, for a specific application.

Due to the fact that the structure of the FCMs resembles this of ANNs, many concepts and procedures from the field of ANNs have been adopted and applied in the case of FCMs. However, while ANNs have been successfully used to classify patterns, the behavior of the FCMs in pattern recognition applications is still unexplored.

Since the major advantage of Fuzzy Cognitive Maps is their ability to describe the behaviour of complex systems, they could be used to model a classifier having appropriate input and output concepts and appropriate interconnections among them. Based on this prospect, the inference procedure of the FCM corresponds to the internal computations the classifier performs in order to map the input data to specific output, to an equilibrium point, according to the stored knowledge.

In order to describe the functionality of the FCM-based classifier in terms of FCMs the following definition has to be declared.

Definition 1. *The Fuzzy Cognitive Map structure used to map a set of input concepts (ICs) to a set of output concepts (OCs), through the execution of the inference formula (1), until an equilibrium point is reached, is called Fuzzy Cognitive Mapper (FCMper).*

The operation of the Fuzzy Cognitive Mapper as a modeling methodology of the dynamics of a classifier can be defined as follows,

Definition 2. *The values of the output concepts (OCs) of the state in which a Fuzzy Cognitive Mapper equilibrates (equilibrium point), when appropriate values of the input concepts (ICs) are applied, correspond to the decision of the classifier being modeled by the FCMper.*

According to the above definitions, a FCM can be used to model the behavior of a classifier, in order to classify patters belonging to specific classes, and its performance depends on several parameters. Some of these parameters are the structure of the FCMper, the nodes' type of activation function, the inference law used to equilibrate the classifier and the learning algorithm used to find the interconnection weights. These factors that affect the performance of a FCMper and determine its classification capabilities are discussed next.

2.1 FCMper – Structure

The first factor that plays important role on the performance of a FCMper, as in the case of ANNs is its structure, which defines the way the input nodes are connected with the output ones. A typical FCMper has the following form (Papakostas et al. 2006, 2008).

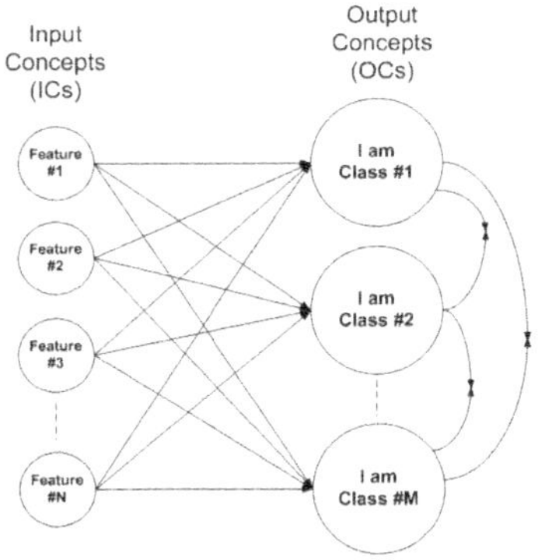

Fig. 1 A typical FCMper – FCMper1

In the above classifier (FCMper1), the input concepts correspond to the features that uniquely describe the patterns and formed by a previous processing step called Feature Extraction Method (FEM), while the output concepts are the classes' labels where the patterns belong. The FCMper1 of Fig.1 is the simplest form of a FCM-based classifier, which present limited performance as compared to the traditional neural classifiers (Papakostas et al. 2008).

More sophisticated hybrid FMCpers structures seem to perform better (Papakostas et al. 2008) in comparison to the previous FCMper1, where one neural classifier and a FCMper1 are operating in a cascade topology, increasing the classification accuracy of the hybrid model. These hybrid structures are illustrated in the following Fig.2 and Fig.3.

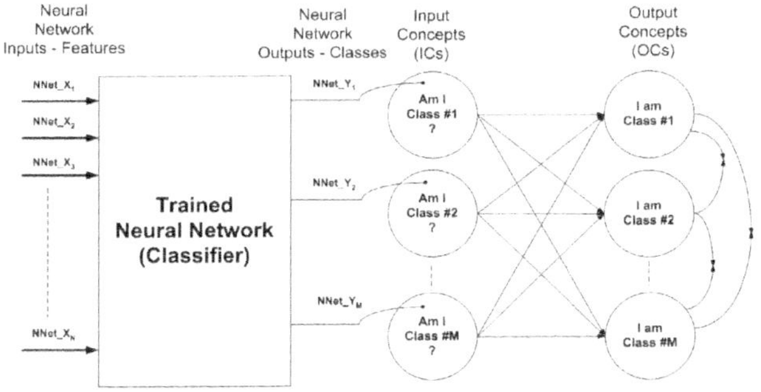

Fig. 2 Hybrid FCMper – FCMper2

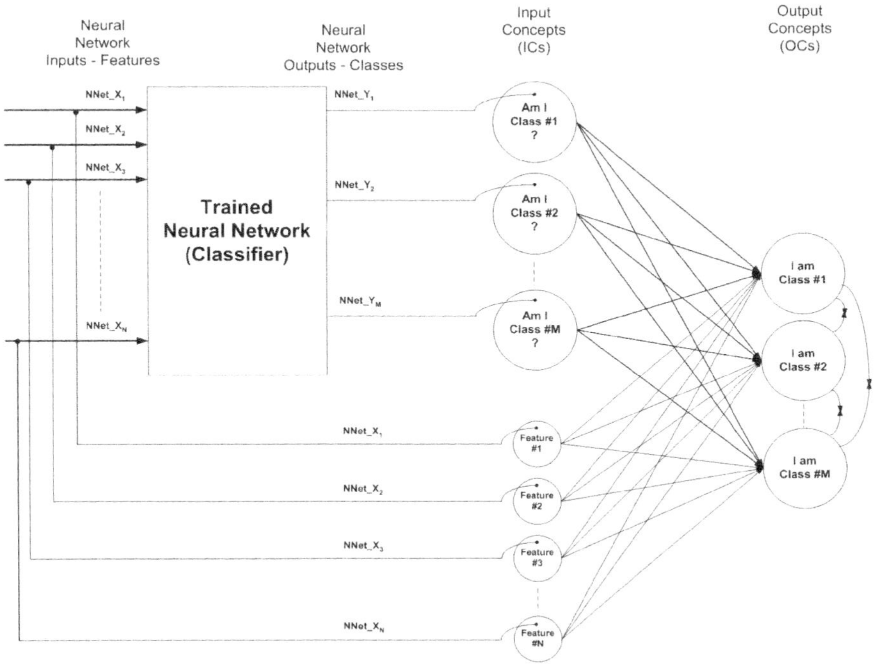

Fig. 3 Hybrid FCMper – FCMper3

The hybrid FCM-based classifiers FCMper2 and FCMper3, take as inputs the problem features (the same as the neural classifier inputs) and the features with the corresponding neural classifier responses respectively and answer to the question which is the appropriate class they belong.

While these hybrid models proved to be superior to the simple FCMper (Papakostas et al. 2008), their complexity make them inappropriate for real-time applications, where a decision about the class belonging of a pattern, has to be obtained in a certain time. This fact makes the usage of simple structures such as the FCMper1 more useful for these applications and therefore new topologies that improve its classification performance need to be developed.

2.2 *FCMper – Activation Function*

The function used to squash the overall impact of the nodes to a specific node, does not play only the role of a barrier function, but also gives a nonlinearity nature in the reasoning mechanism of a FCMper. This nonlinear behavior of a node is of significant importance in the case of difficult classification problems where the decision boundaries consisting by complex hyperplanes of multiple dimensions.

Some interesting conclusions on the affect of the activation functions on the operation of a FCM has been reported in (Koulouriotis et al. 2001c, 2001d, Tsadiras 2008, Bueno and Salmeron 2009, Boutalis et al. 2009), however this affection on the performance of a FCMper is unknown. For this reason three popular functions are selected as activation functions of the FCMper1 nodes, and their classification capabilities in several pattern recognition benchmark problems are studied in the experimental section. The analytical form of these functions which are the *Step, Trivalent* and *Sigmoid* ones are presented in the following equations.

Step Function (F1)
$$f(x) = \begin{cases} 0, & x \leq 0 \\ 1, & x > 0 \end{cases} \tag{2}$$

Trivalent Function (F2)
$$f(x) = \begin{cases} -1, x \leq -0{,}5 \\ 0, -0{,}5 < x < 0{,}5 \\ 1, x \geq 0{,}5 \end{cases} \tag{3}$$

Sigmoid Function (F3)
$$f(x) = \frac{1}{1 + e^{-Ax}} \tag{4}$$

The parameter A in the case of sigmoid function is a parameter that controls the slope of the curve and in the special case where this is equal to 1, the resulted sigmoid functions corresponds to the *logistic sigmoid* function.

2.3 FCMper – Inference Law

The operation of a FCM is based on the repetitive application of a specific inference law, which by taking into account the total impacts of the concepts directly connected to this one, gives the next concept's value, in each iteration. Equation (1) describes an inference law widely used in many FCM applications, but it is not the only one.

Generally, there are four different inference laws regarding the counting or not of the past concept's value or the existence of the self-connection link for each concept node. These inference laws are described in the following formulas and their behavior in classifying patterns by using FCM-based classifiers will analyzed in a next section.

Past & No Self-Connection (Inf1)
$$A_i^{t+1} = f\left(A_i^t + \sum_{i=1, i \neq j}^{N} W_{ji} A_j^t \right) \tag{5}$$

No Self-Connection (Inf2)
$$A_i^{t+1} = f\left(\sum_{i=1, i \neq j}^{N} W_{ji} A_j^t \right) \tag{6}$$

Past & Self-Connection (Inf3)
$$A_i^{t+1} = f\left(A_i^t + \sum_{i=1}^{N} W_{ji} A_j^t \right) \tag{7}$$

Self-Connection(Inf4)

$$A_i^{t+1} = f\left(\sum_{i=1}^{N} W_{ji} A_j^t\right) \tag{8}$$

While in the case of using FCMs in modeling complex systems where the concepts correspond to specific property of the system the selection of the appropriate inference law may be an easy task, in the case of the FCMpers this selection is not straightforward. For this reason, an additional procedure of finding the best inference law that yields better classification rates has to be proceed, as part of the overall calibration stage of the system.

2.4 FCMper – Learning Algorithm

In the early years, the interconnection weights of FCMs are decided by a group of experts which knew the relations in force between the concepts, since FCMs were used to model the behavior of a system, as far as its equilibrium from a certain initial state, is concerned. However, the need to equilibrate a system to a specific final state motivated the scientists to enhance the operation of the FCMs by incorporating an additional procedure of finding suitable weight sets called *learning* in conjunction to the similar operation existing in the ANNs training.

Many researchers in the recent years have tried to enhance the functionality of the FCMs by developing learning algorithms, which find the appropriate set of weight interconnections between the concepts. The proposed algorithms are divided into two types, these which are making use of gradient information and are called *gradient-based algorithms* (Papageorgiou et al. 2003, 2004) and those which are based on evolutionary mechanisms (Koulouriotis et al. 2001a, Stach et al. 2005, 2008, Papageorgiou et al. 2005, Ghazanfari et al. 2007). The former type algorithms have the disadvantage of trapping to local optimum by finding a solution far away from the theoretical optimal, while the latter one give more chances to find the best weight set, due to their stochastic nature.

Until now, the algorithms of both kinds search a weight set which can lead the FCM from a specific initial state to a desired equilibrium point. However, in the case of the FCM-based classifiers the learning algorithm should work quite differently. The learning algorithm for the case of FCMpers, have to find a common weight set, which for initial states of the input concepts that belong to the same class, the FCMpers equilibrate to the same points.

This learning procedure defines some new operational notions for the FCM structures, coping with their ability to form internal representations able to lead them to certain equilibrium points. Moreover, this need highlights the significance of storing enough knowledge to distinct the patterns presented as input concepts and also marks the importance to conduct the capacity of the models.

3 Experimental Study

The experiments are organized in two sections where in the first one the influence to the classifier's performance of the type of activation function and the inference law are investigated. In a second phase the classification capabilities of the FCMpers are compared to those of the traditional ANNs in difficult pattern recognition applications.

3.1 Experimental Part 1

For the needs of the first part experiments a set of well known benchmark pattern recognition datasets widely used in the literature are selected from the UCI repository (UCI-Machine Learning Repository) and their properties are summarized in Table 1.

Table 1 Properties of the datasets used in the experiments

Dataset	Number of Classes	Attributes	Instances
Iris Flowers	3	4	150
Pima Indians Diabetes	2	8	768
Wine	3	13	178
Glass Identification	7	9	214
Hepatitis	2	19	155
Echocardiogram	2	11	132
Breast Cancer Wisconsin	2	9	699
Parkinson's	2	22	195
Mammographic Mass	2	5	961

All the experiments are carried out by using a SGA (Simple Genetic Algorithm) as learning algorithm, without having any advanced genetic operator for preserving the diversity in high levels. This is the simplest form of an evolutionary algorithm, where its simplicity allows its application without any specific calibration procedure.

The MSE (Mean Squared Error) between the desired and the actual values of the FCMper1 output concepts is used as objective function, which has to be minimized and its analytical form is as follows.

$$MSE = \frac{1}{MN}\sum_{i=1}^{M}\sum_{j=1}^{N}\left(OC_{ij}^{Actual} - OC_{ij}^{Desired}\right)^2 \tag{9}$$

where M is the number of training patterns, N the number of the FCMper output concepts and ($OC_{ij}^{Actual} - OC_{ij}^{Desired}$) the difference between the ith output concept and its corresponding desired value (target), when the jth set of input concepts appears to the FCMper's input.

Each experiment is executed 20 times in order to ensure the statistical accuracy of the outcomes and the corresponding mean results for the case of the first three datasets of Table 1 are summarized in the following Table 2, 3 and 4.

In these tables, the activation functions F4 corresponds to the sigmoid function having slope factor different than 1, which the SGA founds during the learning process the same for each node and the function F5 is the same with F4 but in this case each node has its own slope factor.

The performance measures used to study the classification performance of each combination of activation function and inference law, is the MSE measured during the learning procedure, the PMSE (Performance-MSE) which is the MSE measured by using a small fraction of the dataset called testing set (all the datasets are divided into the training 75% and testing 25% sets) and CR the *Classification Rate* defined as the percentage of the correctly classified patterns over the entire testing set.

Table 2 Classification statistics for the case of Iris Flowers dataset

Activation Function	Inference Law	MSE	PMSE	CR(%)
F1	Inf1	0.1730	0.1866	51.78
	Inf2	0.1340	0.1504	57.33
	Inf3	0.1797	0.1926	50.88
	Inf4	0.1314	0.1474	58.88
F2	Inf1	0.1362	0.1563	55.33
	Inf2	0.1384	0.1518	58.66
	Inf3	0.1511	0.1660	54.89
	Inf4	0.1273	0.1444	61.78
F3	Inf1	0.1690	0.1694	66.67
	Inf2	0.1696	0.1701	67.55
	Inf3	0.1712	0.1718	66.89
	Inf4	0.1733	0.1739	66.67
F4	Inf1	0.1078	0.1099	83.33
	Inf2	0.1043	0.1070	82.67
	Inf3	0.1080	0.1104	86.22
	Inf4	0.1097	0.1120	83.78
F5	Inf1	0.1129	0.1140	80.89
	Inf2	**0.1067**	**0.1083**	**89.11**
	Inf3	0.1069	0.1088	88.44
	Inf4	0.1058	0.1081	88.22

Table 3 Classification statistics for the case of Pima Indians Diabetes dataset

Activation Function	Inference Law	MSE	PMSE	CR(%)
F1	Inf1	0.3178	0.3236	56.07
	Inf2	0.3242	0.3238	58.60
	Inf3	0.3229	0.3340	56.11
	Inf4	0.3202	0.3312	55.68

Table 3 (*continued*)

F2	Inf1	0.3049	0.3218	52.44
	Inf2	0.3095	0.3207	54.06
	Inf3	0.3044	0.3244	49.65
	Inf4	0.3017	0.3275	48.43
F3	Inf1	0.2202	0.2169	67.07
	Inf2	0.2197	0.2160	66.77
	Inf3	0.2200	0.2165	67.07
	Inf4	0.2204	0.2174	67.16
F4	**Inf1**	**0.2105**	**0.2062**	**70.61**
	Inf2	0.2113	0.2078	69.74
	Inf3	0.2099	0.2051	69.17
	Inf4	0.2100	0.2071	70.35
F5	Inf1	0.2125	0.2080	70.48
	Inf2	0.2142	0.2102	69.48
	Inf3	0.2116	0.2078	69.43
	Inf4	0.2157	0.2125	69.17

Table 4 Classification statistics for the case of Wine dataset

Activation Function	**Inference Law**	**MSE**	**PMSE**	**CR(%)**
F1	Inf1	0.2403	0.2963	32.96
	Inf2	0.2159	0.2697	33.52
	Inf3	0.2449	0.2901	34.81
	Inf4	0.2645	0.3098	28.33
F2	Inf1	0.3126	0.3808	20.18
	Inf2	0.2325	0.3389	23.52
	Inf3	0.2763	0.3531	33.70
	Inf4	0.3188	0.3605	24.26
F3	**Inf1**	**0.1600**	**0.1600**	**81.67**
	Inf2	0.1541	0.1537	80.74
	Inf3	0.1551	0.1562	80.18
	Inf4	0.1601	0.1624	77.78
F4	Inf1	0.1007	0.1230	76.48
	Inf2	0.0888	0.1064	82.59
	Inf3	0.1145	0.1310	77.22
	Inf4	0.0941	0.1184	78.15
F5	Inf1	0.1225	0.1404	77.22
	Inf2	0.1078	0.1322	76.30
	Inf3	0.1118	0.1292	73.70
	Inf4	0.1225	0.1324	79.81

From the above tables, it is obvious that the Step and Trivalent functions are perform poorly, since they classify wrongly the patters, by giving an average classification rate of 55% for the case of the first two datasets. This fact makes them inappropriate for pattern recognition applications, while their reasoning capabilities are restricted in special cases of the FCM modeling (Tsadiras 2008).

As far as the affection of the inference law is concerned, what is making clear from the above results is that the self-feedback of each node does not play a crucial role in the overall operation of the FCMpe1.

Based on these observations it is useful to study the classification performance of the entire benchmark set only for the cases of F3, F4, F5 with Inf1 and Inf2 as inference laws. The mean classification results are illustrated in the following Table 5.

Table 5 Mean classification statistics for all benchmarks of Table 1

Activation Function	**Inference Law**	**MSE**	**PMSE**	**CR(%)**
F3	Inf1	0.1551	0.1655	72.33
	Inf2	0.1551	0.1643	72.77
F4	**Inf1**	**0.1252**	**0.1521**	**74.68**
	Inf2	**0.1233**	**0.1500**	**74.68**
F5	Inf1	0.1279	0.1558	73.27
	Inf2	0.1259	0.1534	73.91

Table 5, shows that the sigmoid function improves the classification capability of the FCMper1 and in its adaptive form where the slope factor is decided during the learning procedure gives the highest score. Moreover, in the case of using the sigmoid activation function, the type of the used inference law does not play an important role, something which is justified by the fact that the nonlinearity entered into the model, further improves its reasoning operation by making the influence of the inference law less significant.

Consequently, what is concluded by this experimental study is that, the FCM-based classifier the FCMper1, needs a nonlinear activation function in order to construct its internal representations of the patterns consisting the problem. Also, its performance significantly improved when the inference law does not consider self-feedback interconnections, while the past node's value is not important to the overall classification accuracy.

3.2 *Experimental Part II*

In this section, the most high performance configuration of the FCMper1, along with the two hybrid FCM-based classifiers FCMper2 and FCMper3, are used to classify patterns belonging to complex pattern recognition problems, one artificially generated and one real robotic vision application. Their classification performance is compared with that of conventional neural networks of similar structure.

The selected configuration for the FCM part of these classifiers includes the F4 activation function in conjunction with the Inf1 inference law, since with this settings the FCMper1 present good generalization behavior.

Moreover, the experiments' configuration is the same with the previous ones and the neural networks are trained by using the standard *backpropagation* algorithm (Haykin 1999), for all simulations.

3.2.1 The Two Spiral Problem

The two spiral problem is an extremely hard problem for neural networks to solve. The goal of this problem is to learn to discriminate between two sets of training points, which lie on two distinct spirals in the x-y plane (Langlet et al. 2001). Table 6, summarizes the classification performance of a typical multilayer perceptron neural network with 2 inputs, 10 hidden neurons in one hidden layer and 2 outputs, in comparison to the corresponding one of the FCMper structures, through appropriate indices.

Table 6 Classification Statistics in the case of the Two Spiral problem

	Two Spiral			
	Neural Network	***FCM per1***	***FCM per2***	***FCM per3***
Structure	*2-10-2*	*2-2*	*2-2*	*4-2*
MSE	0.1505	0.2485	0.1768	0.1389
PMSE	0.1389	0.2459	0.3158	0.1349
CR	75.27%	56.56%	60.22%	**79.67 %**
Stdv(MSE)	0.0922	0.0012	0.0990	0.1093

The experimental pattern set is consisted of 100 training and 50 testing patterns. In the above table *Stdv* is the *Standard Deviation* of the MSE, as an index of the computational stability of the learning algorithms.

The results presented in Table 6, show that the FCMper3 (last column in bold face), outperforms the other FCMpers, even the neural classifier. More precisely, the FCMper3 having smaller structure (only 16 adjustable parameters) than neural network (52 weights and biases), gives better classification rate of almost 80%, with smaller training error.

3.2.2 A Robotic Vision System

In this experiment a typical classification application, where the Zernike moments are used as discriminative features of the objects be classified, is taken place. Initially some test objects (patterns) are selected. Figure 4b shows a wooden pyramidal puzzle, which is used for robot vision tasks in the Control Systems Lab of DUTH (Democritus University of THrace) (Papakostas et al. 2005). The 9 parts of the puzzle, placed in arbitrary positions, are shown in Figure 4a. The (256x256 pixels) images of these parts are the initial nine patterns of our experiments.

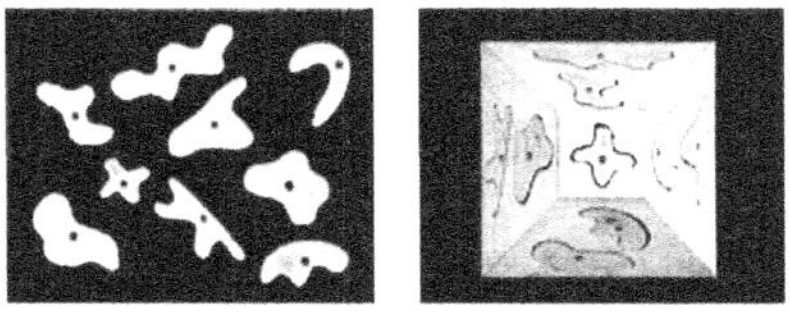

(a) (b)

Fig. 4 The nine work pieces that are placed (a) in arbitrary positions on the table and (b) on a 3-D truncated pyramid

The pattern dataset consists of a 594 images from which 324 are used during the training procedure and the rest 270 are used to test the generalization abilities of the structures being compared. More information about the way the dataset is formed can be found in (Papakostas et al. 2005).

For experimental purposes, the first 12 Zernike Moments of each image are computed and used as features to distinguish the 9 parts. From this set the first two moments are omitted because they are constant for all objects and thus they do not constitute discriminative quantities.

After a typical cross-validation procedure (Haykin 1999) it was decided, the neural classifier to be a typical multilayer perceptron containing 1 hidden layer, with 10 hidden neurons and 9 (equal to the number of classes) output neurons.

This problem differs from the previous ones, since the data set has been formed directly from a real environment, the captured images of the objects. The experimental results of this classification problem are presented in Table 7.

Table 7 Classification Statistics in the case of the robotic vision system

Robotic Vision System				
	Neural Network	*FCM per1*	*FCM per2*	*FCM per3*
Structure	*10-10-9*	*10-9*	*9-9*	*19-9*
MSE	0.0849	0.2415	0.1976	0.0739
PMSE	0.0741	0.2390	0.3395	0.0723
CR	87%	56.2%	58.14%	**89.33%**
Stdv(MSE)	0.0863	0.0150	0.1309	0.0894

The superiority of the FCMper3, is still preserved in this real problem, as it can be seen by Table 7.

4 Conclusions – Discussion

A novel application of Fuzzy Cognitive Maps was presented in this chapter. The appropriate principles that define the basic operation of the FCMs as pattern classifiers were demonstrated and the main functional properties of these FCM-based classifiers were discussed. A FCM-based classifier have the advantage to give the capability to adjust its operation by selecting the appropriate set of free parameters, such as the activation function, the inference law, the structure's connectionist in order to take the highest efficiency depending on the specific problem.

The newly introduced classifiers present high classification capabilities even in their simple form of the FCMper1 and they outperform the traditional neural classifiers when they work in more complicated hybrid forms.

While this study constitutes a first attempt to evolve the FCMs in a way they can be used as classifier modules in modern pattern recognition systems, new directions are raised, regarding the increase of the knowledge capacity by adding dummy concepts and finding activation functions or inference laws, that need more research in the name of high classification performance.

References

Boutalis, Y., Kottas, T.L., Christodoulou, M.: Adaptive etimation of fuzzy cognitive maps with proven stability and parameter convergence. IEEE Transactions on Fuzzy Systems 17, 874–889 (2009)

Bueno, S., Salmeron, J.L.: Benchmarking main activation functions in fuzzy cognitive maps. Expert Systems with Applications 36, 5221–5229 (2009)

Chytas, P., Glykas, M., Staikouras, C., Valiris, G.: Performance measurement in a greek financial institute using the balanced scorecard. Journal of Measuring Business Excellence 10, 87–98 (2006)

Ghazanfari, M., Alizadeh, S., Fathian, M., Koulouriotis, D.E.: Comparing simulated annealing and genetic algorithm in learning FCM. Applied Mathematics and Computation 192, 56–68 (2007)

Glykas, M., Xirogiannis, G.: A soft knowledge modeling approach for geographically dispersed financial organizations. Soft Computing 9, 579–593 (2004)

Haykin, S.: Neural Networks – a comprehensive foundation, 2nd edn. Prentice Hall, Englewood Cliffs (1999)

Khan, M.S., Quaddus, M.: Group decision support using fuzzy cognitive maps for causal reasoning. Group Decision and Negotiation 13, 463–480 (2004)

Kosko, B.: Fuzzy Cognitive Maps. International Journal Man-Machine Studies 24, 65–75 (1986)

Koulouriotis, D.E., Diakoulakis, I.E., Emiris, D.M.: Learning fuzzy cognitive maps using evolution strategies: a novel schema for modeling and simulating high-level behavior. IEEE Congr. Evolutionary Computation, 364–371 (2001a)

Koulouriotis, D.E., Diakoulakis, I.E., Emiris, D.M.: A fuzzy cognitive map-based stock market model: synthesis, analysis and experimental results. In: IEEE International Conference on Fuzzy Systems (FUZZ 2001), vol. 1, pp. 465–468 (2001b)

Koulouriotis, D.E., Diakoulakis, I.E., Emiris, D.M.: Realism in fuzzy cognitive maps: incorporating synergies and conditional effects. In: IEEE International Conference on Fuzzy Systems (FUZZ 2001), vol. 3, pp. 1179–1182 (2001c)

Koulouriotis, D.E., Diakoulakis, I.E., Emiris, D.M.: Anamorphosis of fuzzy cognitive maps for operation in ambiguous and multi-stimulus real world environments. In: IEEE International Conference on Fuzzy Systems (FUZZ 2001), vol. 3, pp. 1156–1159 (2001d)

Koulouriotis, D.E., Diakoulakis, I.E., Emiris, D.M., Antonidakis, E.N., et al.: Efficiently modeling and controlling complex dynamic systems using evolutionary fuzzy cognitive maps. International Journal of Computational Cognition 1, 41–65 (2003) (invited Paper)

Koulouriotis, D.E.: Investment analysis and decision making in markets using adaptive fuzzy causal relationships. Operational Research: an International Journal 4, 213–233 (2004)

Koulouriotis, D.E., Diakoulakis, I.E., Emiris, D.M., Zopounidis, C.D.: Development of dynamic cognitive networks as complex systems approximators: validation in financial time series. Applied Soft Computing 5, 157–179 (2005)

Langlet, F., Abdulkader, H., Roviras, D., Mallet, A., et al.: Comparison of neural network adaptive predistortion techniques for satellite down links. In: Proceedings of International Joint Conference on Neural Networks, pp. 709–714 (2001)

Pajares, G., de la Cruz, J.M.: Fuzzy cognitive maps for stereovision matching. Pattern Recognition 39, 2101–2114 (2006)

Papageorgiou, E.I., Stylios, C.D., Groumpos, P.P.: Fuzzy cognitive map learning based on nonlinear Hebbian rule. In: Australian Conference on Artificial Intelligence, pp. 256–268 (2003)

Papageorgiou, E.I., Stylios, C.D., Groumpos, P.P.: Active Hebbian learning algorithm to train fuzzy cognitive maps. International Journal of Approximate Reasoning 37, 219–249 (2004)

Papageorgiou, E.I., Parsopoulos, K.E., Stylios, C.S., Groumpos, P.P., et al.: Fuzzy cognitive maps learning using particle swarm intelligence. Journal of Intelligence Information Systems 25, 95–121 (2005)

Papakostas, G.A., Karras, D.A., Mertzios, B.G., Boutalis, Y.S.: An efficient feature extraction methodology for computer vision applications using wavelet compressed Zernike moments. ICGST International Journal on Graphics, Vision and Image Processing, Special Issue: Wavelets and Their Applications SI1, 5–15 (2005)

Papakostas, G.A., Boutalis, Y.S., Koulouriotis, D.E., Mertzios, B.G.: A first study of pattern recognition by using fuzzy cognitive maps. In: 13th International Workshop on Systems, Signals and Image Processing (IWSSIP 2006), pp. 369–374 (2006)

Papakostas, G.A., Boutalis, Y.S., Koulouriotis, D.E., Mertzios, B.G.: Fuzzy cognitive maps for pattern recognition applications. International Journal of Pattern Recognition and Artificial Intelligence 22, 1461–1468 (2008)

Stach, W., Kurgan, L., Pedrycz, W., Reformar, M.: Genetic learning of fuzzy cognitive Maps. Fuzzy Sets and Systems 153, 371–401 (2005)

Stach, W., Kurgan, L.A., Pedrycz, W.: Numerical and linguistic prediction of time series with the use of Fuzzy Cognitive Maps. IEEE Transactions on Fuzzy Systems 16, 61–72 (2008)

Tsadiras, A.K.: Comparing the inference capabilities of Binary, Trivalent and Sigmoid fuzzy cognitive maps. Information Sciences 178, 3880–3894 (2008)

UCI-Machine Learning Repository, `http://archive.ics.uci.edu/ml/datasets.html`

Xirogiannis, G., Stefanou, J., Glykas, M.: A fuzzy cognitive map approach to support urban design. Expert Systems with Applications 26, 257–268 (2004a)
Xirogiannis, G., Glykas, M.: Fuzzy cognitive maps in business analysis and performance driven change. IEEE Transactions on Engineering Management 51, 334–351 (2004b)
Xirogiannis, G., Glykas, M.: Intelligent modeling of e-business maturity. Expert Systems with Applications 32, 687–702 (2007)
Xirogiannis, G., Chytas, P., Glykas, M., Valiris, G.: Intelligent impact assessment of HRM to the shareholder value. Expert Systems with Applications 35, 2017–2031 (2008)

Dynamic Fuzzy Cognitive Maps for the Supervision of Multiagent Systems

Aguilar Jose

Abstract. In this work we propose a Dynamical Fuzzy Cognitive Map (DFCM) where its casual relationships are based on fuzzy rules, in a way that the structure of the map changes during the phase of execution (runtime). We propose the modification of the values of the relationships between the concepts through of fuzzy rules derived from the concept states that represent the system modeled by the map. Our DFCM is ideal to build supervision systems for multiagent systems (MAS), in order to study the behavior of the agents community when they fail, use a lot of resource, etc. In this paper, the DFCM is used to build a supervision system for a faults management system based on multiagent systems. Very good results were obtained, demonstrating that the use of these maps as supervisor of multiagent systems is good and reliable.

1 Introduction

In [2, 4] Kosko introduces the Fuzzy Cognitive Maps (FCM) based on the Cognitive Maps of Axelrod [1]. The Fuzzy Cognitive Maps are a tool of causal representation, that are composed by concepts and causal relationships between the concepts, which use the theory of fuzzy logic to describe their structure and to infer answers of the map from input data. In this way, they represent the causal relationship between concepts and analyze inference patterns from a given input to the map. For the design of the structure of these maps, we can use the knowledge of the experts or the historical data of the phenomenon to model, being able to create models of complex systems for which an exact mathematical model is not possible.

These maps have been used for the strategic planning and the analysis of the behavior of the automobile industrial market [4, 7]. Other authors have proposed

Aguilar Jose
CEMISID, Departamento de Computación,
Facultad de Ingeniería, Universidad de Los Andes
Núcleo la Hechicera, Mérida 5101-Venezuela
e-mail: aguilar@ula.ve

M. Glykas (Ed.): Fuzzy Cognitive Maps, STUDFUZZ 247, pp. 307–324, 2010.
springerlink.com © Springer-Verlag Berlin Heidelberg 2010

the use of the FCM to model supervision systems of complex control systems [8, 11] or to control plants [8, 17]. The FCM have been used also to model the behavior of systems in several areas like in political sciences [3, 7, 9], in medical diagnostics [21, 28], failure modes effects analysis and detection [10], software development modeling project [14], for the analysis and decision making [13, 15, 18], for the coordination of cooperative distributed agents [16], and many others. Other works have proposed learning approaches for FCM [19, 22], some of them based on the Hebb rule [20, 21, 26], Genetic Algorithms [23, 25], fuzzy rules [12], among others. Also, new FCM based in neural models have been proposed [5, 7].

An interesting domain of application of the FCM is in the area of supervision of MAS, that is, to study the behavior of a community of agents from initial hypothesis. In this way, we can determine and predict the individual behaviors of the agents according to this hypothesis. That can be very important to improve the design of the MAS, feedback the MAS for its self-organization, etc.

A first version of a dynamic approach of FCM, based on the random neural model, was introduced by Jose Aguilar in [6] (called DRFCM), to allow that the maps change their relationship during their runtime, of such form to be able to model systems of greater complexity than those than can model traditional FCMs. In that approach, random functions were used to provide the dynamic capacity within the maps, based on the random neural networks. In this work, we propose that the dynamics in the DFCM are giving by the modification of the values of the relationship between the different concepts that compose the map, through fuzzy sets derived from concept states of the system modeled by the map. This approach is ideal to build supervision systems for multiagent systems, in order to predict their behaviors when their agents fail, etc. We tested our approach in tasks of Supervision of Multiagent Systems for Faults Management.

This work is organized as follows. In section 2 is presented our proposition of FCM (the DFCM). Section 3 presents the case of study, where a DFCM is used like supervisor of a Faults Management System based on agents. Finally, the conclusions a further works are presented.

2 Dynamical Fuzzy Cognitive Maps

DRFCM are FCM based on the Random Neural Model, where the causal relationships are dynamics [6]. That is, the values of the arcs could be modified during the runtime of the FCM to adapt them to the new environment conditions. The quantitative concepts allowed us develop a feedback mechanism that was included in the causal model to update the arcs. In this way, with the DRFCM we could consider on-line adaptive procedures of the model like real situations. For example, our DRFCM could structure virtual worlds that change with time. The DRFCM does not write down differential equations to change the virtual world. It maps input states to limit-cycle equilibrium. A limit cycle repeats a sequence of events or a chain of actions and responses. Additionally, our DRFCM changed their fuzzy causal web during the runtime using neural learning laws. In this way, our model could learn new patterns and reinforce old ones.

In the DFCM we are going to use fuzzy sets adapted to the system modeled to obtain the dynamics of the causal relationship, unlike the proposition in [6], where a generic form of dynamics of the relationships was defined based on neural learning laws. With this approach we obtain a better adaptation of the FCM to the real system that it models.

2.1 The Causal Relationships Defined as Fuzzy Rules

The relationships are established using a series of fuzzy rules of the type "if the concept t is in C_t then the casual relationship with concept i is w_{ti}", where C_t is one of the possible states of the concept t and w_{ti} will be the value of the casual relationships for this state. In this way, a set of rules defining the value of the relationship is used to determine the relationship value between two concepts.

To define the set of rules, we define a general procedure. For instance, we assume that the state of the concepts in the modeled multiagent system can be located in three zones, according to the following figure:

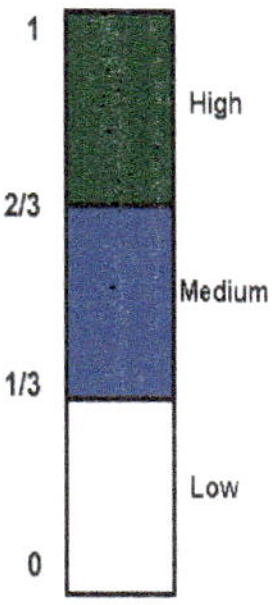

Fig. 1 Identification zones of each concept's state

Using this figure, we can define the state of a concept like:

- A concept has a *high state* (between 2/3 and 1) when it works correctly and contributes substantially with the functioning of the modeled system.
- A concept has a *medium state* (between 1/3 and 2/3) when its functioning must be validated and its contributions to the systems' functioning is not so substantial.
- A concept has a *low state* (between 0 and 1/3) when it does not work and it does not contribute to the functioning of the system.

The state can be defined as a fuzzy variable composed by three fuzzy sets: high, medium and low. Additionally, the possible types of relationships between concepts can be defined like:

Table 1 Possible types of relationships between concepts

Value	Linguistic Variable
1.00	Complete$^+$
0.75	High$^+$
0.50	Medium$^+$
0.25	Low$^+$
0.00	Null
-0.25	Low$^-$
-0.50	Medium$^-$
-0.75	High$^-$
-1.00	Complete$^-$

Also, the type of relationships can be defined as a fuzzy variable composed by nine fuzzy sets: Complete$^+$, High$^+$, etc. Now, we can define the following set of generic fuzzy rules using the concept states and the possible types of relationships defined previously, to define the casual relationships between concepts:

- If the preceding concept is **High** and the consequent one is also **High** then the relationship is **Complete**$^+$(1.0).
- If the preceding concept is **High** and the consequent one is **Medium** then the relationship is **High**$^+$ (0.75).
- If the preceding concept is **High** and the consequent one is **Low** then the relationship is **Low**$^+$ (0.25).
- If the preceding concept is **Medium** and the consequent one is **High** then the relationship is **High**$^+$ (0.75).
- If the preceding concept is **Medium** and the consequent one is **Medium** then the relationship is **Medium**$^-$ (-0.5).
- If the preceding concept is **Medium** and the consequent one is **Low** then the relationship is **High**$^-$ (-0.75).
- If the preceding concept is **Low** and the consequent one is **High** then the relationship is **High**$^-$ (-0.75).
- If the preceding concept is **Low** and the consequent one is **Medium** then the relationship is **Medium**$^-$ (-0.5).
- If the preceding concept is **Low** and the consequent one is **Low** then the relationship is **Complete**$^-$ (-1.0).

The set of generic fuzzy rules follows an adaptation mechanism similar to the hebb learning rule. These rules would be used to determine all the relationships between the different concepts. Thus, every relationship would be determined under the same set of rules, but each one would have a weight defined by the experts that could vary from relationship to relationship. For example, if we take the relationship between Concept 1 and Concept 2, and we assume that Concept 2 has a High state and that Concept 1 has a Medium state, then the relationship resulting from the rules would yield a high$^+$ value (that is, 0.75, if we suppose that we use crisp variables). This value is multiplied by the weight of the relationship defined by the expert (in this example we assume that the weight of this relationship is 0.5, then the final value of the relationship would be 0.375 (see figure 2)).

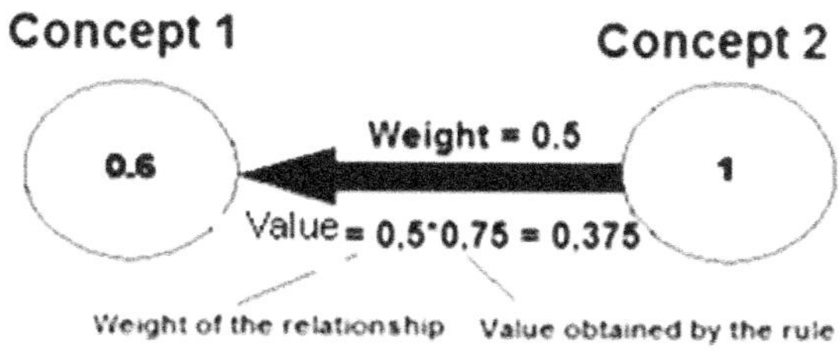

Fig. 2 Example to establish the relationships value between concepts

We can establish equivalence among a concept state and an agent state. In our approach, each agent of a multiagent system is a concept, and the weight of the relationship is defined by the expert according to the relationship that it defines between the agents on the supervised multiagent system.

2.2 General Algorithm of a DFCM

The design of a DFCM is very similar to the design of a DRFCM, which means we need to define the concepts (in or case, the agents) and the initial causal relationships among them. In our case, the initial causal relationships are defined according to the relationships of the agents on the agents' community. Once the DFCM has been designed, the execution algorithm is:

1. Define the Initial agent states $C^0 = \lfloor C_0^0, C_1^0, \ldots, C_n^0 \rfloor$
2. While the system does not converge

   a. Calculate the values of the causal relationship using the procedure defined in the previous section.

   b. Calculate the current states using $C_j^t = \sum_{i=0}^{n} \left(w_{ij} \cdot C_i^{t-1}\right)$, where n is the number of agents.

2.3 Relationships Initial Weight Definition

The initial weights can be defined according to the next procedures:

- Each expert defines its FCM. For the allocation of the weights (opinion of the experts) is used a scale from the 0 to the 1, where 0 represents that the antecedent concept (agent) has not influences over the consequent concept (agent), and a weight of 1 indicates that the consequent concept is sensible to the changes of the antecedent concept.

- We determine a global FCM. For the final calculation of the weight of the concept (global FCM) we can use two formulas:

$$E_{ji}^{G} = \max_{e}\left\{E_{ji}^{e}\right\},\ \forall\ e=1, \text{NE (number of experts)} \tag{1}$$

or

$$E_{ji}^{G} = \sum_{e=1}^{NE} b_e E_{ji}^{e} \,/\, NE \tag{2}$$

Where E_{ji}^{e} is the opinion of the expert e about the causal relationship among agents C_j and C_i, and b_e is the expert's opinion credibility weight.

3 DFCM Like Supervisor of Multiagent Systems

In this section we test the utilization of the DFCM to supervise a reference framework based on MAS to model the Industrial Automation.

3.1 The IDCSA Reference Framework

In this work we study the reference framework "Intelligent Distributed Control system based on Agents" (IDCSA) proposed in [24] to model industrial automation systems. The IDCSA is a multiagent platform designed specifically for control systems (see figure 3).

The IDCSA proposes the next set of agents, which represent the present elements in a loop of control process, with the intention to establish a generic mechanism for the handling of the activities related to the control process:

- *Observer Agent:* this agent will have the mission of gathering the data coming from the repositories of data that can give information about the state of the processes. Also, it can pre-process and/or validate the data, calculate averages and estimations, etc.
- *Controller Agent:* this agent receives the state information emitted by the Observer agent and compares the current conditions of the process with the conditions wished for the same on. In the case where the current conditions are moved away from a certain band of tolerance, it executes control structures, what can be orders of activation of alarms, orders for execution of applications of diagnostic, etc.
- *Actuator agent:* depending on the decisions taken by the controller agent, active alarms and makes them visible for each actor involved with the resolution of the problem (SCADA operators, engineers of optimization, maintenance engineers, etc.), produces changes in the SCADA (for example, it changes the set point), etc.
- *Coordinator agent:* it supervises the operation of the control system and it modifies it if is necessary, changes established values for conditions of normal operation (for example, value nominal for process variables), executes

tests that allow to identify and to locate faults, emits the requirements of services to the specialized agents, etc.

- *Specialized agents:* inside the detection process and diagnostic of faults, maybe is necessary to carry out activities of data mining, mathematical and statistical calculations, predictions, etc. These activities are carried out by specialized agents, each one of them with a specific task to carry out.

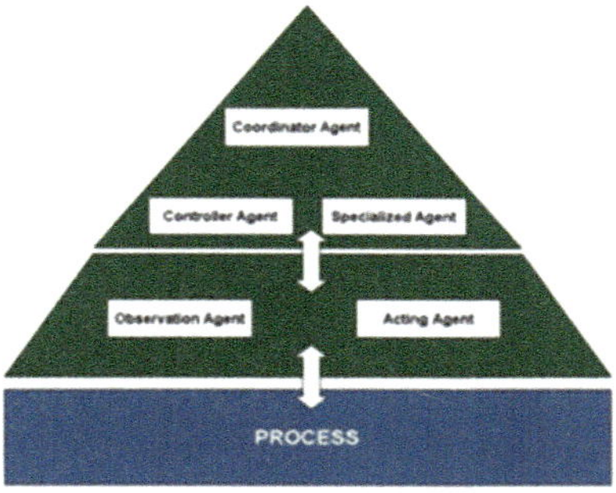

Fig. 3 IDCSA Model

3.1.1 IDCSA and the DFCM

In this work we study the behavior of the agents within IDCSA; for that, we develop a supervisor system based on DFCM. In this way, we can study the relationship between the agents that compose IDCSA by means of the DFCM. In this case, each agent of IDCSA will be a concept (see figure 4).

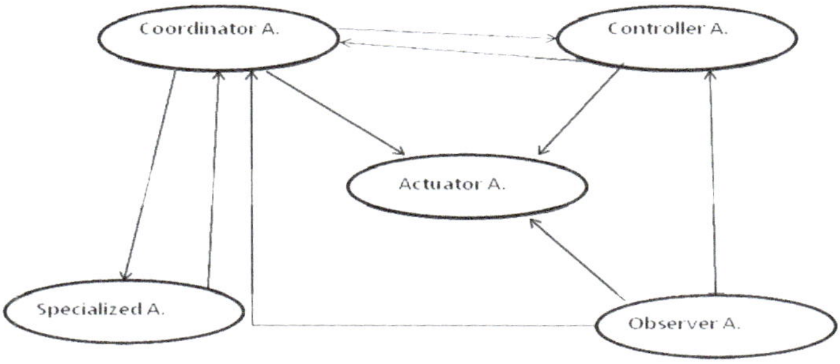

Fig. 4 Example of the DFCM for IDCSA

To test the performance of the DFCM, in the next part we are going to use the modeled of a faults management system developed in [28] using IDCSA.

3.2 *Fault Management System (FMS)*

The FMS proposed in [27] is composed by two subsystems: the first subsystem accomplishes Monitoring and Fault Analysis Tasks (MFAT); the second subsystem accomplishes the Maintenance Management Tasks (MMT). The FMS and the Engineering Management define together the productivity indexes, the human and financial resources, the components stock, among other things. On the other hand, the FMS also interacts with the Fault Tolerant Controlled Process, it being the final receiver of fault detection-diagnosis-decision tasks (see figure 5).

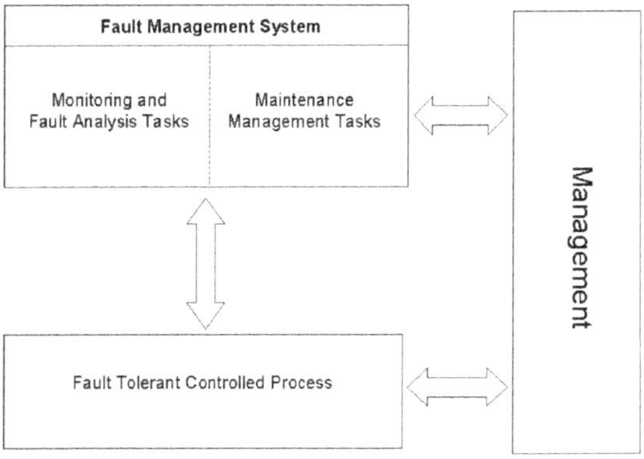

Fig. 5 The FMS

The MFAT subsystem is concerned with the following tasks:

- *Detection:* this task permits the identification of an invalid state in the process. A fault presence is stated from the behavior of significant variables; as a consequence, based on this behavior, the systems may be fault-free, in abrupt failure or incipient fault. The detection task needs detection models and the information from the process.
- *Isolation:* this task estimates the place where is occurring the faulty item.
- *Diagnosis:* This task determines the failure mode related with the fault detection, their causes and consequences. The diagnostic task needs diagnosis models and the information from detection and isolation tasks.

The MMT subsystem is related with the following tasks:

- *Prediction*: this task tries to estimate when an incipient fault goes to a functional failure. This task needs predictive models, and the information from fault detection and diagnosis tasks may be used in order to build these models.
- *Planning*: this task provides a preventive maintenance plan proposing the maintenance tasks that avoid the occurrence of a functional failure. It must also propose a contingency plan in case of abrupt failure, in order to avoid fatal consequences on the process.

- *Maintenance Action:* is concerned to set up and running the maintenance tasks according to the maintenance plan. These tasks are on-condition (detection-isolation-diagnosis) and on-time tasks. Maintenance actions are also related to set up and running contingency plans in case of abrupt or unexpected failures.

3.2.1 MAS-Based Reference Model for FMS

The FMS as MAS should consider the following items:

- Information exchange between the levels of the industrial processes.
- Monitoring and analysis of variables from the lower levels of the industrial processes (Local Control).
- Fault detection, isolation and diagnosis mechanisms and reasoning mechanisms supporting the decision-making process in the preventive maintenance.
- Distributed processing: the FMS activities as fault detection/isolation and estimation of working index should be embedded into a distributed computational model.

Consequently, the FMS works like a system where their modules interact of cooperative form, of way to reach the objective of the process. The FMS provides the following functionalities: Monitoring, faults management system, Detection, Isolation and Analysis of failures, Predictions of failure occurrence, Scheduling of preventive maintenance tasks, Set up and Running of preventive plans and corrective actions.

These functionalities can be embedded into the agents defined in the generic conceptual framework IDCSA [24], which has been adapted to the fault management problem. In this sense, this problem is defined like a generic problem of feedback control system. In this way, using the framework IDCSA eight agents are defined: Detector Agent (specialized agent), Finder Agent (specialized agent), Diagnostician Agent (specialized agent), Predictor Agent (specialized agent), Coordinator Agent, Controller Agent, Actuator Agent, and Observer Agent.

The *Detector* agent identifies if a component is under the presence of an incipient fault. The *Finder* agent looks up for the exact site where the fault is happens in the system, if it is not determined by the Detector agent. The *Diagnostician* agent determines the way of fault, its causes and their consequences. The *Predictor* agent prevents that an incipient fault becomes in a total functional fault in the system.

The *Coordinator* agent gathers the information about the process' items from the specialized agents and, based on this information, it schedules the maintenance tasks. The timeline should be defined according to the item's reliability and the failure effects. These aspects can suggest a decision-making process: take a corrective action or redefine the preventive maintenance plan; the last is also done if a maintenance task is not performed. This timeline can include a long time horizon (Long Term (LT) Plan) and not take into account the human resources and inventory on hand. In this sense, the *Controller* agent takes the preliminary plan provided by the *Coordinator* agent and it proposes a short time horizon plan (Short Term (ST) Plan) satisfying the human resources, inventory on hand and critical claims. That permits to give a daily or weekly plan for easy tracking. On-condition maintenance tasks are notified to the *Coordinator* agent; On-time and corrective maintenance tasks are notified to the *Actuator* agent. The *Observer* agent gathers the information about the process control in order to determine if a functional failure occurs; it also notices about the maintenance state and correlates it with the process state. Maintenance state is related with the foreseen maintenance tasks performance and process state is related with the operational function.

In figure 6 we observe the FMS based on the IDCSA framework (the specialized and the Coordinator agent in the Supervisor level; and the Controller, Observer and Actuator agents in the Process level). In addition, the FMS interacts with the IDCSA of the control process through the Middleware [27].

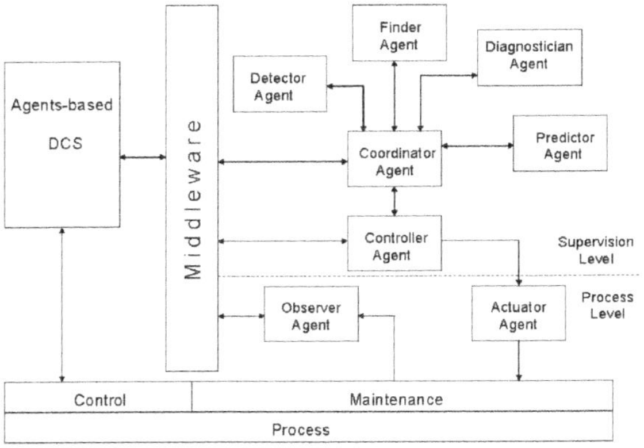

Fig. 6 MAS for the FMS

3.2.2 DFCM for the FMS

We assume that each concept of the map will represent an agent in particular, for that reason we need to define the relationship between the agents of the FMS:

- *Detector Agent:* it work with the coordinator agent informing if the system is in an incipient fault, this task is carried out monitoring variables of the system. A bad detection can cause false alarms and to cause that false plans of maintenance are realized, this entails to a bad operation of the coordinator agent. A good detection can even specify the location of the fault saving therefore the work of the finder agent. The causal relationships that affect to the Detector agent are from the observer agent, which indicates the state of the different variables, and from the process and the middleware.
- *Finder Agent:* Once a fault is detected and its detection does not indicate its location, this agent must locate the place where the fault is happens. If it indicates a false location can produce the accomplishment of unnecessary tasks, and an excellent detection can facilitate the work of the coordinator agent. This agent is affected by the Coordinator and Detector agents and the Middleware.
- *Diagnostician Agent:* this agent determines the type of fault and communicates it to the coordinator agent. A bad diagnostic can entail to plans of maintenance of low quality, the coordinator can generate maintenance plans erroneous if this agent does not determine the type of fault well. This agent is affected by the controller and coordinator agents and the Middleware.

- *Predictor Agent:* this agent avoids that an incipient fault becomes in a total fault of the system. It has a relationship with the Coordinator agent to indicate it if the fault possibly entails to a total collapse of the system. This agent is affected by the Coordinator, the Observer and the diagnostician agents, and the Middleware and the process.
- *Coordinator Agent:* it collects the information of the specialized agents to carry out the plans of maintenance. A bad coordination can waste resources and a total collapse of the system. On the other hand, a good coordinator can try to detect the bad operation of some of the specialized agents and to try to resolve that situation. This agent is affected by the specialized and the Observer Agents, and the Middleware.
- *Controller Agent:* it defines the plan of preventive maintenance to be applied. This agent has relationship with the actuator agent, which executes the maintenance plan, and it is affected by the Coordinator and the observer agents, and the Middleware.
- *Actuator Agent:* this agent executes the maintenance plan and it has relationship with the process. This agent is affected by the Controller and the observer agents, and the process.
- *Observer Agent:* it observes the system and monitors the execution of the maintenance plans, for that reason is affected by the actuator agent and the process.
- *Middleware:* the data of this concept in the map is obtained by the Observer agent, which is the unique one that affects it.
- *Process:* it establishes that so good is the communication of the agents with the process and it is affected by the actuator agent.

The resulting cognitive map is shown in figure 7.

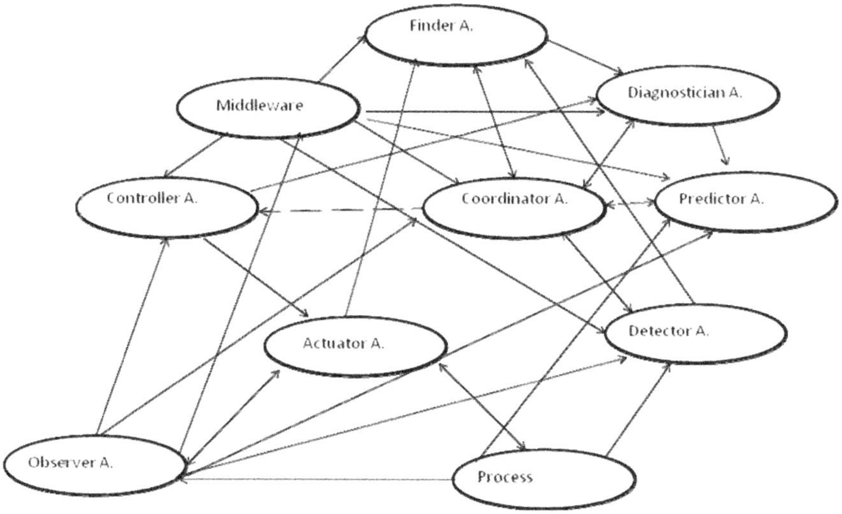

Fig. 7 The DFCM for the FMS

In our case of study, an agent is in *high state* when it fulfills all its functions correctly, it can use new mechanisms of inference successfully, and it does not occupy the resources in an excessive form; in addition, the conclusions and decisions taken by the agent are considered correct in all cases. One agent is in *medium state* when it fulfills its functions but it is not able to use new mechanisms of inference; in addition, the conclusions and decisions taken by the agent must be validated. Finally, an agent is in *low state* when it is not able to use new mechanisms of inference, it excessively occupies the resources of the system, and the conclusions obtained by the agent are considered erroneous and false.

The set of generic fuzzy rules are used to calculate all the relationships between the different concepts (agents) of the DFCM for the FMS. Previously, we have defined the relationship weights according to the experts. We have consulted several experts to define these weights and used the equation 2 to calculate the global FCM, for b_e=1. The global FCM is shown in table 2.

Table 2 Weights of the casual relationships for the DFCM of the FMS

	Coordinator A	Diagnost A	Predictor A	Finder A	Detector A	Controller A	Actuator A	Observer A	Midleware	Process
Coordin. A.	-	0.31	0.27	0.38	0.2	0.28	-	-	-	-
Diagn. A.	0.13	-	0.17	-	-	-	-	-	-	-
Predictor A	0.13	-	-	-	-	-	-	-	-	-
Finder A.	0.13	0.19	-	-	-	-	-	-	-	-
Detector A.	0.13	-	-	0.14	-	-	-	-	-	-
Controller	-	0.12	-	-	-	-	0.60	-	-	-
Actuator A.	-	-	-	-	-	-	-	0.23	-	1.0
Observer A	0.10	-	0.10	-	0.3	0.35	0.20	-	1.0	-
Middleware	0.34	0.38	0.34	0.47	0.3	0.35	-	-	-	-
Process	-	-	0.10	-	0.2	-	0.20	0.77	-	-

3.3 Results Analysis

We have tested the map for different inputs. Particularly, we have observed that there is a large sensitivity of the complete system to the changes in the Middleware, which is the most influential concept of the system. If this concept has a low value the concepts of the rest of the system will be 0 (zero) (see figure 8).

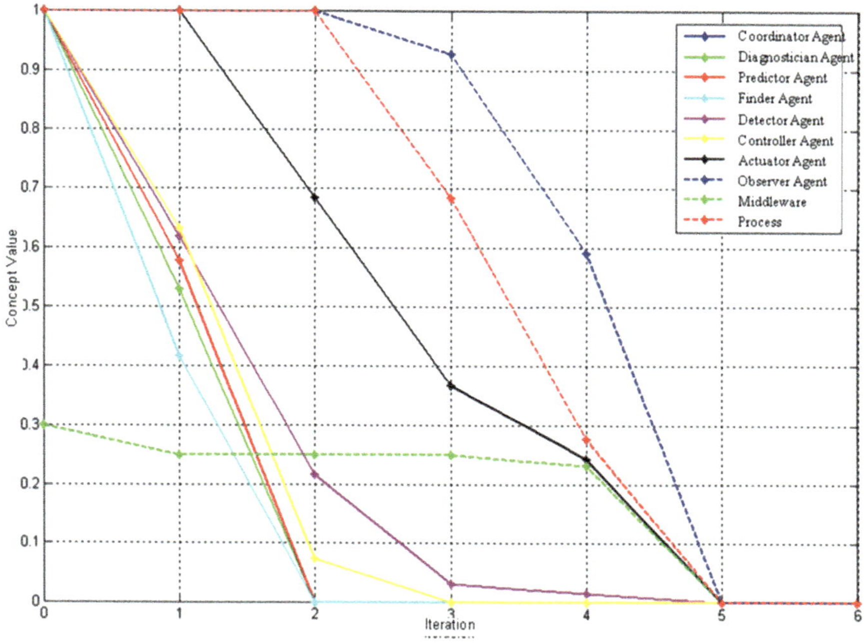

Fig. 8 Evolution of the agents with a Middleware with value of 0.3

We can see that the first concepts that begin to decay are the specialized agents, the coordinator and the controller agents, to a very fast rate. In iteration 1 is observed that the actuator agent begins to very quickly decay due to the fall of the controller agent, and finally, from iteration 2 the observer agent begins to fail, which entails to a total fall of the system. That is due because the agents use the Middleware to collect data to carry out their tasks. In this way, this concept (the middleware) must be in a good state to avoid situations like the predicted in figure 8. When the middleware works badly, the rest of the community of agents begins to handle erroneous information, entailing to a total fault of the system.

Another very important agent in the system is the coordinator agent. In a state of a coordinator agent where it does not work well, the system tends to fall in a bad operation (see the figure 9).

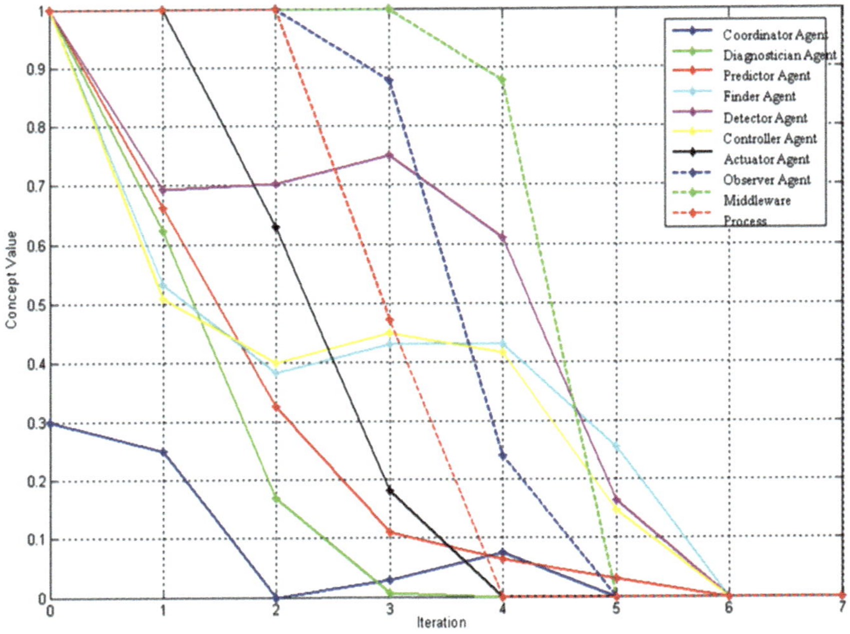

Fig. 9 Evolution of the DFCM for the FMS with a bad coordinator agent

As it is possible to observe in this figure, the specialized agents and the controller agent are themselves affected by the bad operation of the coordinator agent. That is due because these agents use the information of the coordinator agent to carry out their tasks, if the coordinator does not obtain good results and excessively occupies the resources of the system, the specialized agents are affected.

Another interesting test is when the specialized agents were near the threshold between a high and medium state. In this case, we can observe that these agents significantly diminish the operation of all the system (see figure 10).

In the figure 10 we can see that the specialized agents very quickly diminish the behavior of the coordinator agent. Once the coordinator agent does not work absolutely well, the controller agent diminishes the quality of its work since the maintenance plans are generated by the coordinator agent. This type of fault allows that the system continues in operation, but the adjustment must be made in some future time.

Another case of test is when we have a bad actuator agent (see figure 11). A bad actuator agent makes fall the process, falling therefore the quality of the behavior of the observer agent. Once the observer stops working well, the Middleware begins to obtain erroneous data, and as we already saw previously, once this concept begins to fall, the rest of the system fall quickly.

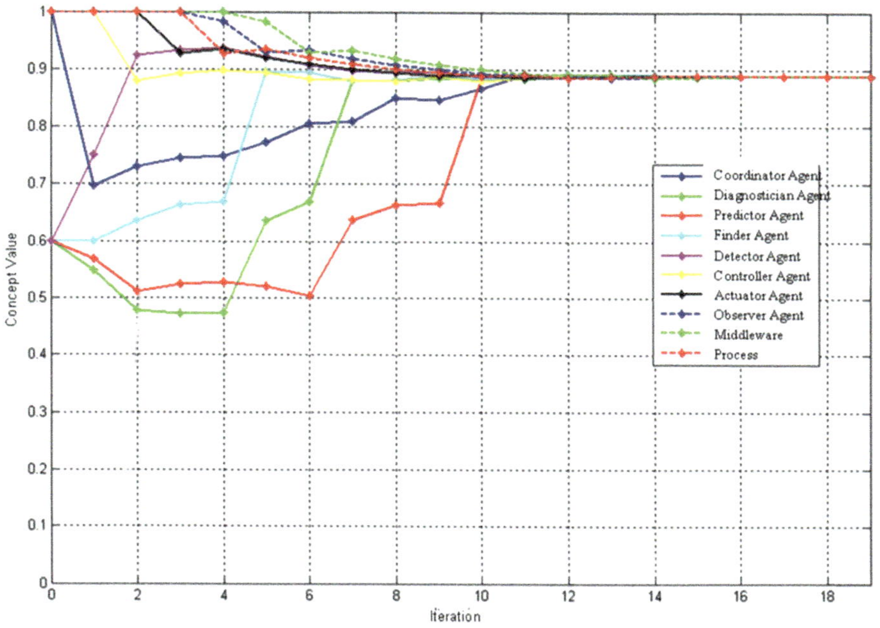

Fig. 10 Evolution of the concepts of the DFCM with specialized agents of medium quality

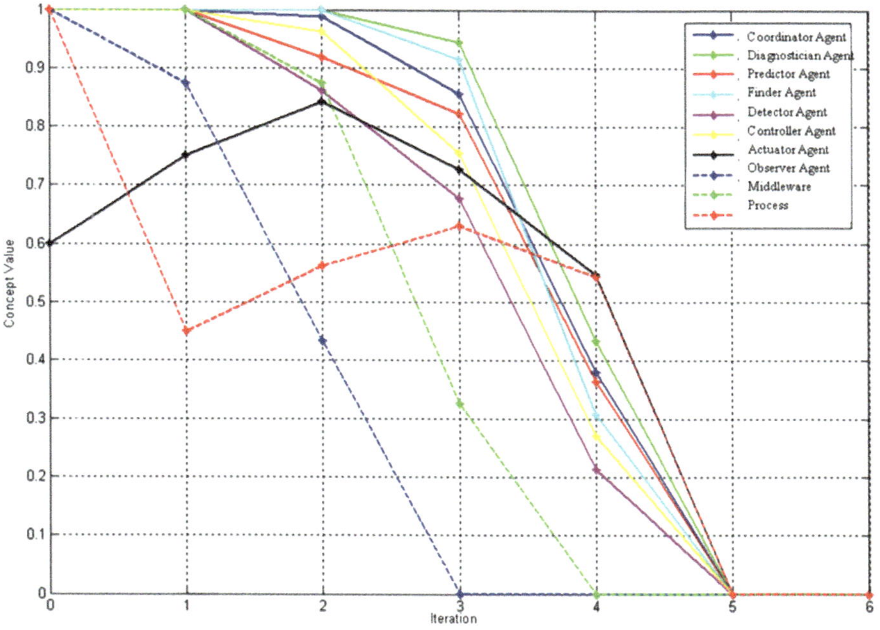

Fig. 11 Evolution of the DFCM with a bad Actuator agent

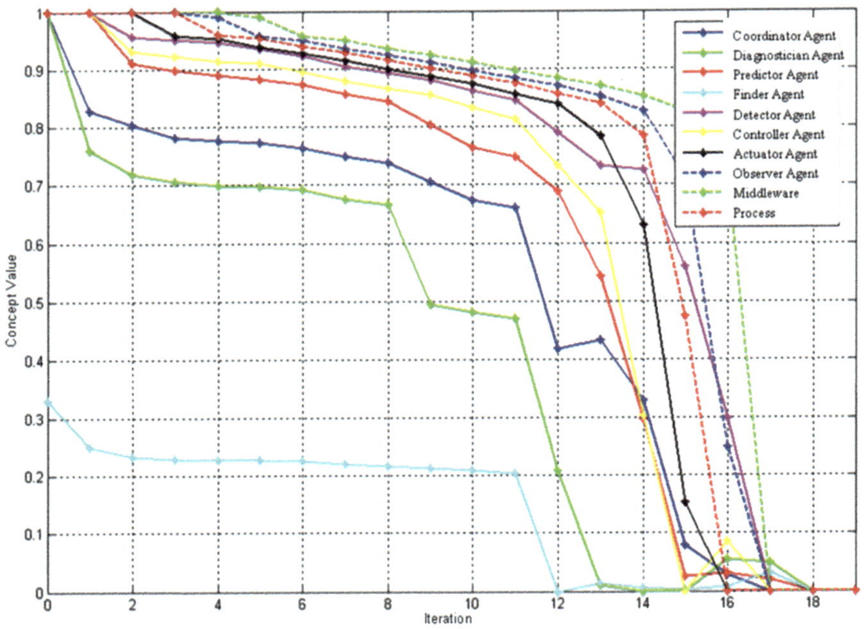

Fig. 12 Evolution of the agents with a bad finder agent

The last test is shown in figure 12, where we can observe that from a state where the finder agent does not work absolutely well, its fault is propagated slowly to the community of agents. Particularly, the diagnoses generated by the diagnostician agent and the plans of maintenance of the Coordinator Agent try to adjust to this situation, before to start to work bad. This type of problem does not entail to a total fault of the system immediately, but in a very near future time.

4 Conclusions

The cognitive maps have demonstrated to be a modeled tool of effective, even more their extensions. The utilization of the DFCM like a supervisor system of MAS demonstrates to be a good tool due to that the behavior of a community of agents can be studied. Specifically, we can study as the bad or good operation of an agent in particular can affect the rest of the community of agents, allowing so make decisions based on the predictions from the map.

Only a previous work has been developed to study MAS using FCM [29], specifically the casual relationship among the agents. In our approach we can model that, additionally we can analyze the behavior of the MAS, predict the behavior of the agents, etc. New extensions to our approach to define different forms of casual relationships can be done. For example, we can define casual relationships like mathematical equations. Other possibility is to define each concept like fuzzy variables, and then the casual relationships are defined like fuzzy relationships among fuzzy variables. Next works will study these approaches.

References

[1] Axelrod, R.: The Structure of Decision: Cognitive Maps of Political Elites. Princeton University Press, Princeton (1976)

[2] Kosko, B.: Fuzzy Cognitive Maps. International Journal of Man-Machine Studies 24, 65–75 (1986)

[3] Eden, C : On the Nature of Cognitive Maps. Journal of Management Studies 29, 261–265 (1992)

[4] Kosko, B.: Fuzzy Engineering. Prentice-Hall, New Jersey (1999)

[5] Aguilar, J.: A Fuzzy Cognitive Map Based on the Random Neural Model. In: Monostori, L., Váncza, J., Ali, M. (eds.) IEA/AIE 2001. LNCS (LNAI), vol. 2070, pp. 333–338. Springer, Heidelberg (2001)

[6] Aguilar, J.: A Dynamic Fuzzy-Cognitive-Map Approach Based on Random Neural Networks. International Journal of Computational Cognition 1, 91–107 (2003)

[7] Aguilar, J.: A Survey about Fuzzy Cognitive Maps Papers (invited paper). International Journal of Computational Cognition 3, 27–33 (2005)

[8] Goto, K., Yamaguchi, T.: Fuzzy Associative Memory Application to a Plant Modeling. In: Proc. of the International Conference on Artificial Neural Networks, pp. 1245–1248 (1991)

[9] Dickerson, J., Kosko, B.: Fuzzy Virtual worlds as Fuzzy Cognitive Maps. Presence 3, 173–189 (1994)

[10] Pelaez, C., Bowles, J.: Using fuzzy Cognitive Maps as a System Model for Failure Models and Effects Analysis. Information Sciences 88, 177–199 (1996)

[11] Stylios, C., Groumpos, P.: The challenge of modelling supervisory systems using Fuzzy Cognitive Maps. Journal of Intelligent Manufacturing 9, 339–345 (1998)

[12] Carvalho, J., Tomé, J.: Rule Based Fuzzy Cognitive Maps – Fuzzy Causal Relations. In: Mohammadian, M. (ed.) Computational Intelligence for Modelling, Control and Automation, pp. 276–281. IOS Press, Amsterdam (1999)

[13] Khan, M., Quaddus, M.: Group Decision Support using Fuzzy Cognitive Maps for Causal Reasoning. Group Decision and Negotiation Journal 13, 463–480 (2004)

[14] Sharif, A., Irani, Z.: Exploring Fuzzy Cognitive Mapping for IS Evaluation. European Journal of Operational Research 173, 1175–1187 (2006)

[15] Zhang, W., Chen, S., Besdek, J.: Pool2: A generic system for cognitive map development and decision analysis. IEEE Trans, Systems Man Cybernet 19, 31–39 (1989)

[16] Zhang, W., Chen, S., Wang, W., King, R.: A Cognitive Map Based Approach to the Coordination of distributed cooperative agents. IEEE Trans. Systems Man Cybernet 22, 103–114 (1992)

[17] Stylios, C., Geogrgopoulos, V., Groumpos, P.: The use of Fuzzy Cognitive Maps in Modeling Systems. In: Proceeding of 5th IEE Mediterranean Conference on Control an System, pp. 518–527 (1997)

[18] Papageorgiou, E., Stylios, C., Groumpos, P.: An integrated two-level hierarchical decision making system based on fuzzy cognitive maps. IEEE Trans. Biomed. Eng. 50, 1326–1339 (2003)

[19] Papageorgiou, E., Parsopoulos, K., Groumpos, P., Vrahatis, M.: Fuzzy Cognitive Maps Learning through Swarm Intelligence. In: Rutkowski, L., Siekmann, J.H., Tadeusiewicz, R., Zadeh, L.A. (eds.) ICAISC 2004. LNCS (LNAI), vol. 3070, pp. 344–349. Springer, Heidelberg (2004)
[20] Papageorgiou, E., Stylios, C., Groumpos, P.: Active Hebbian Learning algorithm to train fuzzy cognitive maps. International Journal of Approximate Reasoning 37, 219–247 (2004)
[21] Papageorgiou, E., Spyridonos, P., Ravazoula, P., Stylios, C., Groumpos, P., Nikiforidis, G.: Grading urinary bladder tumors using unsupervised Hebbian algorithm for fuzzy cognitive maps. Biomed. Soft Comput. Hum. Sci. 9, 33–39 (2004)
[22] Stach, W., Kurgan, L., Pedrycz, W.: A Survey of Fuzzy Cognitive Map Learning Methods. In: Grzegorzewski, P., Krawczak, M., Zadrozny, S. (eds.) Issues in Soft Computing: Theory and Applications, Exit, pp. 71–84 (2005)
[23] Stach, W., Kurgan, L., Pedrycz, W., Reformat, M.: Learning Fuzzy Cognitive Maps with Required Precision Using Genetic Algorithm Approach. Electronics Letters 40, 1519–1520 (2004)
[24] Aguilar, J., Cerrada, M., Mousalli, G., Rivas, F., Hidrobo, F.: A Multiagent Model for Intelligent Distributed Control Systems. In: Khosla, R., Howlett, R.J., Jain, L.C. (eds.) KES 2005. LNCS (LNAI), vol. 3681, pp. 191–197. Springer, Heidelberg (2005)
[25] Stach, W., Kurgan, L., Pedrycz, W., Reformat, M.: Genetic Learning of Fuzzy Cognitive Maps. Fuzzy Sets and Systems 153, 371–401 (2005)
[26] Elpiniki, I., Papageorgioua, E., Stylios, C., Groumpos, P.: Unsupervised learning techniques for fine-tuning fuzzy cognitive. Int. J. Human-Computer Studies 64, 727–743 (2006)
[27] Cerrada, C., Aguilar, J., Cardillo, J., Faneite, R.: Agents-Based design for fault management systems in industrial processes. Computer in Industry 58, 313–328 (2007)
[28] Papageorgiou, E., Spyridonos, P., Glotsos, D., Stylios, C., Ravazoula, P., Nikiforidis, G., Groumpos, P.: Brain tumor characterization using the soft computing technique of fuzzy cognitive maps. Applied Soft Computing 8, 820–828 (2008)
[29] `http://www.damas.ift.ulaval.ca/~fabiola/recherche/`

Soft Computing Technique of Fuzzy Cognitive Maps to Connect Yield Defining Parameters with Yield in Cotton Crop Production in Central Greece as a Basis for a Decision Support System for Precision Agriculture Application

E.I. Papageorgiou*, A.T. Markinos, and T.A. Gemtos

Abstract. This work investigates the yield and yield variability prediction in cotton crop. Cotton crop management is a complex process with interacting parameters like soil, crop and weather factors. The soft computing technique of fuzzy cognitive maps (FCMs) was used for modeling and representing experts' knowledge. FCM, as a fusion of fuzzy logic and cognitive map theories, is capable of dealing with uncertain descriptions like human reasoning. It is a challenging approach for decision making especially in complex environments. The yield management in cotton production is a complex process with sufficient interacting parameters and FCMs are suitable for this kind of problem. The developed FCM model consists of nodes that represent the main factors affecting cotton production linked by directed edges that show the cause-effect relationships between factors and cotton yield. Furthermore, weather factors and conditions were taken into consideration in this approach by categorizing springs as dry–wet and warm-cool. The methodology was evaluated for approximately 360 cases measured over 2001, 2003 and 2006 in a 5 ha cotton field. The results were compared with some benchmarking machine learning algorithms, which were tested for the same data set, with encouraging results. The main advantage of FCM is the simple structure and the easy handling of complex data.

Keywords: Fuzzy cognitive maps, modeling, knowledge representation, fuzzy sets, decision making, cotton, yield.

E.I. Papageorgiou
Department of Informatics and Computer Technology, Technological Educational Institute (TEI) of Lamia, 3rd Km Old National Road Lamia-Athens, 35100 LAMIA, Greece
e-mail: epapageorgiou@teilam.gr

A.T. Markinos · T.A. Gemtos
Laboratory of Farm Mechanization, Department of Agriculture, Crop Production and Rural Environment, University of Thessaly, Volos, Greece
e-mail: markinos@agr.uth.gr, gemtos@agr.uth.gr

* Corresponding author.

M. Glykas (Ed.): Fuzzy Cognitive Maps, STUDFUZZ 247, pp. 325–362, 2010.
springerlink.com © Springer-Verlag Berlin Heidelberg 2010

1 Introduction

Early estimates of agricultural production are of great importance for agricultural policy and trade. Yield prediction based on the combination of the factors affecting it is an important for the management of the farms. Cotton is a very important crop worldwide and especially in Greece, and there is a great need for good estimates of yield and total biomass production. The importance of this factor is more critical for precision agriculture or site specific management of the fields. Precision agriculture is a new farm management practice that takes into account the field and crop variability within a field or a farm. Conventional agriculture used means to manage the farms. A mean yield was considered to be valid for all parts of the fields and the inputs were defined for the mean yield and applied homogeneously in the entire field. With the development of yield sensors in the 90's it was revealed that to assumption of homogeneous yield was wrong. The same applied for the soil properties causing a field spatial variability. Measuring this variability using several new technologies like GPS, GIS and sensors provides with historical data the farmer to help in the management of the fields. The farmer can use the technology of variable rate application (VRA) to vary the inputs according to the real needs of the crops. This technology permits to optimisation of the inputs, the yields and the product quality leading to better economic and environmental results for the farms and agriculture in general. But historical data are not sufficient as a temporal variability was proved to exist in the yield. The potential of predicting yields using historical data is of great importance as management can be defined and applied to improve the viability of the farms and reduce the environmental impact of agriculture. Even prediction of the yield early in the growing season can be of great importance. The correlation of the flowering of apples with yield proved by Aggelopoulou et al. (2009) could permit the early prediction of yield and the appropriate management of the orchard.

Precision farming generates data which, due to their type and complexity, are not efficiently analyzed by traditional methods. The variation of soil properties, yield, product quality and weather should be managed properly to have better cropping results and reduce the adverse effects of agriculture to the environment. A new type of farm management was developed under the name of site specific management (SSM) which requires new analysis tools and model to achieve the maximum of the benefits for the farmer and the environment. The large amount of data collected through several sensors and analysis tools could be handled through the application of this knowledge based approach for cotton management on a qualitative and more quantitative basis. Some research has partly succeeded to do this. In the next paragraphs, a literature review on methods and algorithms for cotton management and yield estimation is presented.

Crop growth models were developed in agriculture by using mean values of input and outputs. Most of them were working satisfactorily usually for the conditions they were developed. But a large number of them encountered problems when site specific data were used. Several crop growth models have been developed (Mathews and Blackmore, 1997, Fraisse et al., 2001, Werner et al., 2000) with questionable value due to large development labor and time. A few cotton management tools have been adopted by researchers that assist them in various aspects; the COTMAN which is computer software used to monitor crop development, enter data easily and generate the reports used to make management decisions (COTMAN, 2007), the GOSSYM/COMAX (McKinion and Wagner, 1994) as a cotton growth expert system and the GRASS geographic information system (GIS) (GRASS, 1999) used to develop a spatial simulation that produces spatially variable outputs. COMAX was proposed to determine spatially variable fertilizer recommendations. GOSSYM was used to simulate cotton growth and yield and to predict spatially variable yield and residual nitrates (McKinion et al., 2001).

A large number of statistical methods and tools have been explored for assessing yield prediction. Many authors have used simple linear correlations of yield with soil properties but the results have varied from field to field and year to year (Drummond et al., 1995, Khakural et al., 1999, Gemtos et al., 2004). Many other studies, using complex linear methods like multiple linear regression, have given similar results (Drummond et al., 1995, Khakural et al., 1999, Kravchenko and Bullock, 2000). Some authors proposed non-linear statistical methods to investigate yield response (Wendroth et al., 1999, Adams et al., 1999).

Artificial intelligence techniques on the other hand have been used by many authors to predict yield in a number of areas, but few studies have focused on knowledge management and spatial analysis in precision agriculture (Liu et al., 2001; Canteri et al., 2002; Miao et al., 2006). Schultz et al. (2000) summarized the advantages of applying neural networks in agro-ecological modeling, including the ability of ANN to handle both quantitative and qualitative data, merge information and combine both linear and non-linear responses. Neural networks have been proposed for identifying important factors influencing corn yield and grain quality variability (Miao et al., 2006), for data analysis (Irmak et al. 2006), for predicting crop yield based on soil properties (Drummond et al., 2003), for setting target corn yields (Liu et al. 2001). Shearer et al. (1999) studied a large number of variables, including fertility, satellite imagery and soil conductivity for a relatively small number of observations in one site–year of data.

The reported work presents the soft computing technique of fuzzy cognitive maps (FCM) to connect yield defining parameters with yield in cotton crop production in Central Greece as a basis for a decision support system for precision agriculture application. FCM was chosen because of the nature of the application as it is a complex process and FCMs have been proved suitable for this kind of problem. FCMs were applied to a large number of diverse application areas such as engineering, medicine, political sciences, earth and environmental sciences, economics and management, etc. (Aguilar, 2005; Kottas et al., 2007). A number of examples of specific applications of FCMs include: political developments (Taber, 1991), analysis of electrical circuits (Styblinski and Meyer, 1991), failure modes effects analysis (Pelaez and Bowles, 1995), B2B ecommerce decision making (Lee and Kwon, 2007) management of relationships among organizational members of airline services (Kang & Lee, 2004), stock investment analysis (Lee & Kim, 1997), ecology and conservation (Ozesmi, 2004; Ramsey and Norbury, 2009), forest management ((Mendoza and Prabhu, 2006), modeling of complex technological systems (Stylios and Groumpos, 2004), modeling of software development project (Stach and Kurgan, 2004, Stach et al., 2004), time-series prediction (Stach et al., 2008), trust dynamics analysis in virtual enterprises (Wei et al., 2008), stock market decision support (Froelich and Wakulicz-Deja, 2007) pattern recognition (Papakostas et al., 2008), business-to-consumer e-commerce web-based systems (Lee and Ahn, 2009), management and organizational learning in support of organizational memory (Irani et al., 2009; Yaman and Polat, 2009), nuclear power reactors (Espinosa-Paredes et al., 2009), agile NPD process (Fekri et al., 2009), identification of critical path in strategic domains (Banerjee, 2009), modeling educational software (Hossain & Brooks, 2008), medical diagnostics (Papageorgiou et al., 2006, 2008, John and Innocent, 2004,), medical decision making (Papageorgiou et al., 2003, Stylios et al., 2008), decision-modeling for assessment of the impact of contemporary human resource management (HRM) practices to the shareholder value and satisfaction (Glykas and Xirogiannis, 2004; Xirogiannis and Glykas, 2007; Xirogiannis et al., 2008), and many others. The scope and range of the applications demonstrate the usefulness of this method and motivate further research.

In the context of knowledge-based systems in the field of agriculture a few trials have been explored by others (e.g. Ambuel et al., 1994, Khan and Khor, 2004) but without considering important yield factors and giving promising results for yield prediction. The modeling approach of FCMs was investigated for the first time by Papageorgiou et al. (2009) to help farmers by making decisions in precision agriculture. Through this work, the Fuzzy Cognitive Maps demonstrate their ability to capture the stakeholders' understanding of the system and their perceptions on the yield requirements of the precision agriculture. The success of precision agriculture depends on accurate and detailed knowledge of yield potential and crop response to specific conditions. The methodology of FCMs is suggested to this work from a different standpoint to accomplish this task including weather factors and conditions and compared with other intelligent benchmarking techniques.

Therefore, this is the first step in the development of an expert system tool that help in decision making process in agriculture, through the design of the knowledge representation and the design of reasoning with FCM to automate the decision. The next section provides the basic aspects of the fuzzy cognitive maps. This is followed by the reasons why the FCM approach was considered appropriate for modeling this particular domain. The third section provides a description of material and methods used to our approach, as well as a brief description on selecting and analyzing weather conditions for the proposed approach. Next the stages in the development of the FCM model that includes weather factors are described, the resulting model presented and its simulation analysis and results discussed. Following a comparison of the results with the benchmark computational intelligent methods and of the limitations of this study, the conclusion is followed by exploring the potential of the FCM as a decision support system in agriculture.

2 Fuzzy Cognitive Maps

2.1 Basic Aspects of Fuzzy Cognitive Maps

Fuzzy Cognitive Maps are appropriate to explicitly encode the knowledge and experience accumulated on the operation of a complex system (Kosko, 1986, 1992). They can be viewed as an extension of Cognitive Maps (Axelrod, 1976), and their main advantages include flexibility and adaptability to a given domain (Aguilar, 2005, Taber et al., 2007). Once constructed for a particular domain, an FCM allows a qualitative simulation of the system. In addition, FCMs represent knowledge in a symbolic manner, encoding the relations between the elements of a mental landscape so that the impact of these elements can be assessed. They are based on knowledge and experience for describing particular domains using concepts (variables, states, inputs, outputs) and the relationships between them, while taking into account the degree of uncertainty that may characterize these relationships in the real world using fuzzy logic (Kosko 1986, 1992). This technique is particularly suitable to model qualitative rather than quantitative systems.

An FCM has the topology of a fuzzy signed directed graph and dynamics similar to feedback non-linear neural networks, as is illustrated in Figure 1. It should be mentioned that all the values in the graph are fuzzy, so concepts take the values in the range between *[0, 1]* and the weights of the interconnections belong to the interval *[-1, 1]*. From simple observation of the weighed causal digraph of FCMs, it becomes clear which concept influences other concepts, showing the interconnection among concepts and it permits thoughts and suggestions for the reconstruction of the graph, i.e., the adding or deleting of an interconnection or a concept.

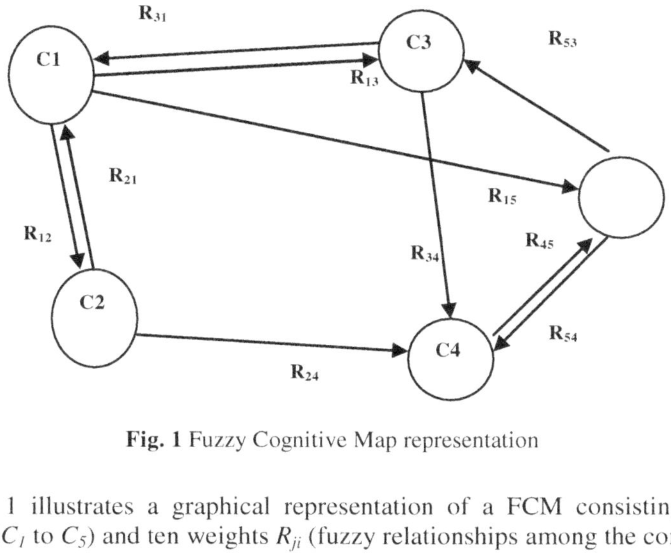

Fig. 1 Fuzzy Cognitive Map representation

Figure 1 illustrates a graphical representation of a FCM consisting of five concepts (C_1 to C_5) and ten weights R_{ji} (fuzzy relationships among the concepts).

The fuzzy causal relationship between two concepts C_j and C_i is described with the weight R_{ji}, taking a value in the range -1 to 1. There are three possible types of causal relationships between concepts:

- $R_{ji} > 0$ which indicates positive causality between concepts C_j and C_i. That is, an increase (decrease) in the value of C_j leads to an increase (decrease) in the value of C_i.
- $R_{ji} < 0$ which indicates negative causality between concepts C_j and C_i. That is, an increase (decrease) in the value of C_j leads to a decrease (increase) in the value of C_i.
- $R_{ji} = 0$ which indicates no relationship between C_j and C_i.

The simulation and inference process of the FCM is based on a mathematical formulation described in equation (2.1). Through the following calculation rule, the values of concepts are calculated at every simulation step k iteratively:

$$A_i^{k+1} = f\left(A_i^{(k)} + \sum_{\substack{j \neq i \\ j=1}}^{N} A_j^{(k)} \cdot R_{ji} \right) \tag{1}$$

where $A_i^{(k+1)}$ is the value of concept C_i at simulation step $k+1$, $A_j^{(k)}$ is the value of concept C_j at simulation step k, R_{ji} is the weight of the interconnection from concept C_j to concept C_i and f is a sigmoid threshold function:

$$f = \frac{1}{1+e^{-\lambda x}} \tag{2}$$

where $\lambda>0$ is a parameter that determines its steepness. In our approach, the value $\lambda=1$ has been used. This function is selected since the values Ai lie within $[0, 1]$.

The values A_i of concepts C_i are initially fuzzy and arise from the transformation of the real values of the corresponding variables to each one concept. Also the values for the weights of the interconnections R_{ji} among concepts are fuzzy. These fuzzy values are converted into numerical values after the defuzzification process of fuzzy logic and are used on the FCM simulation process. The value A_i of the concept C_i expresses the degree of its corresponding physical value. At each simulation step, the value A_i of a concept C_i is calculated by computing the influence of other concepts C_j's on the specific concept C_i following the corresponding mathematical formulation.

Through iteratively multiplying the previous state vector by the connection matrix $\boldsymbol{R}$, using standard matrix multiplication, new state vectors are computed showing the effect of the activated concepts (Peláez and Bowles, 1996). After every multiplication, the values of the state vector are normalized by a non-linear function that allows the vector elements to take a value within a predetermined set of values. Commonly, the functions used allow the variables to take values in $\{0, 1\}$, in $\{-1, 0, 1\}$, or in $[-1, 1]$ (Tsadiras and Margaritis, 1997, Tsadiras 2008). Iteration terminates when it reaches an equilibrium state and stops yielding new data, or when a prearranged iteration count has been reached (Taber, 1991).

2.2 Constructing Fuzzy Cognitive Maps

Fuzzy cognitive maps are constructed mainly by experts' knowledge through the drawing of fuzzy directed graphs. At first, knowledge engineers (used as experts) identify key domain issues or concepts for the specific problem domain. Secondly, they identify the causal relationships among these concepts and thirdly, they estimate causal relationships strengths. Thus, the produced graph (FCM) gathers not only the components and their relations but also the strengths of the corresponding interconnections (Stylios and Groumpos, 2004). FCMs can be constructed either by experts manually or by other source of information computationally (manual FCMs and automated FCMs).

The development process of fuzzy causal graphs is of great importance for its potential to sufficiently model a system. From the literature, most of the methods are dependent on the group of knowledge engineers who operate, monitor, supervise the system and they know its behaviour. This methodology extracts the main knowledge from the knowledge engineers and exploits their experience of the system's model and behavior.

Each one of the knowledge engineers contributes as multiple domain experts; construct the fuzzy causal graphs from their experience. They know the main variables that describe the behaviour of the system; each of these variables is represented by one concept of the fuzzy causal graph. Next, the experts, knowing which elements of the systems influence other elements, determine the negative or positive effect of one concept on the others, with a fuzzy degree of causation for the corresponding concepts. In this way, an expert transforms his/her knowledge in a dynamic weighted graph, the FCM. Each expert, indeed, determines the influence of one concept on another as "negative" or "positive" and then evaluates the degree of influence using a linguistic variable, such as "strong influence", "medium influence", "weak influence", etc.

Usually, a linguistic variable declared as *Influence,* is used to represent the causal interrelationships among concepts. Its term set T (influence), in most case studies, is suggested to comprise seven variables. Using seven linguistic variables, an expert can describe in detail the influence of one concept on another and can discern between different degrees of influence. The seven variables used here are: T(influence)={very very weak, very weak, weak, moderate, strong, very strong and very very strong}. The corresponding membership functions for these terms are shown in Figure 2 and they are: μ_{vvw}, μ_{vw}, μ_{w}, μ_{m}, μ_{s}, μ_{vs} and μ_{vvs}.

The main concepts that represent the model of the system are defined by experts; they describe the structure and the interconnections of the network using fuzzy conditional statements. The fuzzy *IF-THEN* rules that experts use to describe the relationship among concepts assume the following form, where ***A*** and ***B*** are linguistic variables:

IF *value of concept* C_i *is* ***A, THEN*** *value of concept* C_j *is* ***B*** *and thus the linguistic weight* e_{ij} *is* ***C***

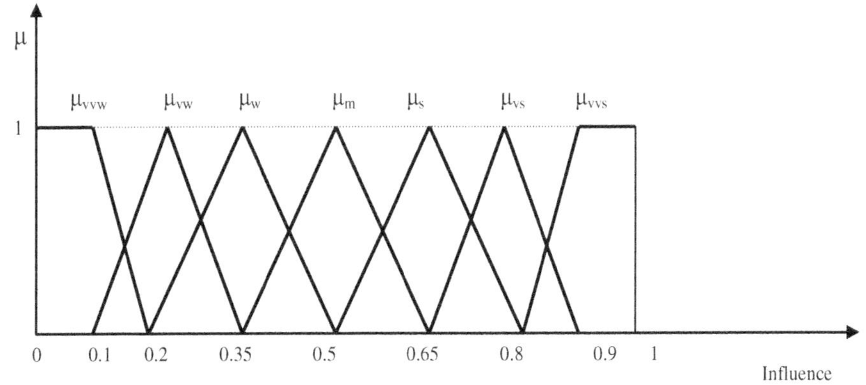

Fig. 2 The 7 membership functions corresponding to each one of the 7 linguistic variables

where ***A***, ***B***, ***C*** are linguistic variables (determined from the previous T(influence)) taking values in the range *[0, 1]*.

Each interconnection described by a fuzzy linguistic variable from the determined set, associates the relationship between the two concepts and establishes the grade of causality between the two concepts. Then, the linguistic variables ***C*** proposed by the experts for each interconnection are aggregated using the SUM method and so an overall linguistic weight is produced through the defuzzification method of the Centre of Area from fuzzy logic. Thus the final numerical weight R_{ij} between concept C_i to C_j is calculated. Using this method, all the weights of the FCM model are inferred (Stylios and Groumpos, 2004).

The main advantage of the FCM development method is that there is no restriction on the number of domain experts or on the number of system variables (Papageorgiou et al., 2005a, 2005b).

3 Materials and Methods

In 2001, an experiment was established in a 5ha commercial field at Myrina, Karditsa prefecture, Central Greece. The field was cultivated with cotton for the two previous years and during the next years till to date. The field was cultivated with cotton (*Gossypium hirsutum L.*) cultivar *Celia* for the last six years. It was managed using spatially uniform applications of inputs. A series of measurements were made each year.

Yield mapping was performed for the years 2001-06 using a *Farmscan*© yield monitor installed on a two row *John Deere*™ cotton picker (Gemtos et al., 2004). After harvesting of a field was completed, a calibration procedure was performed to improve the yield estimation (Markinos et al., 2004).

In February 2002, a 16m x 26m grid was formed in the north part of the field (4.3ha). Overall, 114 soil samples were taken at the grid points at 0-30cm depth. The samples were analyzed for texture, N, P, K, pH, Mg, Ca, Na and organic matter.

In May 2006, a VERIS machine was used to measure the apparent soil EC_a at depths 0-0.30 and 0-0.90m (Lund et al., 1999) from which maps were generated. The machine was pulled through the field at a speed of approximately 7 km/h at a track spacing of 4 m. Data were recorded every 1 s. Figure 3 represents two of the yield maps (for years 2001 and 2003) and some of the soil properties maps, such as pH, Clay, Sand, Organic matter (OM), and Apparent Electrical Conductivity (EC_a).

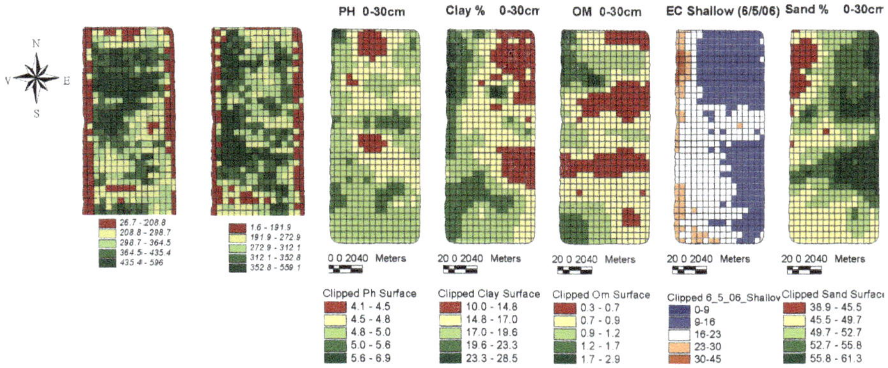

Fig. 3 Yield and Soil Properties Maps for years 2001 and 2003

The *SSToolbox*[TM] *3.61* software was used to store, represent, filter and analyze the acquired field data (SSToolbox, 2004). All the collected data were interpolated in order to produce a map (4.3 ha) on a 10m x 10m grid size that corresponds to a reliable field management unit (cell). The interpolation method of inverse distance was used for yield and EC_a due to dense data sampling, while kriging was used for the soil properties maps of sparse spatial sampling-grid, using *SSToolbox*, 2004, (Markinos et al., 2004).

Data from 20m strips around the field near the edges were filtered and removed to avoid machinery compacted headlands with lower yields. The data of every cell (10mX10m) of filtered maps represent the data to be used as inputs in the FCM model simulations with the yield from each year as output. Every cell of each input map linked to a scalar value in the GIS database. Each particular cell corresponding to the same spatial point represents a vector of scalar values of respective measured soil parameters. The last value in the vector represents the yield at this field point. Every vector constitutes a record in the database extracted from GIS.

Soil and crop data seem to affect the spatial variability of the yield. The temporal variability seems not to be consistent. In the present experiment in the first three years the north and north-west part of the field was giving the higher yield after that a different part of the field was giving the higher yield. Although there are parts of the field giving consistent high or low or variable yields (Markinos et al., 2004) there is always a variation after the first years. Similar results were observed in the long term experiment in wheat in Britain at the beginning of the 1990's.

In our research we have anticipated that the temporal variability can be attributed to the weather variability from year to year. Cotton is sown usually the second 10-days interval of April. Usually by that time the air and soil temperature is higher than a threshold of 15°C. In some years the weather after sowing is warm and dry enhancing the emergence and initial growth of the crop. In others it is cold and wet causing delays in the emergence and the initial growth. The weather conditions interact in this period with the soil. Heavy soils (high content in clay) usually are more fertile than light soils (high content in sand). Heavy soils maintain a higher water content and available water compared to the light soils. Given the heat capacity of the water is about four times that of solids, heavy soils are late in heating in the spring. Therefore when the spring in warm and dry cotton has a good starting in both light and heavy soils and the heavy more fertile ones produce higher yields at the end. In cold and wet years the heavy soils are heated late and delay the emergence and the initial growth compared to light soils. In this case cotton in the light soils start earlier and better and finally out yield the heavy ones.

In the field under study, a variation in the soil texture was observed (Figure 4). The north and north-west parts had higher clay and silt content while the south and south-east part higher sand content. So, it should be expected to show different behaviour in different years, according to the prevailing weather conditions (see Figure 4).

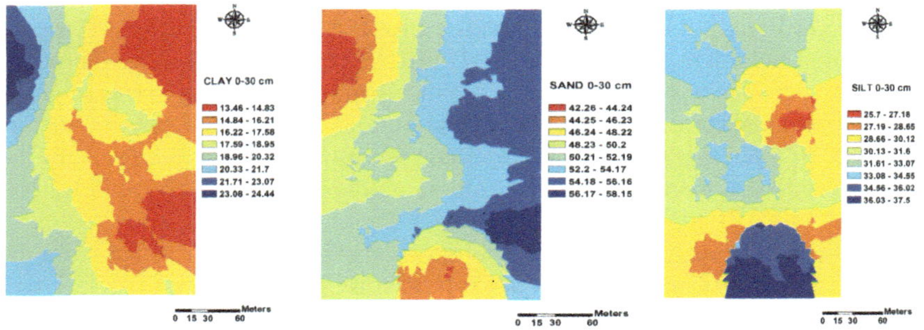

Fig. 4 Clay, Silt and Sand content for 0-30 cm

A number of weather factors and conditions have been taken under consideration in this approach. More specifically, the field has been separated into two zones according to the clay content as described earlier. A general estimation is that:

When the weather is warm and dry in the emergence period (it coincides with the periods of 21-30 of April, and the two first ten days periods of May) then the northern and west part of the field have higher yield and the south and west the lower. The opposite should happen when the weather is cold and wet.

As noted earlier, heavy soils (high content in clay and silt) are generally more fertile. So, when the emergence and initial growth of the crop are favorable then they give higher yields compared to the less fertile sandy or light soils. But when the weather conditions during emergence and initial growth are not favorable (wet and cold) then the less fertile light soil can increase its temperature faster than the heavier soil and this will favour the initial emergence and growth and a higher yield will follow.

In order to analyze the weather data of the area of the experiment, the growing degree days (*GDD*) was used. *GDD* was calculated from the real data of the area with base 15 °C using the formula:

$$GDD = ((T_{\max} + T_{\min})/2) - 15 \qquad (3)$$

where: T_{max} is the maximum air temperature of the day
T_{min} is the minimum temperature of the day

The *GDD* and the total rainfall for the 10 days interval of April and May were calculated. The average values for the six years period (2001-06) were calculated. In studying years with a value above the average are consider warm and wet and below cold and dry.

The results are shown in the following Table 1 for the years of 2001, 2003 and 2006.

Table 1 *GDD* and rainfall for the years 2001, 2003, 2006

Year	Ten days interval	*GDD*	Rain	Comment
2001	3 rd April	11	0	Warm and dry
2001	1st May	31	0	Warm and dry
2001	2nd May	40.5	0	Warm and dry
2003	3 rd April	1.68	3.8	warm and dry
2003	1st May	74.78	0	warm and dry
2003	2nd May	74.20	0	warm and dry
2006	3 rd April	-2.9	16	Cold and wet
2006	1st May	-11.5	1.5	Cold dry
2006	2nd May	32	0	Warm dry

According to the above measurements, we can characterize the conditions for each year as follows:

- 2001: Weather "warm and dry", the north part of the field gave the higher yield
- 2003: Weather "warm and dry", high yield in the west part of the field where the soil is heavier
- 2006: Weather "cold and wet" at the beginning but hot later, we should expect any distribution.

Experts have discriminated the weather conditions for growing months April and May at four categories: "cold and dry", "cold and wet", "warm and dry", "warm and wet". According to the above knowledge, a number of rules have been assigned by experts to be inserted into the FCM model. The rules are gathered at follows:

IF *Clay (C_{11}) is* ***low AND*** *weather at emergence and initial growth (3rd April, 1st and 2nd May) is* ***"cold and wet"*** *(or cold and dry)* ***THEN*** *the yield is* ***very low****.*

IF *Clay (C_{11}) is* ***low AND*** *weather at initial growth (3rd April, 1st and 2nd May) is* ***"warm and dry"*** *(or warm and wet)* ***THEN*** *the yield is* ***high****.*

IF *Clay (C_{11}) is* ***high AND*** *weather at initial growth (3rd April, 1st and 2nd May) is* ***"cold and wet"*** *(or cold and dry)* ***THEN*** *the yield is* ***low****.*

IF *Clay (C_{11}) is* ***high AND*** *weather at initial growth (3rd April, 1st and 2nd May) is* ***"warm and dry"*** *(or warm and wet)* ***THEN*** *the yield is* ***high****.*

The FCM model has been developed based on a raster data GIS approach, i.e. the data are stored in a two-dimensional matrix that represents the spatial distribution of every factor in the field. Each cell of the matrix corresponds to an area of 10mX10m, which is the spatial resolution of the yield data model. The data of every cell of filtered maps (as someone can see in Figure 3) represent the data that will be used as input variables in the proposed FCM model. Each vector of the data record represents the initial concept values of the proposed FCM model that interact through the FCM simulation process till an equilibrium point or decision is reached. Each causal node or factor of the FCM model represents a discrete layer or raster map in the yield data model.

For comparison purposes with FCM tool, the most used machine learning techniques, applied in a large number of scientific fields are used. These algorithms include decision trees, Bayesian networks and neural networks (NNs). In this paper, we assume readers' familiarity with machine learning methods and it is not worthwhile for the paper readability to present them analytically. A brief presentation of each one of the techniques is given in the next paragraphs.

The proposed FCM methodology addresses partly the problem of determining the final spatial variation of cotton yield trend. Although the spatial variability is an important factor especially in precision farming, in the system development, which is mainly a knowledge based approach, it is difficult to consider through its design all the important factors that affect the spatial variability. The spatial variability patterns for each year, have been considered in the construction of fuzzy sets, in the simulation process and threshold determination of FCM tool for cotton yield production as well as on the cross validation of the model and the comparisons with the other benchmarking algorithms. Our aim is not to address all the main issues of precision farming rather to predict the yield considering the experts' knowledge and the availability of data for this system.

4 Fuzzy Cognitive Map Model for Describing Cotton Yield

A FCM model for describing cotton yield spatial variation and management has been previously suggested by Papageorgiou et al. (2009). This model was constructed by three experts following the developing methodology described in the previous section. The three experts were one experienced cotton farmer and two experienced soil scientists, one from Technological Educational Institute of Larissa, Greece and the other from the Laboratory of Regional Soil Analysis and Agricultural Applications of Larissa, Greece. The three experts stated that there are eleven main factors-variables (which represent soil properties) and they added to this work eight more factors regarding weather conditions at initial growth for the ten day's intervals (periods) of 3rd April, 1st and 2nd May, used to determine cotton yield (Table 2 shows all the FCM concepts for this work).

The quantitative values of concepts, available in our database through the performed measurements, were normalized and transformed into the range *[0,1]*. Then, these data were classified into qualitative values, as described in Table 3 and using fuzzy sets theory inserted into the FCM simulation algorithm (Jang et al., 1997).

The set of linguistic variables that every concept can take are depicted in the following Table 3 and the corresponding membership functions for the eleven soil parameters and cotton yield are illustrated in Figure 5.

Then, the experts were asked to describe the degree of influence between the concepts and they determined their inter-relationships using the "*IF-THEN*" rules previously presented to infer a linguistic variable (weight), representing the cause and effect relationship between every pair of concepts. Three linguistic variables have been proposed by the three experts for each inter-connection. These three linguistic weights are aggregated using the SUM method and so an overall linguistic weight is produced which is defuzzified with the Centre of Gravity method and finally a numerical weight for R_{ij} is calculated. The advantage of this methodology is that experts do not have to describe the causality relationships using numerical values, but rather to describe qualitatively the degree of causality between concepts. The fuzzy rule for each interconnection is evaluated using fuzzy reasoning and the inferred fuzzy weight is defuzzified using the Centre of Gravity defuzzification method. Thus the initial weight matrix of the FCM is assigned.

The three experts suggested that the degree of influence between concepts was described by a linguistic variable taking a value in *[0, 1]* and its fuzzy set defined in previous section (shown in Figure 2). It is noticeable that these membership functions have a finer distinction between grades in the lowest and highest end of the influence scale.

Two examples for the specific problem of yield trend description in cotton are given:

IF *a* ***small*** *change occurs in the value of concept C_8 (Organic matter)* ***THEN*** *a* ***small*** *change in the value of concept C_{20} (cotton yield) is caused.*

Table 2 Concepts of the FCM model including weather conditions

Concepts	Description: soil factors measured over 0-300mm soil depth and weather conditions	Type & Number of scaled values
C_1: ShallowEC	Soil shallow electrical conductivity Veris (mS/m)	Five Fuzzy
C_2: Mg	The measured Magnesium in the soil in depth 0-30cm (ppm)	Five Fuzzy
C_3: Ca	The measured Calcium in the soil in depth 0-30cm (ppm)	Five Fuzzy
C_4: Na	The measured Na (Sodium) in the soil in depth 0-30cm (ppm)	Five Fuzzy
C_5: K	The measured Potassium in the soil in depth 0-30cm (ppm)	Five Fuzzy
C_6: P	The measured Phosphorus in the soil in depth 0-30cm (ppm)	Five Fuzzy
C_7: N	The measured NO3 in the soil profile of 0-30cm (ppm)	Five Fuzzy
C_8: OM	The % Organic Matter content in soil profile in depth 0-30cm	Three Fuzzy
C_9: Ph	The pH of the soil in depth 0-30cm	Seven Fuzzy
C_{10}: Sand	The % of the sand in the soil samples in depth 0-30cm	Four Fuzzy
C_{11}: Clay	The percentage % of the clay in samples in depth 0-30cm	Three Fuzzy
C_{12}: cold and wet for 3rd April	Weather is cold and wet for 3rd April ten days interval	Two discrete values (0 or 1)
C_{13}: cold and dry for 3rd April	Weather is cold and dry for 3rd April ten days interval	Two discrete values (0 or 1)
C_{14}: warm and wet for 3rd April	Weather is warm and wet for 3rd April ten days interval	Two discrete values (0 or 1)
C_{15}: warm and dry for 3rd April	Weather is warm and dry for 3rd April ten days interval	Two discrete values (0 or 1)
C_{16}: cold and wet for May	Weather is cold and wet for 1st and 2nd May ten days interval	Two discrete values (0 or 1)
C_{17}: cold and dry for May	Weather is cold and dry for 1st and 2nd May ten days interval	Two discrete values (0 or 1)
C_{18}: warm and wet for May	Weather is warm and wet for 1st and 2nd May ten days interval	Two discrete values (0 or 1)
C_{19}: warm and dry for May	Weather is warm and dry for 1st and 2nd May ten days interval	Two discrete values (0 or 1)
C_{20}: Yield	Seed cotton yield from 1st picking measured by yield monitor (t/ha)	Three Fuzzy

Table 3 Qualitative description (type) of each one of FCM concepts values

C_1		C_2		C_3		C_4	
ShallowEC (mS/m)		**Mg (ppm)**		**Ca (ppm)**		**Na (ppm)**	
Five Fuzzy		Five Fuzzy		Five Fuzzy		Five Fuzzy	
0 – 10	Very Low	< 60	Very Low	< 400	Very Low	< 25	Very Low
10 – 20	Low	60 – 180	Low	400 – 1000	Low	25 – 70	Low
20 – 30	Medium	181 – 360	Medium	1001 – 2000	Medium	71 – 160	Medium
30 – 40	High	361 – 950	High	2001 – 4000	High	161 – 460	High
> 40	Very High	> 950	Very High	> 4000	Very High	> 460	Very High
C_5		C_6		C_7		C_8	
K (ppm)		**P (ppm)**		**N (ppm)**		**OM (ppm)**	
Five Fuzzy		Five Fuzzy		Five Fuzzy		Three Fuzzy	
< 40	Very Low	< 5	Very Low	< 3	Very Low		
40 – 120	Low	5 – 15	Low	3 – 10	Low	< 1.0	Low
121 – 240	Medium	16 – 25	Medium	11 – 20	Medium	1.0 – 2.0	Medium
241 – 470	High	26 – 45	High	21 – 40	High	> 2.0	High
> 470	Very High	> 45	Very High	> 40	Very High		
C_9		C_{10}		C_{11}		C_{20}	
Ph		**Sand (%)**		**Clay (%)**		**Yield (tons/ha)**	
Seven Fuzzy		Four Fuzzy		Three Fuzzy		Three Fuzzy	
<4.5	Very Low						
4.6 – 5.5	Low	< 20	Low	< 15	Low	< 2.5	Low
5.6 – 6.5	Sl. Low	20 – 70	Medium	15 – 37	Medium Texture	2.5 – 3.5	Medium
6.6 – 7.5	Neutral	71 – 80	High	> 37	High	>3.5	High
7.6 – 8.5	Sl. High	> 80	Very High				
8.6 – 9.5	High						
> 9.5	Very High						

– This means that: the influence from concept C_8 to C_{20} is **weak**.

IF *a* ***high*** *change occurs in the value of concept C_1 (ShallowEC)* ***THEN*** *a* ***very high*** *change in the value of concept C_{20} (cotton yield) is caused.*

– This means that: the influence from concept C_1 to C_{20} is **very strong**.

Some rule examples are given relating the weather conditions at initial growth of April and May at ten day's intervals to yield, including the inference for assigning linguistic weights for each condition.

IF *weather at initial growth is* ***"cold and wet"*** *for* ***3rd April*** *interval (C_{12})* ***THEN*** *the yield is* ***very low***.

– *Thus infer: The influence from C_{12} to yield is* ***negatively moderate***.

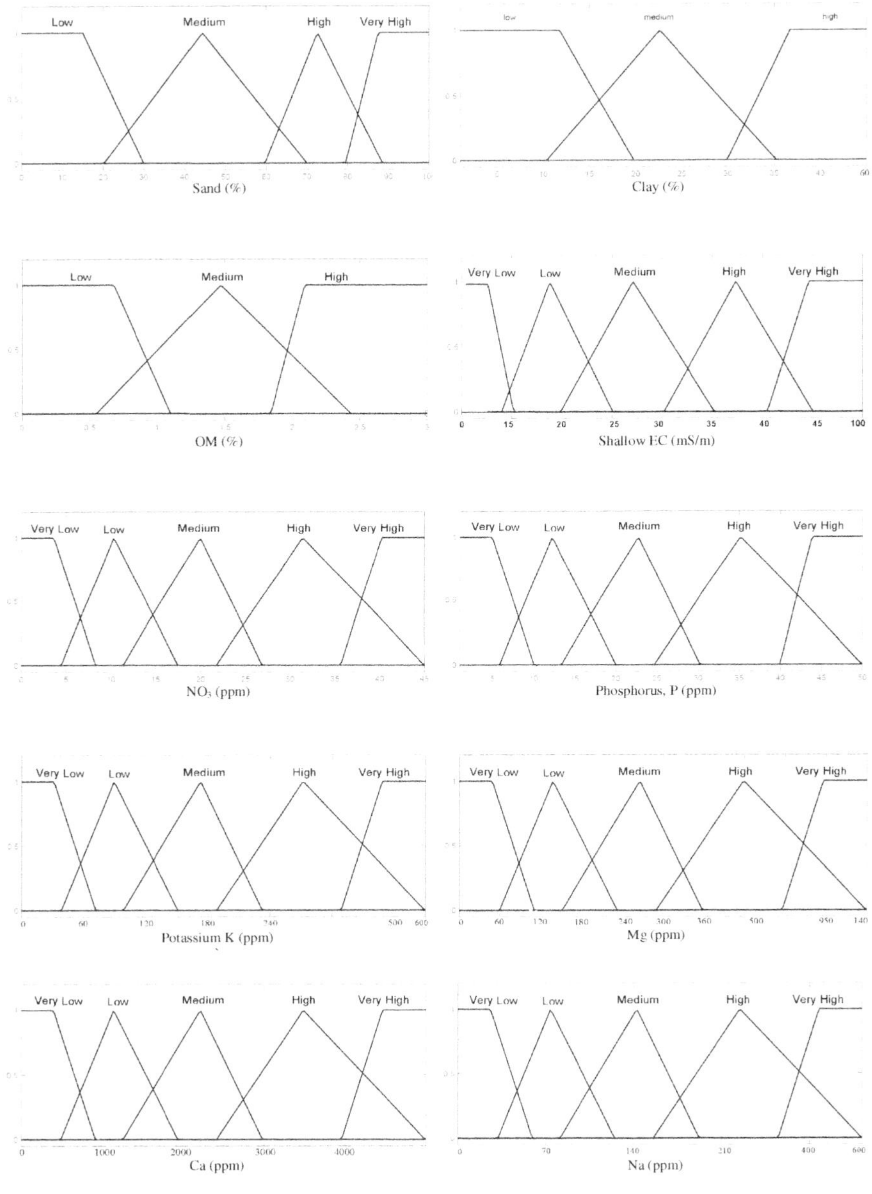

Fig. 5 Membership functions for the eleven soil properties and cotton yield

***IF** weather at initial growth is **"warm and dry"** for **3rd April** interval (C_{15}) **THEN** the yield is **high**.*

– *Thus infer: the influence from C_{15} to yield is **positively strong**.*

***IF** weather at initial growth is **"cold and wet"** for **1st and 2nd May** interval (C_{16}) **THEN** the yield is **low**.*

– *Thus infer: the influence from C_{16} to yield is **negatively strong**.*

***IF** weather at initial growth is **"cold and dry"** for **1st and 2nd May** interval (C_{17}) **THEN** the yield is **low**.*

– *Thus infer: the influence from C_{17} to yield is **negatively weak**.*

Table 4 Fuzzy rules relating yield factors and weather conditions to yield

Factor Concept	**First Expert**	**Second Expert (only where different)**	**Third Expert (only where different)**
	IF EC is VL Then Y is L		
C_1:	IF EC is L Then Y is M		IF EC is L Then Y is L
Shallow EC	IF EC is M Then Y is M	IF EC is L Then Y is L	
(EC)	IF EC is H Then Y is M		IF EC is H Then Y is H
	IF EC is VH Then Y is H		
	IF Mg is VL Then Y is L		
C_2:	IF Mg is L Then Y is L		IF Mg is L Then Y is M
Magnesium	IF Mg is M Then Y is H	IF Mg is H Then Y is M	
(Mg)	IF Mg is H Then Y is H		IF Mg is H Then Y is M
	IF Mg is VH Then Y is M		
	IF Ca is VL Then Y is L		
C_3:	IF Ca is L Then Y is L	IF Ca is M Then Y is M	
Calcium	IF Ca is M Then Y is H	IF Ca is H Then Y is L	-
(Ca)	IF Ca is H Then Y is M	IF Ca is VH Then Y is L	
	IF Ca is VH Then Y is M		
	IF Na is VL Then Y is H		
C_4:	IF Na is L Then Y is M		
Sodium	IF Na is M Then Y is L	-	IF Na is VL Then Y is M
(Na)	IF Na is H Then Y is L		
	IF Na is VH Then Y is L		
	IF K is VL Then Y is VL		
C_5:	IF K is L Then Y is L		
Potassium	IF K is M Then Y is M	IF K is VH Then Y is M	IF K is VH Then Y is M
(K)	IF K is H Then Y is M		
	IF K is VH Then Y is H		

C_6: Phosphorous (P)	IF P is VL Then Y is L IF P is L Then Y is L IF P is M Then Y is M IF P is H Then Y is M IF P is VH Then Y is H	-	IF P is H Then Y is H IF P is VH Then Y is M
C_7: Nitrogen (N)	IF N is VL Then Y is L IF N is L Then Y is L IF N is M Then Y is M IF N is H Then Y is H IF N is VH Then Y is H	IF N is H Then Y is M	IF N is VH Then Y is M
C_8: Organic Matter (OM)	IF OM is L Then Y is L IF OM is M Then Y is M IF OM is H Then Y is M	IF OM is H Then Y is M	IF OM is H Then Y is H
C_9: Ph	IF Ph is VL Then Y is L IF Ph is L Then Y is L IF Ph is SL Then Y is M IF Ph is M Then Y is M IF Ph is SH Then Y is H IF Ph is H Then Y is M IF Ph is VH Then Y is L	IF Ph is SL Then Y is L IF Ph is H Then Y is L	-
C_{10}: Sand (S)	IF S is L Then Y is M IF S is M Then Y is H IF S is H Then Y is L IF S is VH Then Y is L	IF S is M Then Y is M	IF S is L Then Y is H
C_{11}: Clay (Cl)	IF Cl is L Then Y is L IF Cl is M Then Y is M IF Cl is H Then Y is H	IF Cl is H Then Y is M	IF Cl is H Then Y is M
C_{12}:cold and wet April	IF C_{12} is 1 Then Y is VL	IF C_{12} is 1 Then Y is L	IF C_{12} is 1 Then Y is VL
C_{13}:cold and dry April	IF C_{13} is 1 Then Y is VL	IF C_{13} is 1 Then Y is VL	-
C_{14}:warm and wet April	IF C_{14} is 1 Then Y is L	IF C_{14} is 1 Then Y is M	-
C_{15}:warm and dry April	IF C_{15} is 1 Then Y is H	IF C_{15} is 1 Then Y is VH	-
C_{16}: cold and wet May	IF C_{16} is 1 Then Y is VL	-	-
C_{17}:cold and dry May	IF C_{17} is 1 Then Y is L	IF C_{17} is 1 Then Y is VL	-
C_{18}:warm and wet May	IF C_{18} is 1 Then Y is M	IF C_{18} is 1 Then Y is H	-
C_{19}: warm and dry May	IF C_{19} is 1 Then Y is H	IF C_{19} is 1 Then Y is VH	-

***IF** weather at initial growth is **"warm and dry"** for **1st and 2nd May** interval (C_{19}) **THEN** the yield is **very high**.*

- *Thus infer: the influence from C_{19} to yield is **positively strong**.*

All fuzzy rules were pooled by experts for each interconnection of FCM and were gathered in Table 4. In this table, only the different suggestions of the second and third expert were presented according to the first expert fuzzy rules.

Experienced agricultural scientists and farmers (field's experts) suggest that the degree of absorption of each soil nutrient by the cotton plant root depends on the value of some critical soil factors regulating this process. According to the experts opinion, soil pH is the major factor affecting the use by the plant in the soil nutrients. This means that even if there is in a soil a very high content of nutrients the very low pH value, restricts the availability of certain element to the cotton plant. Thus, the initially designed FCM model (Papageorgiou et al., 2009) enhanced by some more fuzzy relationships among concepts of soil parameters such as Mg, K, P, N and Ca towards to pH, suggested by experienced agricultural scientists through the second opinion process. These relationships are shown in Tables 5-9. The output of them presents the real availability of the corresponding element to the cotton plant root that used to the model inputs.

Table 5 Fuzzy logic table of soil parameter Mg to pH

\pH Mg	**VL**	**L**	**SL**	**M**	**SH**	**H**	**VH**
VL	VL	VL	VL	VL	VL	VL	VL
L	VL	VL	L	L	L	L	VL
M	VL	L	L	M	M	M	VL
H	L	M	M	H	H	H	L
VH	L	H	H	VH	VH	H	L

Table 6 Fuzzy logic table of soil parameter K to pH

\pH K	VL	L	SL	M	SH	H	VH
VL	VL	VL	VL	VL	VL	VL	VL
L	VL	L	L	L	L	L	L
M	L	M	M	M	M	M	M
H	M	H	H	H	H	H	H
VH	M	VH	VH	VH	VH	VH	VH

Table 7 Fuzzy logic table of soil parameter N to pH

\pH N	VL	L	SL	M	SH	H	VH
VL	VL	VL	VL	VL	VL	VL	VL
L	VL	VL	L	L	L	VL	VL
M	VL	L	M	M	M	L	VL
H	L	M	H	H	H	M	L
VH	L	H	VH	VH	VH	H	L

Table 8 Fuzzy logic table of soil parameter P to pH

\pH P	VL	L	SL	M	SH	H	VH
VL	VL	VL	VL	VL	VL	VL	VL
L	VL	VL	L	L	L	L	L
M	VL	L	M	M	L	L	L
H	L	M	H	H	M	M	M
VH	L	M	VH	VH	H	H	M

Table 9 Fuzzy logic table of soil parameter Ca to pH

\pH Ca	VL	L	SL	M	SH	H	VH
VL	VL	VL	VL	VL	VL	VL	VL
L	VL	VL	L	L	L	L	VL
M	VL	L	L	M	M	M	VL
H	VL	M	M	H	H	H	L
VH	L	M	H	VH	VH	H	L

The following tables 10, 11 and 12 show the suggested linguistic variables (linguistic weights) by each one of the three experts.

Table 10 Linguistic variables of cause-effect relationships among factor concepts assigned by first expert

Influence	C_1	C_2	C_3	C_4	C_5	C_6	C_7	C_8	C_9	C_{10}	C_{11}	C_{12}
C_1	-	-	-	-	-	-	-	-	-	-	-	WEAK
C_2	-	-	-	-	-	-	-	-	WEAK	-	-	NEG.WEAK
C_3	-	-	-	-	-	-	-	-	WEAK	-	-	MOD
C_4	-	-	-	-	-	-	-	-	-	-	-	NEG.STRONG
C_5	-	-	-	-	-	-	-	-	MOD	-	-	STRONG
C_6	-	-	-	-	-	-	-	-	WEAK	-	-	MOD
C_7	-	-	-	-	-	-	-	-	MOD	-	-	WEAK
C_8	MOD	-	-	-	-	-	-	-	-	-	-	WEAK
C_9	-	-	-	-	-	WEAK	-	-	-	-	-	V. WEAK
C_{10}	NEG.WEAK	-	-	-	-	-	-	-	-	-	-	NEG.STRONG
C_{11}	MOD	-	-	-	-	-	-	-	-	-	-	MOD
C_{12}	-	-	-	-	-	-	-	-	-	-	-	-

Table 11 Linguistic variables of cause-effect relationships among factor concepts assigned by second expert

Influence	C_1	C_2	C_3	C_4	C_5	C_6	C_7	C_8	C_9	C_{10}	C_{11}	C_{12}
C_1	-	-	-	-	-	-	-	-	-	-	-	V. WEAK
C_2	-	-	-	-	-	-	-	-	WEAK	-	-	NEG.MOD
C_3	-	-	-	-	-	-	-	-	MOD	-	-	MOD
C_4	-	-	-	-	-	-	-	-	-	-	-	NEG.V.STRONG
C_5	-	-	-	-	-	-	-	-	STRONG	-	-	STRONG
C_6	-	-	-	-	-	-	-	-	V.WEAK	-	-	STRONG
C_7	-	-	-	-	-	-	-	-	MOD	-	-	WEAK
C_8	WEAK	-	-	-	-	-	-	-	-	-	-	MOD
C_9	-	-	-	-	-	WEAK	-	-	-	-	-	V. WEAK
C_{10}	NEG.V.WEAK	-	-	-	-	-	-	-	-	-	-	NEG.STRONG
C_{11}	WEAK	-	-	-	-	-	-	-	-	-	-	MOD
C_{12}	-	-	-	-	-	-	-	-	-	-	-	-

Table 12 Linguistic variables of cause-effect relationships among factor concepts assigned by third expert

Influence	C_1	C_2	C_3	C_4	C_5	C_6	C_7	C_8	C_9	C_{10}	C_{11}	C_{12}
C_1	-	-	-	-	-	-	-	-	-	-	-	V. WEAK
C_2	-	-	-	-	-	-	-	-	MOD	-	-	NEG.MOD
C_3	-	-	-	-	-	-	-	-	MOD	-	-	MOD
C_4	-	-	-	-	-	-	-	-	-	-	-	NEG.V.STRONG
C_5	-	-	-	-	-	-	-	-	STRONG	-	-	V.STRONG
C_6	-	-	-	-	-	-	-	-	WEAK	-	-	MOD
C_7	-	-	-	-	-	-	-	-	WEAK	-	-	MOD
C_8	MOD	-	-	-	-	-	-	-	-	-	-	MOD
C_9	-	-	-	-	-	WEAK	-	-	-	-	-	V. WEAK
C_{10}	NEG.WEAK	-	-	-	-	-	-	-	-	-	-	NEG.V.STRONG
C_{11}	MOD	-	-	-	-	-	-	-	-	-	-	STRONG
C_{12}	-	-	-	-	-	-	-	-	-	-	-	-

To illustrate how numerical values of weights have been produced, the following example is given. The three experts have described the interconnection between concept C_5 (Potassium) and concept C_{20} (yield) using the following rules:

<u>*1st Expert:*</u>

IF *value of concept* C_5 *is* ***moderate THEN*** *value of concept* C_{20} *is* ***moderate***

– *Infer: The influence from concept* C_5 *towards concept* C_{20} *is* ***moderate***

<u>*2nd Expert:*</u>

IF *value of concept* C_5 *is* ***moderate THEN*** *value of concept* C_{20} *is* ***strong***

– *Infer: The influence from concept* C_5 *towards concept* C_{20} *is* ***strong***

<u>*3rd Expert:*</u>

IF *value of concept* C_5 *is* ***strong THEN*** *value of concept* C_{20} *is* ***very strong***

– *Infer: The influence from concept* C_5 *towards concept* C_{20} *is* ***very strong***

Figure 6 illustrates the three suggested linguistic variables for this particular problem example. These linguistic variables (moderate, strong and very strong) were summed and an overall linguistic weight was produced (also in Figure 5) which, through the defuzzification method of Centre of Gravity (COG), was transformed into the numerical value of $\boldsymbol{R}_{5\text{-}12}$=0.65 (weight value from concept C_5 towards concept C_{12}).

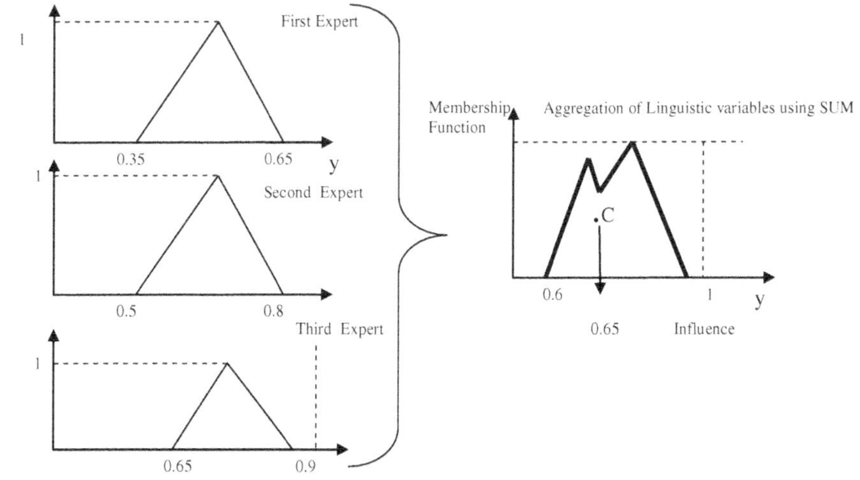

Fig. 6 Aggregation of three linguistic variables using the SUM technique. Point C is the numerical weight after defuzzification using the COG method

The same approach was used to determine all the weights of the FCM model. A weight matrix $\boldsymbol{R}^{initial}$ gathering the initially suggested weights of all the interconnections among the concepts of the FCM model was produced.

$$R=\begin{bmatrix}
0 & 0 & 0 & 0 & 0 & 0 & 0 & 0 & 0 & 0 & 0 & 0 & 0 & 0 & 0 & 0 & 0 & 0 & 0 & +0.25 \\
0 & 0 & 0 & 0 & 0 & 0 & 0 & 0 & 0 & 0 & 0 & 0 & 0 & 0 & 0 & 0 & 0 & 0 & 0 & -0.4 \\
0 & 0 & 0 & 0 & 0 & 0 & 0 & 0 & 0.4 & 0 & 0 & 0 & 0 & 0 & 0 & 0 & 0 & 0 & 0 & +0.5 \\
0 & 0 & 0 & 0 & 0 & 0 & 0 & 0 & 0.3 & 0 & 0 & 0 & 0 & 0 & 0 & 0 & 0 & 0 & 0 & -0.7 \\
0 & 0 & 0 & 0 & 0 & 0 & 0 & 0 & 0 & 0 & 0 & 0 & 0 & 0 & 0 & 0 & 0 & 0 & 0 & +0.6 \\
0 & 0 & 0 & 0 & 0 & 0 & 0 & 0 & 0.6 & 0 & 0 & 0 & 0 & 0 & 0 & 0 & 0 & 0 & 0 & +0.5 \\
0 & 0 & 0 & 0 & 0 & 0 & 0 & 0 & 0.25 & 0 & 0 & 0 & 0 & 0 & 0 & 0 & 0 & 0 & 0 & +0.3 \\
0.4 & 0 & 0 & 0 & 0 & 0 & 0 & 0 & 0.5 & 0 & 0 & 0 & 0 & 0 & 0 & 0 & 0 & 0 & 0 & +0.4 \\
0 & 0 & 0.5 & 0 & 0 & 0.4 & 0 & 0 & 0 & 0 & 0 & 0 & 0 & 0 & 0 & 0 & 0 & 0 & 0 & +0.1 \\
-0.25 & 0 & 0 & 0 & 0 & 0 & 0 & 0 & 0 & 0 & 0 & 0 & 0 & 0 & 0 & 0 & 0 & 0 & 0 & -0.6 \\
0.5 & 0 & 0 & 0 & 0 & 0 & 0 & 0 & 0 & 0 & 0 & 0 & 0 & 0 & 0 & 0 & 0 & 0 & 0 & +0.5 \\
0 & 0 & 0 & 0 & 0 & 0 & 0 & 0 & 0 & 0 & 0 & 0 & 0 & 0 & 0 & 0 & 0 & 0 & 0 & -0.7 \\
0 & 0 & 0 & 0 & 0 & 0 & 0 & 0 & 0 & 0 & 0 & 0 & 0 & 0 & 0 & 0 & 0 & 0 & 0 & -0.25 \\
0 & 0 & 0 & 0 & 0 & 0 & 0 & 0 & 0 & 0 & 0 & 0 & 0 & 0 & 0 & 0 & 0 & 0 & 0 & +0.3 \\
0 & 0 & 0 & 0 & 0 & 0 & 0 & 0 & 0 & 0 & 0 & 0 & 0 & 0 & 0 & 0 & 0 & 0 & 0 & +0.65 \\
0 & 0 & 0 & 0 & 0 & 0 & 0 & 0 & 0 & 0 & 0 & 0 & 0 & 0 & 0 & 0 & 0 & 0 & 0 & -0.65 \\
0 & 0 & 0 & 0 & 0 & 0 & 0 & 0 & 0 & 0 & 0 & 0 & 0 & 0 & 0 & 0 & 0 & 0 & 0 & -0.25 \\
0 & 0 & 0 & 0 & 0 & 0 & 0 & 0 & 0 & 0 & 0 & 0 & 0 & 0 & 0 & 0 & 0 & 0 & 0 & +0.4 \\
0 & 0 & 0 & 0 & 0 & 0 & 0 & 0 & 0 & 0 & 0 & 0 & 0 & 0 & 0 & 0 & 0 & 0 & 0 & +0.65 \\
0 & 0 & 0 & 0 & 0 & 0 & 0 & 0 & 0 & 0 & 0 & 0 & 0 & 0 & 0 & 0 & 0 & 0 & 0 & 0
\end{bmatrix} \quad (4)$$

The developed FCM model is presented in Figure 7 including the concepts that regard the weather conditions.

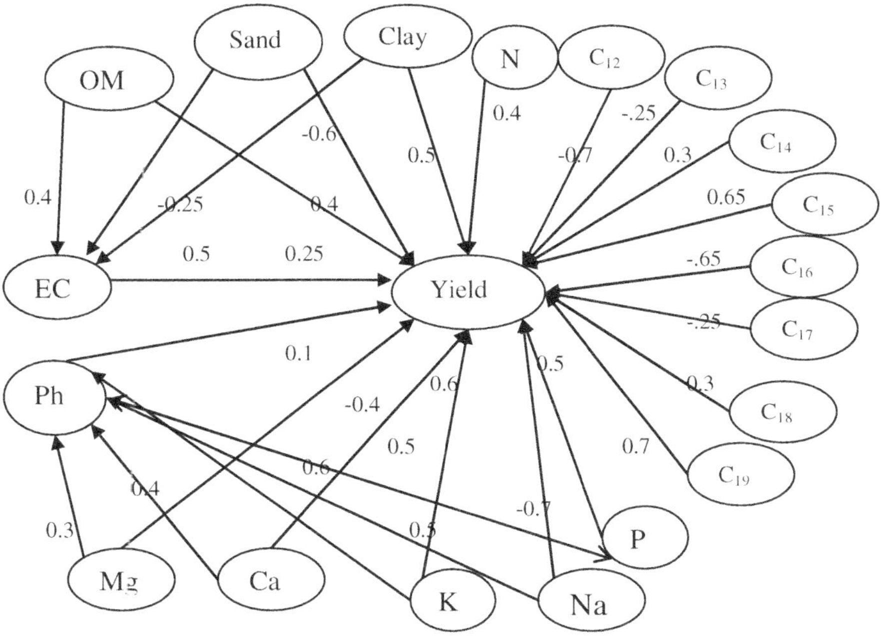

Fig. 7 FCM model for describing the final cotton yield

The aim of the FCM model is to predict the variations in cotton yield with each one of nineteen factors using simulated data. The procedure referred to is based on the determination of the value of output concept "Yield" that represents the percentage of cotton yield from the 1st picking measured with the yield monitor. The concept values correspond to the soil factor values measured over a depth of 0-0.3m.

5 Results

The initial values of concepts were transformed into the range *[0, 1]*, with quantification based on fuzzy sets theory, for the simulation of FCM (Jang et al., 1997). The FCM simulates through eq. (1) and the new values for concepts were calculated till the FCM tool for describing cotton yield reaches an equilibrium point (equilibrium region for the specific set of concepts) where the values of concepts did not change any more from their previous ones. After these limited number of interactions for the FCM convergence, the value A_{20} of concept C_{20} represents the category or the classification degree for the case of cotton yield.

In this point, it is essential to refer that the three experts also determined a threshold value equal to 0.85 to discriminate two different categories for the seed cotton yield production. More specifically, the experts suggested that yield production higher than 85% of desired cotton production could be considered as high cotton production. In our approach this translated as follows:

- If the estimated "yield" value (A_{20}) is less than 0.85 ($A_{20} \leq 0.85$), which means that the yield production is less than the 85% of desired cotton production, then "yield" is categorized as "low".
- If the estimated "yield" value (A_{20}) is higher than 0.85 ($A_{20} > 0.85$), then "yield" is categorized as "high".

Four different case studies were examined for 2001, 2003 and 2006 years to evaluate the proposed methodology based on FCMs for determining category of yield class and assessing the category of cotton production.

<u>First case:</u> A low yield production case study for 2006 weather data ("cold and wet" for 3rd April, "cold and dry" for 1st May and "warm and dry" for 2nd May for year 2006) was considered. The following vector $\boldsymbol{A}^{1}$ gathers the initial values of concepts for this case study (as they were quantified through defuzzification of fuzzy logic into interval $[0,1]$):

$$\boldsymbol{A}^{1} = [0.1\ 0.2\ 0.2\ 0.2\ 0.5\ 0.5\ 0.2\ 0.2\ 0\ 0.5\ 0.2\ 1\ 0\ 0\ 0\ 0\ 0\ 1\ 0\ 0]$$

where initial value of yield production was set equal to zero. The proposed FCM model simulated using these initial values in equations (1), (2) and converged to equilibrium point after a number of iterations. Specifically, after 11 iteration steps, the FCM reached an equilibrium state, where the values did not change any more from their previous ones. This state is:

$$\boldsymbol{A}^{fin\_1} = [0.6602\ \ 0.2000\ \ 0.7514\ \ 0.2000\ \ 0.5000\ \ 0.8093\ \ 0.2000\ \ 0.2000\ \ 0.8865\ \ 0.7044\ \ 0.2000\ \ 1.0000\ \ 0\ \ 0\ \ 0\ \ 0\ \ 0\ \ 1.0000\ \ 0\ \ 0.8241]$$

Figure 8 depicts the subsequent values of calculated concepts for every simulation step. It is observed that the final value of concept A_{20} is 0.8241, which means that, in this region, the yield is less than the threshold value 0.85 which has been considered by experts to discriminate two different categories for the yield based on the 85% of the acceptable seed cotton production. Actually, for this case, the measured yield production was "low" as the initial values corresponded to low cotton production, so that the derived result of the FCM model is the expected one according to the real measurements ("low" yield).

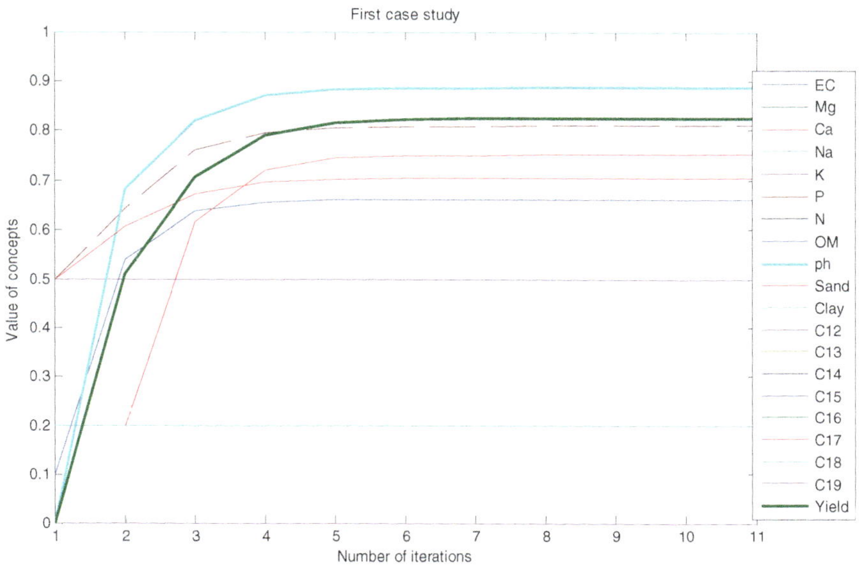

Fig. 8 Subsequent values of concepts for first case till convergence

Second case: Here the previous case's values of yield concepts (for the eleven yield factors) were also considered for the year 2001, but the weather conditions were different: "warm and dry" for April and "warm and dry" for 1st and 2nd May ten day's interval (the same conditions were also for the year 2003). The following vector was assigned for the FCM inference algorithm:

$$\mathbf{A}^2 = [0.1\ \ 0.2\ \ 0.2\ \ 0.2\ \ 0.5\ \ 0.5\ \ 0.2\ \ 0.2\ \ 0\ \ 0.5\ \ 0.2\ \ 0\ \ 0\ \ 0\ \ 1\ \ 0\ \ 0\ \ 0\ \ 1\ \ 0]$$

These values with the corresponding FCM weights were used in equations (1), (2) to calculate the equilibrium region of the process. After 11 iteration steps, the equilibrium region was reached and the final values of concepts were gathered at the following vector:

$$\mathbf{A}^{fin\_2} = [0.6602\ \ 0.2000\ \ 0.7514\ \ 0.2000\ \ 0.5000\ \ 0.8093\ \ 0.2000\ \ 0.2000\ \ 0.8865\ \ 0.7044\ \ 0.2000\ \ 0\ \ 0\ \ 1.0000\ \ 0\ \ 0\ \ 0\ \ 1.0000\ \ 0\ \ 0.9343]$$

Figure 9 gives the subsequent values of calculated concepts. In this case, it is observed that the value of concept A_{20} ("yield") in final state is 0.9649, which means that the yield is "high" according to the previous referred threshold value of 0.85. This means that the weather conditions for the cotton growing periods play a very important role to the yield production and this is clear from the above results.

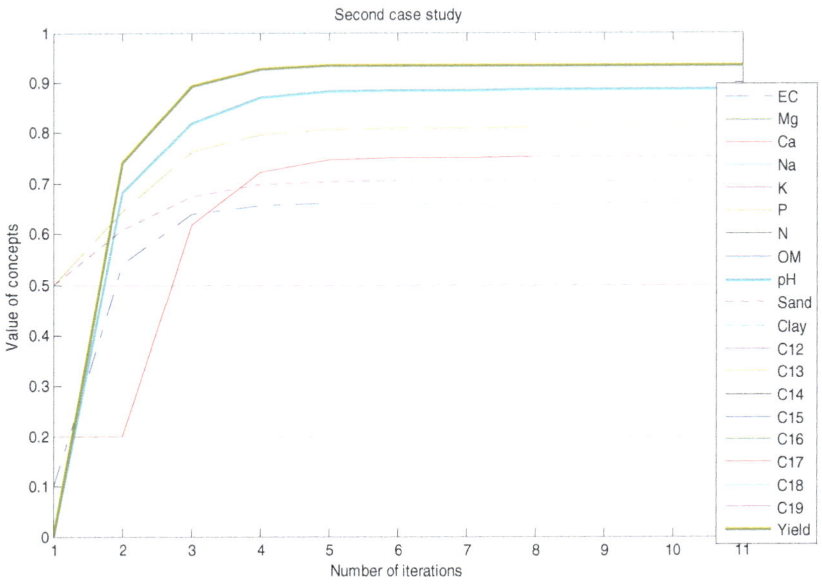

Fig. 9 Subsequent values of concepts for second case till convergence

<u>*Third case:*</u> This case has been characterized by experts and measured data and for the case of "high" Clay (C_{12}=1) as "high" yield for years 2001, 2003 and 2006. The weather data for 2006 were: "cold and wet" for 3rd April, "cold and dry" for 1st May and "warm and dry" for 2nd May. These conditions including the initial eleven factor concepts for this case were considered and gathered at following vector used for the FCM simulation process:

$$A^3 = [0.4\ \ 1\ \ 0.5\ \ 0.2\ \ 0.5\ \ 0.5\ \ 0.2\ \ 1\ \ 1\ \ 0.5\ \ 1\ \ 0\ \ 0\ \ 0\ \ 0\ \ 0\ \ 0\ \ 1\ \ 0\ \ 0]$$

This vector represents the numerical values of measured soil parameters of the physical process, after normalization, transformation and correspondence to numerical values. The weather conditions take only two discrete values, 0 or 1, according to the initial state. Then, using equations (1), (2), the FCM simulates and after 10 iteration steps, the equilibrium region is reached in the following vector:

$$A^{fin\_3} = [0.8287\ \ 1.0000\ \ 0.7552\ \ 0.2000\ \ 0.5000\ \ 0.8512\ \ 0.2000\ \ 1.0000\ \ 0.9280\ \ 0.6082\ \ 1.0000\ \ 1.0000\ \ 0\ \ 0\ \ 0\ \ 0\ \ 0\ \ 1.0000\ \ 0\ \ 0.8837]$$

Figure 10 gives the subsequent values of calculated concepts. Actually, for this case, the measured yield production was "high". It is observed that the calculated value of concept A_{20} in final equilibrium point is 0.8837, which means that the value of yield is higher than the 0.85 threshold value, so that it can be considered as "high" yield. The derived result is the expected one according to the real measurements for high cotton production.

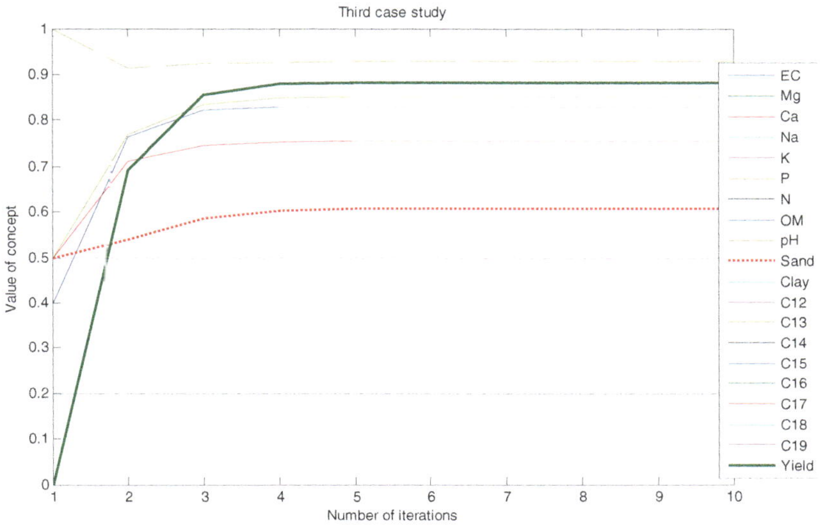

Fig. 10 Subsequent values of concepts for third case till convergence

Fourth case: The same initial values of the above third case for eleven factor concepts and especially for "high" Clay (C_{12}=1) were considered for year 2001, but the weather conditions for this year were different: "warm and dry" for 3rd April and "warm and dry" for 1st and 2nd May ten day's interval (the same conditions were also for the year 2003). The following vector is assigned for the FCM inference algorithm:

$$\mathbf{A}^4 = [0.4\ \ 1\ \ 0.5\ \ 0.2\ \ \ 0.5\ \ 0.5\ \ 0.2\ \ 1\ \ 1\ \ 0.5\ \ 1\ \ 0\ \ 0\ \ 0\ \ 1\ \ 0\ \ 0\ \ 1\ \ 0\ \ 0]$$

This vector represents the numerical values of measured soil parameters of the physical process, after normalization, transformation and correspondence to numerical values. Then, using equations (1), (2), the FCM simulates and after 10 iteration steps, the equilibrium region is reached in the following vector:

$$\mathbf{A}^{fin\_4} = [0.8287\ \ 1.0000\ \ 0.7552\ \ 0.2000\ \ 0.5000\ \ 0.8512\ \ 0.2000\ \ 1.0000\ \ 0.9280\ \ 0.6082\ \ 1.0000\ \ 0\ \ 0\ \ 0\ \ 1.0000\ \ 0\ \ 0\ \ 1.0000\ \ 0\ \ 0.9696]$$

Figure 11 gives the subsequent values of calculated concepts. Actually, for this case, the measured yield production was "high". It is observed that the calculated value of concept A_{20} in final equilibrium point is 0.9696, which means that the value of yield is higher than the 0.85 threshold value so that it can be considered as "high" yield. The derived result is higher than the previous one according to the real measurements for high cotton production.

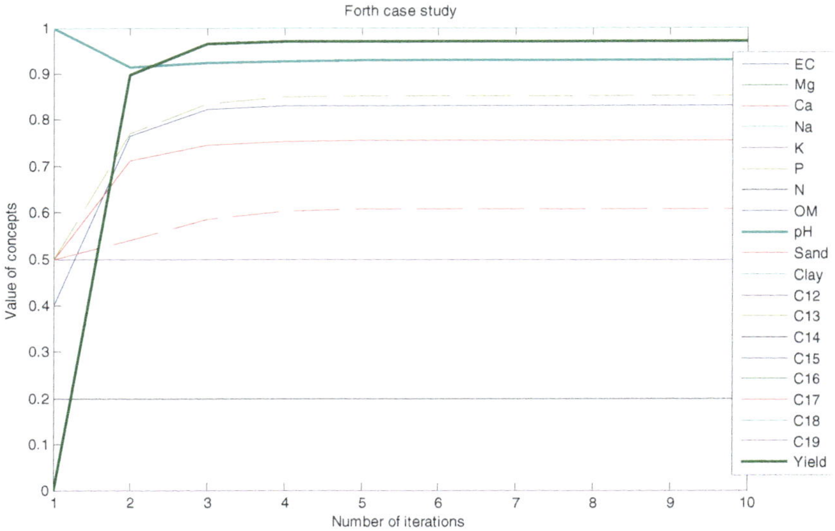

Fig. 11 Subsequent values of concepts for fourth case study till convergence

The FCM tool was evaluated for 360 cases using the measured data of soil parameters of 2001 in order to calculate the average accuracy of the yield production. For these experiments, two categories of "low" and "high" yield respectively were considered. For decision making reasons, the threshold value has been selected equal to 0.85 by the three experts to discriminate the two yield categories-"low" and "high".

This means that if the calculated output values of yield are lower than 0.85 then the produced yield is "low" and vice versa. The estimated value of the output concept-"yield" is essential and it categorizes the cotton yield production as "low" and "high" using a simple discrimination method (Papageorgiou et al. 2006).

For year 2001, the average accuracy is 73.8% which is efficient for this problem using the soft computing technique of FCMs. For the 182 cases of low yield, 135 were characterized as "low" yield and the rest as "high" yield, and for the 178 cases of high yield, 131 were characterized as "high" yield and the others as "low" yield.

The same FCM model was used to predict the accuracy of the yield characterization with the same threshold value for the evaluation procedure to estimate the yield output for the years 2003 and 2006. The results for the three years of 2001, 2003 and 2006, are gathered in Table 13.

Table 13 Average accuracy for three years (2001, 2003 and 2006)

Accuracy/year	2001 (%)	2003 (%)	2006 (%)
Low yield	(135/182):74.18	(126/185):67.57	(103/174):59.20
High yield	(131/178):73.60	(117/175):66.86	(149/186):80.10
Average Accuracy	73.80	67.20	69.65

The average accuracies of years 2003 and 2006 are 67.2% and 69.65% respectively. The yield prediction process in agriculture is depicted on the proposed flowchart (Fig. 12) which describes the procedure for taking the decision about the yield category that actually determines the cotton production according to the available knowledge/data.

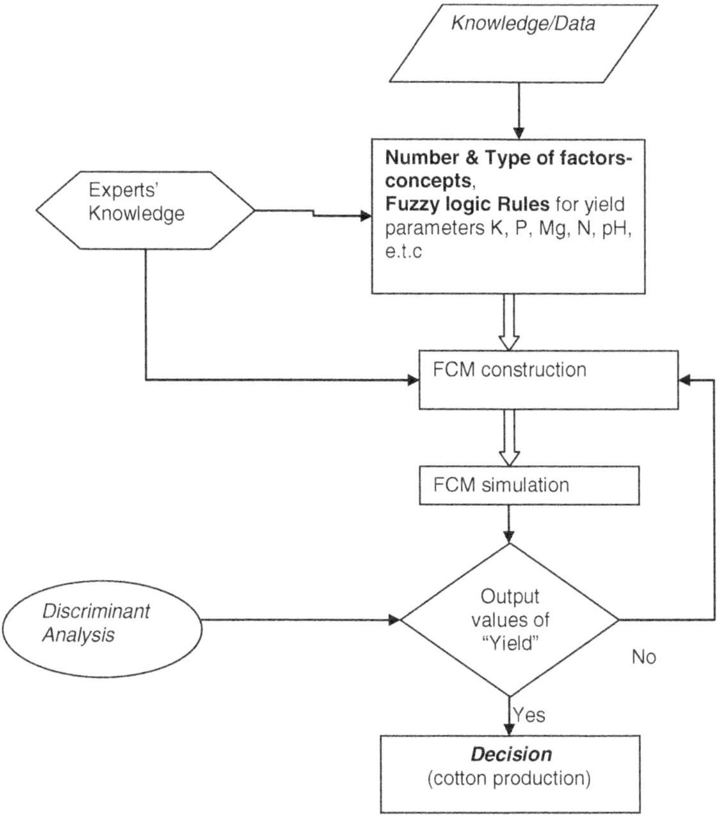

Fig. 12 Flowchart of the yield estimation process in precision agriculture

6 Discussion of Results

Through the literature, more studies have used statistical analysis techniques and only a few have used computational intelligent algorithms to classify and predict cotton yield from large datasets (Liu et al, 2001, Canteri et al., 2002, Miao et al., 2006, Shearer, 1999).

For comparison purposes, decision trees, neural networks with back propagation algorithm and Bayesian networks were employed on the same data set to estimate the classification accuracy into two yield classes (Witten and Frank, 1999,

Jang, 1997, Quinlan, 1990). All methods are machine learning techniques that efficiently accommodate numeric data as well as categorical or symbolic data (Jang, 1997). Other approaches that are used in data mining (Duda et al. 2001), such as statistical classifiers and support vector machines, rely solely on numerical variables and are not suitable techniques for this type of data.

The results of the constructed FCM model, applied to the above set of yield data, were compared with the well-known decision tree learner C4.5 (Release 8) (Quinlan 1993), the neural network training program (Duda et al. 2001), and the Naïve Bayes classifier. The models generated by these programs were compared w.r.t. precision and complexity of the outputs.

For the decision tree algorithms and Bayesian networks, the standard approaches included in WEKA (*Waikato Environment for Knowledge Analysis*) toolbox (WEKA, 2003) were used. WEKA system provides a comprehensive suite of facilities for applying data mining techniques to different type of data (Witten and Frank, 1999). Considering the decision trees, the J48 algorithm, an implementation of C4.5 release 8 (Quinlan 1990, 1993), is used to test data sets and to categorize cotton yield into two categories. For the Bayesian networks, the Naïve Bayes classifier is implemented to categorize the yield data to two output nodes ("low" yield and "high" yield) (Berthold and Hand, 2003).

The classification accuracy was measured by 10-fold cross-validation using a simple discriminant method on the values of output concept "yield" that represents the category of cotton yield. C4.5 was run with standard configuration (Quinlan, 1993). All other experiments were run with 10-fold cross-validation.

The results from the implementation of FCM technique and those obtained by decision trees, multilayer perceptron neural networks and Bayesian networks are shown in Table 14.

Table 14 Comparison of results of FCM tool with some benchmark machine learning algorithms

	FCM	**Decision Trees (C4.5)**	**NNs (backpropagation algorithm)**	**Bayesian**
2001	(135/182)=74.18	(134/182)=73.62	(132/182)=73.07%	(151/182)=82.97
	(131/178)=73.60	(130/178)=73.03	(132/178)=74,15%	(104/178)=58.42
Overall Accuracy	**73.80%**	**73.32%**	**73,61%**	**70.69%**
2003	(126/185)=67.57	(128/185)=69.19	(132/185)=71.35	(124/185)=67.03
	(117/175)=66.86	(110/175)=62.86	(109/175)=62.28	(100/175)=57.14
Overall Accuracy	**67,21%**	**66,03%**	**66,82%**	**62.09%**
2006	(103/174)=59.2	(98/174)=56.32	(96/174)=55.17	(92/174)=52.87
	(149/186)=80.1	(155/186)=83.33	(154/186)=82.79	(148/186)=79.56
Overall Accuracy	**69.65%**	**69.82%**	**68.98%**	**66.22%**

The prediction results of the proposed FCM tool for cotton yield are very encouraging to continue our work towards to this direction; our model achieved prediction of seed cotton yield with nearly 70% overall accuracy for all three years (73.80% for 2001, 67.21% for 2003 and 69.65% for 2006).

It is concluded based on the overall accuracy of each method that the FCM technique performed better than the other benchmark machine learning approaches. The proposed FCM simulation model is able to estimate yield with reasonably high overall accuracy, sufficient for this specific application area. Undoubtedly, more experiments and measurements are required for more accurate results. Additionally, as the experiment in this study was conducted in a certain area and field, the generalization of the results could be untrustworthy.

The developed model offers a prediction of the yield based on data available at the beginning of the growing period. This is very important for precision agriculture because the management of the parts of the field with homogeneous properties (called management zones) can lead to appropriate use of the inputs. Crop yield defines the amount of nutrients taken away from the field. For example, lint cotton contains 3.5% of nitrogen and a yield of 2.5 t/ha would remove 87 kg/ha of nitrogen. This has to be supplied by the soil and by the farmer using chemical fertiliser. In a part of the field with high yield of 3.5 t/ha the nitrogen removed should be 120 kg/ha. The farmer using conventional agricultural practices is usually applying the required chemicals for the higher yield plus 30% for the losses. Using the precision agriculture technologies and the predicted yield by the presented model the farmer can differentiate the N applications with considerable economic and environmental benefits. Economic, as 30 kg of N will be saved in parts of the field and environmental, because the N used above the crop's needs will remain in the soil and with the rains of the next winter will be leached and pollute the underground aquifers. Similar reasoning and practices are valid for all inputs including the water use.

The model can also be used to correct possible reasons for low yield. For example if soil pH is a limiting factor it can be easily corrected by liming the soil with appropriate amounts applied for each management zone. Additional benefits can be obtained from the early prediction of cotton production that is highly affecting the price of the commodity.

Thus, the soft computing technique of Fuzzy Cognitive Maps was explored to model and simulate the complex problem of cotton yield prediction. Fuzzy Cognitive Mapping can handle with complex modeling and knowledge processing problems in decision making tasks. Usually, FCM is used as a front-end tool for knowledge acquisition. The advantages of FCMs derive from their ability to elicit and compare the perceptions of different stakeholder experts, and to unify their viewpoints and understanding of a complex system; in this case, precision agriculture. The method could also be used to integrate perspectives of both lay and expert participants; mobilizing scientific and non-scientific knowledge, values and preferences to evolve a complex and unstructured problem into a range of identifiable, albeit artificial consensus solutions.

Some general limitations of the FCM model could be the absence of any underlying theoretical structure for scoring preferences and explicitly conveying the heuristic values and perspectives of experts. This subjectivity increases where the participants are unfamiliar with the method and the FCMs are not accompanied with additional information. Thus, our recent studies are directed towards this aspect to accompany the construction methodology of FCMs with additional information from data in the form of "*If-Then*" rules by extracting the available knowledge using data mining algorithms (Witten and Frank, 1999, Papageorgiou and Groumpos, 2007, Papageorgiou, 2009).

7 Conclusions

Summarizing the main contribution of the application work, Fuzzy Cognitive Maps, as an efficient methodology for capturing causal knowledge of domain experts, can be a useful tool for capturing the stakeholders' understanding of the system and their perceptions on the cotton yield requirements of the precision agriculture. From the proposed methodology and the presented results the following conclusions can be drawn:

- Fuzzy cognitive maps can be used to capture the stakeholders' understanding of the complex agricultural systems and predict early in the season the yield by using data available at the beginning of the growing season.
- Fuzzy cognitive maps, due to their simplicity, ease of use and less time-consuming for farmers in decision process, can work as a convenient consulting tool in cotton production management.
- FCM model can predict yield variability in the field (spatial variability) and offer explanations of this variability. This characteristic is of great importance for precision agriculture applications in cotton.
- The incorporation of weather data in the model can also explain the temporal variability of the yield with a lot of implications for the crop management, the economics of the farms, the environmental impact of agriculture and the market trends prediction.

This work was not intended to generate novel descriptions and predictions of cotton production but rather to explore a simple, flexible and comprehensible decision support tool to handle efficiently with the knowledge elicitation and processing, as well as with the site specific management behaviour of the crop yield in agriculture. Future work is directed towards the investigation on integrating data mining techniques with experts' knowledge to develop advanced decision support systems in precision agriculture.

References

Adams, M.L., Cook, S.E., Caccetta, P.A., Pringle, M.J.: Machine learning methods in site–specific management research: An Australian case study. In: Robert, P.C., Rust, R.M., Larsen, W.E. (eds.) Proc. 4th Int. Conf. on Precision Agriculture, ASA, CSSA, SSSA, Madison, USA, pp. 1321–1333 (1999)

Aggelopoulou, A.D., Bochtis, D., Koutsostathis, A., Fountas, S., Gemtos, T.A., Nanos, G.D.: Flower spatial variability in an apple orchard. In: JIAC Conference, Wageningen, July 5-8 (2009)

Ambuel, J.R., Colvin, T.S., Karlen, D.L.: A Fuzzy logic yield simulator for prescription farming. Transactions of the ASAE 37(6), 1999–2009 (1994)

Axelrod, R.: Structure of decision, the cognitive maps of political elites. Princeton University Press, Princeton (1976)

Bishop, C.M.: Neural networks for pattern recognition, 1st edn. Oxford University Press, USA (1996)

Berthold, M., Hand, D.J.: Intelligent Data Analysis, 2nd revised and extended edn. Springer, Heidelberg (2003)

Canteri, M., Avila, B.C., Dos Santos, E.L., Sanches, M.K., Kovaleschyn, D., Molin, J.P., Gimenez, L.M.: Application of Data Mining in Automatic Description of Yield Behavior in Agricultural Areas. In: Zazueta, F.S., Xin, J. (eds.) Proceedings of the World Congress of Computers in Agriculture and Natural Resources, Iguacu Falls, Brazil, March 13-15, pp. 183–189 (2002)

Chiu, C., Norcio, A.F., Hsu, C.: Reasoning on domain knowledge level in human–computer interaction. Information Sciences 1, 31–46 (1994)

COTMAN, Cotton Management Expert System Software (2008), `http://www.uark.edu/depts/cotman` (last accessed: 20/9/2008)

Craiger, J.P., Weiss, R.J., Goodman, D.F., Butler, A.A.: Simulating organizational behavior with fuzzy cognitive maps. Int. J. Comp. Intell. Organ 1, 120–133 (1996)

Dickerson, J.A., Kosko, B.: Virtual worlds in fuzzy cognitive maps. In: Kosko, B. (ed.) Fuzzy Engineering, pp. 499–528. Prentice-Hall, Simon & Schuster (1994)

Drummond, S.T., Sudduth, K.A., Birrell, S.J.: Analysis and correlation methods for spatial data. Paper No. 95–1335, ASAE, St. Joseph, MI, USA (1995)

Drummond, S.T., Sudduth, K.A., Joshi, A., Birrell, S.J., Kitchen, N.R.: Statistical and neural methods for site-specific yield prediction. Transactions of the ASAE 46(1), 5–14 (2003)

Duda, R.O., Hart, P.E., Stork, D.G.: Pattern Classification, 2nd edn. John Wiley & Sons, Inc., New York (2001)

Fraisse, C.W., Sudduth, K.A., Kitchen, N.R.: Calibration of the CERES–MAIZE model for simulating site–specific crop development and yield on claypan soils. Applied Eng. Agriculture 17(4), 547–556 (2001)

Froelich, W., Wakulicz-Deja, A.: Learning Fuzzy Cognitive Maps from the Web for the Stock Market Decision Support System. In: Wegrzyn-Wolska, K.M., Szczepaniak, P.S. (eds.) Adv. in Intel. Web, ASC, vol. 43, pp. 106–111 (2007)

Gemtos, T.A., Markinos, A.T., Toulios, L., Pateras, D., Zerva, G.: Precision Farming applications in Cotton Fields of Greece. In: 2004 CIGR international conference, Beijing, China, October 11-14 (2004) (CD-ROM)

Glykas, M., Xirogiannis, G.: A soft knowledge modeling approach for geographically dispersed financial organizations. Soft Computing 9(8), 579–593 (2004)

GRASS GIS (1999) `http://grass.osgeo.org` (Last updated: October 23, 2008)

Groumpos, P., Stylios, C.: Modelling Supervisory Control Systems using Fuzzy Cognitive Maps. Chaos Solit Fract 11(1-3), 329–336 (2000)

Irmak, A., Jones, J.W., Batchelor, W.D., Irmak, S., Boote, K.J., Paz, J.O.: Artificial neural network model as a data analysis tool in precision farming. Transactions of the ASABE 49(6), 2027–2037 (2006)

Jang, J.S.R., Sun, C.T., Mizutani, E.: Neuro-Fuzzy & Soft Computing. Prentice-Hall, Upper Saddle River (1997)

Jang, L.: Soft Computing Techniques in Knowledge-Based Intelligent Engineering Systems: Approaches and Applications. Studies in Fuzziness and Soft Computing, vol. 10. Springer, Heidelberg (1997)

Khakural, B.R., Robert, P.C., Huggins, D.R.: Variability of corn/soybean yield and soil / landscape properties across a southwestern Minnesota landscape. In: Robert, P.C., Rust, R.M., Larsen, W.E. (eds.) Proc. 4th Int. Conf. on Precision Agriculture, ASA, CSSA, SSSA, Madison, USA, pp. 573–579 (1999)

Kang II, Lee, S.: Using fuzzy cognitive map for the relationship management in airline service. Expert systems with applications 26(4), 545–555 (2004)

Khan, M.S., Khor, S.W.: A framework for fuzzy rule-based cognitive maps. In: Zhang, C., Guesgen, H.W., Yeap, W.-K. (eds.) PRICAI 2004. LNCS (LNAI), vol. 3157, pp. 454–463. Springer, Heidelberg (2004)

Kosko, B.: Fuzzy Cognitive Maps. Int. J. Man-Machine Studies 24, 65–75 (1986)

Kosko, B.: Neural Networks and Fuzzy Systems. Prentice-Hall, New Jersey (1992)

Kosko, B.: Fuzzy Engineering. Prentice-Hall, Upper Saddle River (1997)

Kottas, T.L., Boutalis, Y.S., Christodoulou, M.A.: Fuzzy cognitive network: A general framework. Intelligent Decisions Technologies Journal 4, 183–196 (2007)

Kravchenko, A.N., Bullock, D.G.: Correlation of corn and soybean grain yield with topography and soil properties. Agronomy J. 92(1), 75–83 (2000)

Lee, K.C., Kin, J.S., Chung, N.H., Kwon, S.J.: Fuzzy cognitive map approach to web-mining inference amplification. Journal of Experts Systems with Applications 22, 197–211 (2002)

Lee, K.-C., Kwon, S.: CAKES-NEGO: Causal knowledge-based expert system for B2B negotiation. Expert Systems with Applications (2007)

Lin, C.T., Lee, C.S.G.: Neural Fuzzy Systems: A Neuro-Fuzzy Synergism to Intelligent Systems. Prentice-Hall, Upper Saddle River (1996)

Liu, J., Goering, C.E., Tian, L.: A neural network for setting target corn yields. Transactions of the ASAE 44(3), 705–713 (2001)

Liu, Z., Satur, R.: Contextual fuzzy cognitive map for decision support in geographic information systems. IEEE Trans. Fuzz Syst. 5, 495–507 (1999)

Liu, Z.Q.: Fuzzy Cognitive Maps: Analysis and Extension. Springer, Tokyo (2000)

Lund, E.D., Christy, C.D., Drummond, P.E.: Practical applications of soil electrical conductivity mapping. In: Stafford, J.V. (ed.) Proceedings of the 2nd European Conference on Precision Agriculture, pp. 771–779. Sheffield Academic Press, Sheffield (1999)

Markinos, A.T., Gemtos, T.A., Pateras, D., Toulios, L., Zerva, G., Papaeconomou, M.: The influence of cotton variety in the calibration factor of a cotton yield monitor. In: Proc. of the International Conference of the Hellenic Association of Information and Communication Technology in Agriculture, Food and Environment (HAICTA 2004) on Information Systems & Innovative Technologies in Agriculture, Food and Environment, Thessaloniki, Greece, March 18-20, vol. 2, pp. 65–74 (2004)

Markinos, A.T., Papageorgiou, E.I., Stylios, C.D., Gemtos, T.A.: Introducing Fuzzy Cognitive Maps for decision making in precision agriculture. In: Stafford, J.V. (ed.) Proceedings of 6th European Conference on Precision Agriculture (6th ECPA), pp. 77–86. Wageningen Academic Publishers, Netherlands (2007)

Mathews, R., Blackmore, S.: Using crop simulation models to determine optimum management practices in precision agriculture. In: Stafford, J.V. (ed.) Proceedings of the 1st European Conference on Precision Agriculture, pp. 413–420. BIOS Scientific Publishers, Oxford (1997)

McKinion, J.M., Wagner, T.L.: GOSSYM/COMAX: A decision support system for precision application of nitrogen and water. In: Yuanchun, S., Xu, C. (eds.) Integrated Resources Management for Sustainable Agriculture, pp. 32–38. Agricultural University Press, Beijing (1994)

McKinion, J.M., Jenkins, J.N., Akins, D., Turner, S.B., Willers, J.L., Jallas, E., Whisler, F.D.: Analysis of a precision agriculture approach to cotton production. Computers and Electronics in Agriculture 32, 213–228 (2001)

Mendoza, G.A., Prabhu, R.: Participatory modeling and analysis for sustainable forest management: Overview of soft system dynamics models and applications. Forest Policy and Economics 9(2), 179–196 (2006)

Miao, Y., Liu, Z.: On Causal Inference in Fuzzy Cognitive Maps. IEEE Trans. Fuzz Syst. 8, 107–119 (2000)

Miao, Y., Mulla, D.J., Robert, P.C.: Identifying important factors influencing corn yield and grain quality variability using artificial neural networks. Precision Agriculture 7, 117–135 (2006)

Neapolitan, R.E.: Learning Bayesian Networks, 1st edn. Prentice Hall Publishers, Upper Saddle River (2003)

Papageorgiou, E., Stylios, C., Groumpos, P.: An Integrated Two-Level Hierarchical Decision Making System based on Fuzzy Cognitive Maps (FCMs). IEEE Trans. Biomed. Engin. 50(12), 1326–1339 (2003)

Papageorgiou, E.I., Stylios, C.D., Groumpos, P.P.: Active Hebbian Learning to Train Fuzzy Cognitive Maps. Int. J. Approx. Reasoning 37, 219–249 (2004)

Papageorgiou, E.I., Groumpos, P.P.: A weight adaptation method for fine-tuning Fuzzy Cognitive Map causal links. Soft Computing 9, 846–857 (2005a)

Papageorgiou, E.I., Groumpos, P.P.: A new hybrid learning algorithm for Fuzzy Cognitive Maps learning. Applied Soft Computing 5, 409–431 (2005b)

Papageorgiou, E.I., Spyridonos, P., Ravazoula, P., Stylios, C.D., Groumpos, P.P., Nikiforidis, G.: Advanced Soft Computing Diagnosis Method for Tumor Grading. Artificial Intelligence in Medicine 36(1), 59–70 (2006)

Papageorgiou, E.I., Groumpos, P.P.: Neuro-fuzzy, fuzzy decision tree and association rule based methods for fuzzy cognitive map grading process. In: Proceedings of International Conference on Computational Intelligence in MEDicine, CIMED 2007, Plymouth, UK, July 25-27 (2007) (CD-ROM)

Papageorgiou, E.I., Markinos, A.T., Gemtos, T.A.: Application of fuzzy cognitive maps for cotton yield management in precision farming. Expert Systems with Applications 48(1) (2009) (in press)

Papageorgiou, E.I.: A novel approach on designing augmented Fuzzy Cognitive Maps using fuzzified decision trees. In: Cai, Z., et al. (eds.) Proceedings at 4th international Symposium of Intelligence Computations and Applications, ISICA 2009, Computers in Communication and Intelligent Systems-CCIS 51, China, October 23-25, pp. 266–275. Springer, Heidelberg (2009)

Peláez, C.E., Bowles, J.B.: Using fuzzy cognitive maps as a system model for failure modes and effects analysis. Information Sciences 88, 177–199 (1996)

Quinlan, J.R.: Decision trees and decision making. IEEE Trans. System, Man and Cybernetics 20(2), 339–346 (1990)

Quinlan, J.R.: C4.5: Programs for machine learning. Morgan Kaufmann, San Mateo (1993)
Sadiq, R., Kleiner, Y., Rajani, B.: Interpreting fuzzy cognitive maps (FCMs) using fuzzy measures to evaluate water quality failures in distribution networks. National Research Council of Canada (2006), http://irc.nrc-cnrc.gc.ca
Schneider, M., Kandel, A., Chew, G.: Automatic construction of FCMs. Fuzzy Sets and System 93, 161–172 (1998)
Schultz, A., Wieland, R., Lutze, G.: Neural networks in agroecological modeling-stylish application or helpful tool? Computers and Electronics in Agriculture 29, 73–97 (2000)
Shearer, S.A., Thomasson, J.A., Mueller, T.G., Fulton, J.P., Higgins, S.F., Samson, S.: Yield prediction using a neural network classifier trained using soil landscape features and soil fertility data. Paper No. 993042., ASAE, St. Joseph, MI, USA (1999)
Skov, F., Svenning, J.C.: Predicting plant species richness in a managed forest. Forest and Ecological Management 620, 1–11 (2006)
Sowa, J.F.: Principles of Semantic Networks: Explorations in the Representation of Knowledge. Morgan Kaufmann Publishers, San Mateo (1991)
Stach, W., Kurgan, L., Pedrycz, W.: Parallel Learning of Large Fuzzy Cognitive Maps. In: Proceedings of International Joint Conference on Neural Networks, Orlando, Florida, USA, August 12-17 (2007)
SSToolbox for agriculture, User's Guide. N. Country Club Rd., 824, Stillwater, OK, USA (2004)
Stylios, C.D., Groumpos, P.P.: Modeling Fuzzy Cognitive Maps. IEEE Transactions on Systems, Man, and Cybernetics, Part A 34, 155–162 (2004)
Taber, R.: Knowledge processing with Fuzzy Cognitive Maps. Expert Systems with Applications 2, 83–87 (1991)
Taber, R.: Fuzzy cognitive maps. AI Expert 9, 19–23 (1994)
Taber, R., Yager, R.R., Helgason, C.: Quantization Effects on the Equilibrium Behavior of Combined Fuzzy Cognitive Maps. International Journal of Intelligent Systems 22, 181–202 (2007)
Tsadiras, A.: Comparing the inference capabilities of binary, trivalent and sigmoid fuzzy cognitive maps. Information Sciences (2008) Press, Corrected Proof. (Available online: May 23)
Tsadiras, A.K., Margaritis, K.G.: Cognitive mapping and certainty neuron Fuzzy Cognitive Maps. Information Sciences 101, 109–130 (1997)
WEKA, Toolbox (2003), http://www.cs.waikato.ac.nz/~ml/weka (last accessed – 30/11/08)
Wendroth, O., Jurschik, P., Nielsen, D.R.: Spatial crop yield prediction from soil and land surface state variables using an autoregressive state–space approach. In: Precision Agriculture 1999, Proceedings of the 2nd European Conference on Precision Agriculture, JV Stafford ed, pp. 419–428. Sheffield Academic Publishers, Sheffield (1999)
Werner, A., Doelling, S., Jarfe, A., Kuhn, J., Pauly, S., Roth, R.: Deriving maps of yield-potentials through the use of crop growth models, site information and remote sensing. In: Proceedings of Fifth International Conference on Precision Agriculture (CD), Minneapolis, MN, July 16-19, ASA, CSSA, and SSSA, Madison, WI, USA (2000)
Witten, I., Frank, E.: Data mining: Practical Machine Learning Tools and Techniques, 2nd edn. Morgan Kaufmann Publishers, San Mateo (1999)
Xirogiannis, G., Stefanou, J., Glykas, M.: A fuzzy cognitive map approach to support urban design. Expert Systems with Applications 26(2), 257–268 (2004)
Xirogiannis, G., Glykas, M.: Intelligent Modeling of e-Business Maturity. Expert Systems with Applications 32/2, 687–702 (2007)

Analysis of Farmers' Concepts of Environmental Management Measures: An Application of Cognitive Maps and Cluster Analysis in Pursuit of Modelling Agents' Behaviour

Livia Ortolani*, Neil McRoberts**, Nicholas Dendoncker***, and Mark Rounsevell

1 Introduction

The Common Agricultural Policy (CAP) of the European Union (EU) recognises that agriculture is multifunctional and that its multifunctional nature should be promoted. Under this expectation, Europe's agricultural areas are required to provide a diverse mixture of market and non-market goods for the private benefit of agricultural businesses on the one hand, and public good on the other. For reasons that are very well understood, a completely liberalised agricultural market cannot be relied on to provide many of the non-market and public goods expected from a multifunctional agriculture, and, given this, some form of regulation of the market or governmental involvement in it is inevitable. One way to view the CAP, then, is as the mechanism by which the public buys non-market goods from Europe's farmers at a price which is sufficient to have them forego alternative activities which produce no public good. The aim of policy optimisation for the CAP, when viewed in this context, is to allow the public to purchase the maximum amount of good for the least cost and to compensate those farmers who are really contributing to social welfare.

Livia Ortolani · Nicholas Dendoncker · Mark Rounsevell
Center for Environmental Change and Sustainability, School of Geosciences, University of Edinburgh, EH8 9XP, UK

Neil McRoberts
Land Economy & Environment Research Group, Scottish Agricultural College, Edinburgh, EH9 3JG, UK
e-mail: neil.mcroberts@sac.ac.uk

* Present address: Associazione Italiana per L'Agricultura Biologica (AIAB), Via Piave, 14-00178, Roma, Italia
** Corresponding author.
*** Present address: Department of Geography, FUNDP, UC Louvain, 61 Rue de Bruxelles, 5000 Namur, Belgium

M. Glykas (Ed.): Fuzzy Cognitive Maps, STUDFUZZ 247, pp. 363–381, 2010.
springerlink.com © Springer-Verlag Berlin Heidelberg 2010

One important complicating factor in the process of policy evaluation and optimisation is that the people who generate the outcomes of policy (*i.e.* the farmers) are not the people who make the policy, and the farmers' individual and collective objectives may differ from those of the policy makers and, indeed, from each others. Under these circumstances the policy optimisation problem becomes a hierarchical (bi-level) problem and standard mathematical programming techniques are unlikely to produce optimal solutions (Candler *et al.*, 1981). A possible response to this difficulty is to use agent based modelling (ABM) to simulate farmers' responses to potential policy designs. This approach has the advantage of allowing for differences among famers' objectives while also retaining the hierarchical structure of the real world. However, in order to implement an ABM it is necessary to know the rules by which the virtual agents will behave in the model. FCMs may provide a means to describe some of those rules in a formal and structued way, as well as providing additional useful information to the analyst. This chapter is an examintation of these suggestions in the context of specific case study of agri-environment measure uptake by Belgian farmers.

The case study is discussed in later sections. In the remainder of this introduction we briefly describe some general aspects of ABMs and FCMs highlighting a mathematical connection between them; their roots in Markov process models. We then describe some generalised FCMs which have been applied to the study of agricultural intensification and environmental management. The two examples we use, together highlight the importance of the hierarchical problem structure, already noted above, which arises in the agri-environment policy area and we draw readers' attention to the similarity in problem structure discussed in this chapter and the one described by Kafetzis *et al.*, in their chapter in this volume.

1.1 ABMs in Land Use Modelling

As already noted an ABM approach preserves the hierarchical structure of the real world problem by allowing the policy-makers' objectives to configure some aspects of the ABM-world in which the modelled agents interact. The extent to which these objectives are satisfied is then an emergent property of the population of interacting agents (who are, of course, acting according to a set of behavioural rules that will include their own objectives).

ABMs are a class of simulation model which can be traced back to the pioneering work of Thomas Schelling in game theory (Schelling 1967) and the development of cellular automata (CA); starting with John von Neumann's CA models from the 1950's (Burks, 1966). The classical Schelling games and CA models are examples of finite state computational machines in which the interacting components can only take on discrete values from a finite set. In addition, there are typically a finite (and usually constant) number of interacting components and the interactions are typically modelled as discrete time, first order Markov processes (Meyn & Tweedie, 1993, see also Kafetzis *et al.* in this book). The transitions in the state of the modelled system between time points are encoded in sets of rules by which the values of the components are updated. Modern ABMs build on these basic concepts, but rather than modelling agents as cells in a fixed spatial arrangement (as in a CA) they may move over a lattice which can represent abstract or actual physical space (Bous quet & Le Page, 2004; Matthews *et al.*, 2007).

Clearly, ABMs require sets of behavioural rules to encode agent behaviour. One approach to making the rules correspond to farmers' behaviour in the real world, for agricultural systems anal;ysis, is to derive them from empirical data gathered from attitudinal or behavioural surveys. The process of moving from data to rules is recognised to be a problematic area for this type of research and is part of the process of ABM construction that might be improved by using FCMs as an intermediate step between the raw data and the final ABM rules. The rules must specify what the agent's action(s) will be depending on the values of input parameters at each time step. The input values may be from parameters that apply globally to all agents; they may be location specific (so that only agents with particular spatial references are dependent on them), or apply only to certain classes of agent at particular times. These differences essentially reduce to technical issues of implementation which are not the subject of this chapter. The main points to note about ABMs are that they are essentially rule-based and that they are typically Markov processes; these are basic characteristics which they share with *entailment projection* (see Kafetzis *et al.*, in this book) based on FCMs. It is this basic connection between ABMs and FCMs that motivates our investigation of the possibility for combining them.

1.2 FCMs: Generic Features and Specific Applications in Agri-Environment Analysis

FCMs are directed cyclic or acyclic graphs. (Axelrod, 1976) introduced binary cognitive maps and the subsequent development of FCMs was led by Kosko (1986; 1987). In FCMs the graph nodes refer to events, concepts, or quantities in the world while the graph edges represent causal connections between the nodes. A FCM is, then, a graphical summary of a set of entailments, or causal statements. The numerical values of the edges are the fuzzy weights, fuzzy degrees of belief, or fuzzy entailments. In a standard FCM these weights are typically crisp, real numbers in the interval [-1,1], meaning that FCMs are deterministic mathematical models which generate the same output every time a specific set of inputs is supplied. Specifically, as noted in Kafetzis *et al.* in this volume, FCMs are usually employed as first order Markov processes using matrix multiplication (see also Taber, 1981). Matrix Markov process models occur in a wide range disciplines as discussed by Kafetzis *et al.* FCMs, as a specific sub-group of matrix Markov models, have been used for, (among other thing)s: analysis of the sensitivity of the building construction process to different sources of problem (Dissanayake & AbouRizk, 2007); representation of biological processes in bioinformatics (Ettinger, 2002); development of automated supervisory process control systems (Stylios & Groumpos, 1998); medical diagnosis (Froelich & Wakulicz-Deja, 2008, Stylios *et al.*, 2008), analysis of socio-economic problems (Taber, 1981) and; failure mode and effect analysis (Paelaz & Bowles, 1996).

The focus of this chapter is on the value of FCMs as a tool in environmental and ecological modelling. The debate over sustainability and resilience of human activities has led to an increasing awareness of the need to analyse human-environment interactions in an holistic way; as socio-ecological systems (SES) (Gunderson & Holling, 2002; Giampietro, 2004). A feature of SES is that they combine many different types of process, including biophysical, economic and information flows.

This multiplicity of characteristics makes traditional approaches to modelling difficult to implement (Giampietro, 2004). In this situation, the focus which FCMs place only on cause and effect rather than mechanism is an advantage in making headway in understanding the possible dynamics of such systems.

Agriculture is one of the most obvious and important classes of SES, and one in which purposeful human management of the environment is being attempted on a huge scale. In industrialised countries there is a near-universal recognition that intensive agriculture has become unsustainable and active measures are being implemented to attempt to balance food production and environmental health. The balancing act being attempted in the EU, for example, through its CAP (as described above) is one of the most notable examples of these efforts. Aspects of the trade-offs entailed in this balancing act have been studied using FCMs at different scales.

McRoberts *et al.*, (1995), for example, used an FCM derived from a cause and effect diagram drawn by Verijken (1992) to illustrate the factors involved in the intensification of agriculture. The dynamic analysis conducted on the FCM by McRoberts *et al.* (1995) supported Verijken's (1992) original contention that the relationship between agriculture and free markets for food, leads to a closed, positive feedback loop of technological development and intensification. The knock-on effects of this closed loop are a reduction in social welfare resulting from increased unemployment and decreased income, and an increase in pollution. We examine Verijken's (1992) model here in more detail.

Verijken's (1992) cognitive model is reproduced in the upper half of Figure 1 together with the projected state activations when the model is projected as a three-valued cognitive map (i.e. fuzzy weights on the edges take only values of -1, 0, or 1 depending on the direction or existence of causality between concepts) The analysis supports Verijken's (1992) contention that the system is locked in a feedback loop of undesirable effects. In fact, this qualitative result can be inferred directly from the map since it is clear that there is no causal feedback from either of the regulatory policy nodes (*i.e.* Subv and Restr) to the feedback loop at the top of the map. In the lower half Figure 1, we have added such causal links and repeated the projection from the same initial state vector. The added connections are between the community cost and the two regulatory policy nodes and between these nodes and intensification and technology. The interpretation of these changes is to introduce a "polluter pays" feedback into the system: The higher the cost of market driven intensification, the greater the input to subvention and regulatory policies and the more these act to restrict intensification and technology. Note that no direct feedback is applied to the market itself in this analysis. The middle set of projections shows the result of maintaining this pattern of feedback throughout the simulation. The final set of projected data starts with the policy feedback mechanisms just described. In the projection process, thresholding was used so that states with values >0 are depicted as active (solid) while those with values ≤0 are depicted as inactive (open).

After 10 cycles the system enters a stable state in which only subvention policy and community cost nodes are active; the real world interpretation of this result being that imposition of the feedback "polluter pays" policy eventually suppresses the cycle of intensification. Since continuing to enact a costly policy is itself a cost on the community, it would be beneficial if the policy could be de-activated, provided that the system did not return to its previous, damaging configuration. This issue is assessed in the final set of output by using the orginal FCM to project state vector forwards after cycle 10. It can be seen that the damaging positive feedback loop does not become active again. A ranking of the relative cost of these different projections can be made by calculating the proportion of cycles in which the cost node is active (making the assumption that within each cycle, cost is a constant value). The ratio of these cost values, following the sequence from the upper (original Verijken model) to the middle (continuous policy) to lower (discontinued feedback) projections is: 1: 2.14: 1.32.

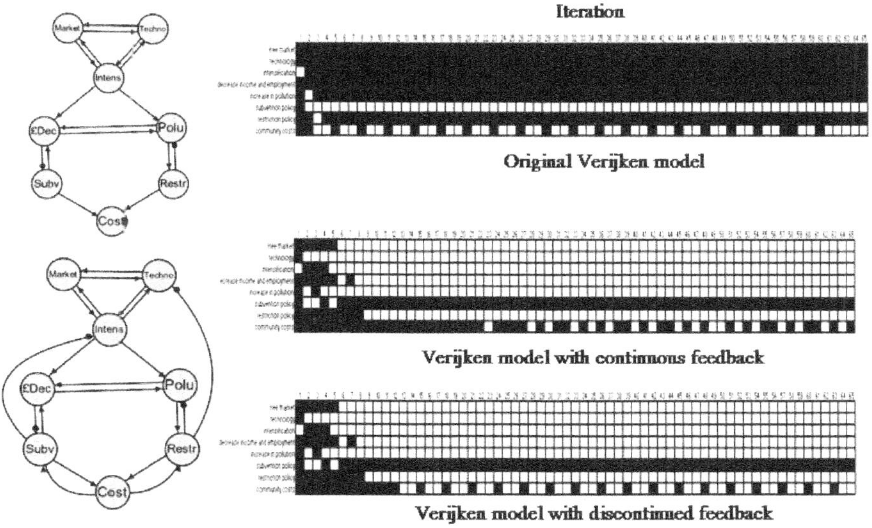

Fig. 1 Simulation of a cognitive map concerned with the relationship between intensive agriculture and negative externalities. The nodes in the map are: Market, free market trade; Techno, technological development in agriculture; Intens, intensification in agriculture; £Dec, decrease in income and employment; Pollu, increase in pollution; Subv, subvention policy; Restr, restrction policy; Cost, community (or social) cost. Filled cells in the projection output indicate active states, empty cells indicate inactive states

The process followed in Figure 1, and described in the preceding paragraph, is essentially one of examining the consequences of internalising the external costs of intensive agriculture. The analysis suggests that given suitable feedbacks these externalities can be controlled, but of course the analysis is conducted at a very broad scale and leaves open the question of how the necessary feedback would be enacted in reality. Answering that question moves us from the broad scale analysis of the previous example, to one concerned with linking actions by individual farmers to policy objectives; this is the FCM analogue of the problem structure (although not the specific detail) which is required for an ABM to explore the issue of agri-environment measures and their uptake by farmers. To begin to explore this issue we examine an hierarchical FCM model of weed control in which domains representing farmers' behavior and policy implementation either do or do not interact.

1.2.1 An Example of FCM Analysis Applied to an Hierarchical Problem Structure

Weeds are an important component of agricultural biodiversity. Plants are the primary mechanism by which energy enters the ecosystem, so much of healthy ecosystem function, and maintenance of biodiversity, depends on sufficient functional plant biodiversity. One of the main impacts of agricultural intensification has been a reduction in both the number and diversity of weeds in agricultural ecosystems (Doyle *et al.*, 2000; Heard *et al.*, 2003). Given the functional importance of weeds in biodiversity and their potentially damaging effects on farmers' financial livelihoods, it is not surprising that the issue of whether weeds can be managed in such a way as to satisfy both farmers and ecosystem service requirements has received a great deal of attention.

The issue can be framed in economic terms by considering the leaving weeds in a crop represents an opportunity cost to farmers, who could utilise the resources captured by the weeds to increase crop yield and thereby increase their income. In contrast, for policy makers, acting to maintain ecological goods and servies for the public, weeds have positive utility. We see in this simple example, then, a specific case of the hierarchical optimisation problem that was mentioned very early in this introduction. The problem has been partially analysed previously by McRoberts & Hughes (2001) using a FCM to illustrate the potential dependence of farmers' willingness to adopt evidence-based weed control measures on a financially-based inducement, expressed in terms of the utility of weeds to the farmers. As with the previous example, we expand here on the original analysis conducted by McRoberts & Hughes (2001).

McRoberts & Hughes (2001) used the implicit assumption that agri-environment payments made from the policy level down to individual farmers could be used by society to buy ecological services from farmers. Here we extend that analysis by making the implicit hierarchical problem structure explicit.

Figure 2, shows an hierarchical FCM which describes the structure of the problem. The problem is seen to comprise two levels. The lower or inner level contains the decision-makers' FCM and is labelled DM-FCM. This is the level at which farmers make decisions about weed control. These decisions depend on the degree to which farmers have a positive perception of weeds. The weed population comprises two parts; the weed plants themselves and a seed bank. The upper or outer level of the model contains the policy-makers' FCM and is labeled PM-FCM. This level contains factors which interact in forming the policy-makers' utility evaluation of weeds and consequently whether or not an agri-environment measure (AEM) to protect weed biodiversity is enacted. Factors such as public pressure and policy cost are present at this level.

The upper set of projection data in Fig. 2 illustrate the behaviour of the system when there is no connection between the upper and lower levels in the hierarchy. It can be seen that the dynamics of the two sub-FCMs are independent and quite different. A real world interpretation of this output is that the policy debate operates in a vacuum because it is not connected to farming or to indicators of the impact of farming on the ecosystem. At the same time, the intrinsic dynamics of farming operate without reference to the policy debate.

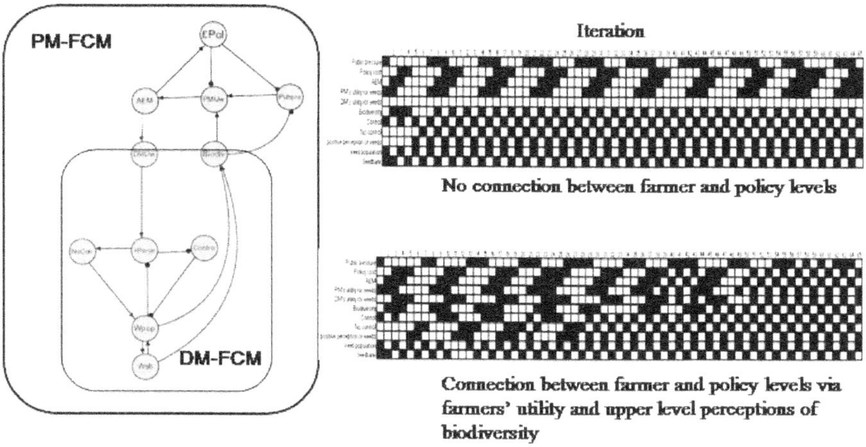

Fig. 2 An hierarchical FCM representing an agri-environment policy issue; in this case how to manage weed biodiversity. The inner or lower part of the FCM represents the farmer (decision-maker, DM) problem (i.e. weed control). The upper or outer level represents the policy-makers' problem (i.e. how to manage the environment). The two levels can be connected through the famers' utility evaluation of weeds, which is influenced from the upper level by agri-environment payments, and the level of biodiversity generated by farming, which influences public pressure on the policy-makers and directly the policy-makers themselves. The two sets of projection data show what happens when the upper and lower systems either are not (upper data set) or are (lower data set) connected

In the lower set of output a long feedback loop between the two levels of the system is activated. Information flows from the upper level into the lower level because the activation of AEM now increases weeds' utility to farmers, increasing the extent to which they are perceived positively and thus reducing the likelihood that they will be subject to control efforts. At the same time, information on the state of biodiversity flows out from the lower level to the upper level and influences public pressure and also policy-makers' utility assessments of agriculture. Although the dynamics of the two parts of the system are initially still somewhat different, eventually (after approximately 50 cycles) the dynamics of the whole system come to resemble those of the lower level. A real world interpretation of this result is that the policy making cycle comes to respond more directly to the dynamics of the system that it refers to. While it might be argued that this is a good thing[1], one might take a more pessimistic view and argue that short cycles imposed on the system would be difficult to deal with in practice; *i.e.* that the policy cycle would be over-responsive to short term effects, resulting in knee-jerk policy formulation. We do not attempt to resolve this issue here, but leave readers to consider it, and how to represent potential solutions, in more detail (in an FCM framework, of course) at their leisure.

The previous example illustrates how an FCM can represent behavioural rules for agents; in this case famers and policy-makers. In the example we invented the rules,

[1] Indeed, it is the basis of evidence-based policy formulation and a central justification for the effort expended on identification of indicators of system properties such as sustainability.

with reference to experience of the subject at hand, simply to illustrate the concept. However, for more realistic models, we have been interested to use FCMs as a means to represent empirical rules extracted from questionnaire/interview data provided by farmers. FCMs have some features which make them particularly attractive in this respect. Not the least of these is that, as will be apparent from other chapters of this book, FCM analysis is far more than simply a means to capture and organise information. Even without its use in developing rule bases for ABM, it can be used for policy analysis in its own right; essentially FCMs can provide a prediction of system behaviour against which the ABM can be evaluated, or they can be used as a precursor to ABM to describe behavioral rules. In the following case study, we describe how we have used FCMs to capture the causal relationships which Belgian farmers recognise among variables associated with environmental management, that reflect their perception of the specific policy options. The case study also provides an illustration of the way in which the structure of the FCMs can be used to provide information on similarities in the mental states of different farmers and on the quality of information collected with respect to the main focus of an interview.

2 Case Study: FCMs and Farmer Types with Respect to Agri-Environment Measures in Belgium

The FCMs analysed here were generated from a database of questionnaire responses and interview transcripts collected as part of a research project on agri-environment measures in Europe. The data allowed us to extract causal relationships between the general concept of agri-environment measures, specific agri-environment measures and a wide range of economic, social and biophysical variables from a group of 20 Belgian farmers. Because the data set was not originally intended to be used in this way, it was not possible reliably to extract causal strengths (i.e. fuzzy weights) of connections identified in the farmers' questionnaire responses; we were restricted to inferences as to whether causal relationships were positive or negative. Thus, our analysis of individual maps is restricted to three-valued maps in which the set of values of possible connections is the set {-1,0,1}. In fact, much of our analysis makes use of the adjacency matrix for the FCMs so only the values [0,1] are employed.

Static, graph theoretic properties of the 20 individual FCMs, based on three-valued weights for the connections, are summarised in Table 1. The number of nodes (*N*) identified by the farmers ranged from 17 to 37, while the number of causal connections (*C*) per map ranged from 19 to 51. In total the 20 farmers identified 183 concepts associated with implementation of agri-environment measures. Figure 3 shows three example FCMs from the analysis. In each case, the shaded node at the centre of the map is the node for agri-environment measures; the three maps were selected to illustrate the range in number of nodes across the sample of farmers.

The main issue facing the policy modeller is how to translate the information captured in the FCMs into something useful for modelling agents' behaviour. Recall that the aim of the modeller is to represent the behaviour of the farmers so that it can be used as part of a policy analysis model.

One approach would be to populate the ABM with as many types of agent as there are FCMs in the empirical data. However, since every FCM is unique, this approach would result in as many types of agent as there are farmers and thus no reduction in the complexity of the real world in constructing the model world. If the model world is not a simplification (or perhaps better put, a generalisation) of the real world it will not be informative as a policy analysis tool; the policy analyst may as well study the real world and dispense with models, but would have to accept that *ex ante* analysis would be very limited in such a situation.

Table 1 Graph theoretic variables calculated from adjacency matrices for cognitive maps elicited from 20 Belgian farmers

Farmer	N[1]	C[2]	T[3]	R[4]	D[5]	h[6]	Deg[7]	[8]var_{in}	[9]var_{out}
1	33	37	12	8	0.03	0.01	1.12	2.42	0.98
2	26	29	9	5	0.04	0.02	1.12	3.79	0.83
3	17	20	8	6	0.07	0.05	1.18	2.78	1.40
4	20	24	10	3	0.06	0.04	1.20	7.01	0.59
5	25	27	6	10	0.04	0.02	1.08	1.83	1.99
6	26	27	11	9	0.04	0.01	1.04	3.40	1.08
7	30	31	8	11	0.03	0.01	1.08	1.69	1.27
8	37	51	23	6	0.04	0.02	1.38	7.74	1.13
9	24	25	6	10	0.04	0.02	1.04.	0.65	1.87
10	22	19	9	9	0.04	0.00	0.86	0.79	0.69
11	21	20	10	6	0.05	0.02	0.95	4.55	0.85
12	20	23	6	4	0.06	0.03	1.15	2.34	0.66
13	18	19	7	4	0.06	0.04	1.06	2.06	0.88
14	28	29	11	9	0.04	0.01	1.04	2.48	1.44
15	23	25	17	14	0.05	0.02	1.09	2.90	0.72
16	25	25	10	8	0.04	0.02	1.00	2.58	0.92
17	31	35	12	9	0.04	0.01	1.13	2.98	1.45
18	25	24	11	10	0.04	0.02	0.96	4.62	2.04
19	17	22	4	6	0.08	0.06	1.29	1.35	2.22
20	20	20	13	14	0.05	0.03	1.00	1.26	1.37

[1]Number of nodes; [2]Number of connections; [3]Number of transmitter nodes; [4]Number of receiver nodes; [5]Map density; [6]Map hierarchy index; [7]mean in (and out) degree – mean indegree and outdegree are equal when calculated on binary matrices; [8]variance of indegree values over nodes; [9]variance in outdegree values over nodes.

FCMs offer two ways of considering agents. In one approach we can look at the *structure* of the FCMs which have been elicited from the farmers. This approach will capture properties of the rule bases which agents can use and, in this way, on the type of information that the farmers give to the interviewer. In the alternative approach, we could analyse the specific cause and effect relationships which are encoded in the FCMs and use these in constructing the agents' rule bases. Following this approach we will be likely to end up with rule bases that deal specifically with the problem posed in the information elicitation phase.

In order to extract this type of information from a set of FCMs the analyst can use a range of multivariate statistical methods which are readily available in most statistical analysis software. In our own research to date we have made use of two different types of methods; hierarchical cluster analysis and principal components analysis. A detailed description of these techniques is beyond the scope of this chapter, but Krzanowski (1988) gives a unified and very clear introdution to this entire field of statistical analysis.

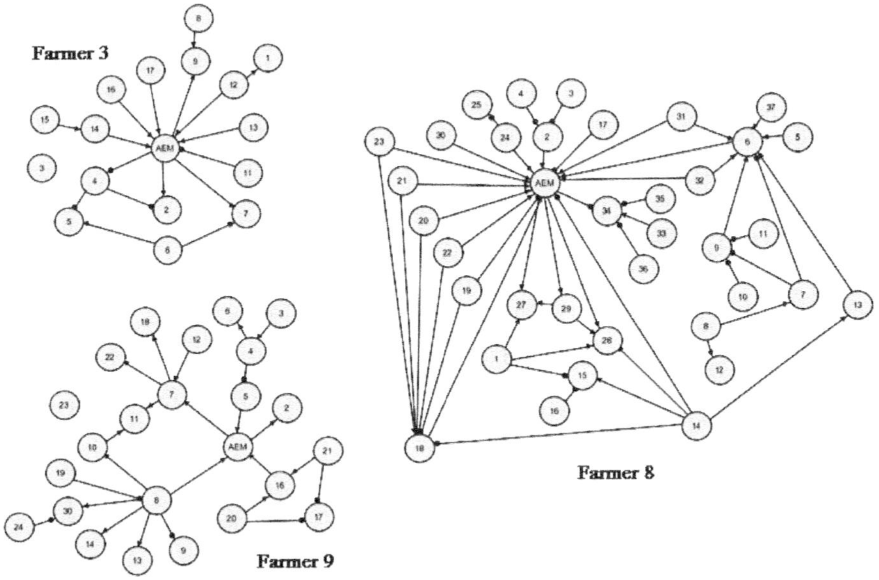

Fig. 3 Three cognitive maps repres0065nting the range of map size recovered from questionnaires and interviews with 20 Belgian farmers on the subject of agri-environment measures. Causal effects are indicated as either • = negative, or ▸ = positive. Note that the similarly numbered nodes in different maps do not necessarily correspond to the same concept

The starting point for both types of method we have used is a two-way array of data, arranged with the individual farmers as the rows and the FCM attributes as the columns. In the similarity analysis, the columns of data are considered to be coordinates for the farmers in a multidimensional data space. The aim of the analysis is, then, simply to calculate the similarities among all the farmers in that space and render an overall summary of the information in a format that can be easily understood. In many applications the usual format is to represent the similarities as lengths in a rooted tree, commonly referred to as a dendrogram (Krzanowski, 1988).

In the second approach, the starting motivation is again to view the data table as giving information about similarities among the farmers in a multidimensional space, but in principal components analysis the aim is to construct a lower dimensional sub-space from linear combinations of the original variables which best represents the distance relationships contained in the initial data. This process reveals not only which FCMs are similar to each other overall (by inspecting which ones end up close to each other in the constructed sub-space), but also provides information on the correlation or co-variance structure among the variables extracted from the FCMS. Both types of information can be inspected together in a graphcial device known as a biplot (Kraznowski, 1988).

2.1 Clustering Based on Map Structure

Graph theoretic variables derived from the FCMs were used to perform an hierarchical cluster analysis of the similarity among the 20 farmers. The variables were all continuous in nature and the similarity matrix among farmers was constructed using a Euclidean distance metric; clustering was performed with a furthest neighbour algorithm implemented in Genstat (Release 11.1, VSN International,. Figure 4 shows the dendrogram generated from the analysis, with an insert showing the relationship between the value of ln(D) against ln(N) for each farmer's FCM (see below).

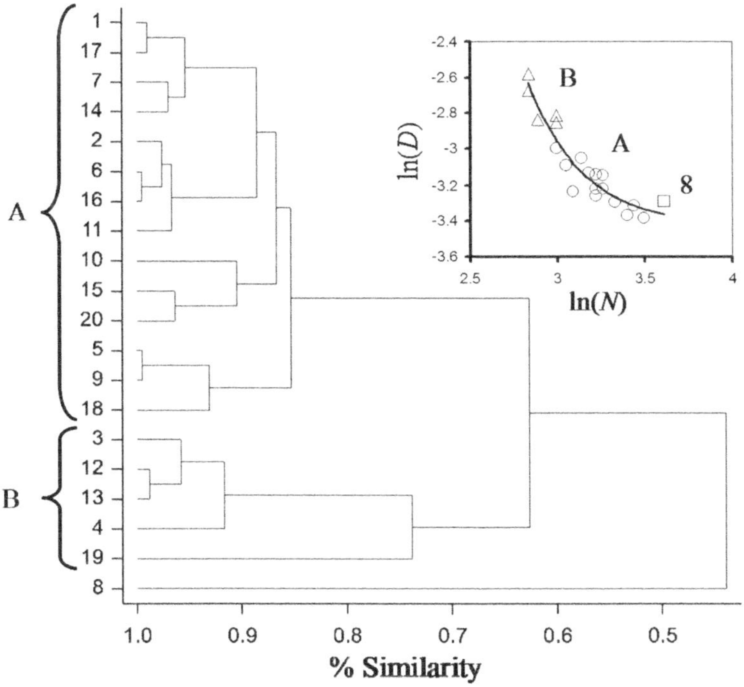

Fig. 4 Main figure: Furthest neighbour dendrogram derived from graph theory variables for 20 cognitive maps elicited from Belgian farmers. Clusters A and B and the unique individual (Farmer 8) are apparent at 75% similarity. Insert: the relationship between the natural logarithm of node number, N, and map density, D, for the same data. Note that cluster B comprises individuals with relatively high density maps

The cluster analysis of map structure suggested that there were two main types of agent, with a third type represented by a single individual. The smaller cluster (B) comprised the 5 individuals with highest map density values. As is well known, map density (*D*) is the proportion of all possible connections which are actually present in a map. Since connections are causal relationships between concepts, higher densities indicate agents whose behavioural response to changes in the world is potentially more variable than those with relatively low *D* values (Özesmi & Özesmi, 2004) For the modeller, the analysis suggests the need to generate rule bases for three different types of agent who will differ in the potential variability of their behaviour in response to changes in the ABM world.

The insert in Figure 4 shows empirical relationship between ln(*N*) and ln(*D*) estimated over the sample of farmers. The statistical relationship in the data is an emergent property of the sample of farmers. In this case a negative exponential relationship between ln(*D*) and ln(*N*) is apparent, with the individuals in cluster B at one end, the unique individual Farmer 8, at the other, and the remaining individuals (comprising cluster A) between these extremes. The emergent relationship can be described mathematically as indicated in equation 1.

$$\ln(D) = \alpha + \beta \cdot r^{\ln(N)} \quad (1)$$

The fitted relationship between ln(*D*) and ln(*N*) was obtained by non-linear regression analysis using maximum likelihood to obtain the parameter estimates. The fitted relationship explained 89% of the variance in ln(*D*); the parameter estimates (s.e. in parentheses) were, α = -3.4 (0.09), β = 6933 (16667), r = 0.041 (0.0357).

One might ask whether the observed trade off between ln(*D*) and ln(*N*) is a particular example arising from a general (and theoretically justified) phenomenon, or is simply an artifact of these data? One potential theoretical justification is as follows.

Since both the number of nodes and the number of connections are components of the complexity of an agent's cognitive model, one might hypothesise that some combination of *N* and *C* will set a limit on the overall complexity which a cognitive model can have. Here we are taking a view of cognitive processing capacity in keeping with the ideas proposed Halford *et al.* (1998), such that "*limits are best defined in terms of the complexity of relations that can be processed in parallel. Complexity is defined as the number of related dimensions or sources of variation*". That is, if there is an upper limit of complexity on the representational models an agent can form, and complexity depends on both *N* and *C*, then one might hypothesise that agents will be able to maximise *N* or *C* in forming a model, but not both. However, other things being equal, we would expect larger maps to have more connections[2], so a complexity limit might be expected to affect not the number of possible connections, C_{max}, but the proportion of possible connections which is filled; this is the map density, *D*. This hypothesis leads to an expectation of a negative correlation between *N* and *D*, as observed.

[2] The number of possible causal relationships in a FCM, C_{max}, increases with the square of N: $C_{max} = N^2$ if self loops are possible, or $C_{max} = N(N-1) = (N^2-N) \cong N^2$ for large N if self loops are not possible.

Generalising, we can say that the trade off between $\ln(D)$ and $\ln(N)$ suggests that people lie somewhere on a continuum between "*cataloguers*" (who maximise N) and "*connectors*" (who maximise D). The trade off means that the number of concepts which a map contains can increase as long as the number of causal connections among concepts decreases. However, without causal connections among its nodes a FCM model will be limited in its usefulness for inferential analysis; tending to be simply a catalogue of potentially relevant concepts.

2.2 Clustering Based on Map Content

The clustering exercise described in the previous section was intended to identify potential types of agent based on the structure of the FCMs elicited from the Farmers. For a specific application to policy questions concerning agri-environment measures it would clearly be useful to know not only broadly what types of individual comprise the population, but also something about the specific concepts and causal relationships they consider. To examine this issue we employed a different approach to detect clustering within the FCMs.

Recall that our context for this analysis is a bi-level problem structure in which the farmers will be represented by model agents at the lower level whose aggregate behaviour will result in outcomes (for example environmental quality) against which policy objectives will be assessed at the upper level. The question for the policy analyst is two-fold: What are the drivers of behaviour which can be affected by pulling policy levers? What are the outcomes that the farmers view as important and against which policy success can be measured?

We examined the FCMs and identified six driving variables and three outcome variables. The drivers were: (*i*) subsidies; (*ii*) presence of less productive land; (*iii*) cereal prices; (*iv*) exterior constraints; (*v*) fixed dates (associated with implementation of particular agri-environment measures); (*vi*) inflexible regulation. The three outcomes were: (*a*) soil erosion; (*b*) public image of farming; (*c*) environmental quality.

To examine relationships among farmers with respect to these drivers and outcomes we constructed a 20×9 binary data matrix by recording whether each driver or outcome was present in the FCM for each farmer. A principal components analysis was used to examine the relationships between the farmers and the driver and output variables by analysing the correlation matrix for the data. Figure 5 summarises the results of this analysis.

For the policy modeller the pattern of vectors and points in Figure 5 suggests (as might be expected) that there are varying degrees of association among the drivers of behaviour and policy outcomes for the set of farmers. Avoiding becoming too engrossed in methodological detail, the analysis suggests that three broad types of agent should be included in the ABM world: a first type who respond most strongly to drivers concerned with commodity market prices and regulatory constraints; a second type who are interested in land productivity and subsidies and who connect these drivers to soil erosion, and a third type who are not strongly concerned with any particular driver but are motivated by the public's

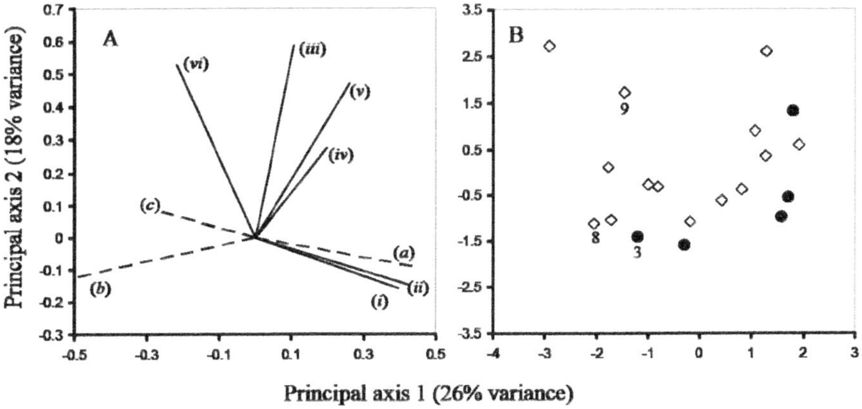

Fig. 5 Ordination diagram showing the output from a principal component analysis of FCM concepts identified as drivers and policy outputs for a set of Belgian farmers. A: vectors representing the drivers (i) – (iv) and outputs (a) – (c) as in the main text. B: Clustering of the farmers in the principal component space. The open symbols are farmers from cluster A in Fig 1., the closed symbols are farmers from cluster B. Farmers 3, 8, and 9 whose FCMs are shown in Fig.1 are identified by numbers

image of farming. The distribution of farmers from the previous clustering exercise in the principal components space suggested that each of the three types just identified should contain a small number of agents whose rule bases are based on relatively high density maps, and a larger number of agents with rule bases reflecting lower density maps.

In all, then, the cluster analyses of FCM structure and content suggested a set of six types of agent for the ABM world. These static analyses give us some idea of what will constitute the agents' behavioural rules, but say nothing about how a world populated by these agents will evolve over time. Some clues as to that issue are available by using the FCMs to iterate any initial set of states over some notional time scale. We illustrate this concept briefly in the next section by examining the responses of three farmers' conceptual models to a policy decision to maintain support for agri-environment measures via two drivers: support for maintaining farming's image with the public and subsidies to farming.

2.3 Anticipating Isolated Agent's Behaviour

Figure 6 shows the output from iterating the FCMs shown in Figure 1 for 20 cycles. In each case the left hand diagram shows the pattern of concept activations when inputs to the map are made at the initiation cycle (cycle 0) and then the FCMs are left to equilibrate without further input. Open cells in the output correspond to inactive concepts, filled cells correspond to active concepts. The right hand diagrams show the pattern of concept activations when inputs to the nodes for "farming's public image" and "subsidies" are made at every cycle. The row in each output corresponding to agri-environment measures (AEM) is indicated.

Fig. 6 Iterated output from three FCMs (see Figure 1) elicited from Belgian farmers on the subject of agri-environment measures (AEM). In each case the left-hand diagram shows the output when activations are made only to the initial cycle, while the right-hand diagram shows the output when input to two driving variables is maintained at each cycle

Several points are worth making about the output from the analysis. First, irrespective of the complexity of the FCM, if no input to driving variables is maintained the systems conceived by all three farmers quickly reach a state with no active concepts. Secondly, in all three cases, maintaining input to support the public image of farming and to subsidies leads to continuous (or near continuous) activation of the AEM node. Thirdly, when input is maintained, we can see that the relative number of *active* nodes per map decreases from the map with highest density (farmer 3, 7/19 concepts active) to the map with the lowest density (farmer 8, 6/37 concepts), with the intermediate density map (farmer 9) between these (8/24 concepts active). A final point of interest, although not one readily apparent from Figure 4, is that while all three FCMs imply that AEM can be maintained in an active state by continuous input to driving variables, none of the specific measures (for example, installing refuges for beneficial insects) was maintained in an active state. This analysis suggests an intriguing disconnection in the minds farmers between the general concept of AEM and the specific mechanisms for delivering environmental goods from farming. Such a disconnection may or may not be important to policy analysis (and hence to the efforts in constructing an ABM) depending on the level of detail at which the policy makers wish to analyse their problem.

3 Conclusions

3.1 General Comments

It is widely recognised that successful, formalised policy analysis must combine elements of objective methodology with realism based on input from relevant stakeholders (Parker *et al.*, 2003). Furthermore, many policy problems are at least bi-level (Candler *et al.*, 1981) so it is necessary for the policy analyst to construct models that take account of objectives at least two scales of resolution. It is clear that ABM has the potential to meet these criteria but that theory and practice are still some distance apart (Schmit & Rounsevell, 2006). Part of the difficulty in successfully applying the ABM approach to policy problems is to obtain sufficient data to construct the models. Because FCM construction focuses on the cause and effect relationships entailed in either qualitative or quantitative data, we believe that it offers a useful methodology for eliciting rule bases from data. Indeed, FCMs have proved useful in studying a wide range of complex problem types (e.g. Glikas & Xirogiannis, 2004; Chytas et al., 2006; Xirogiannis et al., 2008). While the analyses we have presented here are for a single case study and relate to only one type of problem, we believe that the approach of combining FCMs with ABMs will prove to be useful in many analogous arenas.

3.2 Statistical Analysis

Applying statistical analyses to the elicited FCMs from the farmers leads to two general results. First, straightforward listing of the causal relationships captured in the maps provides a catalogue of potential behavioural rules to include in the ABM. Secondly, analysis of the graph-theoretic statistics for the farmers gives us some idea of the types of cognitive agent who are present in the sample; i.e. it gives some guidance to the levels of rule-base complexity to include in the ABM. Finally, analysis of how the pattern of causal relationships among concepts is distributed over the set of FCMs gives some indication of which variables in the ABM should be used to connect the modelled agents with one another and with the larger scale features of the ABM world.

Esssentially the same concepts were suggested by Dickerson & Kosko (1994) when they suggested using FCMs as a means to construct virtual worlds. In their example, rule bases describing elements of the behaviour of predators and prey were linked together in an overall FCM. This optimistic view notwithstanding there are many issues to be overcome in developing the methodology we propose. Not the least of these is how one should handle the philosophical question of what the output of an FCM *means* in relation to other model representations of the same system and indeed the real world.

3.3 Identification of FCM Models with Reality: Lessons from Other Types of Model

A case study chapter may not seem like an obvious place to discuss such a point; one might be tempted to say that such things would be better dealt with in a chapter on theory or methodology. However, we suggest that it is in practical application that such questions become critical and thus ought to be discussed. After all, it matters less if purely abstract entities have physically impossible characteristics in a model than it does if the same is true of model analogues for real entities. This is particularly true when, as in our case, we wish to compare one type of model (FCM) with another (ABM) and use both types to inform decision-makers about potential properties of complex systems embeded in the real world.

If the nodes in an FCM represent real quantities, it is not clear how the process of thresholding and normalisation applied to the iteration outputs should be interpreted. Standard results from the analysis of the eigenvalue structure of constant matrices (May, 1974; Caswell, 2001) mean that the quantities being projected will either grow or shrink exponentially with the number of iterations depending on whether the value of the largest eigenvalue (λ_1) is >1 or <1; in the rare case that $\lambda_1 = 1$, the output will be constant. In a physical system, the eigenvalue results can be interpreted as implying that there either is ($\lambda_1 < 1$) or is not ($\lambda_1 > 1$) a local stable equilibrium in the dynamics. In the analysis of such systems the exponential growth over iterations implied by $\lambda_1 > 1$ is accepted as a limitation of the model structure; the same acceptance does not appear to be a feature of FCM analysis, leaving this potentially useful technique open to the question of what the outputs from FCM models mean. Of course the FCM community is not alone in being open to the criticism of constructing meaningless models. Few books on modelling attempt any sort of philosophical discussion of models. Doucet & Sloep (1992), however, discuss such issues at length (in their Chapter 13 in particular) and provide the invaluable advice that all models should have an identification statement explicitly associated with them. The role of the identification statement is to formally connect the model with the real world by stating of which part(s) of the world it is a model. As Doucet & Sloep (1992) point out without such a statement, the model remains an abstract entity and the logical entailments within it remain un-testable. It is particularly important in real world application, where models are used to inform decision making, that they are clearly identified with the systems they are supposed to represent.

3.4 Future Research

We are actively pursuing the issue of model identification with the real world in our research in FCMs and ABMs. In addition as the FCM and ABM approaches can be used either in conjunction or separately one potential area of future research will be in formal comparison of the quality of model inferences arising from each type of model applied to the same policy problem. Finally, a more practical future objective will be to continue the use of FCMs as a component of participatory research in agriculture.

Acknowledgments. Support for this research was provided by the European Union and the Scottish Government.

References

Axelrod, R.: Structure of Decisions: The Cognitive Maps of Political Elites. Princeton University Press, Princeton (1976)

Bousquet, F., Le Page, C.: Multi-agent simulations and ecosystem management. Ecol. Mod. 176, 313–332 (2004)

Burks, A.W.: Theory of Self-Reproducing Automata, by John von Neumann. University of Illinois Press, Urbana (1966)

Candler, W., Fortunyamat, J., McCarl, B.: The potential role of multi-level programming in agricultural economics. Amer. J. Ag. Econ. 63, 521–531 (1981)

Chytas, P., Glykas, M., Staikouras, C., Valiris, G.: Performance Measurement in a Greek Financial Institute Using the Balanced Scorecard. J. Meas. Bus Ex. 10(2) (2006)

Caswell, H.: Matrix Population Models: Construction, Interpretation and Analysis, 2nd edn. Sinauer Associates, Sunderland (2001)

Dickerson, J., Kosko, B.: Virtual worlds as fuzzy cognitive maps. AI Expert, 25–31 (July 1994)

Diassanayake, M., AbouRizk, S.M.: Qualitative simulation of consrtuction performance using fuzzy cognitive maps. In: Proc. 2007 Winter Sim. Conf., pp. 2134–2140 (2007), `http://www.informs-sim.org/wsc07papers/266.pdf`

Doucet, P., Sloep, P.: Mathematical Modelling for theLife Sciences. Ellis Horwood, New York (1992)

Doyle, C.J., McRoberts, N., Kirkwood, R., Marshall, G.: Ecological Management of Crop-Weed Interactions. In: Masae, S., Hiroshi, K. (eds.) Structure and Function in Agroecosystem Function and Design. CRC Press, Boca Raton (2000)

Ettinger, M.: The complexity of comparing reactions. Bioinform. 18, 465–469 (2002)

Froelich, W., Wakuliicz-Deja, A.: Associational cognitive maps for medical diagnosis supoprt. Intel. Inf. Syst., 387–396 (2008), `http://iis.ipipan.waw.pl/2008/proceedings/iis08-38.pdf`

Giampietro, M.: Multiscale Integrated Analysis of Agroecosystems. CRC Press, Boca Raton (2004)

Glykas, M., Xirogiannis, G.: A soft knowledge modeling approach for geographically dispersed financial organizations. Soft Comp. 9, 579–593 (2004)

Gunderson, L.H., Holling, C.S.: Panarchy: Understanding Transformations in Natural Systems. Island Press, Washington (2002)

Halford, G.S., Wilson, W.H., Phillips, S.: Processing capacity defined by relational complexity: Implications for comparative, developmental and cognitive psychology. Beh. Brain. Sci. 21, 803–831 (1998)

Heard, M.S., Hawes, C., Champion, G.T., Clark, S.J., Firbacnk, L.G., Haughton, A.J., Parish, A.M., Perry, J.N., Rothery, P., Roy, D.B., Scott, R.J., Skellern, M.P., Squire, G.R., Hill, M.O.: Weeds in fields with contrasting conventional and genertically modified herbicide-tolerant crops II. Effects on individual species. Phil. Trans. R. Soc. Lond. B 358, 1833–1846 (2003)

Kosko, B.: Fuzzy cognitive maps. Int. J. Man-Mach. Stud. 24, 63–75 (1986)

Kosko, B.: Adaptive, bidirectional associative memories. Appl. Optics 26, 4947–4960 (1987)

Kosko, B.: Fuzzy Thinking. Harper Collins, London (1993)

Krzanowski, W.J.: Principles of Multivariate Analysis: A User's Perspective. Oxford University Press, Oxford (1988)

Matthews, R., Gilbert, N., Roach, A., Polhil, J.G., Gotts, N.M.: Agent-based land-use models: a review of applications. Lands. Ecol. 22, 1447–1459 (2007)

May, R.M.: Stability and Complexity in Model Ecosystems. Princeton University Press, Princeton (1974)

McRoberts, N., Foster, G.N., Davies, D.H.K., Evans, K.A., McKinlay, R.D., Wale, S.: The role of synoptic models in the development of crop protection for sustainable crop produiction systems. In: BCPC Proceedings 63. Intgrated Crop Protection: Towards Sustainability? BCPC, Farnham, UK, pp. 439–446 (1995)

McRoberts, N., Hughes, G.: Killing or culling? Is it possible to manage weeds as a resource. In: Proceedings BCPC Annual Conference, BCPC, Farnham, UK, pp. 383–390 (2001)

Meyn, S.P., Tweedie, R.L.: Markov Chains and Stochastic Stability. Springer, London (1993)

Özesmi, U., Özesmi, S.: Ecological models based on peoples' knowledge: a mult-step fuzzy cognitive mapping approach. Ecol. Mod. 176, 43–64 (2004)

Parker, D.C., Manson, S.M., Janssen, M.A., Hoffman, M.J., Deadman, P.: Multi-agent systems for the simulation of land-use and land cover change: a review. Ann. Assoc. Amer. Geog. 93, 316–340 (2003)

Pelaez, C.E., Bowles, J.B.: Using fuzzy cognitive maps as a system model for failure modes and effects analysis. Intell. Syst. 88, 177–199 (1996)

Schelling, T.: What Is Game Theory? In: Charlesworth, J.C. (ed.) Contemporary Political Analysis. The Free Press, New York (1967)

Schmit, C., Rounsevell, M.D.A.: Are agriclutural land use patterns influenced by farmer imitation? Agr. Ecosyst. Env. 115, 113–127 (2006)

Stylios, C.D., Groumpos, P.P.: Fuzzy Cognitive Maps in Modeling Supervisory Control Systems. J. Intel. & Fuzzy Syst. Appl. Engi. 8, 83–98 (2000)

Stylios, C.D., Georgopoulos, V.C., Malandraki, G.A., Chouliara: Fuzzy cognitive map architectures for medical decision support systems. Appl. Soft Comp. 8, 1243–1251 (2008)

Taber, R.: Knowledge processing with fuzzy cognitive maps. Exp. Syst. Appl. 2, 83–87 (1981)

Verijken, P.: A methodic way towards more sustainable farming systems. Neth. J. Ag Sci. 40, 209–233 (1992)

Xirogiannis, G., Chytas, P., Glykas, M., Valiris, G.: Intelligent impact assessment of HRM to the shareholder value. Exp. Sys. App. 35, 2017–2031 (2008)

Using Fuzzy Cognitive Maps to Support the Analysis of Stakeholders' Views of Water Resource Use and Water Quality Policy

Alkis Kafetzis, Neil McRoberts*, and Ioanna Mouratiadou

1 Introduction

This chapter is based on two separate case studies concerned with water use and water use policy. One is a study concerning public participation in the Water Framework Directive (WFD) of the European Union (EU) (Mouratiadou & Moran, 2007), and focuses on data collected in the Pinios river basin in Greece. The other is based on previously unpublished research by the authors on trans-boundary river issues in the Maritza river basin shared between Bulgaria, Greece and Turkey. Apart from the obvious similarities of location and their focus on water resources, the two studies are linked by some underlying factors that directly impact on our choice of fuzzy cognitive maps (FCMs) as an analytical method.

The chapter is structured as follows. In this introduction we cover three topics. First, we provide a brief overview of some of the general political, economic and environmental issues concerned with water resources. This discussion leads to the identification of *social scarcity* (Hirsch, 1977) as a key economic issue in water resource management and we introduce some of the important properties of a resource that is subject to social scarcity. This discussion highlights the analytical difficulties which socially scarce resources pose for traditional economic analyses and reveals the suitability of FCMs for capturing the complexity and conflicts inherent in such problems. The introduction then ends, logically, with a review of some of the literature on FCMs and related subjects which are relevant to this particular example. The general introduction is followed by separate analyses of our two case studies, each preceded by some introductory information to provide the reader with a context. The chapter concludes with a synthesis section in which we apply the FCM technique to allow us to make explicit a few generic points which emerge from the separate case studies.

Alkis Kafetzis · Neil McRoberts · Ioanna Mouratiadou
Land Economy and Environment Research Group, Scottish Agricultural College, Edinburgh EH9 3JG, UK
e-mail: neil.mcroberts@sac.ac.uk

* Corresponding author.

M. Glykas (Ed.): Fuzzy Cognitive Maps, STUDFUZZ 247, pp. 383–402, 2010.
springerlink.com © Springer-Verlag Berlin Heidelberg 2010

1.1 Water: Political, Economic and Environmental Issues

Water, and particularly fresh water, is one of the most fundamental resources on Earth. It is a basic necessity of organic life, enabling essentially all biological processes to occur. In addition to being a fundamental necessity, water is also a resource of high value in human societies, since it sustains a number of functions, critical for the progress of the human species. These functions can be classified into five categories: (1) health functions, since clean water is vital for human health; (2) habitat functions, since water supports healthy aquatic ecosystems, which in turn are vital for biodiversity and food supply; (3) two production functions, one linked with biomass production and the other with societal development; for example by facilitating industrial development; (4) two carrier functions, that reveal water's erosive/distributive force as a transporter for solid matter, and its utility as a facilitator of human mobility, which can, in turn, generate problems, and; (5) psychological functions which impact on individual utility, and therefore social welfare, since humans generally exhibit a fundamental affinity with water and a deep aesthetic appreciation for views which include it (Ohlsson, 1995, Espey and Towfique, 2004). Considering these points together, it is clear that water is not only a source of life, but also a malign force, a point that further emphasises the importance of successful water management.

The political and economic significance of water management is further strengthened by its physically finite nature. Increased demands, either because of population growth, or because of rising *per capita* consumption, have depleted water resources globally. Estimations of the world water run-off *per capita*, show a steady decline until 2025, when it will be reduced to 4,692 m^3. The United Nations Environmental Programme (UNEP) has assessed that by that time two thirds of the global population will be under water-stressed conditions. Additionally water resources are characterized by a highly uneven global distribution; a fact that makes water use issues more challenging and links them with questions of equity and fairness (Gleditsch *et al.*, 2006). The statements of Ban Ki-Moon, Secretary General of the United Nations, fore wording the 2009 United Nations World Water Development Report, crystallize these points:

> *"It is well known that water is life......water also means livelihoods. It is the route out of poverty for individuals and communities. Managing water is essential if the world is to achieve sustainable development..... This is important not only for development: it is a matter of security, too. Lack of basic services can contribute to political instability..... There has been a widespread failure to recognize water's vital role in providing food, energy, sanitation, disaster relief, environmental sustainability and other benefits..... We must all work together to address this matter of life and livelihood*s." (UNESCO, 2009)

The UNESCO analysis makes explicit the fact that in the modern world water is a socially scarce resource (Hirsch, 1977, see below) in as much as there is no effective way in which people's water wealth can universally be increased by capturing water provision in an economic activity and "growing" more water. The trade-offs and conflicts which this social scarcity imposes are most obvious in the case of rivers which also, often, introduce the additional complexities of trans-boundary politics into the already complex problem of equitable resource use.

Fresh water traverses the planet in various ways, but rivers and streams are the most crucial vehicle for the utilization of this resource by humans, supplying eighty percent of humanity's needs (Espey & Towfique, 2004). Rivers are intriguing phenomena, with various distinct perceptions regarding their nature. On one side there is the view that rivers are simply water flowing towards the sea, while a contrasting view is that rivers are complex, multilayered ecosystems, that involve many elements and functions (McCully, 1996). Following the second view, rivers are acknowledged as an integral part of earth's physiognomy, shaping surrounding landscapes, and connecting otherwise remote areas. Additionally, they are considered a key component for the ecological integrity of the territories they run through. They form part of a feedback loop or cycle, in which they sustain the functions of ecosystems, which in turn sustain river quality (Sadoff & Grey, 2002).

As a result of their central role in ecosystem function, rivers constitute also a crucial parameter for humanity; a case that is documented through the millennia of human history, as great civilizations tended to flourish in the vicinity of rivers (Caflisch, 1996). The economic, socio-political and cultural significance of these formations is stressed many times. Economically, rivers provide communities with many advantages, as they are a water source that can be used for domestic, agricultural, industrial, energy production and transportation purposes. Socio-politically, they have played an important role in the formation of human societies, as an external pressure compelling communities to join forces in order to "tame" them (Sadoff & Grey, 2002). They continue to shape a large part of today's political reality because, in many cases, they form state borders (Poff, *et al.*, 2003). Last but not least, rivers have shaped many cultures and traditions, often being considered a divine blessing.

Rivers do not always fall neatly inside political boundaries. Many rivers around the globe are shared between a number of co-riparian nations, forming additional issues when managing these resources. Managing a river is by itself a demanding task. It requires the comprehension of an intricate issue, which involves multiple environmental qualities that are linked with various anthropogenic activities. This signifies that river management must address many needs, concerns and perceptions and attempt to pull them all together, in order to reach the appropriate configuration of water demand and supply. These challenges are even more demanding when addressing transboundary rivers. Such a case requires a transboundary management, capable of incorporating parameters accruing from different countries. Furthermore, it must recognise that the long-term success of such endeavours entails the comprehension of a broader set of attributes stemming from the international nature of these rivers.

Water, today, is still regarded as the symbol of life, and its significance continues to affect individual and community perceptions. Ultimately water influences national policies, as exhibited in the second half of the 21st century, when the world's rivers were sites of highly intensive development schemes (Benvenisti,

2002). In conclusion, it would be fitting to say that their significance is evident also in the word "river", that shares a root with the word "rival" (Caflisch, 1996), denoting a resource of a highly complex nature that can also give rise to conflicts.

1.2 Social Scarcity and Water Use

Hirsch (1977) provided the first explicit and sustained exposition of the issue of the social scarcity of goods and the connected idea of the distinction between material goods and positional goods. These ideas may be important to understanding peoples' attitudes to water resources and water use policy and, consequently, we provide a brief summary of them here, to prime readers with the relevant information when the concepts re-appear in our later analyses.

A material good is defined as one whose supply can be increased by increased economic activity such that consumption is essentially not limited by supply. Goods of this type are sometimes referred to as "free" goods in the sense that they are free from limitation. As far as wealth is dependent on possession/consumption of material goods, it can increase for every member of society, and society as a whole, in proportion to the total amount of material goods consumed. In contrast to material goods, Hirsch (1977) defined *positional* goods as those which derive at least part of their value from their scarcity. Such goods are valuable exactly because not everyone can have access to (*i.e.* consume) them in equal degree, and often consumption is restricted to a small minority of wealthy individuals. Typical examples of positional goods are things such as higher educational qualifications and public prestige, but goods such as land ownership have a positional element; an example which raises the point that the social scarcity of a good can be produced by the combination of physical scarcity and an existing unequal distribution of wealth that allows those who are already wealthy better access to physically scarce goods. The implications of the analogy with land ownership for access to a diminishing global supply of water are immediately obvious.

Two further points concerning social scarcity are worth making here. The first is that as material wealth increases in a society, increasing numbers of individuals have their basic material needs met and this brings them into the arena where hopes of gaining positional wealth become, to them, realistic. The second point is that such hopes are largely mistaken; the majority cannot be as wealthy as the wealthiest person is now, even if an economy grows, since growth in material wealth cannot lead to a universal increase in positional wealth. In fact, increased material wealth tends to induce congestion (in the generic sense) which reduces the positional value of the congested goods, leading to unfulfilled hopes of wealth and thus to dissatisfaction. On this basis Hirsch (1977) argued that the existence of positional goods as component of wealth can lead to a decrease in social welfare when the desire for such goods is thwarted[1]. More recent analyses of this issue express the point in the terminology of game theory. Consumption of positional goods is a zero-sum game; the positive amount of a positional good

[1] Welfare is used here in the economic sense of the aggregate of individual utility, where utility itself has the usual definition as the benefit or satisfaction obtained from a good or service.

consumed by one or more individuals is exactly balanced by a negative amount consumed by one or more other individuals (Pagano, 1999).

The preceding explication of social scarcity is relevant to our analysis of attitudes to water use and water use policy in the EU for two reasons. First, policy formulation about water management in the EU has been underpinned by a quasi market-economic approach. Public goods, such as water resources, are viewed as having a value to stakeholders which can be expressed in financial terms. Following elicitation or estimation of that value, the aim of efficient policy formulation is to find a policy that allows the resource to be obtained by would-be consumers at the estimated value. In simple terms, taking clean drinking water as an example, stakeholders are invited to set a value on clean drinking water and then policy is used to distribute the costs and benefits of having clean water among all stakeholders. Thus, the process of water use/management policy development is open to the same criticisms of any market-based provision of goods which are not inherently marketable.

Secondly, we contend that such a conception of water resources is inappropriate because it fails to capture the complex way that water is valued by people (as indicated above). None of the valued attributes of water are available to a market formulation of value (even a quasi market established between the public and private enterprises as competing users of a common resource), because none of them defines water as purely material good; many of them have at least some element of positional value. For example, enjoyment of the leisure opportunities provided by rivers may depend, for many individuals, on a relatively sparse use of the same services by other users with all the inherent problems of congestion noted above (Hirsch, 1977; Pagano, 1999). In such cases, leisure opportunities dependent on water suffer from social scarcity. It will also be apparent that the opportunity to consume this service may be in conflict with opportunities to consume other services; *e.g.* fishing may not be compatible with leisure boating.

Clearly, any methodology applied to the analysis of the issues of water use and policy development, must be capable of capturing conflicts and feedbacks among different components of the system. It, must also be able to capture rather abstract concepts such as utility, which are dependent on subjective opinion. Some of these requirements derive from the type of issue described above while others are imposed by goals of policy analysis.

For example, in both case studies there is a need to incorporate stakeholders into a discussion on the use of a resource to develop sound evidence-based policy. The WFD of the EU, which is the subject of our case study from the Pinios river basin, actually demands of member States that they include stakeholders' views in policy implementation; although the means by which different member states of the EU enact public participation in the WFD is not prescribed. Our analysis in the second case study starts with the accepted premiss that humanity faces daunting challenges regarding present and future availability of water resources. The need to utilize all remaining resources efficiently, yet sustainably, is evident. Meeting this requirement is likely to be rather difficult in the case of transboundary rivers. As these complex ecosystems traverse national boundaries they are subject to various pressures; asymmetries, both of domestic and international nature, market failures, political and administrative inadequacies, multiple and diverse perceptions, rigid state mechanisms and interstate power games that act cumulatively, accelerating their deterioration as resources. Any endeavour to promote an appropriate management must acknowledge the multifaceted nature of transboundary rivers.

The obligation, or desire, to include stakeholders' views on resource use issues into formal policy analysis brings with it two further generic issues for the analyst. First, stakeholders' knowledge and opinion is likely to be expressed in qualitative terms or captured in linguistic quantifiers rather than numerical terms. Secondly, the quantity of data available for building a model for policy analysis is likely to be limited. These two issues set limitations on the type of modelling approach which can be used. Fortunately, FCMs can accommodate both of these limitations.

1.3 Fuzzy Cognitive Maps and Their Analysis

Formally, fuzzy cognitive maps are directed graphs which may be either cyclic or acyclic, although completely acyclic FCMs are rare in practice. The early development of FCMs by Kosko (1986, 1987) from Axelrod's initial concept (Axelrod, 1976) is now well known. An accessible introduction to FCMs is given in chapter 11 of Kosko (1993). In FCMs the graph nodes refer to events, concepts, or quantities in the world while the graph edges refer to causal connections between the nodes. Each FCM is, then, a graphical summary of a set of `IF...THEN` entailments or causal statements. The numerical values of the edges are the fuzzy weights, fuzzy degrees of belief, or fuzzy entailments identified by the presence of each edge on the graph. Note that in a standard FCM these weights are themselves typically crisp real numbers in the interval [-1,1]. This means that FCMs are typically fuzzy not in a sense that is analogous to probabilistic stochasticity, but rather in the sense that edge values are not restricted to the integers 1 and -1, as is the case with the original cognitive maps from which FCMs have been derived. Indeed, in their standard form (including all examples discussed in this Chapter), FCMs fall within the broad class of deterministic mathematical models; *i.e.* for a given set of inputs a standard FCM will always generate the same output. Recognising that FCMs fall within this general class of models, makes for ready comparison with many other fields of analysis which employ the same analytical techniques. A description of these connections, which introduces a useful companion literature for FCM analysts to explore, is best considered by considering FCMs in their matrix rather than graph form.

Every graph can be represented in matrix form. Consider a graph, with $i = 1 \ldots j$ nodes, labelled $n_1 \ldots n_i$. The graph can be represented in a square $i \times i$ matrix E, say, in which the elements, e_{ij}, are the values on the edges between nodes n_i and n_j on the graph. The graph (and corresponding matrix) can represent causal effects that nodes have on each other (as noted above) if the edges are directional and given appropriate weights. Inferences are extracted from the FCM by using it to form a standard matrix product with an i-element state vector whose elements are the initial values of the i concepts at the start of the process. In analogy to the process of population projection in demography, we might refer to this as *entailment projection*. Given the role that the FCM plays in this process it can be viewed as an event or causal effect matrix and the process of entailment projection, then implies an arbitrary (and probably unspecified) temporal scale. Indeed, in some analyses of the nature of time and the relationships between causes and their effects, human perception of time is considered to arise directly from the fact that causes must precede effects (Mellor, 1998, 1999). Entailment projection with the FCM is, therefore, a first order Markov process, and characterised by the deterministic dependence of future states only on the current state and not past states (Doucet & Sloep, 1991).

In a classical Markov process the transition (or projection) matrix contains state transition probabilities (Caswell, 2001). In the case of FCMs, the probabilities are replaced by fuzzy numbers representing degrees and direction of causality. Classical matrix Markov process models occur in a wide range disciplines including, for example; population demography (Caswell, 2001), community ecology (May, 1974; Dambacher *et al.*, 2002), food supply chain analysis (Ledauphin *et al.*, 2006; McRoberts, 2006, 2009), information theory (Shannon, 1948) and financial economics (Bianconi, 2003). In a pure Markov process the transition matrix contains on state transition probabilities. In many areas of application this condition is relaxed and other types of parameter (for example multiplication rates) are also included in the matrix. Thus, for example, the basic structure of interest in much of population demography is the Leslie matrix[2] which describes the growth dynamics of age- or stage-classified populations and has fecundities in the first row and age (stage) class transition probabilities on the main sub-diagonal (Caswell, 2001).

Different areas of science have generally placed emphasis on the different aspects of the set of analytical tools that Markov process theory encompasses. Thus, for example, population demography and community ecology have tended to focus on the eigenvalue structure of the projection matrices and the resulting implications for population growth, age (or stage) structure, community stability and resilience to perturbation. In contrast FCM theory and application has tended to focus on the results of numerical simulation of the process captured in the projection matrix, or the graph theoretic properties of the structure of the FCM.

Conceptually, the use of FCM in entailment projection and inference for abstract problem areas has much in common with the use of life cycle parameter projection matrices in population demography. An interesting analogy here is the focus in both disciplines on **projection** as distinct from **prediction**. To paraphrase Caswell (2001), a projection analysis says what **would** happen given the state of the system now, and supposing that the rules which define its behaviour do not change. In contrast, a prediction analysis says what **will** happen based on the state of the system now and allowing for possibility that the rules defining the system may change.

Turning to specific applications of FCM analysis, a wide range of problems have now been tackled using FCM. For example: analysis of the sensitivity of the building construction process to different sources of problem (Dissanayake & AbouRizk, 2007); representation of biological processes in bioinformatics

[2] Named after the late British population ecologist Patrick Holt Leslie.

(Ettinger, 2002); development of automated supervisory process control systems (Stylios & Groumpos, 1998); medical diagnosis (Froelich & Wakulicz-Deja, 2008), various business and financial management problems (Glikas and Xirogiannis, 2004; Chytas et al., 2006; Xiogiannis et al., 20008) , and; failure mode and effect analysis (Paelaz & Bowles, 1996).

Specifically in the area of environmental impact analysis, McRoberts *et al.*, (1995) used an FCM derived from a cause and effect diagram drawn by Verijken (1992) to illustrate that EU agri-environment policy was likely to subject to oscillatory dynamics as a result of the tension between processes promoting intensification and those promoting environmental protection. McRoberts & Hughes (2000) used a simple FCM to illustrate the potential dependence of farmers' willingness to adopt evidence-based weed control measures on a financially-based inducement expressed in terms of the utility of weeds to the farmers. Özesmi & Özesmi (2004) used FCMs to capture and analyse local stakeholders' knowledge about water in the catchment of the Uluabat lake in Turkey, while Mouratiadou & Moran (2007) used a similar approach to examine stakeholders' attitudes to the implications of the EU Water Framework Directive (WFD) for the Pinios river basin in Greece (see below).

Across many disciplines interest in the use of FCM analysis is steadily increasing as its capacity to represent complex problems and allow an interface between qualitative and quantitative approaches is increasingly recognised.

2 Case Studies

The case studies presented here are not intended to allow direct comparison of FCMs from two different contexts. Rather, we use each study to emphasise slightly different attributes of the FCM methodology and at the same time attempt to draw a synthesis of implications from the studies jointly on future developments in water use policy. We pay particular attention, in drawing this synthesis, to the multi-attribute nature of water as an economic good and consider the implications for future policy and exploitation which arise from at least some of water's value being positional rather than material.

In the first case study we focus on the capacity for FCM analysis to translate qualitative data into quantitative outcomes and the potential, in a policy analysis context, for the method to be used to link with other quantitative analyses. In the second case study we focus more on the utility of the FCM methodology to help in building an organised inventory of important factors in complex policy issues.

3 Stakeholders and the Water Framework Directive in the Pinios River Basin

3.1 Policy Background

Regulation of water use and management within the EU now falls under the WFD which stipulates two elements in the future regulation of water use relevant to our analysis. First, water, or more correctly water use, is viewed as an economic good and the WFD aims to achieve full cost recovery (FCR) for water, by recognising that water use has an opportunity cost[3] and that degradation of water value through individual uses should be managed through the principle of "polluter pays". Secondly, the WFD requires member states to include public/stakeholder participation in the process of developing plans for water use management. This second requirement arises from the belief that successful policy adoption is more likely where stakeholders have been involved in the process of policy formulation and in developing policy measures. The participatory process can be viewed as means for policy makers and stakeholders to arrive at a set of common goals. This avoids the complexities inherent in policy makers and stakeholders having separate goals, which renders the policy optimisation problem a bi-level problem (Candler, 1981). An analogous situation can be observed in the chapter in this book by Ortolani *et al.*, which considers the issue of implementation of agri-environment measures by European farmers.

In addition to increasing the likelihood of acceptance, participation provides an evidence base for policy development. In this case the evidence base will consist, at least partly, of subjective opinion expressed by stakeholders concerning interactions among various issues and factors of concern to them. The question for the policy developer and analyst is then, how best to elicit and record such opinions to allow their use for informing policy development. FCM construction offers one approach to achieve these aims, with the added benefit of simultaneously making available the full range of analytical tools for FCMs.

3.2 Fuzzy Cognitive Map Construction and Analysis

Details of the methodology for data collection and previous analysis for this case study can be found in Mouratiadou & Moran (2007). Briefly, 30 one-to-one interviews were held with stakeholders who were each asked to construct a FCM starting from a pre-defined list of 17 variables and by augmenting the list with variables from their own knowledge base and perceptions. The stakeholders belonged to five different groups: farmers, local residents, water experts, researchers and government officials.

The mean number of concepts per map for each type of stakeholder ranged from just over 20, for farmers and local residents, to just over 23 for government officials, implying that stakeholders added, on average, between 3 and 6 additional concepts to the initial set of 17 provided by the researchers. The number of connections per map ranged from around 45 (for local residents) to more than 55 (for government officials). These results result in map density (D) values which ranged from 0.12 (for residents) to 0.11 (for government officials).

[3] The opportunity cost of an action is the value of the best alternative foregone given the choice that is made. As a simple example, suppose you have a cup of water which you can either drink or use to water your tomato plant. The opportunity cost of drinking the water is a thirsty tomato plant. The opportunity cost of watering the tomato is a thirsty decision maker.

FCMs for individuals were combined to form a social (or consensus) FCM encompassing what was seen as the current state of the issues facing river basin management for the Pinios river. The social FCM contained 22 concepts and 46 causal connections implying a density of 0.10. The concepts with highest centrality in this social FCM were: water deficiency, irrigation and agriculture. The social FCM was reconfigured to reflect a consensus view of what the stakeholders perceived as either desirable or feasible to bring about necessary changes in the management of the river basin. This reconfigured FCM (referred to as the social desirable FCM by Mouratiadou & Moran (2007)) contained 27 concepts and 71 causal connections, giving a D value of 0.10, the same as in the initial social FCM. The three variables with highest centrality in the social desirable FCM were: Water deficiency, State-EU, and agriculture.

It is interesting that in changing from an FCM describing the current state of the world to one describing an idealised configuration, with the intention of resolving conflicts in resource use issues, the stakeholders increased the centrality of state and supra-state (EU) participation in the set of cause and effect relationships. The main reason for the increase in the centrality of State involvement in water use reflects a belief that investment by State and supra-State institutions in local water infrastructure and long-run investment, aiming at the restoration of existing water bodies, augmentation of water quantity, and improvement of water quality, would improve the existing situation and effectively would make people more willing to pay for water services. Also, it is valuable from a policy maker's perspective to note that this result implies either that the participation process was successful in aligning stakeholders' goals with those of policy makers, or that the stakeholders themselves recognise and accept the existing hierarchical goal. Such views reflect a well-founded belief that socially-optimal management of common-pool resources(Ostrom *et al.*, 1994), which may also be subject to social scarcity, requires some degree of control to be imposed from outside (*i.e.* at a higher level than) the system being managed (Hirsch, 1977). We return to these issues in the synthesis section, below.

3.3 Adding Evaluation of Economic Indicators

In asking users to construct a FCM relating to any issue, the analyst is likely to obtain the chance to attach a number of other useful analyses to the activity. In the case of resource use problems which involve environmental goods, one key issue on which useful information can be gained is the acceptability of different mechanisms to attach prices to resources, both in terms of the features of resource use that should be priced and the way in which the cost should be levied. In this case study, an analysis was made of how different categories of costs associated with irrigation and water polluting activities were perceived by different stakeholder groups.

Across the set of stakeholders the most acceptable costs to be recovered were those associated with environmental damage caused by both point and diffuse pollution. Overall, the use of economic instruments for the recovery of environmental costs appeared more justifiable than for recovering resource costs[4]. This reflects a number of stakeholder considerations: 1) the importance of the resource for sustaining the ecological services of a healthy ecosystem; 2) the difficulty in perceiving water as a scarce resource between competitive uses; and 3) the reluctance to treat water as a purely economic good to which different values can be assigned depending on its use. The lowest acceptability was assigned to resource costs imposed on farmers by water pollution and recovered according to a "polluter pays" mechanism, while the recovery of resource costs by direct pricing for irrigation was found to be more acceptable. This difference in preference is mainly related to the difference in value placed on water quantity as opposed to water quality for agricultural water use. The acceptability of the recovery of resource costs imposed by the agricultural sector on urban water users was very low, but equally important between water quantity and water quality.

Overall, even though stakeholders recognised the potential role of economic instruments in regulating water quantity and quality, they were sceptical about their implementation. Most stakeholders identified water pollution as an important problem in the area, but were reluctant about the application of the "polluter pays" principle and taxation mechanisms, due to an unwillingness to impose on farmers a measure of questionable effectiveness for using polluting, albeit "legitimate", production inputs (e.g. fertilisers). The issues of transaction costs and uncertainty in identifying the "polluters" were also raised. Also, demand for both water and fertilisers was found to be rather inelastic, and hence the effect of pricing or taxes was deemed doubtful. The acceptability weights differed between stakeholder groups, with the group of government officials exhibiting higher conformity with the principles and prescriptions of the WFD.

Comparing the results mentioned here with those in the previous section it is apparent that there is a tension among stakeholders between a perception of the need for State and a preference for pricing mechanisms that operate directly rather than through taxation. This tension between regulatory and individualistic behaviours is a subtle but important emergent property in both case studies and one that is important in effective policy implementation for sustainable management of scarce resources.

It is worth noting that here, the process of FCM construction allowed not only the identification and ranking of concepts perceived as important to stakeholders, it also gave guidance as to how financial mechanisms should be included in policy analysis models so as to reflect the preferences of stakeholders for how policy should be enacted.

4 Stakeholders' Perceptions of the Maritza/Meric/Evros Transboundary River

The second case study concerns the Maritza river (also known as the Meric or Evros) which rises in Bulgaria and forms the national boundary between Greece

---

[4] Resource costs were defined as welfare losses that users are confronted with due to misallocation of the resource between high and low value users.

and Turkey for a considerable part of its length. It is not surprising that in addition to the sorts of issues raised by the stakeholders in Greece concerned with the management of the Pinios river basin, a transboundary river such as the Maritza raises additional factors that the analyst must dealt with.

4.1 Policy Background

The UN Convention in the Law of the Non-Navigational Uses of International Watercourses (1997) defines an international watercourse as "a river that flows through or forms a boundary between two or more countries" (Espey and Towfique, 2004). Several other analogous terms are also in use including "shared rivers" and "transboundary rivers". None of these terms is free from contentious connotations and we adopt the term "transboundary rivers" here because it highlights the border and inter-national components in the complex meanings associated with these landscape features. In 1978, the United Nations counted 214 transboundary river basins. This number has risen today to 263, due to political changes and better mapping sources and technology (UN/WWAP, 2003). These basins cover approximately 45% of the planet's land surface (Bernauer, 2002), offering living space to 40% of the world's population (Sadoff and Grey, 2005). A total of 145 nations have transboundary rivers running through their territory. For twenty-one of these, their territory lies entirely within such basins (UN/WWAP, 2003).

The majority of the literature claims that the discourse on transboundary rivers is predominately of political nature. The positioning of the state in the centre of the debate substantiates these claims. Managing transboundary resources is closely linked with state territoriality and sovereignty. Both are founding principles of state structure and function that fragment the holistic approach, underpinning the environmental aspects of river basins (Vlachos, 1999). Additionally, states run the risk of losing legitimacy when they are unable to prevent harm or reduction in the social welfare of their citizens, because the cause of the negative impact is beyond their jurisdiction. In such cases democratic political accountability is proved ineffective, eroding the credibility of state authority. Under circumstances that respect state territoriality, democratic political accountability works properly, as it is able to interconnect the regulatory authorities and both the inflicting and inflicted parties (Mason, 2005, Jessop, 2008).

One of the principal conclusions that can be reached is that transboundary rivers create between co-riparian states some degree of tension, which can have consequences unrelated to the river. In addition these tensions are part of a wider set of factors that constitute the context within which the relations of the co-riparian states are framed. These factors can be environmental, but also economic, cultural and historic. Therefore, in such debates, water concerns are not the only factors at play; rather they are one between many, forming a bundle that is adequately mirrored in politics (Vlachos, 1999, Sadoff and Grey, 2002). The flexibility of FCMs in allowing concepts from different domains to be modelled together was thought to be a key advantage in capturing stakeholders' views about water resource management in the complex context of a transboundary river.

4.2 Data Collection and Map Construction

Face-to-face interviews (or one-to-one telephone conversations) were carried out in July 2009 with stakeholders from Bulgaria, Turkey and Greece. In contrast to the Pinios river basin study, no initial list of concepts was supplied to the interviews, but an example of an FCM from an unrelated subject area was used to familiarise them with the type of information that was being requested. In all, data were collected from 8 stakeholders (2 Bulgarian, 2 Turkish and 4 Greek) who were either government officials, academics or workers for NGOs concerned with water resource use.

Interviewees were asked to consider as wide a range of concept types are they thought relevant including, environmental, economic, social, political, cultural and historical variables. As a result of this open-ended approach to data collection, we were not able to obtain causal weights for most of the relationships elicited from the stakeholders and the analyses mentioned here are based on three-valued maps in which edges take only the values -1, 0 or 1.

4.3 Analysis and Interpretation

Among the 8 stakeholders a total of 248 concepts were elicited. The average number of concepts per map was 49.8, while the average number of causal connections was 64.75; average map density (D) was 0.03. The 12 concepts with highest average centrality values across the 8 stakeholders were: (1) Inter-state cooperation (0.140): (2) Interstate tension (0.076); (3) Flooding (0.074); (4) Water quality (0.064); (5) Army activity (0.053); (6) EU (0.053); (7)Water quantity (0.045); (8) Ignorance of important issues (0.042); (9) Border (0.032); (10) Resource imbalance; (11) Up/downstream issues (0.024); (12) Resource competition (0.018). Informal comparison of this list with the results from the previous case study suggest that there is a similar pattern of core issues in play in the transboundary case, augmented (as would be expected) by some additional complicating inter-state factors.

The stakeholders, some more than others, perceive the transboundary river problem as one that consists of numerous, distinct parameters, all of which must be incorporated in the discussion. A closer examination reveals that the stakeholders did indeed include concepts that stem from different disciplines, including environmental, geographical, economical, political (both domestic and international), administrative, historical and cultural issues. Moreover the environmental issues, while focusing on water related subjects, tend to include issues of land use. All of these parameters are interconnected through cause and effect relationships.

The above reveals the tendency of the stakeholders to perceive the Maritza river basin holistically. They recognise that the Maritza river can be seen only as an ecosystem, which provides many goods and services and thus supports many economic activities. This holistic approach is not constrained to just the environmental and economical aspects; it discloses also the political nature of the issue, connecting river basin management with inter-state cooperation, state sovereignty and multiple indicators of domestic political and administrative inadequacies. As a consequence the successful management of the river basin is seen as a highly complex task, of a multidisciplinary nature, with many interacting factors influencing the status of the system and making the comprehension of the issue in its entirety a formidable challenge. Furthermore this structuring of the problem explicitly contradicts the tendency of simplifying the state, by subtracting many internal qualities, proposed by International Relations theory (Benvenisti, 2002) that in the end can become disadvantageous for the adaption of a holistic approach.

The Maritza river basin is obviously perceived holistically by stakeholders, but there are some issues which hold a more central place in the discourse. The primary focus (and the original motivation to study this area) in the Maritza basin is flooding. This issue is seen in the FCMs of the stakeholders, where floods have a high centrality. However, while flooding may be important as an issue among local people, it is not the most central issue to expert stakeholders. The relatively high concern about flooding may be understood in two ways. The first has to do with the perception of risk events by society. As risk events, in this case floods, have been damaging for all three states, at an increasing rate during the recent years, it is natural that they will shape the cognition of stakeholders, who in the end will become more sensitive towards this issue.

The second way links the increased centrality of floods with administrative incapacity. As administration cannot easily challenge issues of increased complexity, like water quality issues, it will tend to focus more on simpler problems (Kibaroglu *et al.*, 2005). This results in a misconceived centrality of floods that overshadows other, equally important, parameters of the system. In the end, this focus on problems where progress is possible can be an obstacle towards an holistic approach, as management practices are led to adopt a narrow set of goals. On the other hand the overemphasis on floods can have positive effects. For example in the aftermath of the 2005 floods in the Maritza river Bulgaria and Turkey agreed to jointly construct a dam, for protection and irrigation management (Kibaroglu *et al.*, 2005). In conclusion, while high profile concepts such as flooding may have a falsely heightened centrality, which can prove detrimental for the management of the river basin, they can also provide the pressure required for the promotion of state cooperation. The driving force behind this pressure is the transnational public; consisting of individuals who, even though they are not necessarily conationals, are bound together in face of environmental harm (Mason, 2005). As a consequence states have a common interest in regulating the Maritza river; a factor which is identified in the precedence given to inter-state concepts in the elicited FCMs.

5 Synthesis

Although the two case studies summarised here were not carried out as part of an integrated research programme, in bringing them together and comparing their findings, some common themes emerge. In this final section we use FCM as a meta-analytic tool to illustrate these points and help in our synthesis of unified themes in water resource management.

5.1 *Hierarchy: Stimulation and Regulation*

In both case studies the issue of decision-making and policy enactment at a hierarchical level above that of the stakeholders was found to be important. In both cases, a tension was identified between the positive benefits of economic investment by the State or EU in funding necessary governance mechanisms and infrastructure to allow water to be managed effectively, and the detrimental effect that top-down management can have on the long term sustainability of resource management, by inhibiting local involvement and denying stakeholders agency. In the Pinios river basin study, stakeholders expressed the view that willingness to pay (WTP) for water resources would only increase once the State (or EU) had provided water storage infrastructure. In this case, the role of the State or EU was perceived to be as an external mechanism to break a cycle of lock-in which simultaneously prevents the development of good water services and an effective water management plan. These views can be summarised in FCM form, as shown in Figure 1.

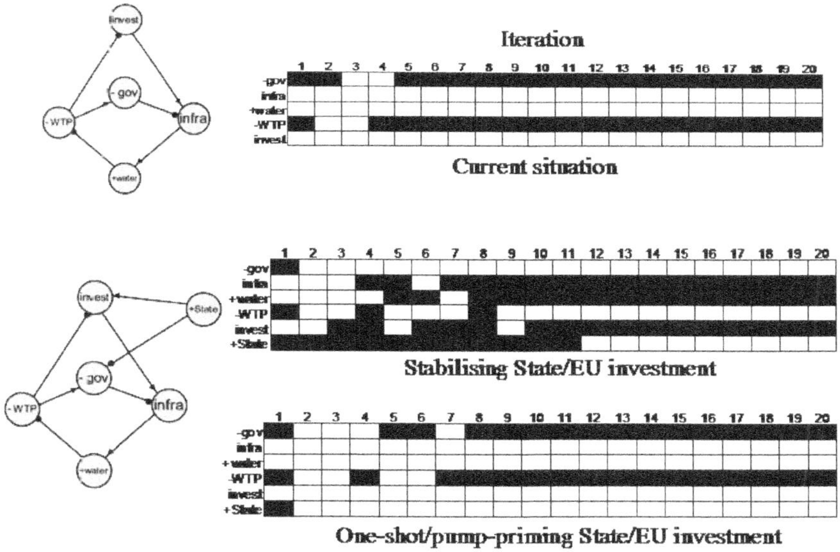

Fig. 1 FCM analysis summarising views about the role of State or supra-State (i.e. EU) input in promoting local involvement and stakeholder agency in water resource management. The upper FCM and projected output depicts the situation identified by Mouratiadou & Moran (2007): Lack of local governance and unwillingness to pay (-WTP) are combined with poor infrastructure, no investment and poor water quality management; the system is locked in a vicious stable state. The lower portion illustrates two different outcomes when State/EU input serves to stimulate local governance and investment. In the first projection State/EU input is maintained until a stable pattern of investment, local govenrance , WTP and water quality emerges at iteration 11 at which point Stare/EU investment is removed but the system remains in a stable configuration. In the second projection a pump-priming State/EU input is made only in the first cycle. This initially stimulates WTP and removes obstacles to local governance but the system falls back into the same configuration as the current situation without any improvement in water quality and quantity

The output of the FCM representing the current situation (as expressed by stakeholders) appears to support their contention that the situation is locked into a self-enforcing state in which low water quality and poor infrastructure discourages acceptance of water charges which prevents investment in improved infrastructure and quality improvement. Adding a concept representing stimulus of the local economy by the State or EU has interesting implications for both local stakeholders and policy makers. For the policy maker there is a tension between the objective of improving water quality and sustainability of supply, and the cost of achieving those objectives.

The local stakeholders' position might be paraphrased as a challenge to the policy makers: "If you pay to improve water quality and supply infrastructure, we'll start to value it and pay for these services which will mean that your financial input isn't open-ended". Clearly from the policy makers' perspective, the fear is that the financial commitment will be both large and open-ended. It is worth noting that in both case studies while stakeholders identified the need for State or EU input to stimulate desired effects, they also considered that on-going State/EU involvement would eventually stifle the agency of local stakeholders leading to sub-optimal solutions being imposed on the system. The FCM analysis presented in the lower portion of Figure 1, suggests that all of the stakeholder and policy-maker goals can be met. Provided input from the State or EU is maintained long enough to fix the system in a state in which there is a positive feedback between the local value of water and WTP it can then be withdrawn avoiding an open-ended commitment for policy-makers and suppression of local agency. The effect identified in the analysis associated with Figure 1 is associated with a cross-scale effect in the system (Giampietro, 2004). A similar effect comes into play when FCM is used to explore solutions to the problem of social scarcity in water goods.

5.2 Examining the Effect of Regulation on a System with Socially Scarce Goods

The need for regulation to modify individual objectives to deliver socially optimal resource use was discussed at length in the context of social limits to growth by Hirsch (1977) and has remained something of an ungrasped nettle in the ecological economics literature. Pooling the themes which emerged from the two case studies we believe one representation of the problem can be captured in the FCM shown in Figure 2.

The situation without regulation, as depicted in the upper part of Figure 2, can be described as follows: Unregulated development allows access to common resources (such as water) based on initial opportunity. Without regulation, the costs of exploiting the resource can be externalised for the user leading to maximisation of individual use at the expense of social welfare. This pattern of behaviour leads to perceptions of inequality and increased demand as more people attempt to access resources in order to stake a claim and increase their consumption. This increased competition leads to congestion and increased costs for would-be consumers. These features can be inferred from the analysis of both river systems. The presence of transboundary complexities in the second case study can be

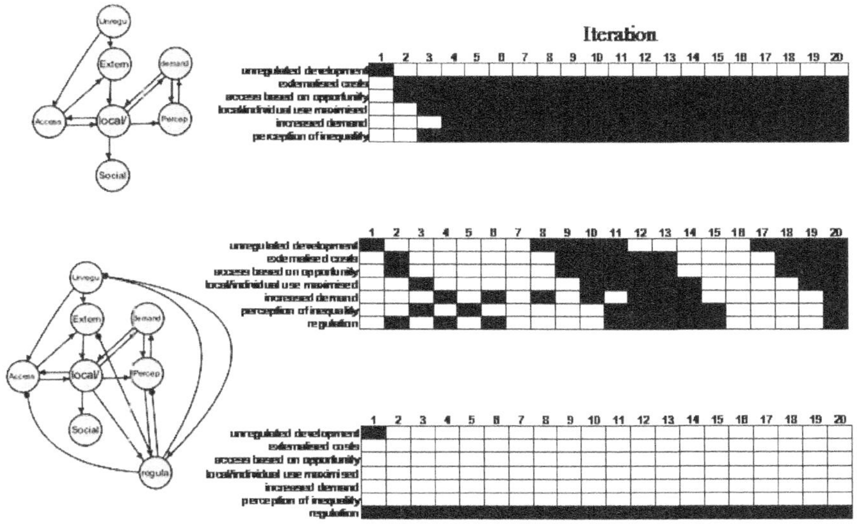

Fig. 2 A simple illustration of the impact of external regulation on the behaviour of a model system representing a liberalised market approach to use of a shared and socially scarce resource. In the upper portion the system contains only positive feedback loops (indicated by causal connections with standard arrow heads) and after initiation from unregulated development, exhibits an unsustainable pattern of constant resource depletion. In the lower portion, regulation is introduced (indicate by the causal connections with rounded heads). If regulation is itself subject to feedback from the system (as depicted in the second set of projection data) a limit cyclic behaviour is imposed on the system dynamics that include periods of improved social use of resources. In contrast, constant regulation, without feedback from the system, results in suppression of all unwanted state activations

thought of as exacerbating the tendency for local or individualistic behaviour by increasing the sense of competition with "the other" that characterises liberalised markets. It can be seen that any system represented by the first network in Figure 2 is inherently unstable because all of the causal connections are positive; the system contains only positive feedback loops. This is demonstrated in the iteration output which shows that all of the nodes of the graph (expect the upper one representing unregulated development) are active in every cycle. In contrast, the lower portion of Figure 2 illustrates the behaviour of the system when regulation is imposed. In this case it is assumed that regulation acts to reduce unregulated development, externalisation of costs, and perceptions of inequality. In turn the presence of regulation is stimulated by unregulated development, perceptions of inequality and localised or individualistic consumption of resources. If the occurrence of regulation is dependent on feedback from the system in the way just described, the net result is to cause the system to cycle between periods of individualistic, unregulated behaviour and regulated behaviour with more socially optimal outcomes. Imposition of constant regulation (as depicted in the final set of projection data in Figure 2) leads to suppression of the unwanted effects of positional competition for resources. The presence of transboundary issues in this case is assumed to be subsumed into the question of whether or not they can be overcome in order to achieve active regulation in the first place. Clearly, the international nature of transboundary rivers adds another level of the hierarchy of interactions to the system; that of international politics. We are actively researching the use of FCMs to model such 3-level water resource systems.

Acknowledgments. SAC is a Main Research Provider for the Scottish Government. The FCM analysis of Mouratiadou & Moran (2007) was undertaken using 'Fuzzy Cognitive Mapping Integrated Environment 1.16' software package, which has been developed in collaboration with the Information Technology Company Interplanetic.

References

Axelrod, R.: Structure of Decisions: The Cognitive Maps of Political Elites. Princeton University Press, Princeton (1976)

Benvenisti, E.: Sharing International Resources: International Law and Optimal Resource Use. Cambridge University Press, Cambridge (2002)

Bernauer, T.: Explaining Success and Failure in International River Management. Aq. Sci. 64, 1–19 (2002)

Bianconi, M.: Financial Economics, Risk and Information. World Scientific, Hackensack (2003)

Pagano, U.: Is power an economic good? Notes on social scarcity and the economics of positional goods. In: Bowles, S., Franzini, M., Pagano, M.U. (eds.) The Politics and Economics of Power, Routledge, New York, NY (1999)

Caflisch, L.: Emerging Rules on International Waterways: the Contribution of the United Nations. Pol. Geog. 15, 273–285 (1996)

Candler, W., Fortunyamat, J., McCarl, B.: The potential role of multi-level programming in agricultural economics. Amer. J. Ag. Econ. 63, 521–531 (1981)

Caswell, H.: Matrix Population Models: Construction, Interpretation and Analysis, 2nd edn. Sinauer Associates, Sunderland (2001)

Chytas, P., Glykas, M., Staikouras, C., Valiris, G.: Performance Measurement in a Greek Financial Institute Using the Balanced Scorecard. J. Meas. Bus Ex. 10(2) (2006)

Dambacher, J.M., Li, H.W., Rossingnol: The relevance of community structure in assessing indeterminacy of ecological predictions. Ecol. 83, 1372–1385 (2002)

Doucet, P., Sloep, P.: Mathematical Modelling for theLife Sciences. Ellis Horwood, New York (1992)

Diassanayake, M., AbouRizk, S.M.: Qualitative simulation of consrtuction performance using fuzzy cognitive maps. In: Proc. 2007 Winter Sim. Conf., pp. 2134–2140 (2007), `http://www.informs-sim.org/wsc07papers/266.pdf`

Espey, M., Towfique, B.: International Bilateral Water Treaty Formation. Water Resources Res. 40, 1–8 (2004)

Ettinger, M.: The complexity of comparing reactions. Bioinform. 18, 465–469 (2002)

Froelich, W., Wakuliicz-Deja, A.: Associational cognitive maps for medical diagnosis supoprt. Intel. Inf. Syst., 387–396 (2008), `http://iis.ipipan.waw.pl/2008/proceedings/iis08-38.pdf`

Giampietro, M.: Multiscale Integrated Analysis of Agroecosystems. CRC Press, Boca Raton (2004)

Gleditsch, N.P., Furlong, K., Hegre, H., Lacina, B., Owen, T.: Conflicts over shared rivers: resource scarcity or fuzzy boundaries? Pol. Geog. 25, 361–382 (2006)

Glykas, M., Xirogiannis, G.: A soft knowledge modeling approach for geographically dispersed financial organizations. Soft Comp. 9, 579–593 (2004)

Hirsch, F.: Social Limits to Growth. Routledge & Kegan Paul, London (1977)

Kibaroglu, A., Klaphake, A., Kramer, A., Scheumann, W., Carius, A.: Cooperation on Turkey's Transboundary Waters (online), Berlin. The German Federal Ministry for Environment, NatureConservation and Nuclear Safety (2005), `http://www.bmu.de`

Jessop, B.: State Power: a Strategic-Relational Approach. Polity Press, Cambridge (2008)

Kosko, B.: Fuzzy cognitive maps. Int. J. Man-Mach. Stud. 24, 63–75 (1986)

Kosko, B.: Adaptive, bidirectional associative memories. Appl. Optics 26, 4947–4960 (1987)

Kosko, B.: Fuzzy Thinking. Harper Collins, London (1993)

Ledauphin, S., Pommeret, D., Quannari, E.-M.: A Markovian model to stufy products shelf lives. Food Qual. Pref. 17, 598–603 (2006)

Mason, M.: The New Accountability: Environmental Responsibility Across Borders. Earthscan, London (2005)

McCully, P. (1996) Silenced Rivers: the Ecology and Politics of Large Dams. Zed Books Ltd.McCully, London (1996)

McRoberts, N.: Karpography: A generic concept of quality for chain analysis and knowledge transfer in supply chains. Acta Hort 712, 153–158 (2006)

McRoberts, N.: Introduction. In: Zude, M. (ed.) Optical Methods for Quality assessment of Fresh and Processed Products in the Agricultural Supply Chain. CRC Press, Boca Raton (2008)

McRoberts, N., Foster, G.N., Davies, D.H.K., Evans, K.A., McKinlay, R.D., Wale, S.: The role of synoptic models in the development of crop protection for sustainable crop produiction systems. In: BCPC Proceedings 63. Intgrated Crop Protection: Towards Sustainability?, BCPC, Farnham, UK, pp. 439–446 (1995)

McRoberts, N., Hughes, G.: Killing or culling? Is it possible to manage weeds as a resource. In: Proceedings BCPC Annual Conference, BCPC, Farnham, UK, pp. 383–390 (2001)

Mellor, D.H.: Real Time II, 2nd edn. Routledge, London (1998)

Mellor, D.H.: The Facts of Causation. Routledge, London (1999)

Mouratiadou, I., Moran, D.: Mapping Public Participation in the Water Framework Directive: A Case Study of the Pinios River Basin. Greece. Ecol Econ. 62, 66–76 (2007)

Ohlsson, L.: Hydropolitics: conflicts over water as a development constraint. Zed books, London (1995)

Ostrom, E., Gardiner, R., Walker, J.: Rules, Games and Common-Pool Resources. University of Michigan Press, Ann Arbor (1994)

Özesmi, U., Özesmi, S.: Ecological models based on peoples' knowledge: a mult-step fuzzy cognitive mapping approach. Ecol. Mod. 176, 43–64 (2004)

Pelaez, C.E., Bowles, J.B.: Using fuzzy cognitive maps as a system model for failure modes and effects analysis. Intell. Syst. 88, 177–199 (1996)

Poff, L.N., Allan, D.J., Palmer, M.A., Hart, D.D., Richter, B.D., Arthington, A.H., Rogers, K.H., Meyer, J.L., Stanford, J.A.: River Flows and Water Wars: Emerging Science for Environmental Decision Making. Front.Ecol. Env. 1, 298–306 (2003)

Sadoff, C.W., Grey, D.: Beyond the River: the Benefits of Cooperation on International Rivers. Water Pol. 4, 389–403 (2002)

Sadoff, C.W., Grey, D.: Cooperation on International Rivers: A Continuum for Securing and Sharing Benefits. Water Int'nl. 30, 1–8 (2005)

Shannon, C.E.: A Mathematical Theory of Communication. Bell Syst. Tech. J. (July 1948), http://faculty.kfupm.edu.sa/RI/abdallah/041/ee406/shannon.pdf

Stylios, C.D., Groumpos, P.P.: Fuzzy Cognitive Maps in Modeling Supervisory Control Systems. J. Intel. & Fuzzy Syst. Appl. Engi. 8, 83–98 (2000)

UN/WWAP (United Nations/World Water Assessment Programme). In: 1st UN World Water Development Report: Water for People, Water for Life (online book) Paris, UNESCO and Berghahn Books (2003), http://www.unesco.org/

United Nations Educational Scientific and Cultural Organization. In: The 3rd United Nations World Water Development Report: Water in a Changing World (online book), London, Earthscan (2009), http://www.unesco.org/water/

Verijken, P.: A methodic way towards more sustainable farming systems. Neth. J. Ag Sci. 40, 209–233 (1992)

Vlachos, E.: Practicing Hydrodiplomacy in the 21st century. Water Resources Update 111, 76–82 (1999)

Xirogiannis, G., Chytas, P., Glykas, M., Valiris, G.: Intelligent impact assessment of HRM to the shareholder value. Exp. Sys. App. 35, 2017–2031 (2008)

Fuzzy Cognitive Map to Support Conflict Analysis in Drought Management

R. Giordano and M. Vurro

Abstract. Empirical investigations in scientific literature have highlighted the differences between the stakeholders' perceptions of a given drought phenomenon's severity and the results of scientific – technical evaluation. This means that there can be several perceptions of the phenomenon, and the scientific models used to assess the drought's severity do not consider these differences. Facing a drought phenomenon, stakeholders adopt different mental models to assess its severity, taking into account additional elements, other than just water availability and climatic conditions. This, in turn, could have a strong negative impact on the effectiveness of drought mitigation strategies. In fact, if the mitigation actions are selected without considering the stakeholders' perceptions of the drought, then the actions themselves could be considered as unsatisfactory by the stakeholders or, even worst, not acceptable at all. If the degree of acceptability is low, then the implementation of the mitigation actions could be hampered by strong opposition from the stakeholders. Therefore, an in depth analysis of the potential conflicts and the definition of effective negotiation strategies could be really useful. In this perspective, we propose a methodology based on a Fuzzy Cognitive Map (FCM) to support the elicitation and the analysis of stakeholders' perceptions of drought, and the analysis of potential conflicts. The method was applied to a drought management process in the Trasimeno Lake area (Umbria Region) in order to analyze potential conflict.

Keywords: Drought management, drought perception, Fuzzy Cognitive Map, Conflicts analysis.

1 Introduction

Drought is considered as one of the water scarcity phenomena. It is defined as a natural and temporary imbalance of water availability, consisting of a persistent

R. Giordano · M. Vurro
Water Research Institute – I.R.S.A., Italian National Research Council – C.N.R., Bari, Italy
e-mail: raffaele.giordano@cnr.it

M. Glykas (Ed.): Fuzzy Cognitive Maps, STUDFUZZ 247, pp. 403–425, 2010.
springerlink.com © Springer-Verlag Berlin Heidelberg 2010

lower than average precipitation (Pereira et al., 2009) which is of uncertain frequency, severity and duration. Moreover, drought is difficult to predict in terms of its beginning, ending and severity. Therefore, management strategies aimed only at increasing the seasonal availability of water through a merely technological and infrastructural approach are not sufficient. It is widely acknowledged that coping with drought requires the development of a risk management plan to support the timely implementation of mitigation measures.

Nevertheless, policy making to plan for drought is hindered by the lack of clearly agreed definitions of drought, which makes it difficult to implement preparedness measures, to apply timely mitigation measures when a drought occurs, or to adequately evaluate drought impact (Pereira et al., 2009b). Ohlsson stated that indicators of water scarcity – and, thus, of drought – are "not fixed stars" (Ohlsson, 2000 – pag. 215), but they show what has been postulated as important in the analysis of the phenomena. The prevailing technical dimension of drought management imposes indicators for drought analysis which are mainly based on the amount of precipitation and water availability.

Empirical investigations in scientific literature have highlighted the differences between the stakeholders' perceptions of drought phenomena and the results of scientific – technical evaluation (Noemdoe et al., 2006). Thus, characterizing drought simply as a departure from normal precipitation and as a reduction of the amount of water available provides only a one-dimensional definition of drought (Noemdoe et al., 2006). There is no unique definition of the problem, but each individual has his/her own perspective in defining and interpreting it (Lane and Oliva 1998). A distinction is needed between hard and soft system thinking, where the former adopts an "objectivist" stance that sees problems as independent of individuals' views and beliefs. Soft system thinking, on the other hand, requires a "subjectivist" stance that recognises the importance of participants' perceptions (Rosenhead and Mingers, 2001). Facing a drought phenomenon, stakeholders adopt their own mental models to assess its severity, taking into account additional elements other than just water availability and climatic conditions. Mental models influence an actor's perception of a problematic situation by influencing both his/her observation of the world and his/her conclusions based on that observation (Pahl-Wostl 2007). They can be considered as the windows through which people view the world (Timmerman and Langaas 2004). Mental models determine what information the actors perceive in the real world and what knowledge the actors derive from it (Kolkman *et al.* 2005).

The perception of drought is influenced by the main impacts of drought on a stakeholder's perceived environment (Slegers, 2008) and on the related water use activities. For these reasons, on the one hand, a farmer quickly recognizes the onset of a drought due to soil water deficit (agricultural drought) because this drought process is among the first to be detected. On the other hand, an urban citizen may not perceive the drought until water is not available at home, i.e., in the last stage of a drought because water supply drought, which is due to a surface storage deficit, is the last drought process to be detected (Pereira et al., 2009b).

Thus, different stakeholders can perceive a drought's severity differently, and, moreover, drought can be perceived at different times. These differences result in ambiguity in the definition of the problem. Ambiguity implies that a problematic situation can be approached and interpreted in different ways (Hommes et al., 2009), leading actors to act in different ways (Checkland, 2001), and, consequently, to judge the actions taken by the others according to different criteria.

The ambiguity in drought definition could have a strong negative impact on the effectiveness of drought mitigation strategies. In fact, if mitigation actions are selected without considering the stakeholders' perceptions of the drought then the actions themselves could be considered as unsatisfactory by stakeholders or, even worst, not acceptable at all. The latter case could occur when the mitigation actions are expected to have a negative impact on the main elements of stakeholders' perceptions.

If the degree of acceptability is low, then the implementation of the mitigation actions could be hampered by strong opposition from stakeholders. This would lead to a reduction in the effectiveness of the mitigation actions, particularly in the case of actions to be implemented by the stakeholders (e.g. a reduction in crop irrigation by farmers). In the worst cases, the low level of acceptability makes the implementation of mitigation actions impossible, resulting in an increase in the drought's impact and the cost of drought management. Therefore, sound methodologies to elicit, structure and analyze the stakeholders' perceptions of a drought are required to support effective drought management.

In this work, a methodology based on a Problem Structuring Method (PSM), and, in particular, Fuzzy Cognitive Map (FCM), is applied in order to identify similarities and differences among stakeholders' perceptions of drought phenomena. The methodology was experimentally implemented analyzing drought perception in the Lake Trasimeno area, situated in the Umbria region (Central Italy).

The remaining part of this article is structured as following. Section 2 summarizes a review of the literature on the potential of PSMs in supporting the resolution of complex and unstructured problems. Section 3 describes the approach adopted and the results obtained in the case study. Section 4 summarizes the lessons learned.

2 Problem Structuring Methods for Environmental Management

Environmental management problems are characterized by the existence of multiple actors, multiple perspectives, conflicting interests and key uncertainties (Mingers and Rosenhead, 2004). These characteristics result in lack of consensus about values and norms to be considered in problem analysis and resolution, and in an uncertain knowledge base (Hommes et al., 2009). Therefore, the most demanding and troublesome task in environmental management often consists in defining the nature of the problem, rather than its solutions (Rosenhead and Mingers, 2001).

Problem Structuring Methods (PSMs) start from the basic assumption that problem formulation cannot be separated from problem solution (Hommes et al, 2009). PSMs support the elicitation of the different perceptions of the problematic situation and facilitate the debate in which assumptions about the world are teased out, challenged, tested and discussed (Checkland, 2001). During the debate, participants become aware of each other's perspectives and key interests. The objective of the debate is the establishment of a common understanding, which supports information exchange and co-operation.

PSMs do not aim to create a linear process through which an unstructured problem becomes structured. PSMs aim to identify, confront and integrate different views with respect to a given problem situation (Hommes et al., 2009).

Mostly, PSMs have been used to facilitate group work within business organizations. New approaches are attempting to apply these methodologies in more complex shared decision processes such as participatory natural resource management (e.g., Hjorsto, 2004; Ozesmi and Ozesmi, 2003). In fact, PSMs recognize and integrate participants' subjective perspectives, the importance of mutual learning, iterative process design and adaptive decision making. Comparing these characteristics to those of environmental management approaches indicates that PSMs may provide a feasible platform for organizing public participation in environmental management (Hjorsto, 2004).

Among the different PSMs, this work focuses on cognitive mapping methodologies. Two different interpretations seem to emerge in scientific literature about what a cognitive map (CM) represents. On the one hand, it can be seen as a model which is as close as possible to the cognitive representation made by decision makers. Thus the model can be considered as a "mirror" of the causes and effects that are inside the mind of decision makers (Montibeller et al., 2001). On the other hand, the constructivist view of knowledge assumes that in order to understand reality knowledge must change dynamically. According to the constructivist approach, a CM is a construct that can be useful to help the decision maker reflect on the problem. Thus the decision maker is involved in the iterative psychological construction of the real world, rather than the perception of an objective world (Eden and Ackermann, 2001).

Fuzzy Cognitive Maps (FCMs) can be included in the first group of CMs. In fact, FCMs can simulate the cause – effect relationships between the main variables in the model. The FCM have been largely used to analyze system dynamics in the business domain (e.g. Xirogiannis and Glycas, 2007; Glykas and Xirogiannis, 2004). Kang et al. (2004) developed a FCM tool to analyze the complex causal relations among conflict, communication, balanced power, shared values, trust, and cooperation in order to enhance the management of relationships among organizational memers in airline services. Xirogiannis et al. (2007) developed a decision modelling tool based on FCM' intellingent computing characteristics able to support stratetic – level shareholders decisions. The FCM are increasingly applied in spatial and environmental planning are increasing. Ozesmi and Ozesmi (2004) used FCM to analyze the perceptions about an ecosystem held by people in different stakeholder groups. De Kok et al. (2000) adopted a FCM approach qualitative to integrate social science concepts in a quantitative modeling for water management scenarios development. Xirogiannis et al. (2004) proposed an FCM – based approach to model experts' decision mechanisms in the field of urban area management.

Given the aims of this work, the potentialities of FCM to support environmental management are particularly interesting. To this aim, we should consider the two main phases of a decision process, i.e. the divergent and the convergent thinking phases (Montibeller et al., 2001). From decision analysis point of view, during the divergent thinking stage, the issue is disclosed, different views are encouraged and proposed, alternatives are generated, objectives are defined and the boundaries of the problem definition are discussed during the debate among the decision makers. Thus, CMs as suggested by Eden and Ackermann (2001) can be useful during divergent thinking phase because Cognitive Mapping supports creative definition of the problem's characteristics and the identification of alternatives. It can be used to clarify what interests are involved in the discussion and to facilitate the debate.

During the convergent thinking phase, criteria are defined to measure the performances of alternatives on the objectives, data about these performances are gathered, compensations between criteria are stated, alternatives are ranked, and the 'best' alternative is selected and implemented (Montibeller et al., 2001).

FCMs can be used to support the convergent thinking phase given their potentialities to simulate, even qualitatively, the impact of the different management actions on the main elements of the stakeholders' perceptions.

In this work, a methodology based on the sequential implementation of Cognitive Mapping and Fuzzy Cognitive Mapping is proposed in order to support divergent and convergent thinking for drought management as described in the next sections.

3 The FCM to Support Drought Management

The methodology adopted in this work aims to elicit and analyze the different perceptions of drought, and to investigate the links between the ambiguity in drought perception and the emerging of conflict among actors in drought management. To this aim, a multi-step cognitive mapping approach was implemented.

The main steps are:

- elicitation of the stakeholders' drought perceptions;
- assessment of drought management acceptability.

The description of the results in the case study are used here to lead the narration of the adopted methodology.

3.1 Description of the Case Study

The methodology developed was applied to elicit and analyze drought perceptions in the area of Lake Trasimeno, located in the Umbria region (Central Italy) (fig.1).

The Trasimeno Lake covers a surface area of 128 km^2. The lake has unusual hydromorphological conditions, characterized by the absence of substantial inlet and outlet rivers. The tributary catchment of the lake covers a limited area. Moreover, the depth of the lake is around 4 m, with a maximum of 6 m. These conditions make the lake particularly vulnerable to drought phenomena. Therefore, the

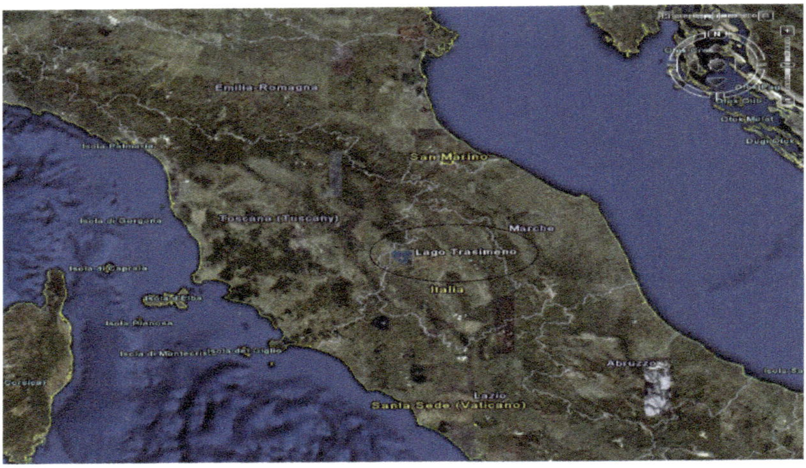

Fig. 1 Trasimeno Lake

amount of water in the lake is strongly influenced by climatic conditions. Evaporation during sunny and windy days in a normal summer period can significantly reduce the level of the lake.

Drought increases the effects of already adverse climatic conditions. Drought is quite recursive in this area as shown in fig. 2.

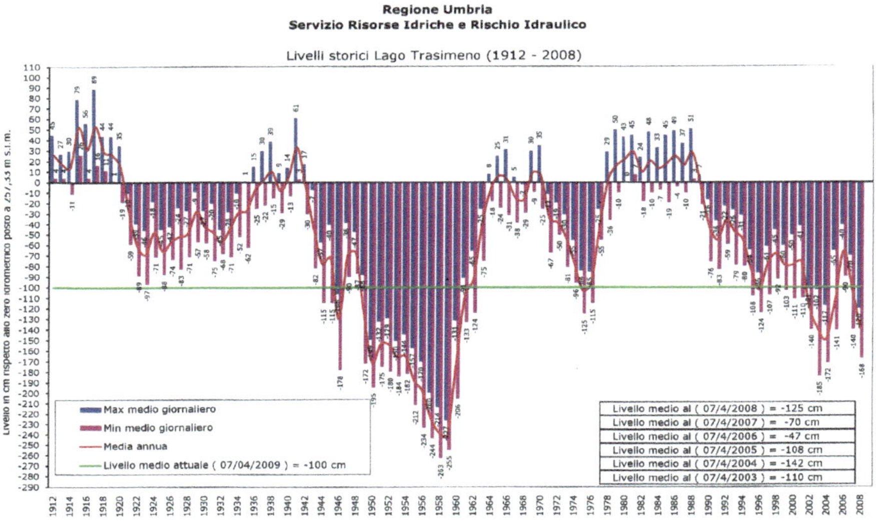

Fig. 2 Level of the lake Trasimeno from 1912 to 2008

The last strong drought phenomenon initiated in 2002 and finished in 2006. During this period, the drought had a strong negative impact on the local socio –

economic conditions. In fact, most of the economic activities were strongly influenced by the state of the lake. Farmers used to withdraw water for irrigation directly from the lake. Therefore, the reduction of the level significantly decreased the water available for irrigation. Moreover, the reduction in the level of the lake had a strong negative impact on the touristic industry in the area.

The drought management strategies adopted in the past were mainly based on the reduction in permits to withdraw water for irrigation directly from the lake. This led to conflicts between the different users. In many cases, farmers did not accept this strategy and continued to use the water from the lake for irrigation. This not only reduced the effectiveness of the drought management strategy, but also increased the perception of the negative role played by farmers in drought mitigation.

An analysis of the conflicts which have emerged in the past due to drought phenomena allowed the authors to identify the main stakeholders to be involved in this study. The list of participants is as follows:

- the Umbria Regional Authority;
- the Local Irrigation System Management (EIUT);
- the local Municipalities;
- the Local Development Support Association (GAL);
- the local Farmers Association;
- the Regional Environmental Protection Authority (ARPA);
- the local Tourist Industry Association.

The first three actors are the decision makers, while the others can be considered as stakeholders who are influenced by decisions about drought management. The decision makers were involved in the first step of the process, in order to collect information about potential drought management strategies. The results of this step are described below:

- Emergency planning: this action concerns the limitation of permits to withdraw water for irrigation directly from the lake. This is the most common action taken by the Regional Authority in the first stages of drought phenomena.
- Reuse of wastewater: this action aims to increase the availability of water for irrigation by improving the use of treated water. This is a management strategy rather than an emergency decision.
- Technical support to farmers: this action aims to reduce the negative impact of drought on farmers' income by supporting them in the adoption of technical innovations.
- Changes in agricultural practice: this management strategy aims to decrease the quantity of water-demanding crops grown in the area, in order to reduce their impact on water resources.

This information was used as basis for the conflict analysis for drought management, as described in the next sections.

3.2 Elicitation of Drought Perceptions

The first step of the adopted approach was aimed at eliciting and structuring the mental frames used by each stakeholder to perceive the drought. Any kind of drought assessment, and thus even the perceptual analysis made by the stakeholders, could concern the beginning of the phenomenon, its termination, or its severity. Since the aim of this work is to support drought management, the focus is on the perception of drought severity.

Therefore, the first step of the Cognitive Mapping process was aimed at eliciting and structuring the stakeholders' perceptions about the severity of a? drought and to identify the elements they used to make this assessment. In order to analyze similarities and differences among perceptions, the stakeholders were interviewed individually. A round of semi-structured interviews was carried out involving the stakeholders mentioned in the previous section. As stated by Slegers (2008), the stakeholder's perception of a drought is influenced by previous drought experiences. Therefore, the interviews were aimed at eliciting stakeholders' understandings about both direct and indirect drought impact on the perceptual environment. That is, the part of the whole environment which is closest to the stakeholder and in which she/he operates and makes decisions about how to respond and to behave (Slegers, 2008). Moreover, the stakeholders were required to specify elements which can either increase or decrease the negative impact of a drought. Some of the CMs developed from these interviews are shown below:

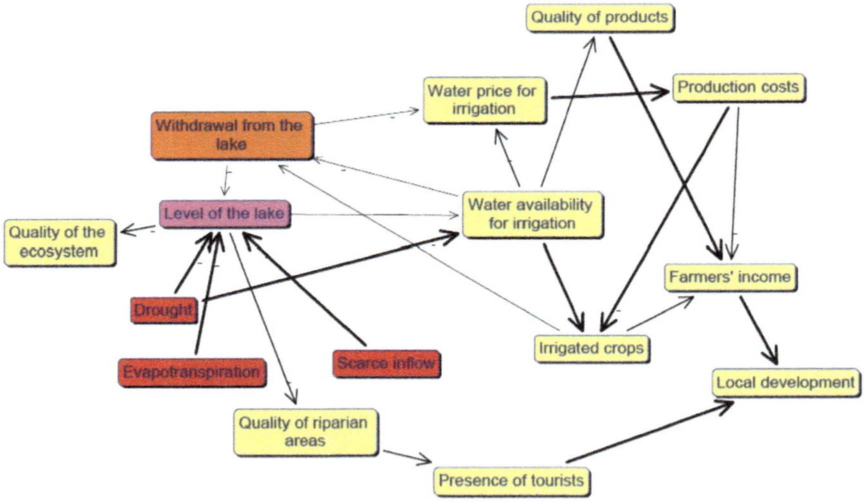

Fig. 3 Cognitive Map of Farmers' Association

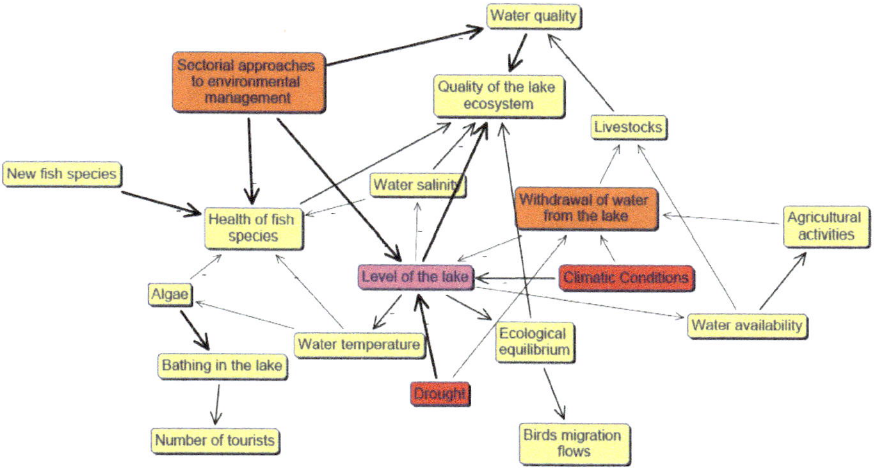

Fig. 4 Cognitive Map of the ARPA

The CMs developed were used to identify the most important elements in the stakeholders' perceptions of a drought, that is the "nub of the issue" (Eden, 2004). The basic assumption in assessing the degree of importance of the concepts contained in the CM is that the more central the concept in the CM, the more important the concept is in the stakeholder's perception (Giordano et al., 2007). Taking into account that the meaning of a concept in a CM depends on its explanations and consequences (Eden and Ackermann, 2001), the centrality of each concept can be assessed analyzing the complexity of the surrounding causal chains. Eden (2004) introduced the domain analysis, which calculates the total number of in-arrows and out-arrows from each concept. In this work, the weighted extended domain analysis was applied. This method extended the domain analysis by adding successive layers of concepts, and giving a decreasing weight to each layer according to a decay weight function (Eden, 2004).

In the present work, the authors used this method to identify the most important elements of the stakeholders' CM. Table 3 shows the results of this analysis.

A second round of meetings with stakeholders was organized in order to validate both the CM and the assigned degrees of importance. The stakeholders were quite satisfied with the results obtained and, thus, no changes were required.

Drought perception depends on the impact of a drought on the perceptual environment. Thus an analysis of the stakeholders' CMs allowed the perceptual environment for each stakeholder to be structured. Next, the analysis of drought perception was completed by assessing and comparing the drought impact on the main elements of each stakeholder's environment.

To this aim, the CMs were used as basis for the development of the Fuzzy Cognitive Map (FCM) (Axelrod, 1976; Ozesmi and Ozesmi, 2004; Xirogiannis et al., 2004). A weight and a polarity were assigned to each link considering the results of the interviews with the stakeholders. A positive link between two variables A and B means that, according to the stakeholder's understanding, an increase in A results in an increase in B. A negative link between the same variables means that a change in A in one direction implies a change in B in the opposite direction. The strength of a link between two concepts indicates the intensity of the relationship between them, that is to say, how strong is the influence of one concept over the other according to the stakeholder's understanding. The strength can assume values in the interval [-1; 1]. The relationships between concepts can be represented in an adjacency matrix. In the FCM, this matrix allows the overall effects of a change on the elements in the map to be inferred qualitatively.

The initial system state represents the value of the elements in the FCM at the beginning of the simulation process. The values in the square matrix represent the strength of the impact between the elements of the FCM. The adjacency matrix allows the propagation of the change in one variable in the FCM to be simulated, considering the system of causal relationships.

The impact of a drought on the main elements of the FCM was analyzed by comparing the state of the variables without drought (the drought value is 0 in the initial state vectors) and the system state in case of drought (the drought value is changed to 1 to simulate the effects of this phenomenon). In the first state, the "climatic conditions" is the only active variable. In fact, as ARPA said, the drought effects on the level of the lake are added to the already existing effects of the current climatic conditions in the study area.

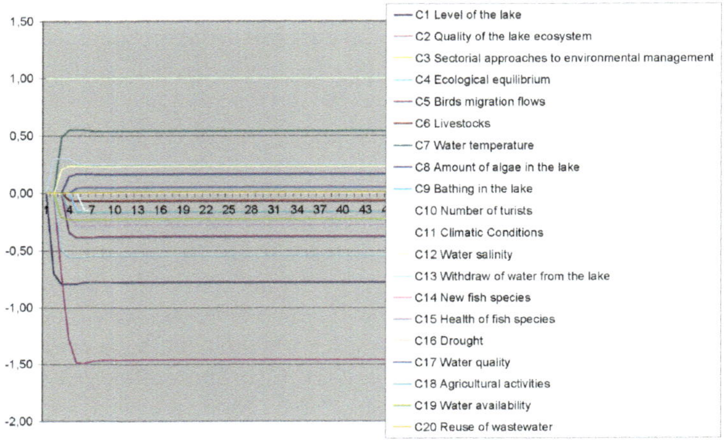

Fig. 5 State of system before beginning of drought according to the Arpa's FCM

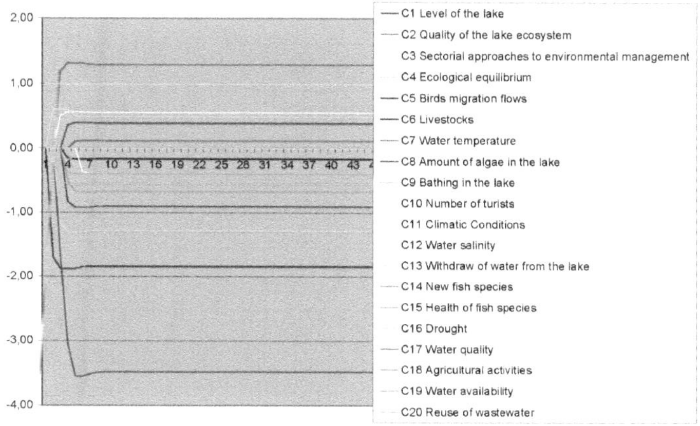

Fig. 6 State of system after beginning of drought according to the Arpa's FCM

The comparison between the system states is done taking into account the stable states, that is the state achieved by the system at the end of the simulation processes.

The degree of change for each element in the FCM due to the beginning of the drought phenomenon was assessed using the fuzzy linguistic variable shown in fig. 7,

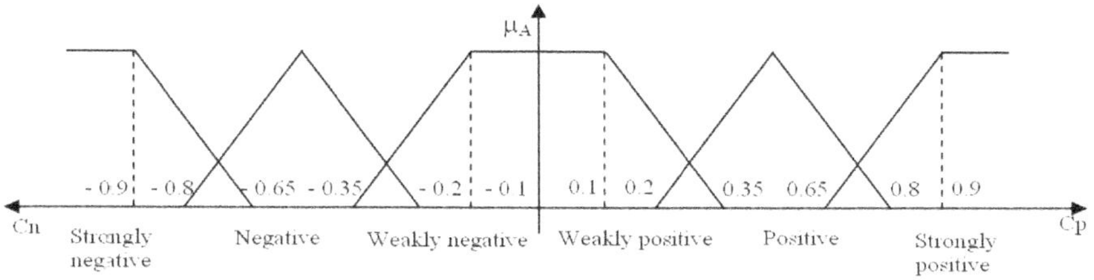

Fig. 7 Fuzzy function to describe the degree of change due to drought initiation

where C_n represents negative changes due to drought. A negative change can occur either when a negative element (e.g. water salinity) increases or when a positive element (e.g. quality of the lake ecosystem) decreases. C_p represents positive impacts.

The degree of change was normalized to 1 as a ratio between the change of the i-th element due to the action a, and the maximum change due to the same action. The normalized value is the degree of impact (table 1).

Table 1 Impact of drought according to the Arpa's perception

Variable	Degree of Importance	Degree of Impact
Level of the lake	Very important	Negative (decrease)
Quality of the lake ecosystem	Very important	Strongly negative (decrease)
Ecological equilibrium	Very important	Negative (decrease)
Health of fish species	Important	Weakly negative (decrease)
Water quality	Important	Weakly positive (increase)

The aggregation of the degree of importance and the degree of impact allowed the elements with the highest impacts on stakeholder's perception of drought to be identified. The aggregation was carried out considering that the more important the element, and the more negative the drought impact, the stronger is the influence of the element on the stakeholder's perception of drought. Fuzzy *if...then* rules were defined. The degree of influence on drought perception was assessed applying the centroid defuzzification method.

Fig. 8 shows the defuzzification process for "quality of the lake ecosystem".

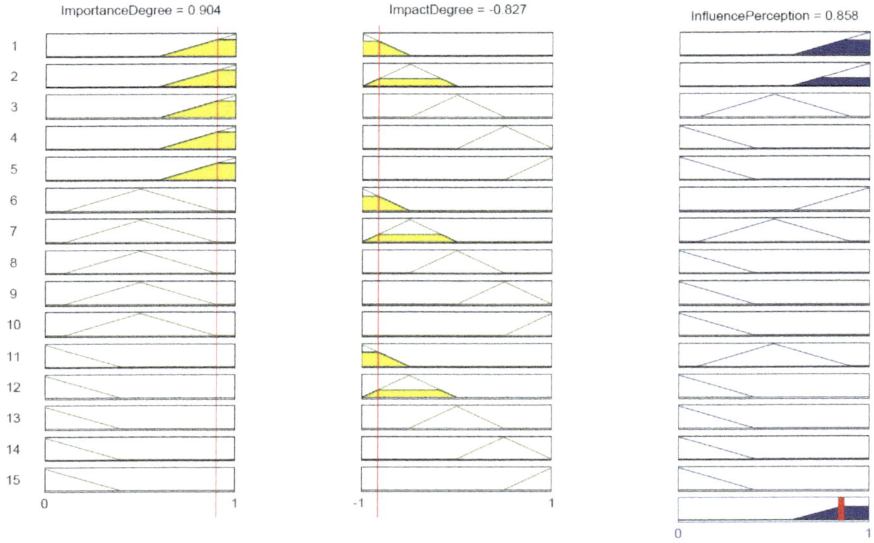

Fig. 8 Impact of "quality of ecosystem" on the stakeholder's perception of drought

The results of this analysis for the ARPA's perception are shown in Table 2.

Table 2 Influence on drought perception of the elements in the ARPA's FCM

Variable	Influence on drought perception
Level of the lake	0,72
Quality of the lake ecosystem	0,86
Ecological equilibrium	0,76
Health of fish species	0,38
Water quality	0,14

The same analysis was carried out for all the involved stakeholders. Table 3 summarizes the results obtained during this first step of the approach.

Table 3 Results of FCM analysis for all the involved stakeholders

Stakeholder	Variable	Degree of importance	Degree of Impact	Perception of Influence
ARPA	Level of the lake	Very important	Negative	0,72
	Quality of the lake ecosystem	Very important	Strongly negative	0,86
	Ecological equilibrium	Very important	Negative	0,76
	Health of fish species	Important	Weakly negative	0,38
	Water quality	Important	Weakly positive	0,14
Trasimen National Park	Level of the lake	Very important	Strongly negative	0,92
	Farmers' income	Very important	Negative	0,72
	Local Economic Development	Important	Negative	0,43
	Touristic sector income	Important	Strongly negative	0,62
	Quality riparian area	Important	Negative	0,4
GAL	Level of the lake	Very important	Strongly negative	0,82
	Withdrawal of water from the lake	Very important	Strongly negative	0,8

Table 3 (*continued*)

Stakeholder	Variable	Degree of importance	Degree of Impact	Perception of Influence
	Touristic sector income	Very important	Negative	0,55
	Local Economic Development	Important	Negative	0,47
	Farmers' income	Important	Strongly negative	0,66
Touristic sector manager	Level of the lake	Very important	Strongly negative	0,91
	Touristic sector income	Very important	Strongly negative	0,9
	Withdrawal of water from the lake	Important	Negative (increase)	0,42
	Water quality	Important	Weakly negative	0,3
Farmers	Level of the lake	Very important	Strongly negative	0,9
	Water availability for irrigation	Very important	Strongly negative	0,9
	Farmers' income	Very important	Strongly negative	0,93
	Irrigation Water price	Important	Negative (increase)	0,48
	Production costs	Important	Negative (increase)	0,46

Concerning the element "level of the lake", there is a high consensus among the participants about its importance to assess the impact of a drought phenomenon. In fact, as discussed with stakeholders during the FCM development phase, the level of the lake is the first recognizable effect of the drought, and it has the most important impact on local activities.

It is interesting to note that there is no consensus for two elements directly linked with agricultural activities in the area, i.e. "Withdrawal of water from the lake" and "Water availability for irrigation". While some of the stakeholders seemed to consider irrigation as a factor which exacerbates the impact of a drought on the lake, other stakeholders considered the impact of agricultural activities as negligible, if compared with the effects of climatic conditions. A third group, instead, considered the agricultural irrigation as the main victim of drought rather than one of the most important causes.

The "influence on perception" values were used to assess the degree of acceptability of potential drought mitigation actions, as described in the following section.

3.3 Assessment of Drought Management Acceptability

This step of the work was aimed at supporting the identification of the most consensual drought management strategies. The basic assumption was that if the consensus degree was high, then the proposed management action was considered acceptable by most of the stakeholders. This would facilitate the implementation of the drought management action.

The acceptability of actions was assessed considering their impact on the main elements of stakeholders' perception of drought. That is, the acceptability was based on the analysis of the impacts of each management action on the stakeholder's FCM.

To this aim, a third round of meetings with stakeholders was organized to define the expected impact of the set of potential drought management actions defined during the first round of interviews with the decision makers:

- Reuse of wastewater;
- Technical support to farmers;
- Changes in agricultural practices;
- Emergency planning.

At the end of this round of interviews, the drought management actions were integrated in the stakeholders' FCM. Fig. 8 shows the expected impact of "reuse of wastewater" on the Arpa's FCM.

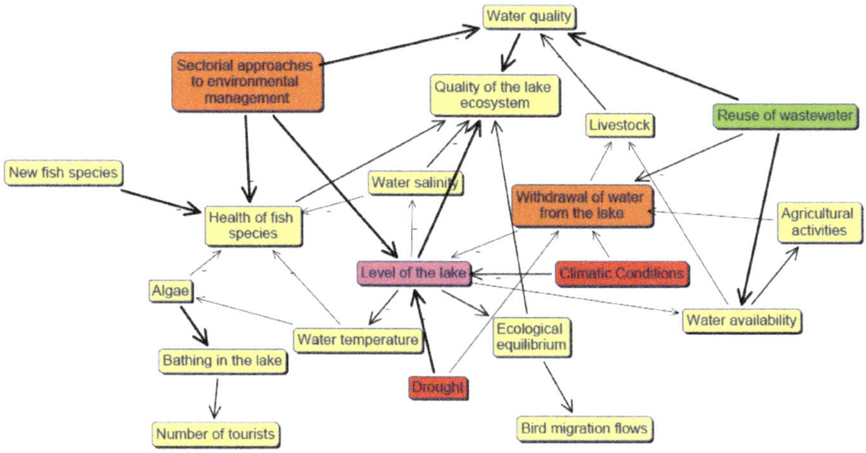

Fig. 9 In the Arpa's opinion, the reuse of wastewater will increase the water availability for irrigation, reduce withdrawals from the lake and will have a positive impact on the water quality

The adjacency matrix of this FCM allowed the impact of the proposed action on the main elements to be simulated (fig. 10).

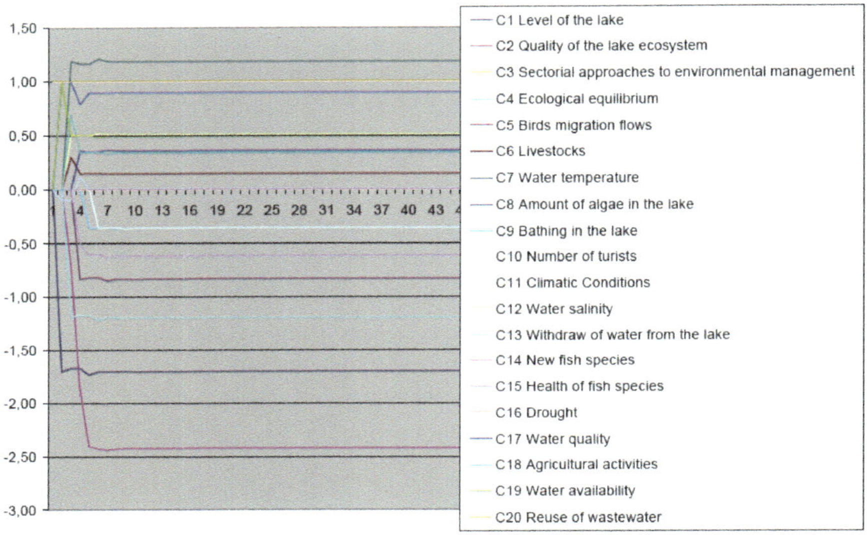

Fig. 10 Simulation of the impact of "reuse of wastewater" on the elements of the Arpa's FCM. Although the negative impact of drought cannot be avoided, this action would allow the impact on important elements such as "Quality of lake ecosystem" to be mitigated

Table 4 summarizes the impact of the proposed action on the main elements of the Arpa's perception of drought. The impact was calculated by comparing the results of the FCM simulation in case of drought and the results of the FCM with the action. That is, these two elements were activated (value = 1) in the system state vector. The influence on the stakeholder's perception is reported in brackets. The overall degree of acceptability was assessed combining the impact on each element and taking into account the influence on perception.

Table 4 Impact of "reuse of wastewater" on the main elements of the Arpa's drought perception

Level of the lake (0,72)	**Quality of the lake ecosystem (0,86)**	**Ecological equilibrium (0,76)**	**Health of fish species (0,38)**	**Water quality (0,14)**	**Degree of Acceptability**
Weakly positive	Positive	Positive	Weakly positive	Weakly positive	Acceptable

The proposed action was considered acceptable because it was perceived to have a positive impact on the three elements with the strongest influence on the drought perception.

The same action was integrated in the farmers' FCM (Fig.11).

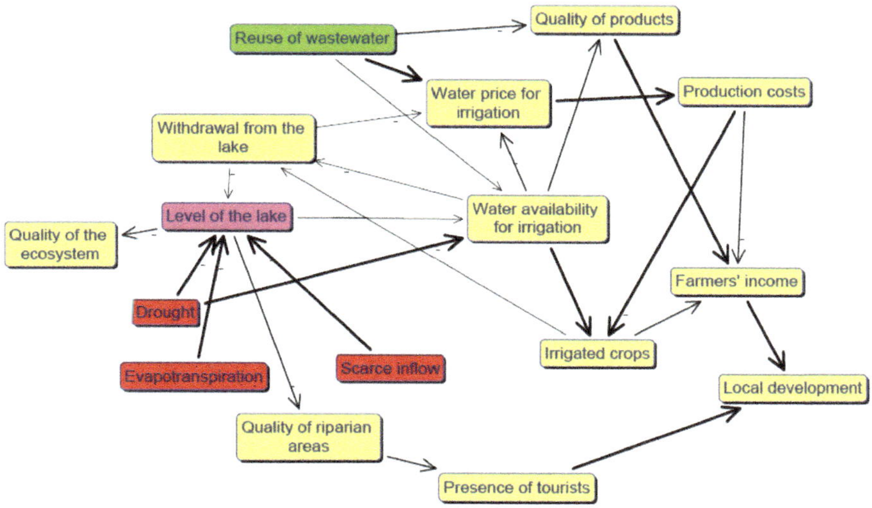

Fig. 11 Farmers' FCM with the introduction of the "reuse of wastewater"

In the farmers' opinion, although the proposed action could increase the water availability for irrigation, it would result in a great increase in production costs. The overall impact of "reuse of wastewater" on the farmers' FCM is shown in Fig. 12.

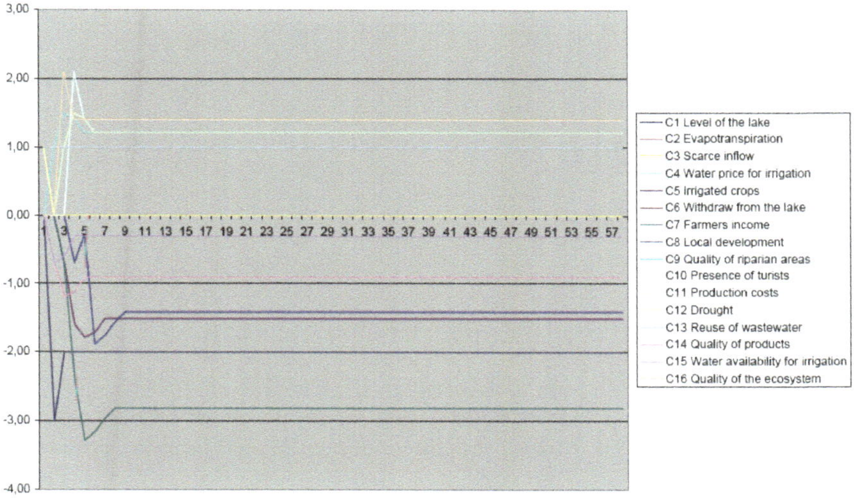

Fig. 12 Simulation of the impact of "reuse of wastewater" on the elements of the farmers' FCM

Table 5 Impact on the most important elements of the farmers' perception of drought

Level of the lake (0,90)	**Water availability for irrigation (0,90)**	**Farmers' income (0,93)**	**Irrigation water price (0,48)**	**Production costs (0,46)**	**Degree of Acceptability**
Weakly positive	Positive	Strongly negative	Strongly negative (increase)	Strongly negative (increase)	Not acceptable

The acceptability degree of "reuse of wastewater" is low for farmers because of the strongly negative impact on production costs and, consequently, on farmers' income.

The analysis of the degree of acceptability of this action was carried out for each stakeholder. A similarity measure was then assessed to compare their opinions. To this aim, the degree of similarity between the degrees of acceptability expressed by each stakeholder was assessed using the following formula (Munda, 1994):

$$S_d(1, 2, x_i) = 1 - |\mu_1(x_i) - \mu_2(x_i)|$$

where, $S_d(1, 2, x_i)$ defines the degree of similarity between stakeholders 1 and 2 on the action x_i (in our case, reuse of wastewater); $\mu_1(x_i)$ expresses the opinion of 1 regarding the acceptability of action x_i and $\mu_2(x_i)$ expresses the opinion of 2 regarding the acceptability of the same action. The results of the degree of similarity assessment were then used to develop the similarity matrix to specify the differences between the actors.

Table 6 Similarity matrix concerning "reuse of wastewater"

	ARPA	Tras. Park	GAL	Tourism	Farmers
ARPA	-	0,83	0,54	0,23	0,16
Tras. Park	0,83	-	0,67	0,33	0,12
GAL	0,54	0,67	-	0,78	0,35
Tourism	0,23	0,33	0,78	-	0,8
Farmers	0,16	0,12	0,35	0,8	-

This table shows high similarity between the farmers' and tourist operators' opinions. In fact, neither did the latter accept the proposed action because of the potential negative impact on water quality and, consequently, on the presence of tourists in the area.

The data contained in the similarity matrix were used to assess the degree of consensus among the participants. To this aim, the stakeholders were clustered according to degree of similarity. The procedure to assess the degree of consensus is described in Giordano et al., 2007. This methodology allows the degree of consensus to be assessed using three factors, i.e. the number of clusters created considering the degree of similarity, the distribution of the stakeholders in the different clusters and the semantic distances between the clusters.

Using this methodology, the degree of consensus was calculated for each of the proposed drought mitigation actions. Table 7 shows the results of this step. The degree of consensus can assume values between 0 and 1; the higher the value, the higher is the consensus among stakeholders.

Table 7 Degree of consensus for the proposed drought management actions

The analysis of the FCM allowed the expected negative impact of the proposed actions on the stakeholders' perceptions of drought to be defined. Therefore, FCM can be used to identify the main reasons behind conflicts.

According to the obtained results, the action with the lowest consensus degree is the "emergency planning", that is the decision to reduce the seasonal amount of water available for irrigation in cases of drought. Although this action is currently considered as effective by some of the involved stakeholders – i.e. the tourism sector – this decision seems highly controversial due to its negative impact on the local development in the FCMs of the farmers' and the GAL. This could result in a strong opposition from these two stakeholders.

The "reuse of wasterwater" for irrigation has a medium level of conflict. This is because of the potential opposition of farmers (expected negative impact on farmers' income and on the quality of products) and weak opposition from the tourism sector (expected weak negative impact on water quality).

This information could then be used by water managers to initiate a negotiation process with stakeholders in order to reduce the level of conflict. This step is not discussed in this work.

4 Discussion

The adopted approach was discussed with the involved stakeholders in order to identify benefits and weaknesses. The lessons learned from this analysis are described in this section. Firstly, the strong points of the system are presented, highlighting the expected positive impact of the system. Secondly, some weaknesses are discussed, and suggestions for improvements and future developments are made. The analysis of the suitability of the adopted approach concerned its ability to analyze different drought perceptions and to support decision makers in dealing with conflict in drought management.

Concerning the first issue, participants stated that one of the positive results of the adopted methodology is its ability to make explicit differences in drought perception. A significant strength of Fuzzy Cognitive Mapping was that the modelling was similar to natural language, which reflected the ways stakeholders were used to talking and thinking about the issues considered. The adoption of a descriptive approach enhanced the comprehensibility of the FCM and, consequently, the sharing of information.

The results of the FCM analysis of the influences on perception were discussed with the stakeholders involved. Thus, they became more aware of the interests and concerns of the other participants about drought impact and drought management. In the opinion of the participants, as expressed at the end of the process, this information allowed them to reflect about divergences and similarities of problem perceptions. The methodology allowed participants to identify, confront the different perceptions and to start the debate about the integration of the divergent views of the same problems. These are actually the main aim of a Problem Structuring process.

The capabilities of FCM to structure the cause – effects chains of stakeholders' understanding of the problem at hand have an important benefits, compared with other approaches adopted to analyze drought perceptions and to support drought management. As we learned during the feedbacks phase with stakeholders, FCM analysis – i.e. the assessment of the "perception of influence" – suggested important elements which were not immediately in participants minds but which were acknowledged as important during the discussion about the obtained results. Therefore, FCM supports participants to avoid anchoring to the first ideas, as it could happen applying methods based on the elicitation of participants memories and experiences in past drought situations (Slegers, 2008; Dagel, 1997).

From the decision makers' point of view, as they stated, the main benefits are related to the ability to make the reasons for potential conflict about drought management explicit. This information can be used by them to identify the most consensual management strategies. Moreover, when the implementation of a strategy cannot be avoided, the information obtained can be used to identify "side-measures" to be implemented together with the identify strategy in order to reduce the level of conflict. For example, the information about the farmers' concerns over the negative impact of treated wastewater on the quality of agricultural products suggested to decision makers that enhancing technical support for farmers could be an effective action.

For what concerns the consensus degree of the drought mitigation options, the proposed methodology is based on the assumption that the consensus is an iterative process which can be monitored defining a consensus measures. Several methods are described in the scientific literature (e.g. Fedrizzi et al., 1999; Herrera-Viedma et al., 2002; Herrera et al., 1996; Szmidt and Kacprzyk, 2003). These methods are based on the comparison of the explicit opinions of the participants. The proposed methodology, based on the implementation of FCM, aims to support participante to formulate their opinions about drought mitigation actions by simulating their impacts, either positive or negative, on the main participants' concerns and interests.

One of the drawbacks highlighted during the analysis of the results concerns the qualitative nature of the results of FCM simulation. As described previously in the text, FCMs are used to assess also the potential impact of the different actions on the elements of the map. Nevertheless, during the presentation of the results to the decision makers it was important to highlight the fact that the results should be interpreted as a change in the state of the element rather than as an exact value. This represented a weakness of the system according to the decision makers, who are familiar with quantitative assessment. Thus, for them, qualitative results could be considered as not completely reliable. An important improvement in the system could be made by coupling the FCM with some quantitative models in order to increase the reliability of the results for the decision makers. To this aim, research activities are currently in progress to integrate a quantitative analysis of drought and the effects of drought management with qualitative perceptions of the phenomenon.

5 Conclusions and Future Developments

The complexity and unstructured nature of drought management issues originates from uncertain knowledge about the phenomenon and from the existence of divergent perceptions among the local actors. Scientific investigations are trying to enhance the knowledge base to address these issues. Particularly, many efforts are currently in progress to define an effective monitoring and early warning system able to make short term drought predictions more reliable.

Nevertheless, dealing with these complex and unstructured problems is not only a matter of knowledge production. It is also a problem of ambiguity. The ambiguity in drought perceptions and definition strongly influences the effectiveness of drought management actions. Therefore, methods and tools to support the elicitation and comparison of the different perceptions are required.

A Problem Structuring approach based on the use of Fuzzy Cognitive Maps is described in this work. The proposed method was able to identify the main elements of the stakeholders' perception of drought and to make this information accessible and easily understandable for both decision makers and stakeholders. Thus, it increased the awareness in each actor of the other members' interests and concerns over drought management. The sharing of this information allowed the decision makers to become aware of potential conflict due to the implementation of certain drought management actions. Moreover, the availability of information on the reasons for conflict allowed them to define a negotiation strategy. Currently, the negotiation process has not yet started.

In future resarch activities, the FCM methodology will be integrated in a Group Decision Support System able to facilitate the collaborative decision making concerning drought management. The capabilities of FCM analysis to identify the main concerns and interests for each participants will allow to select those having an interest in the topic to be discussed. This means that participants will not run the risk of being involved in a discussion far from their interests. This could have a positive impacts on the actors' willingness to take part in collaborative decision making.

References

Axelrod, R.: Structure of Decision - The Cognitive Maps of Political Elites. Prince-ton University Press, Princeton (1976)

Checkland, P.: Soft System Methodology. In: Rosenhead, J., Mingers, J. (eds.) Rational Analysis for a Problematic World. John Wiley and Sons, Chichester (2001)

De Kok, J.L., Titus, M., Wind, H.G.: Application of fuzzy sets and cognitive maps to incorporate social science scenarios in integrated assessment models. Integrated Assessment 1, 177–188 (2000)

Eden, C.: Analyzing cognitive maps to help structure issues and problems. European Journal of Operational Research 159, 673–686 (2004)

Eden, C., Ackermann, F.: SODA – The principles. In: Rosenhead, J., Mingers, J. (eds.) Rational Analysis for a Problematic World. John Wiley and Sons, Chichester (2001)

Giordano, R., Uricchio, V.F., Passarella, G., Vurro, M.: Integrated conflict analysis and consensus reaching in a decision support system for water resources management. Journal of Environmental Management 84, 213–228 (2007)

Glykas, M., Xirogiannis, G.: A soft knowledge modeling approach for geographically dispersed financial organizations. Soft Computing 9(8), 579–593 (2004)

Hiorsto, C.N.: Enhancing public participation in natural resource management using Soft OR-—an application of strategic option development and analysis in tactical forest planning. European Journal of Operational Research 152, 667–683 (2004)

Hommes, S., Vinke de Kruijf, J., Otter, H.S., Bouma, G.: Knowledge and Perceptions in Participatory Policy Processes: Lessons from the Delta-Region in the Netherlands. Water Resources Management 23, 1641–1663 (2009)

Kang, I., Lee, S., Choi, J.: Using fuzzy cognitive map for the relationship management in airline service. Expert Systems with Applications 26, 545–555 (2004)

Kolkman, M.J., Kok, M., van der Veen, A.: Mental model mapping as a new tool to analyse the use of information in decision-making in integrated water management. Physics and Chemistry of the Earth 30(4-5), 317–332 (2005)

Lane, D.C., Oliva, R.: The greater whole: Towards a synthesis of system dynamics and soft system methodology. European Journal of Operational Research 107, 214–235 (1998)

Mingers, J., Rosenhead, J.: Problem structuring methods in action. European Journal of Operational Research 152, 530–544 (2004)

Montibeller, G., Ackermann, F., Belton, V., Ensslin, L.: Reasoning Maps for Decision Aid: A Method to Help Integrated Problem Structuring and Exploring of De-cision Alternatives. In: ORP3, September 26-29 (2001)

Munda, G.: Multicriteria evaluations in a fuzzy environment. Phisica Verlag, The Hague (1995)

Noemdoe, S., Jonker, L., Swatuk, L.A.: Perceptions of water scarcity: The case of Genadendal and outstations. Physics and Chemistry of the Earth 31, 771–778 (2006)

Ohlsson, L.: Water Conflicts and Social Resource Scarcity. Physics and Chemistry of the Earth (B) 25(3), 213–220 (2000)

Ozesmi, U., Ozesmi, S.: A Participatory Approach to Ecosystem Conservation: Fuzzy Cognitive Maps And Stakeholders Groups Analysis in Uluabat Lake, Turkey. Environmental Management 31(4), 518–531 (2004)

Pahl-Wostl, C.: The implication of complexity for integrated resources management. Environmental Modelling and Software 22, 561–569 (2007)

Pereira, L.S., Cordery, I., Iacovides, I.: Copying with water scarcity. Springer Science + Business Media (2009)

Rosenhead, J., Mingers, J.: A New Paradigm for Analysis. In: Rosenhead, J., Mingers, J. (eds.) Rational Analysis for a Problematic World. John Wiley and Sons, Chichester (2001)

Slegers, M.F.W.: "If only it would rain": Farmers' perceptions of rainfall and drought in semi-arid central Tanzania. Journal of Arid Environments 72, 2106–2123 (2008)

Timmerman, J.G., Langaas, S.: Conclusions. In: Timmerman, J.G., Langaas, S. (eds.) Environmental information in European transboundary water management. IWA Publishing, London (2004)

Xirogiannis, G., Stefanou, G., Glykas, M.: A fuzzy cognitive map approach to support urban design. Expert Systems with Applications 26, 257–268 (2004)

Xirogiannis, G., Chytas, P., Glykas, M., Valiris, G.: Intelligent impact assessment of HRM to the shareholder value. Expert Systems with Applications 35(4), 2017–2031 (2008)

Zimmermann, H.J.: Fuzzy Sets, Decision Making, and Expert Systems. Kluwer Academic Publisher, Dordrecht (1987)

Author Index

GPSR Compliance

The European Union's (EU) General Product Safety Regulation (GPSR) is a set of rules that requires consumer products to be safe and our obligations to ensure this.

If you have any concerns about our products, you can contact us on ProductSafety@springernature.com

In case Publisher is established outside the EU, the EU authorized representative is:

Springer Nature Customer Service Center GmbH
Europaplatz 3
69115 Heidelberg, Germany

Batch number: 10755461

Printed by Printforce, the Netherlands